BIOCHEMICAL PHARMACOLOGY

BIOCHEMICAL PHARMACOLOGY

MICHAEL PALMER
University of Waterloo

with chapters by

ALICE CHAN
University of Alberta

THORSTEN DIECKMANN
University of Waterloo

JOHN HONEK
University of Waterloo

WILEY

A JOHN WILEY & SONS, INC., PUBLICATION

Published by John Wiley & Sons, Inc., Hoboken, New Jersey.
Published simultaneously in Canada.

Library of Congress Cataloging-in-Publication Data:

Palmer, Michael, 1961 May 4–
Biochemical pharmacology / Michael Palmer ; with chapters by Alice Chan, Thorsten Dieckmann, John Honek. — 1st ed.
p. ; cm.
Includes bibliographical references and index.
ISBN 978-0-470-17445-6 (hardback)
I. Title.
[DNLM: 1. Drug Interactions. 2. Pharmacokinetics. 3. Drug Delivery Systems. 4. Pharmaceutical Preparations. QV 38]

615.7'045—dc23 2011041617

Printed in the United States of America.

10 9 8 7 6 5 4 3

CONTENTS

Preface ix

Acknowledgments xi

1 Introduction 1

1.1 Origin and preparation of drugs 3
1.2 Drugs and drug targets 4
1.3 Drug molecules may or may not resemble the physiological ligands of their receptors 7
1.4 Strategies of drug discovery and development 11
References 14

2 Pharmacodynamics 15

2.1 Molecular features of drug-receptor interaction 15
2.2 Theory of drug-receptor binding 17
2.3 Dose-effect relationships in signaling cascades 25
2.4 Potency and efficacy 28
2.5 Beneficial and toxic drug effects 28
2.6 Appendix 30
Study questions 35
References 36

3 Pharmacokinetics **37**
3.1 Anatomical barriers to drug transport 38
3.2 Solute transport across cell membranes 40
3.3 Drug application 49
3.4 Drug distribution 53
3.5 Drug elimination via the kidneys 56
3.6 Quantitative experimental methods in pharmacokinetics 63
Study questions 65
References 65

4 Drug metabolism **67**
4.1 Overview: Possible functional outcomes of drug metabolism 68
4.2 Phase I and phase II reactions 70
4.3 Cytochrome P450 71
4.4 Reductive drug metabolism 74
4.5 Conjugation reactions 75
4.6 Enzyme induction 83
4.7 Metabolism-related toxicity as a therapeutic target 84
Study questions 85
References 85

5 G protein-coupled receptors **87**
5.1 Overview 87
5.2 GPCR structure 89
5.3 Structural GPCR families 89
5.4 Activation of GPCRs 90
5.5 GPCR dimerization and oligomerization 96
5.6 G proteins 98
5.7 GPCR phosphorylation, endocytosis, and G protein-independent signaling 102
5.8 Appendix 103
Study questions 107
References 108

6 Pharmacology of cell excitation **111**
6.1 Ions, pumps, and channels 114
6.2 ATP-driven active ion transport 116
6.3 Voltage-gated channels and the action potential 116
6.4 Channels controlled by intracellular ligands 120
6.5 Transient receptor potential (TRP) channels 120
6.6 Voltage-gated channels of nerve cells as drug targets 121
6.7 Synaptic transmission 127
6.8 Pharmacology of individual transmitters 136
6.9 Appendix 153
Study questions 160
References 161

7 Hormones 163
7.1 Hormone receptors 164
7.2 The hypothalamus and the pituitary gland 164
7.3 Thyroid gland hormones 170
7.4 Steroid hormones 175
7.5 Endocrine control of bone mineralization 183
Study questions 188
References 189

8 Pharmacology of nitric oxide 191
8.1 Characterization of nitric oxide as a biological signaling molecule 192
8.2 Nitric oxide synthase and its isoforms 192
8.3 Biochemical mechanisms of NO signaling 195
8.4 The biological function of iNOS 200
8.5 NO-releasing drugs 200
8.6 NOS inhibitors 204
8.7 Phosphodiesterase inhibitors 204
Study questions 205
References 205

9 Eicosanoid mediators and related drugs 207
9.1 Biosynthesis of eicosanoids 208
9.2 The cyclooxygenase reaction 211
9.3 Cyclooxygenase isoforms and inhibitors 211
9.4 Phospholipase A_2 inhibitors 217
9.5 Derivatives of prostaglandin H_2 and related drugs 217
9.6 Lipoxygenases, leukotrienes, and related drugs 218
9.7 Eicosanoids synthesized by cytochrome P450 219
9.8 Endocannabinoids and related drugs 219
9.9 The role of polyunsaturated fatty acids in eicosanoid signaling 223
Study questions 224
References 224

10 Intermediate metabolism, diabetes and atherosclerosis 227
10.1 Hereditary enzyme defects 227
10.2 Gout 235
10.3 Diabetes mellitus 238
10.4 Atherosclerosis 241
Study questions 251
References 252

11 Chemotherapy of infectious diseases 255
11.1 Pathogenic microbes: Diversity and selective toxicity 256
11.2 Pharmacokinetic considerations 257

11.3 Resistance to antimicrobials 258
11.4 Antibacterial chemotherapy 259
11.5 Chemotherapy of fungal infections 277
11.6 Chemotherapy of parasite infections 279
11.7 Antiviral chemotherapy 284
Study questions 291
References 292

12 Tumor chemotherapy 295
12.1 Some principles of tumor biology 296
12.2 Cell-type-specific antitumor drugs 304
12.3 Drugs that target specific oncoproteins 306
12.4 Cytotoxic antitumor drugs 309
Study questions 320
References 320

13 Ribonucleic acids as drug targets and drugs 323
13.1 RNA as drug target 324
13.2 RNA as a therapeutic agent 336
Study questions 339
References 339

14 Drug delivery 343
14.1 Improving intestinal drug absorption 343
14.2 Improving drug distribution 348
14.3 Targeted drug delivery 352
14.4 Kinetically controlled drug release 358
14.5 Controlling drug toxicity 363
14.6 Delivery of nucleic acids 367
Study questions 368
References 368

15 Drug discovery 373
15.1 Target selection and validation 373
15.2 Screening of candidate compounds 375
15.3 Computational screening 381
15.4 Phenotypic screening 384
15.5 Compound acquisition 384
Study questions 389
References 389

Answers to study questions 393
References 400

Index 403

PREFACE

> Those who have taken upon them to lay down the law of nature as a thing already searched out and understood, . . . have therein done philosophy and the sciences great injury. For as they have been successful in inducing belief, so they have been effective in quenching and stopping inquiry.
>
> Francis Bacon

This book aims to explain the biochemical principles of drug action, to students of biochemistry and of related scientific disciplines, at the advanced undergraduate or beginning graduate level. For each drug that we discuss, we cover the biochemical concepts needed to understand its function. We explain some of the experimental methods used to investigate drugs and their receptors, along with examples of the evidence that these methods provide. Clinical and physiological aspects are outlined where this seems useful to appreciate the significance of a drug's action. Finally, we have tried to heed Bacon's advice and point out some of the open questions that will need to be solved by future researchers.

In selecting the topics for a teaching text, one has to balance breadth and depth. We chose to forgo comprehensive coverage in favor of treating selected drugs in some more detail. Therefore, many drugs that are important in clinical medicine are not covered here. Similarly, topics from fundamental biochemistry are limited to the context required for the chosen example drugs. While such limited treatment of biochemistry may seem natural, it is important to realize that it will miss out on recent discoveries that may provide the foundations for future drugs; the biochemistry of today is the pharmacology of tomorrow. This point is well illustrated in Chapter 13, contributed by Thorsten Dieckmann, which discusses emerging applications of ribonucleic acids as drugs and drug targets.

Among the other chapters, the first five are devoted to general principles of drug action. Chapters 6–12 discuss a number of mostly proven drugs, but also some experimental ones, grouped along biochemical and pathophysiological lines. Chapter 14, by Alice Chan, covers drug delivery, and chapter 15, by John Honek, discusses drug discovery. While these topics are not strictly part of pharmacology, they provide valuable context for this subject nevertheless.

The book has evolved from courses taught at the University of Waterloo that met with lively interest among our students. We envision that the book may facilitate similar course offerings at other institutions. To aid in the preparation of such classes, the figures contained in this book are available as slides to instructors through this book's web page at `wiley.com`. We would love to hear from our readers and appreciate any corrections and suggestions for improvement.

ACKNOWLEDGMENTS

First and foremost, I would like to thank my coauthors. I enjoyed our collaboration and discussions, and I feel that their chapters strengthen this book substantially.

Jonathan Rose of John Wiley & Sons suggested the idea for this book and saw it through to the end. I am grateful for his friendly advice and for his patience when at times the pace of my progress resembled Achilles' race with the tortoise. Alice Chan, Michael Peinecke, Eric Brefoh-Mensah, Kate Wood and Bruce Reed read parts of the manuscript and gave suggestions for improvement.

Finally, I thank my wife, Shenhui Lang, for the encouragement, understanding and support she showed me during this work.

M. P.

CHAPTER 1

INTRODUCTION

This book deals with drugs and their biochemical mechanisms of action. The term "drug" is used here in an inclusive sense, and we will neglect the following possible distinctions: (1) Some drugs are legal, others are not. This difference will not matter for the purpose of this book—we will look at examples of either group. (2) One might argue that drugs, at least the legal ones, should be distinguished from poisons by the beneficial nature of their effect. Yet most drugs turn into poisons if overdosed, and many poisons make useful drugs if properly diluted.

The relationship between drugs and poisons was first, or at least most famously expressed by the Swiss physician Paracelsus in the 16th century:

> All things are poison and nothing is without poison; only the dosage makes it so that something is not a poison.[1]

Given our inclusive definition of the term, do drugs have anything in common at all? Let's start with a simple criterion: size. Figure 1.1 shows the smallest of all drugs—lithium. Neglecting the pharmacologically inert counterion, the atomic mass of this drug is just 7 Da. On the other end of the scale, we have rather large

[1]German: "Alle Ding' sind Gift und nichts ohn' Gift; allein die Dosis macht, dass ein Ding kein Gift ist." Latin short version of spurious origin: "Dosis sola facit venenum." While Paracelsus certainly deserves credit for recognizing the general principle, rumor has it that he frequently erred in applying it to himself.

Biochemical Pharmacology. By M. Palmer, A. Chan, T. Dieckmann, J. Honek

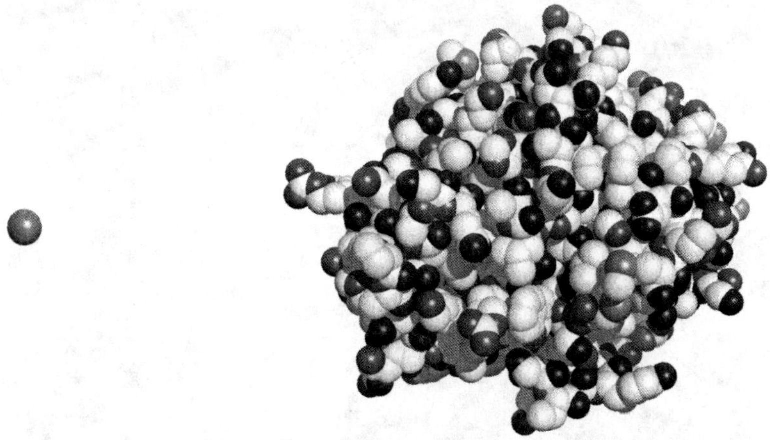

Figure 1.1 The smallest drug (Li^+, left) and a very large drug molecule (urokinase, right). Lithium salts are used in the treatment of bipolar disorder. Urokinase, a protein, can be used to dissolve blood clots in thrombosis and stroke.

molecules—proteins. Also shown in Figure 1.1 is the structure of urokinase, a human protein that can be used to dissolve blood clots. Urokinase has mostly been superseded by another protein of similar size and function, namely, tissue plasminogen activator (tPA). The concentration of tPA in human tissues is very low, but by means of recombinant expression in cell culture it can be produced in clinically useful amounts. Recombinantly expressed tPA is now the gold standard in the thrombolytic therapy of brain and myocardial infarctions. Other proteins that are used for therapeutic purposes are coagulation factor VIII, which is missing and must be substituted in hemophilia; albumin, which is used in the emergency treatment of acute loss of a large volume of blood; and the ever more popular botulinum toxin. These proteins have molecular weights of tens or hundreds of kilodaltons.

While lithium and protein drugs illustrate the wide possible size range and diversity of drug molecules, both of them are fairly atypical examples. A few arbitrarily selected drug molecules with more typical sizes are shown in Figure 1.2. Like these examples, most practically useful drugs are organic molecules, with molecular weights of roughly 100–2000, mostly below 1000 Da. Interestingly, this also applies to many natural poisons, although on average they are probably somewhat larger. Are there reasons for this?

Large drug molecules are usually more difficult to obtain than small ones. Moreover, drug molecules need to reach their target sites in the human body, typically by diffusion. Diffusion through solution and particularly across anatomical barriers becomes increasingly difficult with size.

A lower size limit is imposed by the requirement for specificity. There are numerous examples of low-molecular weight poisons—probably the better part of the periodic table is poisonous. The toxicity of the elements varies with their reactivity, but their lack of structural features makes it difficult for them to discriminate between individual macromolecular targets. A few drugs, however,

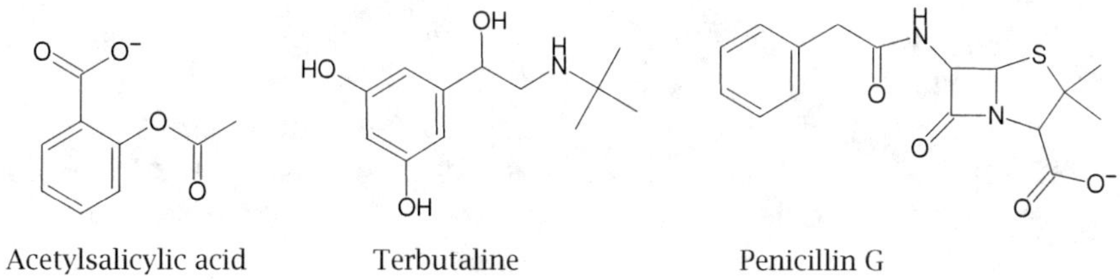

Figure 1.2 Some organic drug molecules of typical size. Acetylsalicylic acid inhibits cyclooxygenase, terbutaline stimulates β_2-adrenergic receptors, and penicillin G is an inhibitor of bacterial muramyl transpeptidase.

seem to be doing fine without pronounced structural features, such as lithium and the inhalation anesthetics chloroform, diethylether, or xenon.[2]

Apart from its size, the ability of a drug molecule to reach its target is strongly influenced by its *polarity*. Ionic or polar moieties in drug molecules strongly hinder their diffusion across cell membranes, which is required for migration across anatomical barriers and for access to intracellular targets. A case in point are antisense oligonucleotides. These are single-stranded nucleic acids or nucleic acid analogs that inhibit the expression of specific viral or cellular genes through complementary base pairing with the mRNA transcripts. While this approach is in principle very elegant and versatile, the combination of relatively large size and strongly polar nature of oligonucleotides poses a great challenge to their actual use in clinical medicine. The only such drug in clinical use is fomivirsen, an inhibitor of cytomegalovirus replication, works because it can be topically applied to the site of infection (the eye). Effective methods for systemic delivery of oligonucleotide drugs are still being investigated.

If your ambition is to develop drugs that are useful in practice, you should take such practical constraints into account right from the start. We will consider them in more detail in Chapters 3 and 14.

1.1 ORIGIN AND PREPARATION OF DRUGS

A great number of drugs have been isolated from natural sources—plants, fungi and bacteria, many of which were observed to have toxic or medically useful activity long ago. Examples are:

1. Atropine, the essential component of the plant known as *deadly nightshade*. It widens the pupils and reddens the cheeks, and thus enhances one's appearance. It also induces hallucinations and so may enhance perception of one's surroundings and company; this, too may come in handy if the object is to fall in love.[3]

[2]This is valid only to a degree. Lithium seems to interact with multiple target proteins, and it is not yet clear which one of these is responsible for the therapeutic effect. The molecular targets of inhalation anesthetics are similarly still somewhat shrouded in mystery (see Section 6.7.3).

[3]The various names of the plant—*Atropa belladonna* (*bella donna* is Italian for "pretty woman"), *Tollkirsche* (German), meaning "craze cherry", and *deadly nightshade*—suggest that preferred applications and dosage regimens may vary with locality.

2. Ergotamine is one of the active compounds found in the ergot fungus (*Claviceps purpurea*). It promotes contraction of smooth muscle cells both in the womb and in blood vessels. In the womb, ergotamine will induce labor; this application of the drug has been abandoned. Contraction of blood vessels makes ergotamine useful in the treatment of migraine. At higher concentrations, however, it leads to *ergotism*, in which excessive constriction of blood vessels deprives tissues of their blood supply altogether, causing their demise. Ergotism has occurred historically as a rather disastrous form of food poisoning caused by contamined grain.

3. Antibiotics, which are chemical weapons used by fungi or other soil-dwelling microorganisms to ward off their microbial competition.

In very many cases, natural compounds that are useful as drugs comply with Paracelsus' maxim by being poisons—see atropine and ergotamine, above. Most antibiotics are toxic for humans, too, and are either not used at all in medicine or sometimes as cytotoxic agents in cancer therapy. Penicillin G (Figure 1.2) has an exceptionally low level of toxicity for humans. In severe infections, it can be used in quantities of tens of grams per day, usually without causing any side effects whatsoever. It still is a poison, but its poisonous effect happens to be highly selective for bacteria. Penicillin embodies the principle of *selective toxicity*, which is the holy grail of anti-infectious chemotherapy.

While nature still is an important source of novel compounds and of inspiration, organic synthesis became the most prominent source of new drug molecules in the last century. It is now also used to prepare many of the drugs originally obtained as natural compounds; it is so generally used that in the following text we will not even mention the fact that a drug has been so obtained. The confidence that almost any molecule that is stable can also be synthesized is the basis for the modern concept of drug design. Nevertheless, even today, some drugs are still obtained from natural sources, where that is easier and more economical than total synthesis. An example is the production of morphine for both legal and illegal use from poppy seed; other ones are β-lactam antibiotics such as penicillins and cephalosporins, which are semi-synthetically derived from fermentation products of *Penicillium notatum* and related fungi.

1.2 DRUGS AND DRUG TARGETS

Drugs need to bind to drug targets. Given our current knowledge of biochemistry, this insight seems trite; one may therefore be surprised to find that it is only about one hundred years old. The idea of distinct target molecules was conceived by Paul Ehrlich. He called them *receptors*, which name has stuck ever since. Ehrlich worked on a variety of microbes and microbial toxins. He observed that many dyes used to highlight specific structures within microbial cells for microscopic studies also exerted toxic effects on the microbes. This observation inspired him to systematically try every new dye he could get a hold of—and new dyes were all the rage in the late 19th century!—on his microbes. After screening more than 600 compounds, Ehrlich and his coworker Sahachiro Hata finally settled on arsphenamine (Figure 1.3), an organic arsenic

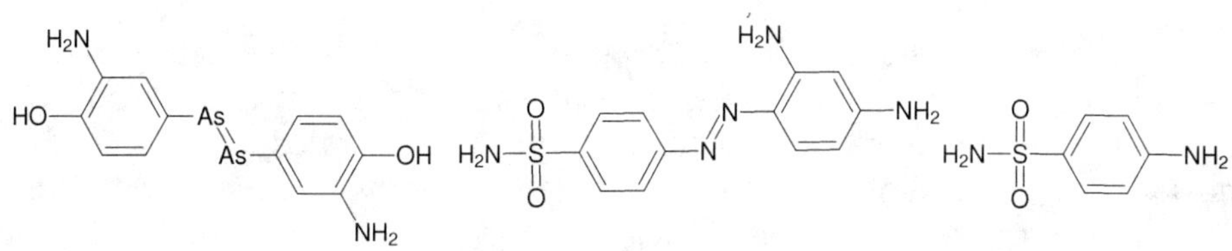

Arsphenamine Sulfamidochrysoidine Sulfanilamide

Figure 1.3 Structures of arsphenamine ("salvarsan"), and of sulfamidochrysoidine ("prontosil rubrum") and its antibacterially active metabolite sulfanilamide, which is released through azoreduction by enzymes of the host.

compound that was dubbed "salvarsan" and clinically used to treat syphilis for several decades, until penicillin became available.

In keeping with his enthusiasm for colors and dyes, Ehrlich is credited with having possessed one of the most colorful lab coats of all times; he also had one of the most paper-jammed offices ever. His Nobel lecture [4] is an interesting read—a mix of brilliant thoughts and ones that would today seem utterly naive, and that make it startlingly clear how very little was known in biochemistry and cell biology only a century ago.

Most drug receptors are proteins and fall into one of the following functional categories: (1) enzymes, (2) hormone and neurotransmitter receptors, (3) ion channels, (4) membrane transporters, and (5) cytoskeletal proteins.

Most enzymes—especially biosynthetic ones, such as HMG-CoA reductase, a key enzyme in cholesterol synthesis, and cyclooxygenase, which has a central place in the synthesis of prostaglandins—are intracellularly located. Some regulatory enzymes, such as the proteases that control blood clot formation and dissolution, are found extracellularly.

Hormone and neurotransmitter receptors may be located on the cell surface or inside the cell. Many of the most widely used drugs such as antihistamines, opioids, or β-blockers act on *G protein-coupled receptors* (GPCRs), which are located on the cell surface. This is the largest single family of drug targets, and its significance is likely to grow as more of the many "orphan" receptors that have been uncovered by the sequencing of the human genome can be assigned ligands and physiological functions.

Another functional class are the *nuclear hormone receptors*. These are intracellular proteins that, when activated by their cognate hormones, will bind to specific sequence motifs on the DNA and thereby set in motion the hormones' characteristic effects on gene expression. Examples are the receptors of steroid hormones such as cortisol and aldosterone, of thyroid hormones, and of retinoic acid, all of which are drug targets in practical pharmacotherapy.

Drugs that bind to enzymes are virtually always inhibitors, or *antagonists*. In contrast, drugs that act on hormone or neurotransmitter receptors may be either inhibitors or *agonists*, that is, activators. The same also applies to drugs that act on ion channels. These come in two varieties: (1) ligand-gated channels and (2) voltage-gated channels. Ligand-gated channels open or close in response to binding of specific agonists. They are prominently involved in synaptic transmission between nerve cells: A neurotransmitter released from the first or presynaptic cell binds to a ligand-gated channel on the second or postsynaptic cell and causes it to open, thereby triggering a flux of specific ions. This may increase or decrease the electrical potential across the membrane of the postsynaptic cell. Prominent examples of ligand-gated channels are the receptors for acetylcholine, glutamate, and γ-aminobutyric acid (GABA). The $GABA_A$ receptor channel is the target of many widely used drugs such as ethanol, barbiturates, and benzodiazepines.

Voltage-gated channels are controlled by changes in the surrounding electric field. When they open, this will, in turn, change the local electric field and so affect other voltage-gated channels in the vicinity. Therefore, these channels amplify membrane potential undulations and propagate them along the surfaces of nerve and muscle cells. Physiologically, the regulation of voltage-gated channels by ligand binding plays at most a secondary role. Nevertheless, there are many natural poisons and synthetic drugs that manage to interfere with the activity of these channels by arresting them in their open or closed states. Some such drugs are therapeutically used for local anesthesia and in the treatment of epilepsy.

Membrane transporters are proteins that facilitate the transport of solutes, including ions, metabolites and transmitters, across cell membranes. Examples of drugs that act on membrane transporters are diuretics, cocaine, and proton pump inhibitors; the latter have become a mainstay in the therapy of gastric ulcers. Cytoskeletal proteins are important in cell division and mobility. Inhibition of the polymerization and depolymerization of actin and tubulin is used in the treatment of inflammatory diseases and of malignant tumors.

The predominance of proteins as drug targets is, of course, due to the fact that proteins fulfill most cellular functions. However, some essential functions are brought about mainly by RNA. A key example is ribosomal protein synthesis, and ribosomal RNA is indeed the target of several classes of antibiotics that are used in antimicrobial therapy. In addition, RNA mediates and regulates gene expression, and in this capacity is increasingly attracting interest both as a drug and as a drug target. Chapter 13 explores various strategies of RNA-related pharmacology.

While drugs that bind RNA tend to do so in a highly sequence-specific manner, most drugs in current use that react with DNA do so without significant selectivity for specific sequences. These are mostly DNA-alkylating agents, which unsurprisingly are highly toxic yet of use in tumor therapy; some examples are discussed in Chapter 12.

Structures other than proteins or nucleic acids also may function as drug targets. Examples are the antibiotics vancomycin, which binds to an intermediate substrate of bacterial cell wall synthesis, and polymyxin, which binds to lipids in the outer membrane of Gram-negative bacteria. The antimalarial drug chloroquine is thought to bind to heme that is released when malaria parasites feed on hemoglobin, and thereby interfere with mechanisms that the parasite uses to reduce the toxicity of the free heme. Finally, some drugs do not bind to a receptor at all but act simply via their osmotic activity. Examples include laxatives (e.g., sodium sulfate), blood plasma volume expanders (dextran, hydroxyethyl-starch), and diuretics (mannitol). Naturally, these drugs are not very interesting from a biochemical viewpoint, so we will not discuss them any further.

Returning to the subject of drug-receptor interaction, the next question to consider is: Are there any general, useful rules and laws governing this interaction? Indeed there are; we will consider them in some detail in the next chapter.

1.3 DRUG MOLECULES MAY OR MAY NOT RESEMBLE THE PHYSIOLOGICAL LIGANDS OF THEIR RECEPTORS

Many types of drug receptors discussed above have binding sites for cognate ligands that occur physiologically, and drugs acting on those receptors very often resemble these physiological ligands in structure. With synthetic drugs, this resemblance is of course not coincidental; the physiological ligand is already known to bind the target avidly and specifically, so it is a logical starting point for designing drugs targeting this receptor. This rationale has been successfully applied over and over, both for receptor agonists, that is, drugs that bind and activate the receptor much like the physiological ligand, and for antagonists, which occupy the same binding site but do not activate the receptor.

1.3.1 Histamine receptor agonists and antagonists

As an example, we may consider the development of antagonists for the H_2 histamine receptor. The H_2 receptor, a GPCR, occurs in the parietal cells of the stomach mucous membrane. Stimulation of the H_2 receptor by histamine controls the secretion of gastric acid, and H_2 blockers can be used in the treatment and prevention of gastric and duodenal ulcers (Figure 1.4). The H_2 antagonist cimetidine was derived from histamine in several successive steps. The first derivative that displayed a strong reduction of receptor activation—while still binding to the receptor, of course—was *N*-guanylhistamine. The guanidino group is retained in cimetidine, which improves on *N*-guanylhistamine with respect to receptor affinity. Cimetidine was the first clinically useful H_2 receptor blocker. It represented a major improvement in ulcer therapy at the time and is still in use today, although more modern drugs are now more commonly used.

Apart from H_2 receptors, there are also H_1 receptors. These are relevant as drug targets in their own right, since they are involved in triggering allergic reactions such as hay fever. While both H_1 and H_2 receptors are stimulated by histamine, cimetidine is selective for H_2 receptors, and H_1-selective antagonists exist as well.[4] Subtype-selective drugs are of both practical and theoretical interest; they often permit more targeted pharmacotherapy with fewer side

[4]In addition to H_1 and H_2 receptors, there also are H_3 receptors, which occur in the brain, and H_4 receptors, which like H_1 receptors contribute to allergic reactions.

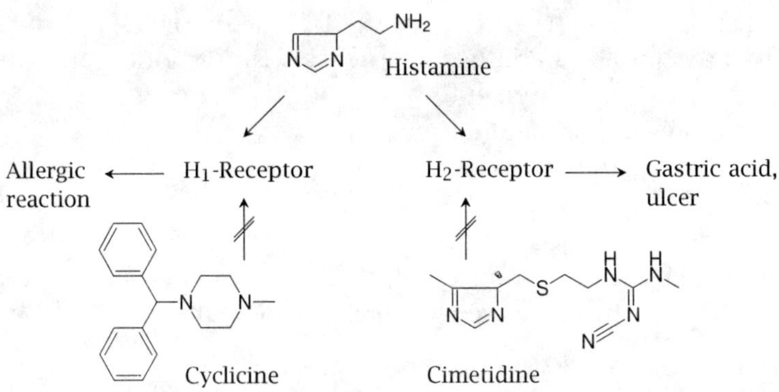

Figure 1.4 Histamine and two subtype-selective receptor blockers. H_1 and H_2 histamine receptors are GPCRs that occur in different cell types and trigger different reactions. Cyclizine selectively inhibits H_1 receptors and is used to treat allergy, whereas cimetidine blocks H_2 receptors and is used to reduce gastric acid secretion.

effects, and they also are very useful in the identification and functional characterization of new receptor subtypes.

An early example of a synthetic drug that enabled the experimental characterization of receptor subtypes was isoproterenol. The existence of distinct adrenergic receptor subtypes, now known as α and β receptors, had been surmised initially because of the different physiological responses to the natural catecholamines. Among these, norepinephrine acts on α receptors only, while epinephrine acts on both subtypes. The development of isoproterenol, which is highly selective for β receptors, made it possible to clearly distinguish the two receptor subtypes.

Different receptor subtypes for a given physiological mediator are quite often found in different anatomical locations and distinct functional roles, as is illustrated by the example of histamine H_1 and H_2 receptors given above. In the physiological setup, there is little interference, because the mediator is typically released in proximity to its receptor and is degraded or at least greatly diluted before it gets a chance to interact with different receptor subtypes. However, this is difficult to reproduce in drug therapy, which most often applies drugs systemically. Chemical selectivity of a drug for a receptor subtype may then compensate for the lack of targeted application.

1.3.2 The renin-angiotensin system

Varying the structure of physiological ligands is also a viable approach to the development of enzyme inhibitors. For illustration, we may consider some enzymes from the human renin-angiotensin system, which is important in the regulation of blood pressure (Figure 1.5). Angiotensinogen is a plasma protein that is synthesized in the liver. From this protein, the peptide angiotensin I is cleaved by the specific protease renin, which is found in the kidneys. Angiotensin I is cleaved further by angiotensin converting enzyme, which is present in the blood plasma. This second cleavage releases angiotensin II, which is a very powerful vasoconstrictor. Angiotensin II activates a specific

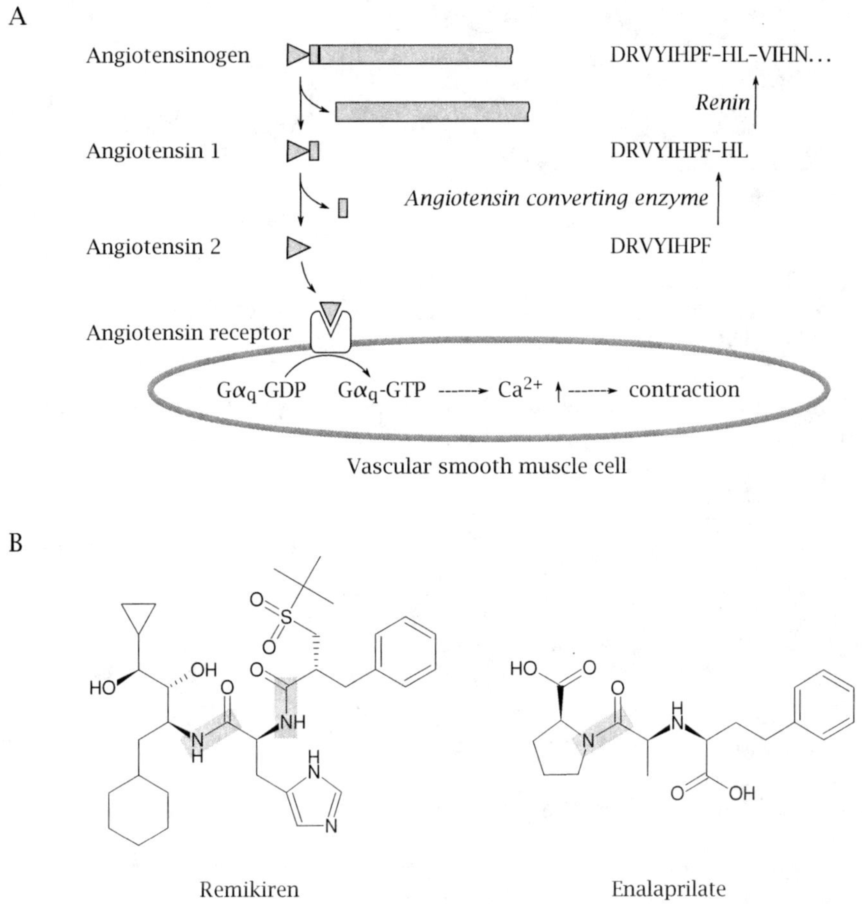

Figure 1.5 Enzyme inhibitors in the renin-angiotensin system. **A:** Function of the renin angiotensin system. Angiotensinogen, a plasma protein, is cleaved by the protease renin. The N-terminal fragment is angiotensin I. Angiotensin converting enzyme removes two more residues to release angiotensin II, which activates a receptor on smooth muscle cells in blood vessels, inducing contraction and raising the blood pressure. **B:** Structures of two enzyme inhibitors. Remikiren inhibits renin, whereas enalaprilate inhibits angiotensin converting enzyme. Peptide bonds in both molecules are highlighted.

receptor on the surface of vascular smooth muscle cells. Through a cascade of intracellular events, this receptor triggers contraction of the muscle cell, which leads to constriction of the blood vessels and an increase in blood pressure.

Increased activity of the renin–angiotensin system is frequently observed in diseases of the kidneys. To counter the ensuing *renovascular hypertension*, several elements in the renin–angiotensin system can be addressed. The first one is renin itself. An inhibitor of renin is remikiren (Figure 1.5B). Its structure is modeled on the polypeptide sequence in the vicinity of the proteolytic cleavage site. Remikiren is effective but has several shortcomings, such as low *bioavailability*, which means that the drug does not enter the systemic circula-

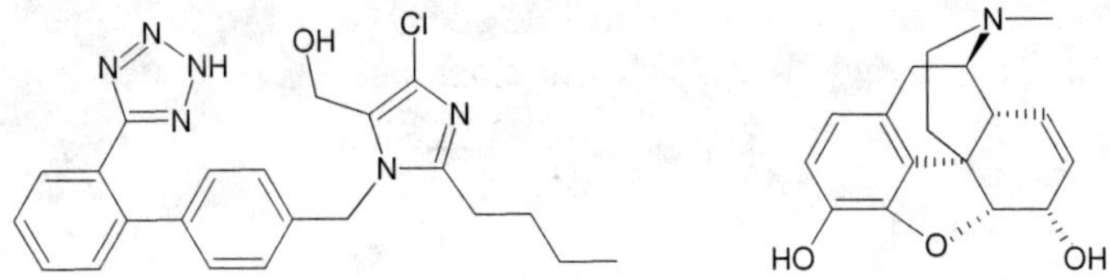

Losartan Morphine

Figure 1.6 Nonpeptide inhibitors of peptide receptors. The synthetic antagonist losartan blocks the angiotensin AT_1 receptor, whereas the natural compound morphine blocks several opioid receptors.

tion efficiently after oral uptake. Of course, oral drug application is important in the treatment of high blood pressure, which mostly occurs as a long term condition and thus requires long term treatment. Generally speaking, a major cause of low drug bioavailability is metabolic inactivation in the intestines and the liver. Remikiren contains two peptide bonds; these are susceptible to cleavage by peptidases, which are abundant in the small intestine.

The drug enalaprilate inhibits angiotensin converting enzyme. It is modeled on the structure of the peptide substrate, too, but it is smaller than remikiren and has only one peptide bond, which is also more sterically protected than those found in remikiren. These features give it better bioavailability. Enalaprilate is indeed sufficiently stable and suitable for the oral therapy of patients with hypertension.

Another checkpoint in the renin–angiotensin system is the angiotensin receptor itself. A receptor antagonist that is derived in a straightforward fashion from angiotensin II is saralasin. This is a synthetic peptide, in which three amino acid residues of angiotensin II have been replaced:

Angiotensin sequence: `Asp-Arg-Val-Tyr-Ile-His-Pro-Phe`
Saralasin sequence: `Sar-Arg-Val-Tyr-Val-His-Pro-Ala`

`Sar` represents sarcosine (N-methylglycine). It lacks the negative charge of the aspartate residue that it replaces, which breaks interactions between peptide and receptor that are important for receptor activation. Since saralasin is a relatively large drug molecule and contains multiple peptide bonds, it will not surprise you that it is unsuitable for oral therapy. An inhibitor of the receptor that *is* orally bioavailable is losartan (Figure 1.6). This synthetic molecule has no obvious structural similarity with angiotensin whatsoever, and its lack of peptide bonds and much smaller size make it useful in oral therapy. Nevertheless, it fits into the same binding pocket on the receptor molecule as angiotensin and engages some of the same amino acid residues. This illustrates that it is not mandatory for an inhibitor to closely resemble the covalent structure of the physiological agonist. The same holds true for morphine, which avidly and selectively binds to receptors for endogenous opioid-like peptides, the *endorphins.* The existence of morphine actually enabled the discovery of first the opioid receptors and later of the endorphins themselves.

Like the adrenergic and the histamine receptors, the angiotensin receptor and the opioid receptors belong to the class of G protein-coupled receptors (GPCRs). Among the great number of as yet uncharacterized GPCRs, there most likely are quite a few peptide receptors. As the examples of losartan and morphine show, such receptors can be targeted with drugs that, unlike the physiological peptide ligands, are sufficiently small and stable to be practical for pharmacotherapy. Peptide receptors among GPCRs have great potential for both fundamental research and drug development.

Finally, drugs may also bind to target sites that don't seem to have any physiological ligand whatsoever. Examples among the types of receptors mentioned so far include the voltage-gated sodium channel, which is inhibited by local anesthetics like lidocaine, and tubulin, whose polymerization is inhibited by the anticancer drug vinblastine.

1.4 STRATEGIES OF DRUG DISCOVERY AND DEVELOPMENT

An instructive account of the discovery of a drug can be found in a letter by the Reverend Edmund Stone of Chipping-Norton, Oxfordshire, to the Right Honourable George Earl of Macclesfield, President of the Royal Society, from June 1763 [7]:

> Among the many useful discoveries, which this age hath made, there are very few which, better deserve the attention of the public than what I am going to lay before your Lordship.
>
> There is a bark of an English tree, which I have found by experience to be a powerful adstringent, and very efficacious in curing anguish and intermitting disorders.
>
> About six years ago, I accidentally tasted it, and was surprised at its extraordinary bitterness ... As this tree delights in a moist or wet soil, where agues chiefly abound, the general maxim, that many natural maladies carry their cures along with them, or that their remedies lie not far from their causes, was so apposite to this particular case, that I could not help applying it; and that this might be the intention of Providence here, I must own had some little weight with me. ...

The compound responsible for the bitter taste of the bark of the willow tree is salicylic acid, which indeed has anti-inflammatory activity and is still in use today, mostly as a component of acetylsalicylic acid. What lessons can we learn from this story? Several:

1. Chance, if not Providence, has played and continues to play an important part in drug discovery.
2. If you want to make a contribution to drug discovery, you very likely are in the wrong profession—find a job that lets you get out more!
3. Unsound reasoning may, at least on occasion, be conducive to success, since it encourages an optimistic and exploratory attitude. So, don't permit your enthusiasm to be stifled by the caustic and cynical comments of your superiors.

Commendable though as they may be, the reverend's exploits are hardly representative of today's methods of drug development, which rely less on

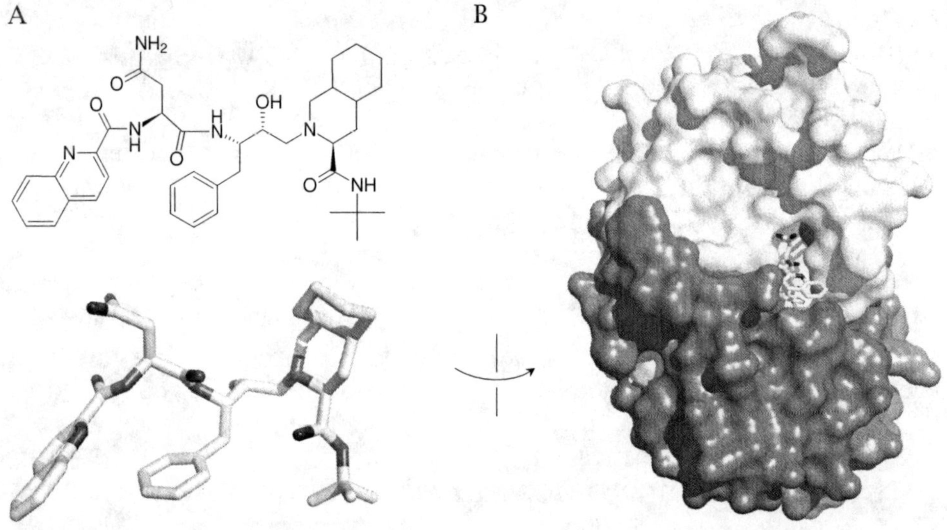

Figure 1.7 The HIV protease inhibitor saquinavir fits tightly into the enzyme's active site. **A:** Structural formula of saquinavir, and stick model of its conformation within the active site. **B:** The inhibitor within the active site of the protease (rotated by approximately 90°). The two subunits of the enzyme, which together enclose the inhibitor, are shown in different shades. Rendered from 1fb7.pdb [6].

mere chance but instead usually combine rational design and systematic brute-force screening of large numbers of compounds in varying proportions.

Rational design does, of course, work best if the drug target is known in a great deal of detail. A good example is the development of inhibitors for the protease of the human immune deficiency virus (HIV), the causative agent of AIDS. HIV protease cleaves the viral polyprotein—the initial translation product of the viral genome, which contains all viral proteins hanging together like pearls on a string—into its individual components; this cleavage is essential for the maturation of virus particles. The crystal structure of HIV protease was used to design synthetic molecules that would snugly fit and avidly bind within the active site. Figure 1.7 shows the inhibitor saquinavir bound to the enzyme. HIV protease inhibitors have become one of the mainstays of HIV therapy; used in combination with the previously introduced reverse transcriptase inhibitors, they have brought to pass a very significant increase in the life expectancy of HIV patients.

An important success of the brute-force approach was the discovery of "prontosil rubrum" (sulfamidochrysoidine, Figure 1.3), the first antibacterial drug of the sulfonamide type [2]. *Rubrum* means "red" in Latin, so this is another dye turned drug. The biochemical mechanism of action was completely unknown at the time, but the drug proved very active against a considerable range of bacterial species [3]. The discovery of sulfonamides in the 1930s may have contributed to the delay in the development of penicillin, the effect of which was discovered by Fleming in 1928 but which was not available for clinical use before 1942.

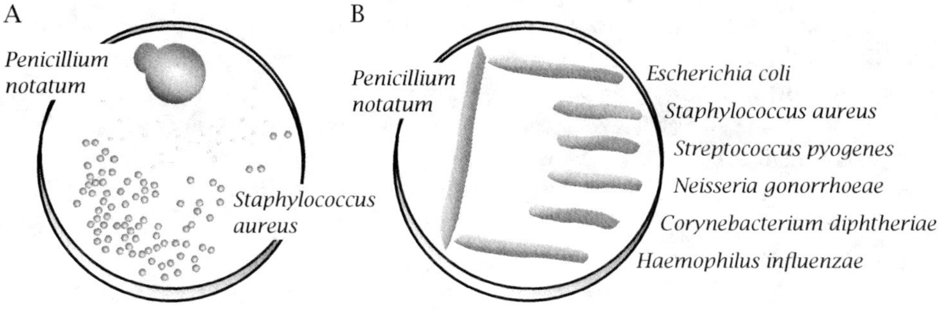

Figure 1.8 The discovery of penicillin. **A:** Sketch of Fleming's original petri dish, with a large contaminating colony of the mold *Penicillium notatum* and fading *Staphylococcus aureus* colonies in the vicinity. **B:** Cross-streak of *Penicillium* and various bacterial species. The Gram-negative bacteria *Escherichia coli* and *Haemophilus influenzae* are resistant to penicillin and grow right up to the mold, whereas several Gram-positive species are inhibited. The only Gram-negative species to be inhibited is *Neisseria gonorrhoeae*. (Drawn after the original photographs in Ref. [5].)

The brute-force approach is, of course, very labor-intensive, particularly if the screening process involves animal experiments. This was the case with sulfamidochrysoidine, since the drug molecule itself has no antibacterial activity. Its antibacterial component is sulfanilamide, which is only released through reductive metabolism in the animal's or patient's liver. Sulfanilamide competitively inhibits the utilization of *p*-aminobenzoic acid by dihydropteroate synthase, which occurs in the bacterial synthesis of folic acid. Since mammals don't synthesize folic acid but instead require it as a vitamin, they are not affected by this inhibition.

The amount of labor required is one important reason why drug development is now largely done by major pharmaceutical companies and not in academia. In fact, sulfamidochrysoidine was discovered at the biggest pharmaceutical company of the era, the German IG Farben.[5] To cut down on the amount of work required, in vitro assays and cell culture are preferred over animal experiments, where that is possible; high throughput formats of assays and analytical instrumentation are an important current focus of research (see Chapter 15).

An example of rational design and systematic screening being used in combination is the development of histamine H_2 receptor antagonists (see above, page 7). While histamine, the agonist, served as the starting point, the structure of the receptor and the binding site was unknown, which required a wider net to be cast at the stage of structural variation. Structural biology has made great strides in recent years, and the structures of many soluble proteins are now available; however, membrane proteins, including G protein-coupled receptors, are lagging behind, and therefore the need for screening large numbers of compounds remains.

[5] *Farbe* is German for dye, paint, or color. The IG Farben was dismantled after World War II because of its involvement with the production of poisons used in the holocaust.

While rational design and systematic screening are certainly the dominant paradigms today, mere chance and an open mind for taking advantage of it remain important. The most striking example that comes to mind is the discovery of penicillin (Figure 1.2). Here is a summary of this strategy: (1) forget to properly cover your petri dish and wait for it to be contaminated by a vagrant mold, (2) observe that the mold kills bacteria (Alexander Fleming, 1929; Figure 1.8), (3) purify the active ingredient from cultures of the mold (Howard Florey and Ernst Chain, 1942).

While the discovery of penicillin was fortuitous, Darwin's theory suggested that the production of antimicrobial poisons might not be limited to this particular mold, *Penicillium notatum*, but rather be common among competing microoorganisms.[6] Selman Waksman took up this idea by systematically screening soil microorganisms for compounds with activity against pathogenic bacteria. He discovered a large number of antibiotics, mostly from member species of the bacterial genus *Streptomyces.* The first one, actinomycin D, is toxic for both prokaryotic and eukaryotic cells. It therefore cannot be used in bacterial infections, but it is useful in tumor chemotherapy. Other important discoveries of his are streptomycin, tetracycline, and chloramphenicol, which can be used systemically and still have their place in the treatment of bacterial infections today.

REFERENCES

[1] J. Brunel, "Antibiosis from Pasteur to Fleming," *J. Hist. Med. Allied Sci.* 6 (1951), 287–301.

[2] G. Domagk, "Chemotherapie der Streptokokken-Infektionen," *J. Mol. Med.* 15 (1936), 1585–1590.

[3] G. Domagk, "Weitere Untersuchungen Über die Chemotherapeutische Wirkung Sulfonamid-haltiger Verbindungen bei Bakteriellen Infektionen," *J. Mol. Med.* 16 (1937), 1412-1418.

[4] P. Ehrlich *Partial Cell Functions,* 1908 URL: `http://nobelprize.org/nobel_prizes/medicine/laureates/1908/ehrlich-lecture.html`.

[5] A. Fleming, "On the antibacterial action of cultures of a *Penicillium*, with special reference to their use in isolation of *B. influenzae*," *Br. J. Exp. Pathol.* 10 (1929), 226–236.

[6] L. Hong et al., "Crystal structure of an in vivo HIV-1 protease mutant in complex with saquinavir: insights into the mechanisms of drug resistance," *Protein Sci.* 9 (2000), 1898–1904.

[7] E. Stone, "An Account of the Success of the Bark of the Willow in the Cure of Agues," *Philos. Trans. R. Soc.* 53 (1763), 195–200.

[6] Fleming was indeed not the first one to observe antibacterial activity of fungi or other bacteria. Several similar preceding observations are described in [1]. In view of the considerable time that elapsed after those observations, and even between the observation of penicillin and its purification, it appears that the state of the art in chemistry hindered a systematic exploitation of those previous findings.

CHAPTER 2

PHARMACODYNAMICS

Pharmacodynamics is concerned with general principles of drug action.[1] The effect of a drug may be divided into two aspects: (1) the immediate interaction of the drug with its receptor and (2) the function of the receptor, as influenced by the drug, in the context of the cell and the organism.

If a protein[2] binds a drug without undergoing any distinct, specific change in its function, this protein does not qualify as a receptor. It is not uncommon for drugs to bind to one or more sites other than their receptors. For example, many drugs bind to the plasma protein albumin, which has multiple binding sites for free fatty acids and other solutes of a hydrophobic nature. Such binding may have consequences for the retention and distribution of a drug within the organism, as we will see in Chapter 3. Yet, albumin binding will not by itself trigger any significant physiological effect, and therefore it does not make albumin a receptor of these drugs.

2.1 MOLECULAR FEATURES OF DRUG–RECEPTOR INTERACTION

Binding of a drug to its receptor can be noncovalent or covalent. Noncovalent binding may involve all the usual suspects among the molecular forces, that is, hydrogen bonds, electrostatic and van der Waals forces, hydrophobic and

[1]When applied to a specific drug, the term simply refers to that drug's mode of action.

[2]We have seen in Chapter 1 that most drug receptors are proteins. We will confine the discussion to this case in the current chapter.

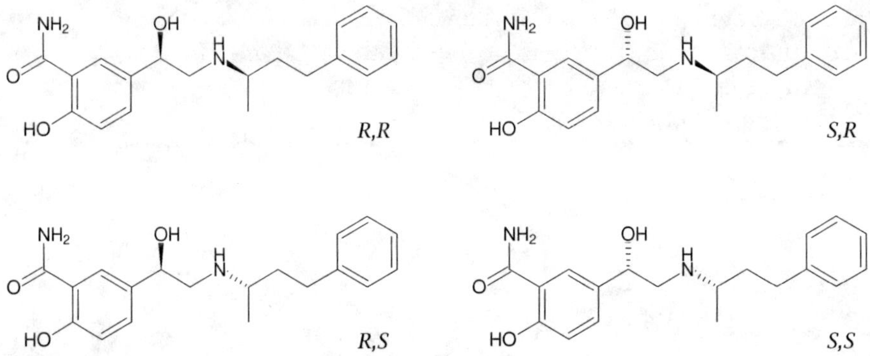

Figure 2.1 Labetalol as an example of stereoselective drug action. Labetalol, an α- and β-adrenergic antagonist, is a racemic mixture of four stereoisomers. The *R,R* isomer carries most of the β-adrenergic blocking activity, whereas the *S,R* isomer carries most of the α-blocking activity. The *R,S* and *S,S* isomers are dead freight [5].

cation–π interactions. High affinity of a drug for its receptor requires a combination of multiple noncovalent bonds. This means that multiple functional groups in the drug molecule need to interact favorably with those in the receptor's binding site. Just as with a substrate that binds to the active site of an enzyme, this requires a high degree of steric complementarity between receptor and drug, which is often reflected in the structural resemblance between the drug and the corresponding physiological ligand. To some extent, steric complementarity is also required with drugs that bind covalently, because prior to the covalent reaction they still need to be steered to their proper binding sites by noncovalent forces.

If there are multiple drugs available that bind to the same site on the same receptor, it will often be possible to derive a consensus set of essential molecular features that support an avid interaction. Each feature can be expressed as a vector that comprises the type of binding force and its orientation in space. A set of consensus features expressed in this quantitative form is called a *pharmacophore* and can be used in molecular simulation calculations to evaluate potential new derivatives. If a high-resolution structure of the receptor is available, it may even be possible to construct pharmacophores from scratch, without any prior experimental binding data. Chapter 15 will discuss this subject in greater detail.

Again as with enzyme-substrate interactions, the asymmetric nature of the receptor proteins implies that binding of ligands with asymmetric centers will be stereoselective. Physiological metabolites or mediators and natural compounds that are asymmetric often occur in one stereoisomeric form only. In contrast, synthetic drugs are usually produced as racemic mixtures. In such cases, the pharmacological effect is often due to only one of the stereoisomers, although with some drugs multiple isomers may contribute. A case in point is labetalol [5]. This drug has two asymmetric centers and has antagonistic activity on both α- and β-adrenergic receptors. These two activities are associated with two of the four stereoisomers (Figure 2.1).

Stereoisomers that do not contribute to the desired effect of a drug may still contribute to its side effects. Selective synthesis or purification of stereoisomers may be involved and costly; whether it is required must be decided for each drug individually. If you look at the molecular structures in this book, you will see that many have one or more asymmetric centers; however, we will generally not discuss the differential activity of their stereoisomers.

2.2 THEORY OF DRUG-RECEPTOR BINDING

The simplest possible case of drug-receptor binding is also a very common one, so it is a logical starting point. In this scenario, binding is (1) noncovalent, (2) stoichiometric (that is, each receptor molecule binds a single drug molecule), and (3) subject to the law of mass action. There are exceptions to each of these rules, some of which will be considered later. The major conclusions will be presented here with only a modicum of equations; some more mathematical detail is given in the appendix of this chapter.

2.2.1 Drug-receptor binding and the law of mass action

We assume that the binding of the drug L to its receptor R to form the drug-receptor complex LR is governed by a single equilibrium constant K:

$$K = \frac{[\mathrm{L_{free}}][\mathrm{R_{free}}]}{[\mathrm{LR}]} \tag{2.1}$$

As shown in Section 2.6.1, this yields the following expression for the *receptor occupancy* Y, which is a key parameter in drug action:

$$Y = \frac{[\mathrm{LR}]}{[\mathrm{R_{total}}]} = \frac{[\mathrm{L}]}{[\mathrm{L}] + K} \tag{2.2}$$

Figure 2.2 shows three different graphical representations of mass action binding kinetics. The linear plot of receptor occupancy versus ligand concentration has the same shape as the Michaelis-Menten law of enzyme activity, for the simple reason that the latter is also based on the law of mass action. In the semilogarithmic plot, an increase in the dissociation constant—equivalent with a lower affinity—shifts the receptor saturation curve to the right but doesn't change its shape. The semilogarithmic plot can also accommodate a very wide range of ligand concentrations.

Binding data can also be plotted using the Scatchard transformation [11], which is explained in Section 2.6.4. If binding is indeed subject to a single equilibrium, all data points will fall on a straight line in the Scatchard plot. If they don't, it indicates a more complex situation, such as the presence of two or more binding sites with different affinities.

What degree of receptor occupancy should a drug achieve in practice, and with what affinity should it bind? To obtain a stable and reproducible level of drug activity, we will frequently want to saturate the receptor completely. Most

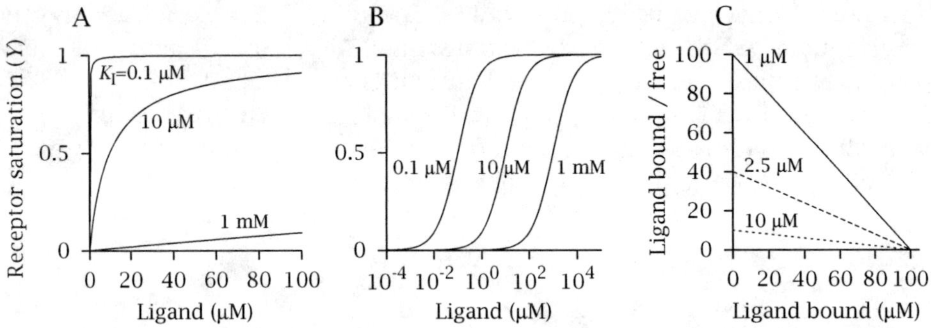

Figure 2.2 Drug-receptor binding with mass action kinetics plotted in different formats. **A:** Linear plot of receptor saturation versus ligand concentration for various dissociation constants. **B:** Semilogarithmic plot of the same. **C:** In the Scatchard plot, the ratio of bound ligand to free ligand is plotted against the concentration of bound ligand, which yields a straight line as long as all binding sites have the same affinity.

drugs are applied in dosages of several milligrams to several hundred. With typical molecular weights of a few hundred daltons and a human body weight in the range of 50–100 kg, this works out to low micromolar concentrations in the tissues at the most. Therefore, to reliably achieve receptor saturation, dissociation constants should be in the nanomolar range. Note, however, that for a maximal effect downstream of the receptor complete saturation is not always required. This is explained in Section 2.3.

2.2.2 Competitive binding

Frequently, a drug binds to its receptor within the same site as the receptor's physiological ligand, which it thereby prevents from binding, without however evoking any receptor activity itself. This mode of action is known as *competitive inhibition.* It applies to most clinically used enzyme inhibitors, and it is also very common with receptor antagonists of hormones or neurotransmitters such as norepinephrine, histamine, or serotonin.

The extent of competitive inhibition is measured by the reduction in the occupancy of the receptor with its physiological ligand. With K_L and K_I as the equilibrium constants for the ligand L and the antagonist I, respectively, the apparent equilibrium constant K' will be given by

$$K' = K_L \left(1 + \frac{[I]}{K_I}\right) \tag{2.3}$$

At a fixed inhibitor concentration $[I]$, this is tantamount to an increased dissociation constant, or reduced affinity, for the physiological agonist L. Therefore, the inhibitor will cause a dose-dependent right shift of the receptor occupancy curve (Figure 2.3C).

One consequence of this relationship is that the effect of even very large amounts of a noncovalent, competitive inhibitor can still be overridden by sufficient quantities of physiological ligand. Likewise, the effect of an excessive dosage of a physiological agonist can be neutralized by applying an equally excessive amount of antagonist. A historical illustration of this latter principle was the treatment of parathion poisonings with atropine. Parathion, an insecticide that is no longer in use, blocks the cleavage of acetylcholine by cholinesterase (see Section 6.8.3). Consequently, acetylcholine piles up to enormous concentrations in cholinergic synapses. This condition could be treated with atropine, a competitive antagonist of acetylcholine. However, the milligram dosages of atropine prepackaged for more conventional applications were totally ineffective. Gram amounts were necessary, and the job of cracking open the scores of milligram ampules needed commonly fell to the youngest resident.[3]

2.2.3 Covalent or irreversible inhibition

Some inhibitory drugs not only competitively displace their corresponding physiological ligands, but also attach to their receptors covalently and therefore irreversibly.[4] In this case, there will be no dynamic equilibrium, but a stable state will be reached once all available drug molecules have covalently reacted with the receptor or with alternative reactants such as glutathione, or have been eliminated otherwise. Thereafter, the unreacted fraction of the receptor molecules will continue to equilibrate between the free and ligand-associated states. The equilibrium constant for the physiological ligand will be unchanged, and therefore the horizontal position of the receptor occupancy curve will remain the same. The curve will, however, be vertically compressed in a dose-dependent fashion; the decrease in height corresponds to the fraction of receptor molecules that were covalently modified (Figure 2.3C).

2.2.4 Example: Reversible and irreversible inhibition of α-adrenergic receptors

For an experimental illustration of the foregoing, let us look at two inhibitors of α-adrenergic receptors. These receptors are stimulated by epinephrine and norepinephrine; stimulation will increase the tension of blood vessel walls and therefore enhance blood pressure. α-Adrenergic receptors are very numerous in the spleen. The spleen has a sponge-like tissue structure and stores about half a liter of blood, which upon adrenergic stimulation will be squeezed out and injected into the circulation.[5] This extrusion of blood is brought about by the contraction of smooth muscle cells that are embedded in the spleen tissue. Accordingly, if we take a fresh slice of spleen and bathe it in solutions

[3]Parathion poisonings were not infrequent and occurred mostly with suicidal intention. At some point, extra large dosages of atropine became available for the very purpose of treating these patients.

[4]Covalent and irreversible inhibition are often but not always equivalent. Examples of slowly reversible covalent inhibition are the acylation of muramyltranspeptidase by β-lactams (Section 11.4.2) and the carbamoylation of cholinesterase by carbachol (Figure 6.26). Examples of noncovalent yet almost irreversible inhibitors are reserpine (Figure 6.23) and Cbz-Phe-Val-Phe phosphonate (Figure 15.2).

[5]A kind of endogenous blood doping—epinephrine and norepinephrine are often referred to as "fight-or-flight hormones".

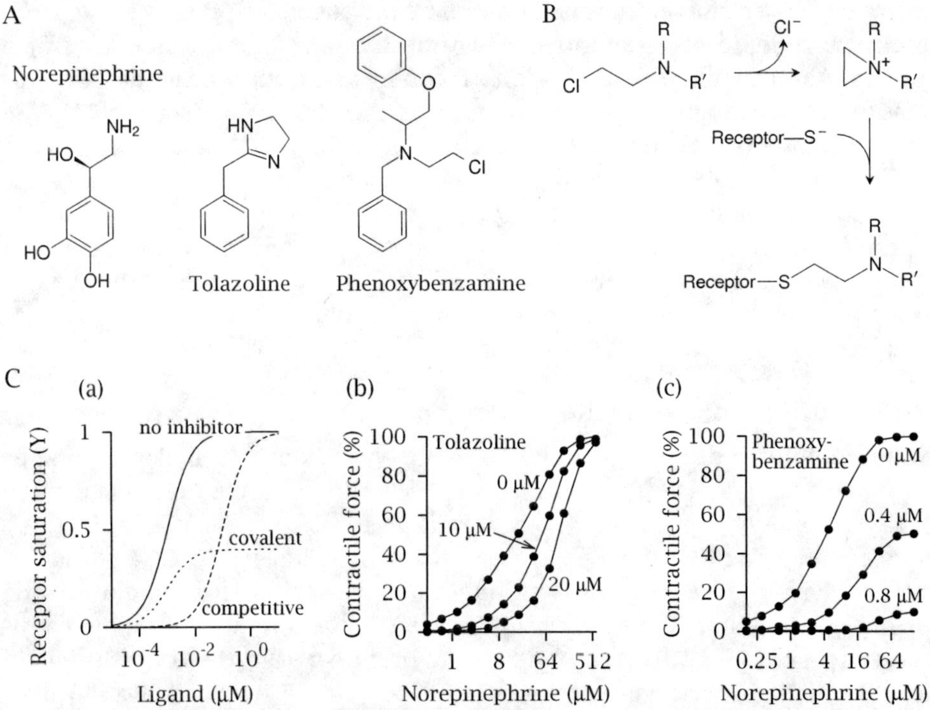

Figure 2.3 Competitive and covalent inhibition of α-adrenergic receptors. **A:** Structures of norepinephrine, the competitive antagonist tolazoline and the irreversible antagonist phenoxybenzamine. **B:** Phenoxybenzamine undergoes spontaneous cyclization to the aziridine derivative, followed by covalent reaction with a cysteine residue in the receptor. **C:** Dose–response curves in the presence of inhibitors. (a) Numerical examples. Competitive inhibition causes a right shift, covalent inhibition a vertical compression of the receptor occupancy curve. (b,c) Effects of tolazoline and phenoxybenzamine on the contraction of spleen smooth muscle induced by norepinephrine. (Panels (b) and (c) redrawn from original data in Ref. [2].)

of mediators or drugs, we can measure its mechanical tension to quantify the extent of α-adrenergic stimulation. Figure 2.3 shows the force of contraction developed by such spleen tissue strips in response to varying concentrations of norepinephrine, alone or in the presence of tolazoline or phenoxybenzamine, respectively. By comparing these experimental data to the theoretical plot on the left, you will be able to decide which inhibitor is the reversible one and the irreversible one, respectively.

Let us consider the molecular principles behind the two modes of inhibition. Figure 2.3A shows the structures of norepinephrine and of the two inhibitors. With some imagination, one can spot the similarity between physiological agonist and inhibitors, so that it is understandable that all of them bind to the same site on the α-adrenergic receptor. Tolazoline has no obvious reactive groups, and it will therefore bind noncovalently and reversibly. Phenoxybenzamine, on the other hand, has a reactive chloroethyl group attached to the nitrogen that will spontaneously form an aziridine intermediate and then react with a cysteine residue on the receptor.[6] The drug therefore becomes covalently attached to the receptor and irreversibly inactivates it (Figure 2.3B).

Several things are notable about the action of phenoxybenzamine. Firstly, the initial reaction, which forms the aziridine ring, is rather slow; this causes the pharmacological action to lag behind the plasma concentration of the drug. On the other hand, receptor blockade will persist long after any excess drug has been eliminated. With most drugs that act by noncovalent association with their receptors, plasma levels correlate much more closely with the intensity of drug action. Secondly, while the benzylamine moiety of phenoxybenzamine targets it to the α receptor, the chemical reactivity of the ethyleneimine group is rather non-selective. Molecules not bound to the receptor are prone to react in random locations, potentially causing harm including DNA damage. Because of this, phenoxybenzamine is not used in most clinical applications of α receptor blockade. It is, however, the drug of choice in one particular disease called *pheochromocytoma.* This is a tumor of the adrenal glands that produces and intermittently releases very large amounts of epinephrine and norepinephrine, which then cause dangerous spikes in blood pressure. The advantage of phenoxybenzamine in pheochromocytoma is a direct consequence of its covalent mode of binding: The covalently inactivated receptor cannot be reactivated by whatever amounts of hormones released. A competitive inhibitor such as tolazoline, in contrast, could be overridden in this particular situation.

2.2.5 The two-state model of receptor activation

In the preceding sections, we have assumed the interaction of a drug with its receptor to be governed by a single equilibrium. This assumption is adequate with a competitive inhibitor, since it is at least conceivable that the receptor retains the same conformation with or without the ligand bound to it. However, if the ligand is an *agonist*, meaning that it activates the receptor, binding must impose some conformational change on the receptor molecule. This conformational change will then cause another site of the receptor molecule to initiate interaction with some downstream element, such as a G protein in the case of the α-adrenergic receptor, or ATP and a substrate protein in the case of a receptor tyrosine kinase. If the receptor is a ligand-gated channel, the conformational change will result in the opening or closing of the passageway that controls the flow of ions across the membrane.

The effect of receptor agonists is entirely analogous to the regulation of enzymes by allosteric activators: In both cases, the protein may adopt two distinct conformations, one active and the other inactive, and binding of a ligand influences the transition between them. However, the nomenclature is a little different: Agonists or antagonists that bind within the same binding site as the physiological ligand of the receptor are called *orthosteric* ligands; the term *allosteric* is applied only to those ligands that bind elsewhere and

[6]Reaction with a cysteine has been demonstrated with α_2 receptors [4]. The cysteine in question is conserved in α_1 receptors also.

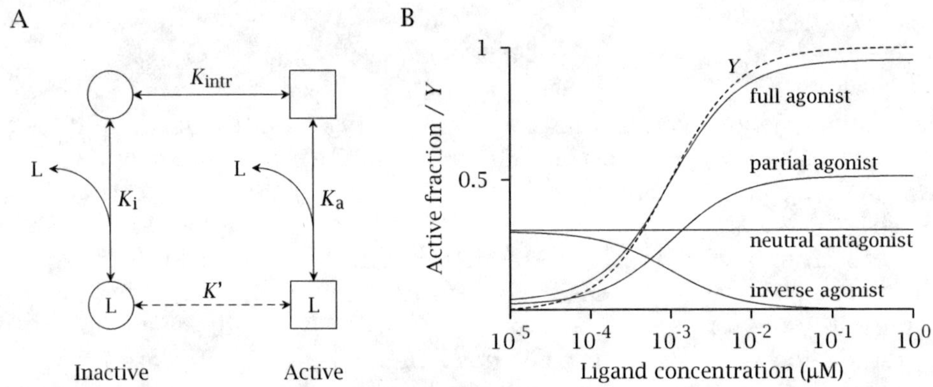

Figure 2.4 The two state model of receptor activation. **A:** The receptor is controlled by three linked equilibria. The transition between the active and the inactive conformations of the free receptor is subject to K_{intr}. Ligand binding to the active and inactive conformation is subject to K_a and K_i, respectively. Together, these three constants also determine the equilibrium constant for the conformational transition of the bound receptor, K'. **B:** Numerical examples for receptor active fraction as a function of ligand concentration, for various types of ligands. Each curve represents a different set of values for K_{intr}, K_a and K_i, but all of them correspond to the same receptor saturation curve (dashed line).

may therefore bind the receptor simultaneously with the physiological ligand or other orthosteric ligands.

The two-state model describes the effect of agonists and antagonists on a receptor by means of three linked equilibria. In the absence of any ligand, the equilibrium between the inactive state of the receptor R_i and the active state R_a is controlled by the intrinsic constant K_{intr}:

$$K_{intr} = \frac{[R_i]}{[R_a]} \tag{2.4}$$

Since the two receptor states have different conformations, they will bind the ligand with different affinities, which are given by the equilibrium constants K_a and K_i, respectively. The functional effect of a drug on the receptor then depends on the ratio of these two equilibrium constants, and it is manifest in the *active fraction* (f_A) of the receptor, which is a function of the ligand concentration. In contrast to the simple mass action model, f_A is no longer identical with the receptor occupancy (see Figure 2.4). We can distinguish the following cases:

1. $K_a \ll K_i$ —the ligand has a very strong preference for binding to the active form. At saturating concentration, it will convert almost all receptor molecules to the active conformation. A ligand with this behavior is called a *full agonist.*

2. $K_a < K_i$ —if the difference between K_a and K_i is less pronounced, some receptor molecules will remain in the inactive state even at saturating ligand concentrations. Drugs with this behavior are called *partial agonists.* The ex-

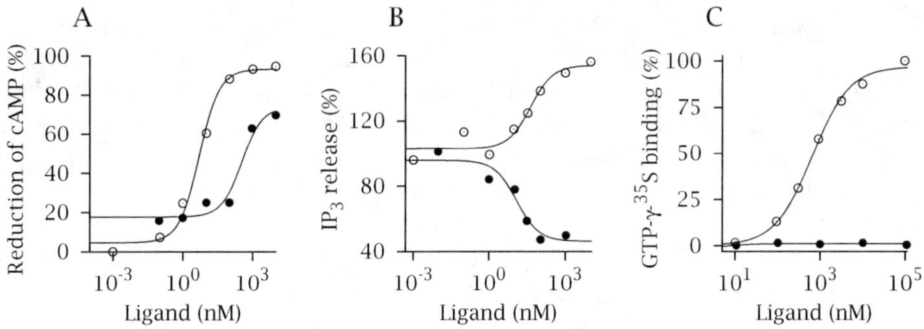

Figure 2.5 Activities of the synthetic drug aripiprazole (•) and of the physiological ligands (⊙) on serotonin (5-HT) and dopamine receptors. Aripiprazole is a partial agonist at the 5-HT_{1A} receptor (**A**) and an inverse agonist at the 5-HT_{2B} receptor (**B**). It also inhibits the dopamine D_{2L} receptor (**C**). Since the latter receptor has no detectable basal activity, inverse agonism of aripiprazole is not detectable either. (Figure redrawn from original data in Ref. [12].)

act balance of active and inactive receptors will also depend on the intrinsic conformational equilibrium constant K_{intr}, as is discussed in Section 2.6.5.

3. $K_a = K_i$ —the ligand binds with equal affinity to both the active state and the inactive state (this will only rarely be strictly true). In this case, f_A will be unaffected by the ligand concentration. Ligands with this behavior are strictly neither agonists nor antagonists. Nevertheless, they are conventionally called *neutral antagonists.* This is due to the fact that most well-behaved receptors only turn up the volume when asked explicitly to do so, that is, in the absence of any ligands their f_A is very small. A neutral antagonist will compete against an agonist if both are applied simultaneously and thus keep receptor activity at its low intrinsic level.

4. $K_a > K_i$ —the ligand binds preferentially to the inactive form. A drug with this behavior is a true antagonist, also called an *inverse agonist.*

If f_A is close to zero to begin with, even an inverse agonist will have very little functional effect when applied alone. However, there are receptors—occurring either as wild-type forms or as *constitutively active mutants*—that in the absence of ligand have an active fraction significantly greater than zero. With such receptors, the shift toward the inactive state in response to an inverse agonist will be significant and result in a functional response. Inverse agonism thus is a property of the receptor as much as of the ligand (Figure 2.5).

In the two-state model, the conformational change of the receptor that results in its activation requires an input of energy. This energy will be taken out of the free energy of ligand binding. Consider a hypothetical drug molecule with three features that equally contribute to the overall binding energy. Removal of any one of these features may then lower the binding energy to such an extent that it no longer suffices to drive the activating conformational change. Therefore, the two-state model suggests that drug activity may be quite sensitive to even small changes of molecular structure. It also suggests that, if the receptor should undergo a mutation to higher intrinsic activity, its affinity for the agonist should increase, which has indeed been observed experimentally [10].

In the same way that enzymes may possess more than one allosteric binding site and thus bind more than one allosteric effector simultaneously, it is also possible for one receptor to bind two different ligands at the same time. A particularly interesting case may arise if one of these is the physiological ligand, whereas the other is a drug that binds to an alternate site that does not overlap the orthosteric binding site. Depending on the vigor of its conformational impact, the drug may then act as a modulator of receptor activation, amplifying or attenuating the signal conveyed by the physiological mediator rather than supplanting it [7]. A case in point is the $GABA_A$ receptor, which contains several allosteric binding sites that bind different classes of drugs and seem to differ in the maximal degree of receptor response (see Section 6.7.3).

2.2.6 Receptor behavior not covered by the simple two-state model

The relatively simple, straightforward two-state model laid out in the preceding section readily describes a large number of experimental observations. However, some kinds of receptor behavior are not adequately accounted for.

Cooperative behavior of multimeric receptors. Many receptors do not occur as monomeric molecules but as complexes of two or more subunits. While this does not a priori exclude the possibility that ligand binding and receptor activation follow the simple rules discussed so far, most multimeric receptors behave *cooperatively*, meaning that the conformational transitions are coupled between their subunits. As is illustrated in Figure 2.6, this results in steeper dose-response curves for both occupancy and activation, and in more favorable signal-to-noise ratios than those observed with simple mass action kinetics.[7]

Section 2.6.5 takes a closer look at the relationship between the simple two-state model and the cooperative model of receptor activation.

Multiple active states. The two-state model only allows for a single active state of any given receptor. This is a straightforward assumption with receptors that perform just one kind of activity. However, many G protein-coupled receptors activate several different G proteins that control different biochemical cascades inside the cell. With some receptors, it has been observed that different agonists may preferentially activate different subsets of these G proteins. This behavior is called *agonist-specific coupling* or stimulus trafficking. An example is shown in Figure 5.5.

While it is readily conceivable that any receptor might preferentially activate one G protein over another, in the two-state model the order of preference should not depend on the agonist, since all agonists are supposed to induce the very same active conformation of the receptor. Therefore, the observation of agonist-specific coupling contradicts the two-state model, and it requires us to assume as many distinct active conformations as there are patterns of G protein

[7]If mass action kinetics were the only mechanism of receptor activation, the signal-to-noise ratio would degrade in each successive binding event along a signal transduction cascade. Cooperative proteins make complex signal processing possible.

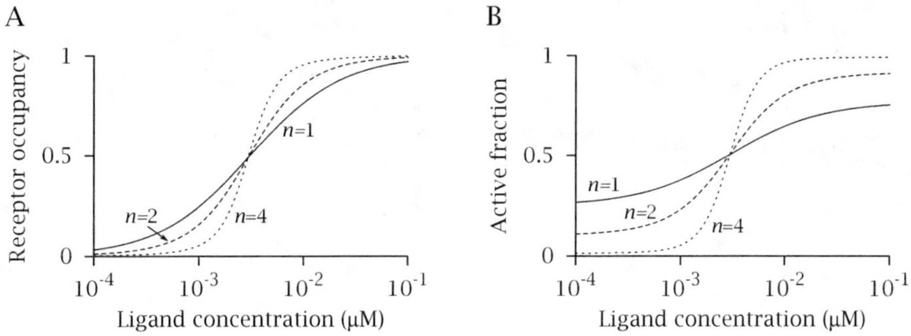

Figure 2.6 Ligand saturation (left) and active fraction (right) of a set of fictitious receptors with one (n=1), two and four subunits, respectively. Each subunit behaves according to the same set of two-state model parameters (see Section 2.2.5). The conformational transitions in the dimeric and tetrameric receptors are assumed to be fully cooperative.

selectivity. Agonist-specific coupling is not only theoretically interesting but it also offers a prospect for the development of drugs with enhanced selectivity. We will come back to it in Chapter 5.

Refractory states. In the two-state model, the receptors will switch back and forth between an inactive state and an active state. This does not adequately describe the behavior of many ion channels, which instead cycle through a sequence of three distinct states. After switching from the inactive or closed state to the active or open state, they convert to a distinct refractory state. In both the inactive and the refractory states, the channel is closed, but the two differ in that the refractory state cannot directly revert to the open state. Therefore, the receptor has to move on from the refractory state to the inactive state before it is ready to participate in another round of activation.[8] This behavior applies to both voltage-gated and ligand-gated channels; it will be considered in more detail in Chapter 6.

2.3 DOSE-EFFECT RELATIONSHIPS IN SIGNALING CASCADES

In the example of reversible and irreversible α receptor blockade, we noted the similarity between empirical dose-effect relationships and theoretical plots (see Figure 2.3). This needs to be qualified in two ways: (1) while the theoretical plots modeled receptor occupancy, the experiment measured muscle tension, and (2) the resemblance of theoretical and experimental dose-response curves is not perfect. These two observations are indeed related. A perfect similarity of theoretical and experimental plots could be expected only in case of a strictly linear relationship between the occupancy of α receptors with norepinephrine and muscle contraction. Considering that muscle contraction is triggered several steps downstream of receptor activation, there are numerous possible factors that may distort this linearity; in reality, no linear relationship

[8]To the extent that this cycle is really unidirectional, it requires an input of energy; it is interesting to think about the different kinds of energy that are involved.

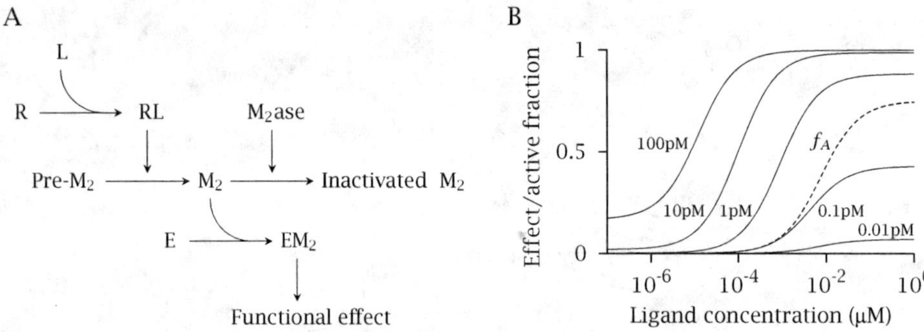

Figure 2.7 Dose-response relationship in biochemical cascades. **A:** A simple model cascade, consisting of a receptor R, a second messenger M_2, and an effector E. **B:** The influence of varying receptor abundance on the dose-response curve in our model cascade, measured in terms of the functional effect. The dashed curve depicts the active fraction (f_A) of the receptor, which does not change with receptor abundance. See text for further details.

will ever be observed if drug target and drug effect are separated by intervening biochemical cascades. It thus turns out that the shape of a dose-effect relationship will depend strongly on the functional proximity of the drug receptor molecule and the choice of experimentally observed parameter.

If the observable effect is directly associated with the receptor, we may indeed expect a linear relationship between receptor occupancy and function. There are viable examples of this. If the receptor is an enzyme, we can measure its turnover rate; if the receptor is an ion channel, we may observe its conductivity by patch-clamping. On the other hand, very often the functional drug effect is observed a long way downstream of the receptor, as in our example of smooth muscle contraction and α-adrenergic blockers. Other examples are the inhibition of prostaglandin synthesis by acetylsalicylic acid, with perceived pain relief as the functional readout, or the activation of nuclear hormone receptors by synthetic androgens, with a readout downstream of transcriptional regulation such as increased muscle mass or inhibition of sperm production. In cases like these, there will be numerous possible reasons for deviations from linearity in the dose-effect relationship. Such deviations might, of course, take any shape. Nevertheless, some useful generalizations are possible.

2.3.1 A model cascade

Let us consider a very simple hypothetical model cascade, with the following elements: (1) a ligand binds to and activates a receptor; (2) the activated receptor causes the formation of a second messenger; (3) the second messenger binds to and activates an effector molecule, which causes some final, quantifiable effect; and (4) the second messenger is subject to enzymatic inactivation, as for example, cyclic AMP that is cleaved by phosphodiesterase.[9]

Figure 2.7 illustrates the relationship between the ligand concentration and the functional effect in this system. The dose-effect curves show the influence

[9]The quantitative assumptions of the model are more thoroughly explained in Section 2.6.6.

of varying the abundance of receptor $[R_{total}]$. Also shown is the active receptor fraction f_A. Note that even at saturating ligand concentrations f_A does not reach 1, which means that our hypothetical ligand is a partial agonist. We can make the following observations:

1. At sufficiently high total receptor concentration, the effector will become maximally activated, even though the receptor is not.
2. The threshold of activation is shifted to lower ligand concentrations with increasing values of $[R_{total}]$. In fact, the downstream effector may become more sensitive to the ligand than the receptor itself.
3. At very high total receptor concentrations, effector activity may be observed even in the complete absence of ligand.

When the receptor concentration is high and the functional effect precedes receptor saturation, we can knock out a substantial fraction of receptors using an irreversible inhibitor, yet still maximally activate the downstream effector with a sufficient dosage of agonist. The fraction of receptors that can be removed before the maximum effect becomes diminished is referred to as *receptor reserve* or *spare receptors.*

2.3.2 Experimental cascade effects

The effects of variable receptor concentration apparent in our hypothetical model cascade can also be observed in pharmacological experiments. An example is the stimulation of β-adrenergic receptors in the heart by epinephrine. Half-maximal increase of heart muscle contractility in response to epinephrine is observed at 2% receptor occupancy, and the effect will saturate long before the receptor does [3]. Overexpression of β-adrenergic receptors in transgenic mice can activate the heart muscle to the same extent as otherwise observed only with maximal stimulation by isoproterenol [8]. Interestingly, β receptors, like other G protein-coupled receptors, are subject to regulation by phosphorylation and reversible removal from the cell surface. Either regulation may reduce the sensitivity of the system to epinephrine but leave the maximum response observed at receptor saturation unchanged.

The model cascade discussed here is simpler than most real signaling cascades, yet it does resemble some of them. For example, acetylcholine may activate certain ion channels through a cascade with the following elements: (1) acetylcholine binds to and activates the M_2 muscarinic acetylcholine receptor, and (2) the ligand-bound, active receptor catalyzes the activation of G protein molecules, the $G\beta\gamma$ subunits of which act as the second messengers and open the channels. If we measure channel conductance as the system's response to acetylcholine, we can expect a signal that is linearly related to the occupancy of the effector with the second messenger.

The conclusions from our simple model cascade are also relevant for more complex cascades. With several successive intermediate messengers, the left shifts of the overall dose-response curves will be cumulative. The possibility of cascade effects must be kept in mind when working with recombinantly expressed receptors in cell culture, which is very common nowadays in experimental pharmacology. Receptor overexpression may change both the basal activity and the sensitivity to agonist stimulation [13].

2.4 POTENCY AND EFFICACY

Two concepts that at this point should present no more difficulty are those of *potency* and *efficacy* of a drug. The potency measures how much of a drug is required for its specific effect; it is expressed as the inverse of the EC_{50}, that is of the concentration required to achieve 50% of its maximal effect. In contrast, the efficacy measures the maximum strength of the effect itself, at saturating drug concentrations. Thus, in Figure 2.8, the drug Dashed Mite™ exceeds the competing drug Solid Forte® in potency, while the opposite is true of the efficacy.

Two drugs that act on different receptors will typically differ in efficacy. For example, drugs that block adrenergic receptors or activate cholinergic ones will both lower the heart rate, but the maximal strength of that effect will likely differ. Two drugs that act on the same receptor may also differ in efficacy if one is a full agonist and the other is a partial agonist.

Both potency and efficacy of a drug may be affected by the existence of spare receptors if the observed effect is indirectly related to receptor occupancy, as discussed in Section 2.3. For full agonists, increasing the number of receptors will lower the EC_{50} and thus increase the potency. With partial agonists, a larger number of occupied receptors may compensate for partial receptor activation and therefore also increase the efficacy, possibly to the same level attained by a full receptor agonist.

2.5 BENEFICIAL AND TOXIC DRUG EFFECTS

As an anticlimactic finale to this theory-heavy chapter, let us consider the relationship between beneficial effects and toxic effects of drugs. Possible forms of this relationship are schematically depicted in Figure 2.9. In some cases, toxicity arises simply as an extension of the beneficial or therapeutic effect. As an example, consider warfarin, an indirect inhibitor of the posttranslational γ-glutamyl carboxylation and concomitant activation of several blood coagulation factors (see Section 10.4.4). While warfarin helps to prevent thrombosis and stroke when used in low amounts, an excessive drug effect will be highly dangerous, leading to events such as spontaneous hemorrhage into the brain. Complete inhibition of γ-glutamyl carboxylation is outright deadly, and indeed drugs of this type are also used as rat poisons. Therefore, warfarin is very clearly an exception from the rule stated above that we typically want to achieve receptor saturation.

Generally speaking, toxicity as an extension of therapeutic action is usually associated with a small *therapeutic index*, which is simply the ratio of the toxic concentration of the drug in the blood plasma over the therapeutic concentration:

$$\text{Therapeutic index} = \frac{[\text{Drug}]_{\text{toxic}}}{[\text{Drug}]_{\text{therapeutic}}} \tag{2.5}$$

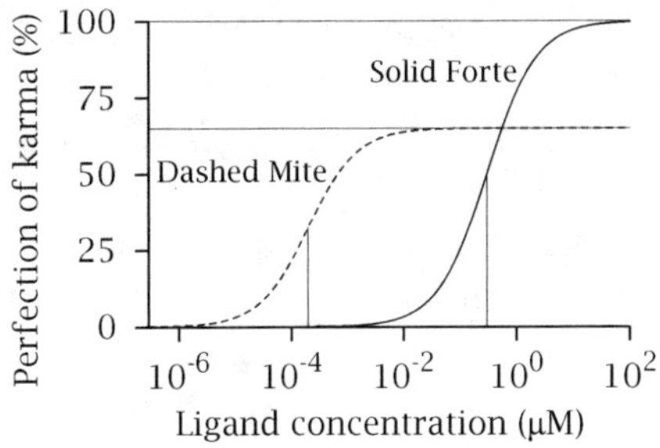

Figure 2.8 Potency and efficacy. Vertical lines indicate the potencies, horizontal lines the efficacies of the fictitious drugs Dashed Mite and Solid Forte.

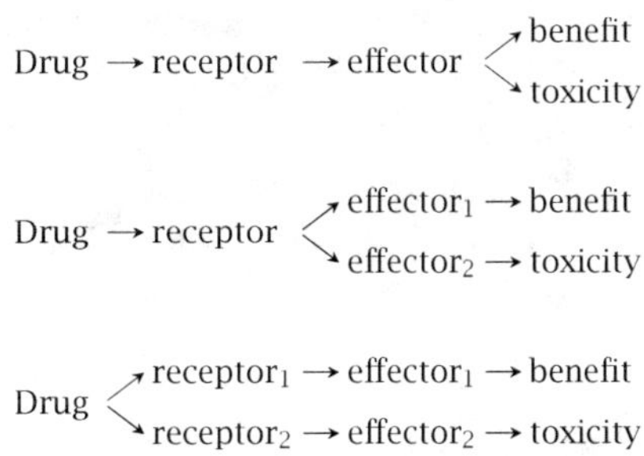

Figure 2.9 Therapeutic and toxic drug effects. Toxicity may be inseparable from the beneficial effect (top) or occur through separable mechanisms (center, bottom).

The dosage of a drug required to achieve a therapeutic concentration may vary quite substantially between any two patients, even if they have the very same body weight and we apply the same dosage at the same intervals; this is due to individual variations in drug metabolism and excretion (see Chapter 3). Such variations may easily cause the plasma concentration to either exceed the toxicity threshold, or drop below the minimum amount required for the therapeutic effect. To avoid this, the concentration or the functional effect of a drug with a small therapeutic index must be monitored. Accordingly, in our example, patients receiving warfarin treatment must have their blood clotting activity measured at regular, frequent intervals.

In the third case depicted in Figure 2.9, the therapeutic and the toxic effects are triggered from different receptors, both of which respond to the same drug. The challenge then will be to find drugs that will selectively act on the receptor responsible for the therapeutic action. For an example, we may again turn to β-adrenergic receptors. The blockade of β receptors in the heart is used in the treatment of hypertension and of heart disease. On the other hand, blockade of β receptors in the bronchi will promote bronchoconstriction and may aggravate the symptoms of asthma. The β receptors in the heart mostly belong to the β_1 subtype, whereas in the bronchi we mainly find β_2-receptors. β_1-Selective adrenergic antagonists ("cardioselective β blockers") are available and preferred in heart patients that also suffer from asthma.

In the second case in Figure 2.9, the drug binds to a single receptor alright, but the pathways leading to the therapeutic and toxic effects, respectively, diverge at a point upstream of the ultimate effectors. Inhibitors of monoamine oxidase provide an example. Monoamine oxidase (MAO) is responsible for the degradation of the *biogenic amines*, that is, epinephrine and norepinephrine, dopamine, serotonin, and histamine. Inhibition of this enzyme will therefore have a broad range of effects.[10] To find better therapeutic options, it would not

[10]Subtypes of MAO exist and show preferential activity on some substrates, which can be exploited with subtype-selective inhibitors. However, these still do not match the selectivity of modern receptor antagonists.

seem very useful to find better inhibitors; instead, one should try to find drugs that act downstream of the branching point. In fact, MAO inhibitors have now largely been superseded by drugs that act selectively on the receptors for the different mediators.

In this scenario, it may also be possible to improve the therapeutic index by taking advantage of agonist-specific coupling (Section 2.2.6). However, so far there are no clear-cut clinical applications of this interesting receptor behavior.

2.6 APPENDIX

This section expands on the derivation of the equations and on some of the conclusions presented in the preceding sections, and it also discusses some additional procedures for the analysis of drug-receptor interaction. It is not necessary for an understanding of the subsequent chapters of this text.

2.6.1 Mass action equilibrium

Equation 2.2 can be obtained from equation 2.1 with these additional considerations: For both the drug L and the receptor R, we can write simple mass balance equations

$$[L_{free}] + [LR] = [L_{total}] \tag{2.6}$$

$$[R_{free}] + [LR] = [R_{total}] \tag{2.7}$$

Assuming that the concentration of the drug is far larger than that of the receptor, we can neglect the depletion of free drug by receptor binding:

$$[L_{free}] = [L_{total}] = [L] \tag{2.8}$$

Therefore, equation 2.1 becomes

$$K = \frac{[L]([R_{total}] - [LR])}{[LR]} \tag{2.9}$$

which can be rearranged to equation 2.2.

2.6.2 Competitive inhibition

If an inhibitory drug I binds competitively, association of the receptor R with the drug and the physiological ligand L are mutually exclusive. The three possible states of the receptor—free, bound to the antagonist, or bound to the physiological ligand—will be subject to two competing equilibria:

$$K_I = \frac{[I][R_{free}]}{[IR]}, \; K_L = \frac{[L][R_{free}]}{[LR]} \tag{2.10}$$

as well as to the mass balance

$$[R_{total}] = [R_{free}] + [LR] + [LR] \tag{2.11}$$

From these equations, we obtain the receptor occupancy:

$$Y = \frac{[\mathrm{LR}]}{[\mathrm{R_{total}}]} = \frac{[\mathrm{L}]}{[\mathrm{L}] + K_\mathrm{L}\left(1 + \frac{[\mathrm{I}]}{K_\mathrm{I}}\right)} \tag{2.12}$$

Equation 2.3 is obtained by comparing Equation 2.12 to Equation 2.2.

2.6.3 Covalent inhibition

The extent of inhibition of a receptor by a covalent inhibitor I will depend on a large number of empirical variables such as the drug's reactivities with the receptor and other targets, duration of exposure and so on. However, once the reaction has run its course, the remaining activatable fraction of receptor $\mathrm{R'_{total}}$ is simply given by

$$[\mathrm{R'_{total}}] = [\mathrm{R_{total}}] - [\mathrm{IR}] \tag{2.13}$$

Then, if we add an agonist L, the receptor occupancy relative to the total number of receptors corresponds to

$$Y = \frac{[\mathrm{LR}]}{[\mathrm{R_{total}}]} = \frac{[\mathrm{L}]}{[\mathrm{L}] + K_\mathrm{L}}\left(1 - \frac{[\mathrm{IR}]}{[\mathrm{R_{total}}]}\right) \tag{2.14}$$

Comparison to equation 2.2 shows that this corresponds to a vertically compressed version of the receptor saturation curve that applies in the absence of inhibitor.

2.6.4 The Scatchard equation

In the foregoing, we have always considered ligands interacting with a single, homogeneous class of binding sites only. In reality, drugs or other ligands may bind to several distinct receptors and possibly other kinds of binding sites as well. In order to determine whether there is a single class of binding sites or several classes with distinct affinities, one needs to measure binding data at various ligand concentrations. These data are usually obtained using a radiolabeled ligand, which allows for the facile measurement of both free (unbound) and bound ligand across a wide range of concentrations. Equation 2.1 can then be re-stated as

$$K = \frac{\mathrm{L_{free}}\,(\mathrm{R_{total}} - \mathrm{L_{bound}})}{\mathrm{L_{bound}}} \tag{2.15}$$

which in turn can be rearranged to

$$\frac{\mathrm{L_{bound}}}{\mathrm{L_{free}}} = \frac{\mathrm{R_{total}}}{K} - \frac{1}{K}\mathrm{L_{bound}} \tag{2.16}$$

In the Scatchard plot (Figure 2.10), $\mathrm{L_{bound}}/\mathrm{L_{free}}$ is plotted as a function of $\mathrm{L_{bound}}$. If there indeed is only one homogeneous population of binding sites, the graph

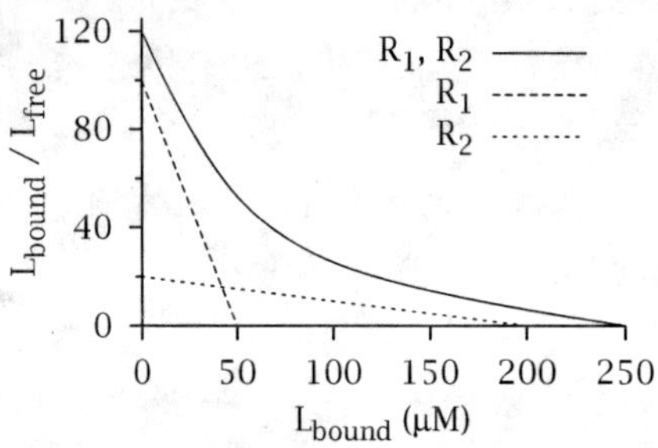

Figure 2.10 Scatchard plots for two hypothetical receptors (R_1, R_2) that each obey the law of mass action but differ in abundance and in affinity, occurring either alone or in combination.

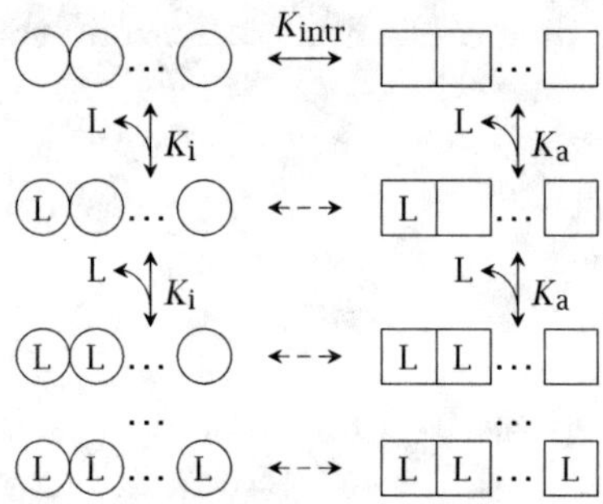

Figure 2.11 WMC model for cooperative behavior of multimeric receptors. All subunits bind the ligand with the same affinity and change conformation synchronously.

will be a straight line. The total number of binding sites can be obtained from the x-axis intercept, and the dissociation constant from the slope. Deviations from linearity may indicate the presence of two or more classes of binding sites; however, another possible reason for deviation from linearity is a cooperative mode of ligand binding.

While linear transformations like the Scatchard plot are appealing for displaying data, they have the disadvantage of introducing unequal weighting of the experimental error associated with individual data points. For quantitative analysis, it is preferable to apply nonlinear fitting methods to the untransformed underlying equation—in this case, for example, Equation 2.2.

2.6.5 Two-state and cooperative receptor behavior

While cooperative behavior of receptors and other multimeric proteins may be infinitely complex, theoretical analysis usually makes some simplifying assumptions. Of the two classical models, the sequential model devised by Koshland et al. [6] is more general, while the Wyman-Monod-Changeux (WMC) model [9] is simpler. The key axiom of the WMC model is that all subunits of the multimer change their conformation at the same time, or in other words that the conformational transition between the inactive and the active states is fully cooperative.[11]

The behavior of the individual subunit within the multimer can be described in terms of the two-state model (see Section 2.2.5, and Figure 2.11). We further assume the three characteristic equilibrium constants (K_{intr}, K_i and K_a) to be the same for all subunits. Then, for both simple two-state and cooperative behavior, it is straightforward to determine the abundance of any state of conformation and ligand saturation, relative to the unbound, active state, by multiplying [L], K_{intr}, and K_i or K_a as many times as appropriate. Expressions

[11] Hybrid states, in which some subunits are in the active, others in the inactive conformation, are allowed but penalized in the Koshland model. The WMC model can be considered a limiting case of the Koshland model, with infinite penalties placed on the hybrid states.

describing the overall receptor occupancy and active fraction are then obtained by summation of terms as appropriate. For the simple two-state model, the active fraction f_A works out to

$$f_A = \frac{[R_a] + [LR_a]}{[R_{total}]} = \frac{K_a + [L]}{K_a(1 + K_{intr}) + [L])\left(1 + K_{intr}\frac{K_a}{K_i}\right)} \tag{2.17}$$

In the absence of ligand ([L] = 0), this simplifies to

$$f_A = \frac{1}{1 + K_{intr}} \tag{2.18}$$

whereas for saturating concentrations of ligand ($[L] \gg K_a, K_i$), we obtain

$$f_A = \frac{1}{1 + K_{intr}\frac{K_a}{K_i}}, \tag{2.19}$$

The receptor occupancy function will resemble equation 2.2, with the global dissociation constant K' given by

$$K' = \frac{(1 + K_{intr})K_i K_a}{K_a + K_{intr}K_i} \tag{2.20}$$

The expressions describing occupancies and active fractions for cooperative receptors obtained in the same way are not given explicitly but are illustrated by example in Figure 2.6.

2.6.6 The model cascade

In the cascade model illustrated in Section 2.3, we have made the following quantitative assumptions: (1) The receptor R behaves according to the simple two-state model (see Section 2.2.5). (2) The concentration of the second messenger M_2 is proportional to the concentration of activated receptors: $[M_2] = r f_A [R_{total}]$. This condition will be met if the rate of formation of M_2 is proportional to the concentration of activated receptors, and the inactivation of M_2 is a first-order process. (3) The second messenger M_2 binds to the effector E according to simple mass action kinetics, and the strength of the observed effect is directly proportional to the concentration of the EM_2 complex.

This model is essentially the same as the one discussed by Strickland and Loeb [14], except that we here treat the receptor according to the two-state model rather than to simple mass action kinetics. A significant consequence of this change is that, at sufficiently high receptor expression levels, some effector activity will be observed even in the complete absence of agonist.

2.6.7 The Schild plot

So far, we have considered the behavior of ligands in systems with well-defined receptors and effectors. In experimental pharmacology, however, it is also necessary to examine the function of new drugs in complex systems, for example in tissue culture or in an animal model. Even if we know that a new drug binds to a known receptor, it might also bind to other, similar receptors or even to entirely unrelated ones. In addition, the signal measured in the experiment may occur downstream of a biochemical cascade, which will cause the dose-response curves for the signal and for receptor binding to look quite different (see Section 2.3).

A useful approach to address these problems is to measure the interaction of the new drug with a well-known drug that interacts specifically with the same receptor that we expect our new drug to bind to, yet has the opposite functional effect. One drug must be an agonist and the other one an antagonist; it is not important which one is the new drug. The experiment is carried out as follows: (1) Using incremental doses of the agonist, the signal is titrated to its half-maximal level. (2) A defined amount of the inhibitor is added, which will decrease the signal. (3) Agonist is added until the response is again restored to its half-maximal level. (4) Steps 2 and 3 are repeated to collect a sufficient number of data points. In each iteration, the total amounts of agonist and inhibitor added to the system are recorded.

The underlying assumption of this approach is that, even though their relationship may be non-linear, the same signal intensity should still always correspond to the same degree of receptor occupancy with the agonist. An experimental strategy that, like the present one, achieves cancellation of systematic errors by measuring all data points at the same signal strength, is called a *null method.*

The data points are then used to test the hypothesis that the two drugs do indeed bind to the same receptor in a competitive fashion. Using the assumptions that (1) all data points were acquired at the same level of receptor saturation with agonist, and (2) the inhibitor acts indeed by competition, we can write, for the first data point (inhibitor concentration $[\mathrm{I}]_0 = 0$, agonist concentration $[\mathrm{A}]_0$) and any other data point (inhibitor concentration [I], agonist concentration $[\mathrm{A}]_\mathrm{I}$)

$$Y = \frac{[\mathrm{A}]_0}{[\mathrm{A}]_0 + K_\mathrm{A}} = \frac{[\mathrm{A}]_\mathrm{I}}{[\mathrm{A}]_\mathrm{I} + K_\mathrm{A}\left(1 + \frac{[\mathrm{I}]}{K_\mathrm{I}}\right)} \tag{2.21}$$

from which we obtain

$$[\mathrm{A}]_0\left(1 + \frac{[\mathrm{I}]}{K_\mathrm{I}}\right) = [\mathrm{A}]_\mathrm{I} \tag{2.22}$$

and hence

$$\frac{[\mathrm{A}]_\mathrm{I}}{[\mathrm{A}]_0} - 1 = \frac{[\mathrm{I}]}{K_\mathrm{I}} \quad \Leftrightarrow \quad \log\left(\frac{[\mathrm{A}]_\mathrm{I}}{[\mathrm{A}]_0} - 1\right) = \log[\mathrm{I}] - \log K_\mathrm{I} \tag{2.23}$$

Therefore, a plot of $\log([\mathrm{A}]_\mathrm{I}/[\mathrm{A}]_0 - 1)$ against log[I] should produce a straight line with a slope of 1 if the two drugs are in fact antagonists of each other. This plot is known as the *Schild plot* [1]. We can distinguish the following possible cases:

1. The Schild plot yields the expected straight line, and the unknown drug is the inhibitor. We can assume that the unknown drug binds to the same receptor as the known agonist and obtain its binding constant from the x axis intercept, which occurs at $-\log K_I$.

2. The Schild plot yields the expected straight line, and the unknown drug is the agonist. We then still can assume that it binds to the same receptor as the known antagonist; however, we will not know its binding constant. The latter could be determined by comparison with known agonists.

3. The Schild plot does not yield the expected straight line. Then, we have to conclude that the two drugs do not bind competitively and may interact with different receptors.

STUDY QUESTIONS

2.1 The plot below shows the contraction of smooth muscle isolated from pig small intestine in response to the drug carbamoylcholine, without pretreatment (control), after pretreatment with inhibitor A alone (A), or with both inhibitor A and inhibitor B (A+B).

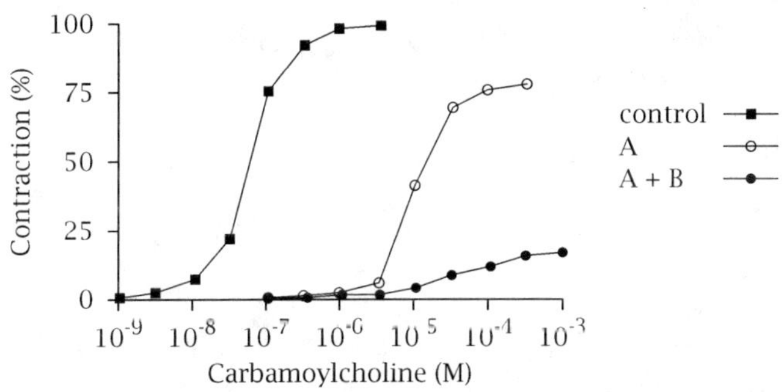

Let's assume that the amounts of A and B used were saturating, that is, increasing their dosages further would *not* change the observed response. How can we explain the observed behavior of this system? Is the inhibition exerted by A reversible or irreversible? What about B?

2.2 The effect of a combination of two drugs can be examined using a so-called *isobologram*. The concentrations of each of the two drugs is independently varied, and the strength of the observable effect is measured. Curves connect points with equal strength of effect (see sketch below).

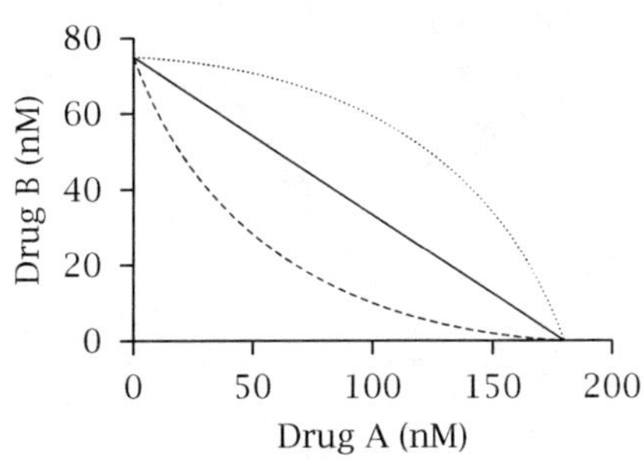

Questions: (1) If the two drugs A and B are both competitive inhibitors of the same enzyme or receptor, what will be the shape of the curve in the isobologram? (2) What would be the significance of the other two curve shapes in the sketch above?

✎ **2.3** Find out about the stereoisomers of quinine and their different pharmacological activities.

REFERENCES

[1] O. Arunlakshana and H. O. Schild, "Some quantitative uses of drug antagonists," *Br. J. Pharmacol. Chemother.* 14 (1959), 48–58.

[2] R. K. Bickerton, "The response of isolated strips of spleen to sympathomimetic drugs and their antagonists," *J. Pharmacol. Exp. Ther.* 142 (1963), 99–110.

[3] L. Brown et al., "Spare receptors for β-adrenoceptor-mediated positive inotropic effects of catecholamines in the human heart," *J. Cardiovasc. Pharmacol.* 19 (1992), 222–232.

[4] H. Frang et al., "Phenoxybenzamine binding reveals the helical orientation of the third transmembrane domain of adrenergic receptors," *J. Biol. Chem.* 276 (2001), 31279–31284.

[5] E. H. Gold et al., "Synthesis and comparison of some cardiovascular properties of the stereoisomers of labetalol," *J. Med. Chem.* 25 (1982), 1363–1370.

[6] D. E. Koshland, G. Némethy, and D. Filmer, "Comparison of experimental binding data and theoretical models in proteins containing subunits," *Biochemistry* 5 (1966), 365–385.

[7] K. Leach, P. M. Sexton, and A. Christopoulos, "Allosteric GPCR modulators: taking advantage of permissive receptor pharmacology," *Trends Pharmacol. Sci.* 28 (2007), 382–389.

[8] C. A. Milano et al., "Enhanced myocardial function in transgenic mice overexpressing the β_2-adrenergic receptor," *Science* 264 (1994), 582–586.

[9] J. Monod, J. Wyman, and J. P. Changeux, "On the Nature of allosteric transitions. A plausible model," *J. Mol. Biol.* 12 (1965), 88–118.

[10] P. Samama et al., "A mutation-induced activated state of the β_2-adrenergic receptor. Extending the ternary complex model," *J. Biol. Chem.* 268 (1993), 4625–4636.

[11] G. Scatchard and E. S. Black, "The effect of salts on the isoionic and isoelectric points of proteins," *J. Phys. Colloid Chem.* 53 (1949), 88–99.

[12] D. A. Shapiro et al., "Aripiprazole, a novel atypical antipsychotic drug with a unique and robust pharmacology," *Neuropsychopharmacology* 28 (2003), 1400–1411.

[13] P. A. Stevens and G. Milligan, "Efficacy of inverse agonists in cells overexpressing a constitutively active β_2-adrenoceptor and type II adenylyl cyclase," *Br. J. Pharmacol.* 123 (1998), 335–343.

[14] S. Strickland and J. N. Loeb, "Obligatory separation of hormone binding and biological response curves in systems dependent upon secondary mediators of hormone action," *Proc. Natl. Acad. Sci. U. S. A.* 78 (1981), 1366–1370.

CHAPTER 3

PHARMACOKINETICS

Whatever the actual mechanism of action of a drug may be, we will always want to know: Will the drug actually reach its site of action, and if so, for how long will it stay there? These questions are the subject of pharmacokinetics. While it might seem merely a distraction to someone chiefly interested in biochemical mechanisms of drug action, pharmacokinetics is a crucial and often limiting factor in any form of pharmacotherapy, and it will be wise for anyone aspiring to develop novel therapeutic approaches to take it into consideration right from the beginning.

A drug's site of action may be a single anatomical location, such as a single organ, a localized infection, or a solid tumor. On the other hand, it may also be diffusely distributed throughout the entire body. This is the case, for example, in diabetes—insulin is required in all kinds of tissues—or in a systemic infection. If the site of action is confined, it may be possible to apply the drug topically, that is, directly and exclusively to this site. This will minimize any unnecessary exposure of organs that are not afflicted by the disease. Topical application also permits the use of drugs that would be too toxic for systemic application, such as iodine-based antiseptics, which can be used in the treatment of skin infections.

The more common case, however, is that the drug must be applied systemically, because the site of action is either diffuse or inaccessible. In this case, the success or failure of a drug to reach its target is governed by these processes:

Biochemical Pharmacology. By M. Palmer, A. Chan, T. Dieckmann, J. Honek

(1) *absorption*, which means the uptake of the drug from the compartment of application into the blood plasma, (2) *distribution*, which means the equilibration between the blood plasma and the rest of the organism, usually including the target site, and (3) *excretion* or *metabolic inactivation*, which may prevent the drug from reaching its target in the first place or else will limit the duration of its action. Organs with prominent roles in drug elimination are the kidneys and the liver.

Anatomical barriers may impede both the absorption and the distribution of a drug and divert it toward elimination. In addition, en route to its destination or as part of the elimination process, the drug may be enzymatically modified and inactivated.

3.1 ANATOMICAL BARRIERS TO DRUG TRANSPORT

While historically pharmacological research has relied mostly on animal experiments, cell culture has become an important tool, particularly in early screening and high-throughput experiments. It is also widely applied in mechanistic studies, often using molecular genetics to recombinantly express or knock out drug target molecules. Such studies can precisely define the roles of specific receptors in the mode of action of a drug. Cell cultures expressing specific drug-metabolizing enzymes help to elucidate the metabolic activation or inactivation of drugs. Yet, cell cultures do not capture some key problems of pharmacokinetics, which is one of the reasons why animal experiments remain an indispensible part of drug development. In a cell culture, all cells are in immediate contact with the medium to which the drug is applied. If the receptor is located at the cell surface, there is no barrier at all between the drug and its target. If the receptor is located in the cytosol, it will be separated from the drug only by a single cell membrane, any hindrance by which may be overridden in the experimental setup by the application of dosages higher than those possible in vivo.[1]

A drug that is applied systemically *in vivo* will have to cross multiple anatomical barriers at all phases of transport, that is, during absorption, distribution, and elimination. These anatomical barriers are composed of two basic structural elements: cell membranes and basal membranes.

3.1.1 Cell membranes and basal membranes

A cell membrane consists of a lipid bilayer with embedded proteins; it controls the scope, rate and direction of solute exchange between the cell and its environment. Solutes, including drug molecules, may negotiate a cell membrane in different ways: (1) They may cross the lipid bilayer by distributing into its hydrophobic core and emerging again on the other side. This mode of transport is restricted to small, lipophilic molecules. (2) A drug may be translocated by a specific transport protein if it resembles the physiological cargo molecule of that transporter. There also exist transport proteins that have a relatively

[1] Another limitation of cell cultures is the absence of complex physiological functions—how do you measure blood pressure or observe sleep in a petri dish?

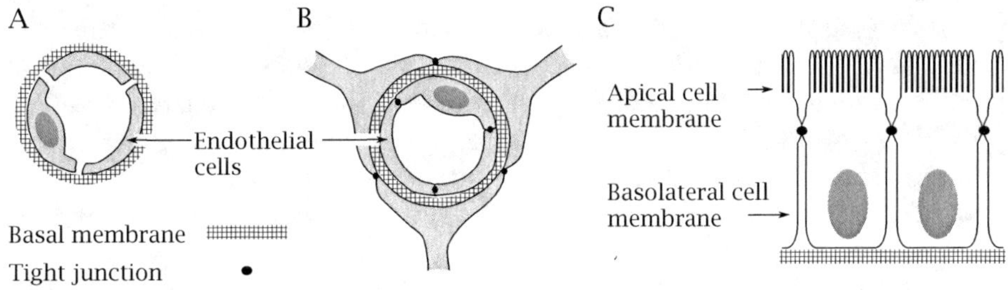

Figure 3.1 Cell layers as barriers to drug uptake and distribution. **A**: Ordinary capillaries. Gaps between and fenestrations across endothelial cells allow passage of large and small molecules; the supporting basal membrane spans these gaps and restricts passage of macromolecules. **B**: Capillaries in the brain have two continuous cell layers, sealed by tight junctions, which form the blood–brain barrier. **C**: Epithelia in the intestine and kidney tubules have one continuous cell layer, again sealed by tight junctions between cells.

broad substrate specificity and may act on drugs that do not closely resemble any physiological solute. (3) A drug may adsorb to the cell surface and then be taken up by phagocytosis. This is important mostly with larger molecules or particles.

Cell membranes form the most stringent barriers to drug transport; we will consider their transport properties in more detail below. Basal membranes restrict drug transport to a far lesser degree than cell membranes do. They consist of a meshwork of proteins and proteoglycans. The spaces in this meshwork are water-filled; therefore, drugs do not have to traverse a different medium to get across. The main criterion of selectivity is size; the exclusion limit of the basal membrane filter is approximately 10 kDa. Therefore, basal membranes will grant free passage to drugs and other solutes that are small molecules, as long as these are not bound to proteins or other macromolecules. Basal membranes always occur in close association with cell layers. Depending on the structure of the cell layer, either the basal membrane or the cell membranes may be rate-limiting for the transport of drug molecules; this varies with the anatomical location.

3.1.2 Epithelia and endothelia

Drugs are often applied orally and have to cross the mucous membranes of the gastrointestinal tract, usually the small intestine. The surface of the mucous membrane is formed by an epithelial cell layer that sits atop a basal membrane. The epithelial cells are connected to each other by *tight junctions* (Figure 3.1), and their apical and basolateral cell membranes essentially function as two continuous barriers connected in series. The basal membrane acts in series as well, but in this case does not add significantly to the much more stringent selectivity imposed by the cell membranes.

A drug that has made its way into the systemic circulation usually needs to reach some target outside of it, which means that it has to cross the capillary *endothelium*. The endothelial cells in a typical capillary are again supported by a basal membrane. In contrast to the epithelial cells in the intestine, they are separated by gaps and even are perforated by fenestrations that facilitate solute exchange between the blood plasma and the interstitial, extracellular fluid of the tissues. The two compartments are in effect only separated by the molecular sieve of the basal membrane, and therefore small drug molecules will cross this barrier with ease.

3.1.3 The blood–brain barrier

While the capillary endothelium in most organs conforms to the permissive structure described above, the capillaries in the central nervous system are distinctly different. Here, no fenestrations are present, and the endothelial cells are interconnected by tight junctions, just like the epithelial cells in the intestine and other epithelial organs. In addition, the outer circumference of the basal membrane is lined by protuberances of glia cells,[2] which are again joined to each other by tight junctions. Therefore, all told, solutes that pass between the brain tissue and the blood plasma need to cross four cell membranes arranged in series. This protective structure is referred to as the *blood–brain barrier*.[3] The barrier function of the cell membranes is enhanced by transport proteins, which belong to the ABC and the SLC families (see below). Many drugs that may penetrate the endothelial cells are exported back into the bloodstream by such transporters.

The effectiveness of the blood–brain barrier is strikingly illustrated by the experiments, performed by Ehrlich and others [8], that first demonstrated its existence. Injection of the organic dyes trypan blue or methylene blue into the general circulation caused most tissues to be stained but not the brain, whereas injection into the cerebrospinal fluid resulted in staining of the brain only.

The blood–brain barrier is of great significance for practical pharmacotherapy, since it excludes many drugs that effectively distribute into other tissues. If the drug target is in the periphery, this is a good thing, as it protects the brain from side effects; however, it can be a problem if the drug target is in the central nervous system.

3.2 SOLUTE TRANSPORT ACROSS CELL MEMBRANES

My high school chemistry teacher used to summarize the traditional relationships between individuals and the authorities in different countries as follows:

> In Russia, everything is forbidden, even if explicitly allowed.
> In Germany, everything is forbidden unless explicitly allowed.
> In England, everything is allowed unless explicitly forbidden.
> In France, everything is allowed, even if explicitly forbidden.

On this scale of permissiveness, cell membranes are a lot like Germans—they will only let those solutes pass that have an explicit permit. The default refusal

[2] *Glia* is the collective name for cells other than nerve cells in brain tissue. Glia cells are more numerous than neurons, which may have given rise to the silly myth that "humans only utilize 10% of their brain cells".

[3] The name has also been applied by anesthesiologists to the drapery that separates their own workspace from that of the surgeons at the operation table.

of transit is imposed by the lipid bilayer, which forms the matrix of any cell membrane. This control is very tight, as illustrated by the following facts:

1. All the major cations—sodium, potassium, and calcium—are distributed across the membrane in a highly asymmetric way. The concentration gradient is steepest for calcium, for which it amounts to 4–5 orders of magnitude, with only nanomolar concentrations inside the cell. These ion concentration gradients cause an electrical potential across the membrane that amounts to approximately 80 mV or—using a membrane thickness of 4 nm—a field strength of approximately 20,000 V/mm. This potential is fundamental to the function of excitable cells, which we will consider in Chapter 6.

2. Water, which is a small molecule and present in very high concentrations, can pass the lipid bilayer with a certain efficiency; however, cell membranes that need to enable a high rate of water transport contain specific water transport proteins called *aquaporins* for this purpose.

While the lipid bilayer establishes the default characteristics of transport across the cell membrane, there are numerous and diverse transporter proteins that facilitate transport of solutes unable to cross the lipid bilayer. Like other solutes, drugs may be transported either directly through the lipid bilayer or by membrane proteins. We will consider both modes of transport in turn.

3.2.1 Drug transport by diffusion across lipid membranes

Most of the various lipid species that occur in cell membranes have an amphiphilic structure, illustrated in Figure 3.2 for phosphatidylcholine. The lipids form a bilayer, the interior of which comprises the hydrophobic fatty acyl chains of the lipid molecules, while the hydrophilic headgroups face the aqueous phase.

While it is true that the lipid bilayer effectively restricts the diffusion of most physiological solutes, this is due in part to the mostly polar, often ionic nature of the latter. The metabolites of glycolysis may serve as an example. In the very first step, glucose is phosphorylated to glucose-6-phosphate, and all downstream metabolites retain at least one ionic group, which effectively pins them down inside the cytosol. There are, however, some physiological solutes that diffuse across lipid bilayers quite well. An example is the steroid hormone cortisol. This molecule is fairly lipid-soluble, which enables it to partition into the lipid bilayer and then emerge on the other side.

The lipid solubility of an organic molecule can be estimated by its octanol-water partition coefficient, which is influenced in predictable ways by the functional groups it contains. It is usually reported as the logarithm of the partition coefficient (logP). For example, a logP of 3 means that the concentration in octanol is 10^3 times higher than in water.[4]

Charged and polar moieties will reduce lipid solubility, and therefore render the drug molecules less membrane-permeant. In principle, the lipid solubility of a drug that is poorly absorbed or distributed could be improved by removing polar or ionic groups. However, those groups are often required for the

[4]Computational approaches for estimating this parameter have been developed [5].

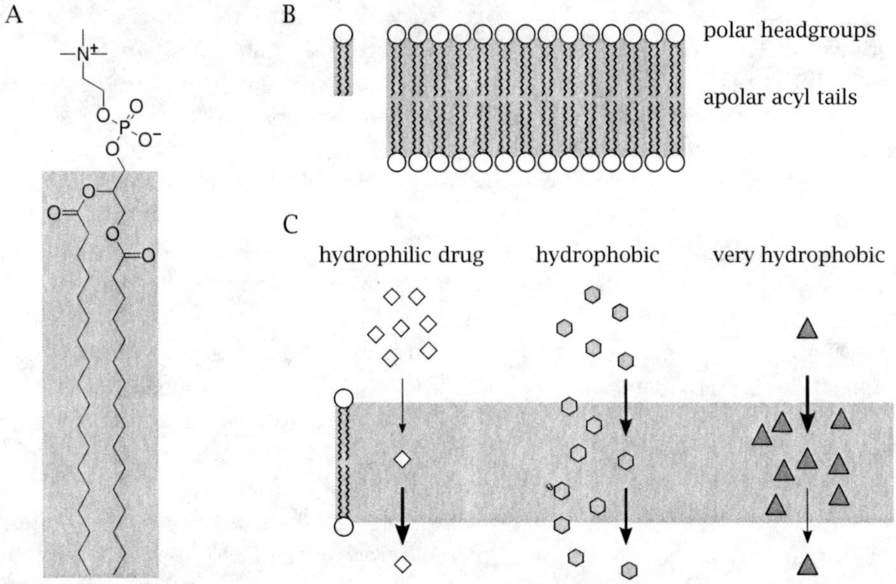

Figure 3.2 Lipid bilayers as permeability barriers for drugs. **A:** Structure of phosphatidylcholine, a major component of biological membranes. Like other phospho- and glycolipids, phosphatidylcholine contains hydrophobic alkyl or acyl chains (gray) and a hydrophilic headgroup. **B:** In a lipid bilayer, the lipid molecules align such that the acyl chains form a continuous hydrophobic phase, while the hydrophilic headgroups face the aqueous phase. **C:** Permeation of hydrophobic and hydrophilic drug molecules across lipid bilayers. Permeation of hydrophilic drugs is limited by partition into the hydrophobic phase; very hydrophobic drugs do not partition out of it. Moderately hydrophobic drugs partition both into and out of the hydrophobic phase and therefore permeate the membrane most rapidly.

pharmacological activity of a drug, which limits the possible extent of this pharmacokinetic optimization. A strategy that can sometimes be applied in such circumstances is to convert the drug to a *prodrug*, a typically inactive derivative of the drug in which some hydrophilic or ionizable groups are masked by hydrophobic ones. An example is the *resorption ester* bacampicillin, which is a derivative of the antibiotic ampicillin (Figure 3.3, left). Esterification of the carboxylic acid in ampicillin facilitates its uptake from the gut lumen. Intracellular esterases in the intestinal epithelium will cleave the ester and release ampicillin, which is then passed on into the circulation.

Masking hydrophilic groups also enhances the uptake of drugs across the membranes of the blood-brain barrier into the brain. A good example is heroin, which is the diacetylated derivative of morphine (Figure 3.3). It penetrates the brain more efficiently than morphine and therefore is more strongly inebriating and addictive.[5]

Another strategy to improve the membrane permeation of a drug is based on the effect of *nonionic diffusion*. An example is provided by the two *ganglion*

[5]Ironically, heroin was initially conceived as an improved, nonaddictive morphine. As you may know, methadone was later invented to treat heroin addiction.

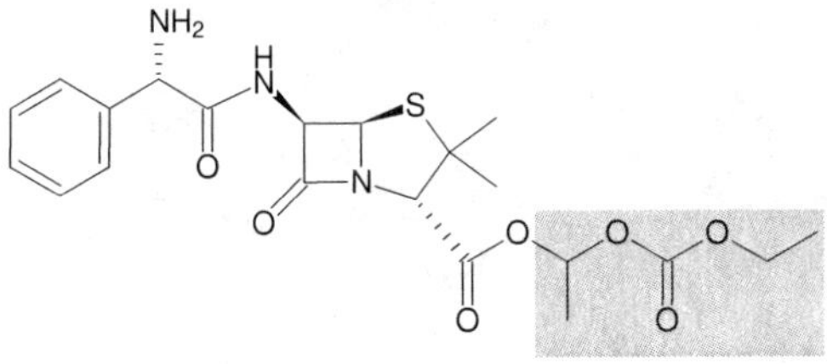

Bacampicillin

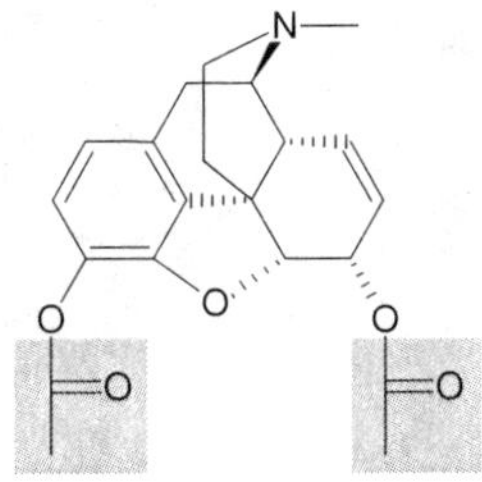

Heroin

Figure 3.3 Resorption esters facilitate drug absorption and distribution. The cleavable ester groups are highlighted. The prodrug bacampicillin is cleaved to release ampicillin after uptake into the intestinal epithelium. Heroin is morphine with two added acetyl groups, which mask two polar groups and facilitate distribution into the central nervous system. Heroin undergoes cleavage by esterases, which releases morphine.

blockers hexamethonium and mecamylamine (Figure 3.4), which act as antagonists at certain receptors of the neurotransmitter acetylcholine and, once upon a time, were used as antihypertensive agents (see Section 6.9.5). Acetylcholine is a quaternary amine; the positive charge is important to its interaction with its receptors. Hexamethonium, similarly, is a dual quaternary amine; as such, it is not able to traverse lipid bilayers and thus is not taken up from the intestine after oral application. In contrast, mecamylamine is a secondary amine and can adopt an uncharged form that traverses lipid bilayers much more readily. The deprotonated form itself is pharmacologically inactive; however, having reached its target site, mecamylamine can revert to its protonated, pharmacologically active form and bind to the receptor. Mecamylamine therefore can be orally applied.

Nonionic diffusion can also produce unwanted effects, as in the case of acetylsalicylic acid (aspirin). In the acidic milieu of the stomach, this molecule will be protonated and thus uncharged, which promotes its diffusion into the cells of the stomach mucous membrane. Inside the cell, the pH is very close to neutral, which will deprotonate the carboxylic acid group. Diffusion of the deprotonated, charged form out of the cell will be much slower than entry of uncharged form, so that aspirin will accumulate inside the cells to concentrations considerably higher than in the stomach lumen. When compared to other inhibitors of cyclooxygenase (see Chapter 9), aspirin has a stronger tendency to trigger side effects such as gastritis and gastric or duodenal ulcers.

Molecular size is another factor that is relevant to the ease of membrane permeation. This may be illustrated by comparing dimethylether to polyethylene glycol (PEG), which may formally be considered a linear polymer of dimethylether. Dimethylether crosses membranes readily, whereas PEG is quite efficiently excluded by membranes, particularly in its higher molecular weight varieties. It needs to be said, however, that PEG is not only larger than dimethylether but, for subtle reasons, is also somewhat more polar.

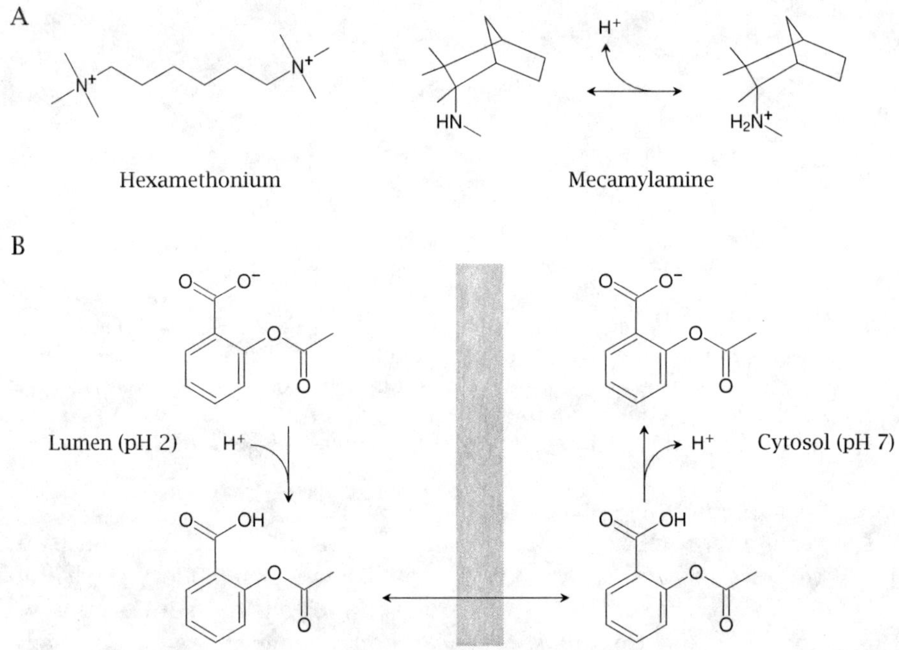

Figure 3.4 Nonionic diffusion in drug transport. **A:** Hexamethonium has two permanent charges and therefore cannot cross lipid bilayers by diffusion. In contrast, mecamylamine can cross bilayers in its unprotonated form and then bind a proton to acquire the charge necessary for receptor binding. **B:** Because of the proton gradient that exists across the apical membrane of the gastric epithelium, nonionic diffusion of acetylsalicylic acid leads to its accumulation inside the epithelial cells.

3.2.2 Drug transport by membrane proteins

Transport proteins in membranes are relevant to pharmacology for two reasons: (1) they may accelerate or impede drug transport across anatomical barriers and drug uptake into target cells, and (2) they may themselves be targets of drug action.

The most fundamental distinction concerning protein-mediated solute transport across membranes is that between active and passive transport. Active transport always proceeds in a fixed direction, which is a built-in property of the transporter protein. This direction is typically opposite to the solute's concentration gradient across the membrane, and active transport therefore requires energy. In primary active transport, this energy is supplied by a linked chemical reaction such as ATP hydrolysis. Secondary active transport derives its energy from simultaneous transport of another solute down its concentration gradient. The two coupled transport processes may occur in parallel or antiparallel directions. These two cases are referred to as *symport* and *antiport*, respectively.

Very many secondary active transport processes are directly or indirectly driven by sodium ions; Figure 3.7 shows an example. The extracellular concentration of Na^+ is about tenfold higher than the intracellular one; an opposite gradient exists for the other major cation, K^+. Both gradients are sustained by the Na^+/K^+-ATPase, which is the largest single consumer of ATP in a typical cell.

In passive transport, substrate translocation is not coupled to any other energetically relevant process; therefore, the transport proteins do not impose a direction but simply facilitate diffusion of their substrates downhill their concentration gradients. If a drug happens to be a substrate of a passive transporter protein, it will overcome anatomical barriers more easily. In contrast, an active transporter may either help or, more commonly, impede the progress of a drug on its way to the target.

Like enzyme molecules, transport proteins have limited substrate turnover rates and occur in limited numbers; therefore, the kinetics of transport follow the saturation pattern familiar from Michaelis-Menten enzyme kinetics. Saturation may be relevant with physiological solutes present at high concentration. For example, the transport capacity for glucose reuptake in the kidneys is tuned fairly closely to the range of glucose plasma concentrations occurring in healthy persons, and therefore glucose readily appears in the urine of diabetic patients. Drugs, however, are typically present at concentrations much lower than the K_M values of their transport proteins, so that their transport rates will vary linearly with concentration.

3.2.3 ABC transporters

Membrane transporters account for a significant share of all proteins encoded in the human genome, and quite a few of these transporters have actual or potential roles in drug transport. This is particularly true of the family of *ATP-binding cassette transporters*, or ABC transporters for short. ABC transporters are found in both prokaryotic and eukaryotic organisms, and most of them act as extrusion pumps; that is, they pick up substrate molecules in the cytosol and transport them out of the cell, at the expense of ATP cleavage. The most widely studied such transporter in humans is P-glycoprotein or MDR-1 (MDR is an acronym for *multidrug resistance*). It was discovered in tumor cells, which developed resistance to anti-cancer drugs through overexpressing this transporter. This resistance applied to a wide range of structurally unrelated drug molecules, indicating a remarkably broad substrate specificity of MDR-1. Similar observations have also been made with ABC transporters that cause antibiotic resistance in bacteria.

The molecular structures of MDR-1 and of a few members of the ABC family have been elucidated. Taken together, these structures cover both the inward and the outward facing conformation of the transporters, which gives some insight into the conformational changes that accompany the transport process (Figure 3.5). However, the mechanisms of coupling between substrate binding, ATP cleavage and membrane translocation are not yet fully understood.

MDR-1 and the related BCRP (breast cancer resistance protein) transporter are strongly expressed in the intestinal epithelium and at the blood-brain barrier, that is, at anatomical boundaries that significantly restrict drug uptake

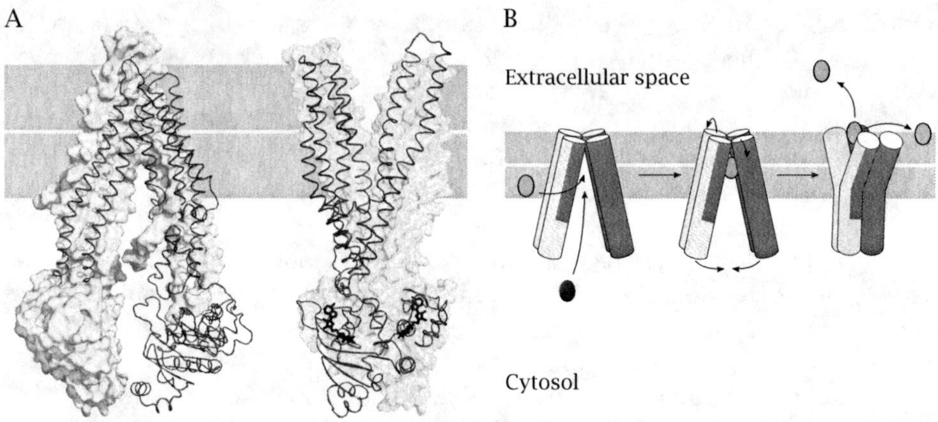

Figure 3.5 Structure and function of ABC transporters. **A:** Structures of the inward facing conformation of P-glycoprotein (MDR-1) [2] and of the outward facing conformation of the bacterial ABC transporter Sav1866 [6]. One of the two homologous functional domains (MDR-1) or separate monomers (Sav1866) is rendered in cartoon mode, while the other is shown in surface representation. The outward facing conformation contains two molecules of a nonhydrolyzable ATP analog inside the binding sites at the interface of the two subunits (shown in stick representation). **B:** Functional cycle. Substrates are accepted from the cytosol or from the inner leaflet of the cell membrane. The ATP-powered conformational change first sequesters the substrate inside the transporter and then expels it into the outer membrane leaflet or the cytosol. A colored version of this figure is shown in the insert in the middle of this book.

and distribution. They are also strongly expressed in the liver and the kidneys, where they participate in secretion of drugs and their metabolites into the bile and urine, respectively. This applies both to unmodified drug molecules and to drug metabolites that have undergone conjugation with glucuronic acid or glutathione. In keeping with this functional relatedness, the expression of several ABC transporters is coordinated with that of drug-metabolizing enzymes (see Section 4.6).

As suggested by the context of their discovery, overexpression of ABC transporters is a major cause of resistance acquired by tumors in the course of treatment with anticancer agents. ABC transporters found in bacteria play a major role in bacterial resistance against antibiotics. Both of these challenges have inspired efforts to develop inhibitors of ABC transporters, which would be administered together with antitumor or antibacterial drugs. Apart from facilitating entry into resistant tumor cells, inhibitors of human ABC transporters will also facilitate traversal of the blood–brain barrier (Figure 3.6). While this may be beneficial in tumors of the brain, it would be prone to cause side effects in all others.

Inhibition of MDR-1 or bacterial ABC transporters is also observed with drugs that primarily act on other targets, such as quinidine, verapamil, and reserpine.[6] Such drugs may facilitate the absorption and distribution of other

[6]Intriguingly, the primary established targets of these three drugs are all transport proteins that are not ABC transporters.

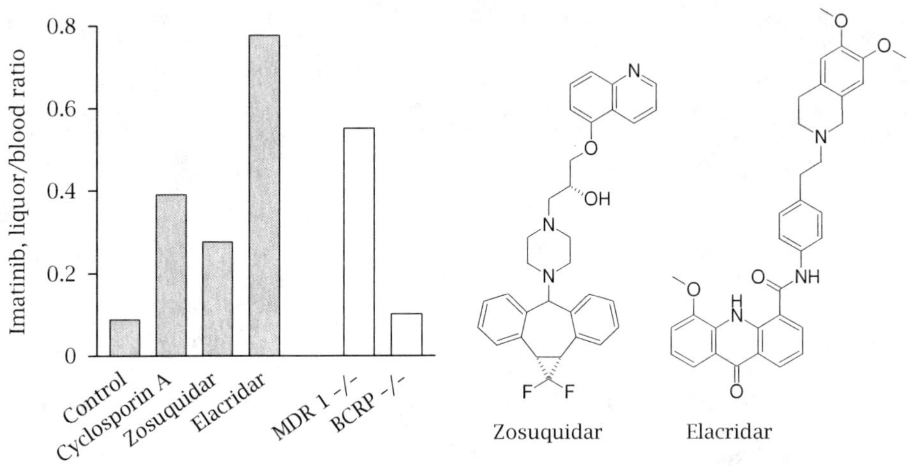

Figure 3.6 Penetration of the antitumor drug imatinib through the blood-brain barrier: Effect of ABC transporters. When applied alone, the concentration of imatinib in the liquor is 90% lower than in the blood. Cyclosporin A and zosuquidar, which inhibit MDR-1, significantly increase the penetration of imatinib. A stronger effect is observed with the drug elacridar, which inhibits MDR-1 as well as BCRP. Genetic knockout of MDR-1 (MDR-1 -/-) but not BCRP (BCRP -/-) increases CNS penetration of imatinib, which suggests that the strong effect of elacridar may be due to inhibition of additional transporters. (Figure redrawn from original data in Ref. [3].)

drugs that are applied simultaneously. This interaction may result in toxicity caused by the drug applied in combination, as illustrated by a case of CNS toxicity leading to paralysis in a patient treated simultaneously with verapamil and colchicine [9].

Some proteins in the ABC transporter family have evolved to fulfill functions other than transport. An interesting example is the sulfonylurea receptor, which acts as a sensor of intracellular ATP and is important in the regulation of insulin release from pancreatic β cells and of smooth muscle tone (see Section 6.4).

3.2.4 The SLC superfamily of membrane transporters

The SLC (solute carrier) transporter superfamily is substantially larger and more diverse than the ABC family. Within this group, we find both active and passive transporters, and the solutes or substrates that are transported include ions, metabolites, neurotransmitters, and drugs. SLC transporters again occur in both prokaryotes and eukaryotes; within eukaryotes, they occur both in the plasma membrane and in intracellular membranes such as neurotransmitter vesicles, mitochondria, and the endoplasmic reticulum.

Several SLC transporters are important as drug targets, in particular those that mediate transport of neurotransmitters in synapses (see Chapter 6). Of particular relevance to pharmacokinetics are the groups of the so-called organic anion transporters (OAT) and organic cation transporters (OCT). These are found in many cells, but unsurprisingly, expression is most prominent in the

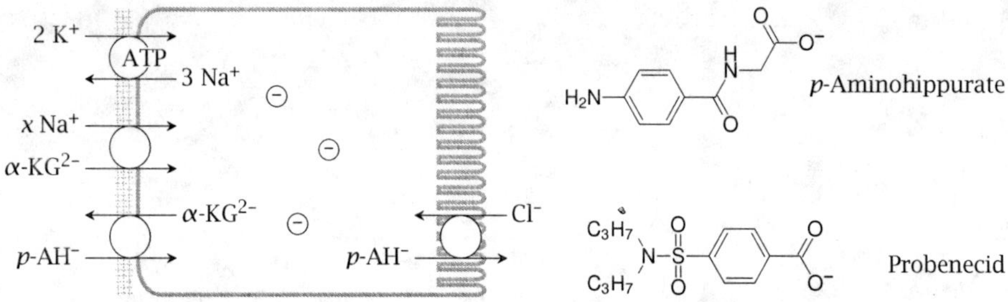

Figure 3.7 Tubular secretion of *p*-aminohippurate (*p*-AH) in the kidneys. Import of *p*-AH across the basolateral membrane is driven by α-ketoglutarate (α-KG) antiport; the α-KG gradient in turn is maintained by Na^+-symport, and the Na^+ gradient maintained by Na^+/K^+-ATPase. Excretion of *p*-AH across the apical membrane into the tubular lumen is driven by Cl^-- or HCO_3^--antiport; it can be inhibited by probenecid. The free energy of solute transport is affected not only by concentration gradients but also by the negative-inside membrane potential.

organs with the highest rates of solute transport, namely, the kidneys, the liver and the intestinal epithelium.

Both OAT and OCT have several subtypes with different preferential location and apparent substrate specificity. However, these transporters often have low specificity. For example, OAT subtype 1 transports not only anions but also the basic drug molecule cimetidine, which at the same time is also a substrate for the cation transporter OCT1. As with ABC transporters, substrate specificity cannot be pinned down to distinct structural features of the drug but rather seems to be correlated with global properties such as overall charge, hydrophobicity, and molecular weight.[7]

OAT and OCT transporters may engage in passive or active transport; the mode of transport is not always known with certainty. However, cellular substrate uptake via OAT is often powered by α-ketoglutarate antiport. α-Ketoglutarate has a lower concentration outside the cell than inside. In addition, it carries two negative charges, so that its exchange for a singly charged anion will result in the net export of one negative charge from the cytosol—the transport is *electrogenic*. Since the cytosol is electrically negative relative to the extracellular space, the export of a negative charge is exergonic. To make this process sustainable, α-ketoglutarate must be brought back into the cell, which is accomplished by a separate transporter through cotransport of sodium ions. With cations, the negative-inside membrane potential makes uptake inherently exergonic, and indeed most OCT transporters that mediate cation uptake don't seem to require co- or antiport of auxiliary substrates.

[7]Given that an organism has only a limited number of transporters to cope with a large and unpredictable variety of noxious substances in the environment, the low degree of substrate specificity seems logical. However, the apparent ability of this array of nonspecific transporters to discriminate between poisons that should be rejected and excreted on one hand, and useful metabolites that should be absorbed and retained on the other hand, is somewhat mystifying. The situation is similar with the very broad specificity of many drug-metabolizing enzymes.

Both OAT and OCT are important in the secretion and reabsorption of solutes at the tubule epithelia in the kidney. In these processes, solutes are first taken up by the epithelial cells on one side and then extruded at the other side. As an example, Figure 3.7 depicts the transport processes involved in the tubular secretion of *p*-aminohippuric acid. Since the free energy of uptake is different from that of extrusion, it is not surprising that two different transporters are involved; both are members of the OAT family. With other substrates, the secretion step may be mediated by ABC transporters instead.

So far, we have introduced some general concepts of drug transport. We will now see how these concepts apply to specific organs and to the various stages of pharmacokinetics.

3.3 DRUG APPLICATION

You are certainly aware that drugs can be applied by various routes. The choice of route depends on the stability and other properties of the drug, the urgency of the treatment, and the importance of quantitative uptake.

3.3.1 Oral drug application

This is the most commonly used mode of application, mostly because of convenience. However, the drugs so applied must overcome multiple and formidable barriers before admittance into the circulation; optimizing drug molecules for oral application is a common challenge in drug development.

Inside the digestive tract, drug molecules encounter a fairly aggressive chemical milieu. The acidic pH (≤ 2) in the stomach and the presence of proteases and nucleases in the small intestine preclude the application of proteins, nucleic acids, and other chemically labile molecules. The apical membranes of the gut epithelial cells present a barrier that many drugs are unable to diffuse across. Those drugs that manage to enter the epithelial cells are faced with transport proteins that will extrude a broad variety of compounds right back into the gut lumen. The cells also possess enzymes such as esterases and oxygenases that modify and thereby inactivate many drug molecules.

For those drugs that make it past these barriers, the liver presents another formidable barrier. All blood drained from the intestines, as well as the spleen and the pancreas, is first passed through the liver via the portal vein before being released into the general circulation. This *portal circulation* is schematically depicted in Figure 3.8. A notable feature of the liver tissue is its lack of real capillary walls that would separate the blood from the liver epithelial cells. Therefore, the latter get into intimate contact with the blood and can easily extract from it any compound they see fit. Solute uptake into the liver cells is facilitated by an array of OAT and OCT type transporters. The liver also is packed with enzymes and is capable of chemically modifying numerous substrates, including drugs, in many ways and at great speed. In fact, many drugs cannot be orally applied at all, because even during the initial passage the liver extracts them quantitatively from the blood it receives via the portal vein. This phenomenon is called the *first-pass effect.*

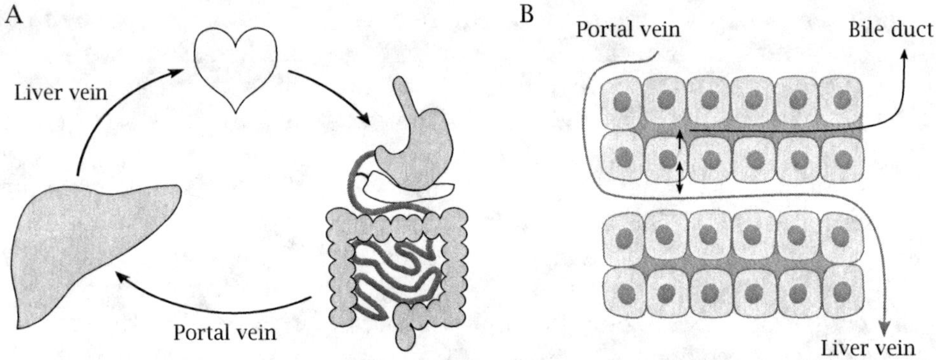

Figure 3.8 The portal circulation. **A:** The intestinal organs—stomach, pancreas, spleen, small and large intestines—receive arterial blood from the left heart. Their venous blood is drained by the portal vein and passed through the liver before reaching the systemic circulation again via the liver vein. **B:** Within the liver, blood received from the portal vein flows around the liver epithelial cells. Solutes are exchanged with the cells at their basolateral membranes. At their apical sides, the cells secrete solutes into the bile.

An example of a drug that undergoes a substantial first-pass effect is propranolol, which blocks β-adrenergic receptors and is commonly used in patients with cardiovascular disease (Figure 3.9). Only about 30% of the propranolol ingested actually shows up in the systemic circulation; the rest is either not absorbed or metabolized in the liver during the first passage. The extent of this first-pass effect shows considerable variation between individuals, which means that the required dosage may vary accordingly and has to be empirically determined with each patient. The fraction of the drug that does reach the systemic circulation (30% in our example) is called its *bioavailability*. In the case of propranolol, the bioavailability is actually considered acceptable, and the drug is indeed applied orally; this goes to show how vigorous the first-pass effect can be with those drugs that are deemed unfit for oral use.

If and when an orally applied drug has passed all these barriers, it will at least have taken some time to do so. Because of this and of the other inherent vagaries of absorption, oral application is not suitable for treatment in emergencies.

3.3.2 Intravenous application

Next to oral application, intravenous drug application is most widely used. If we define absorption as the process of uptake into the bloodstream, then in this case it is both quantitative and instantaneous. This is valuable if drug action is needed immediately, as is the case in emergency therapy. Intravenous application is the only means for delivering large drug molecules, in particular proteins such as tissue plasminogen activator (tPA) or coagulation factor VIII, which needs to be substituted in patients with hemophilia.

We have seen above that drugs for oral application need to be designed with stability in mind. In contrast, intravenous application is feasible with

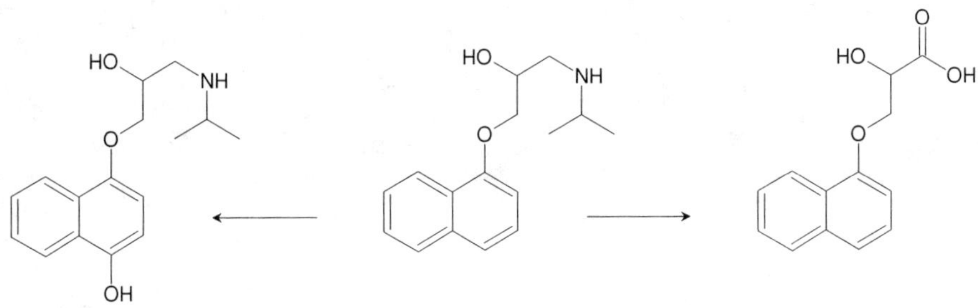

Figure 3.9 Propranolol and the first-pass effect. During the first passage through the intestine and the liver, propranolol undergoes metabolic transformation to an extent of >50%. The metabolites shown here are formed, or their formation initiated by cytochrome P450 enzymes. 4-Hydroxypropranolol has a pharmacological activity similar to propranolol, whereas naphthyloxylactate is inactive.

short-lived drugs as well. This is sometimes advantageous in emergency care treatment. The drug in question is applied by continuous infusion. Since it survives in the circulation for only a couple of minutes, the plasma concentration follows the volume flow rate of infusion very closely; the infusion rate can thus be continuously varied such as to titrate the desired clinical effect. Examples of this strategy are (1) the application of succinylcholine, which is rapidly degraded by plasma cholinesterase, to induce muscle relaxation during systemic narcosis, (2) the control of blood pressure in hypertonic crisis with sodium nitroprusside, and (3) the application of insulin and glucose in the acute treatment of patients with derailed diabetes mellitus.

Intravenous application also has its disadvantages. It is involved and dangerous if not performed properly, therefore it requires trained professionals. Like therapeutic drug effects, adverse reactions will be more instantaneous and vigorous; an example is the intravenous application of penicillin with patients allergic to the drug, which may easily have a fatal outcome.

3.3.3 Intramuscular and subcutaneous application

Intravenous, intramuscular, and subcutaneous injection are collectively referred to as *parenteral* application. They all bypass the obstacles associated with enteral (intestinal) uptake. However, intramuscular and subcutaneous injection will deposit the drug *outside* the blood vessels; it will thus have to be absorbed across the walls of the capillaries. The rate of uptake into the general circulation depends on the size of the drug molecules, and on the rate of blood perfusion; subcutaneous tissue is more slowly perfused than muscle tissue. Very large molecules such as proteins, such as vaccines, will not be able to permeate the capillary walls at all but will instead be drained by the lymphatic vessels. Along this passage, they will be subject to immunological processing within the lymph nodes.

Intramuscular and subcutaneous injection can be used with drug molecules that are not suitable for oral application but nevertheless should become

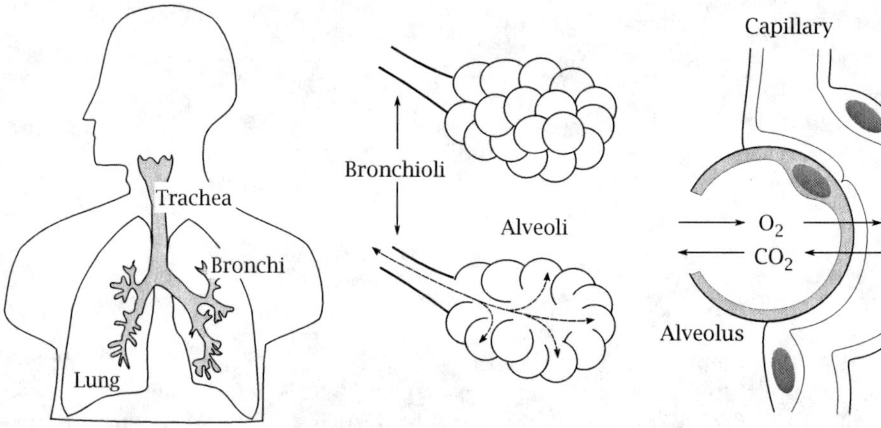

Figure 3.10 Overview of lung anatomy. The trachea branches out into bronchi, which, in turn, branch into bronchioli that end within alveoli. The alveoli collectively have a very large surface (~80 m^2), and the distance between the air-filled space and the blood in the surrounding capillaries is very short. The combination of large active surface and short diffusion distance enables very rapid gas exchange.

systemically available in a sustained, protracted fashion. A good example is the subcutaneous application of insulin in the long-term therapy of diabetes mellitus.

3.3.4 Pulmonic application

The effect of drug application by inhalation depends on the aggregation state of the drug. Gaseous drugs reach the alveoli and partition across the cell membranes of the alveolar endothelium and into the bloodstream; therefore, they will be systemically available. An adult's lung has a full 80 m^2 of exchange-active area (Figure 3.10). This makes the pulmonic absorption of gaseous drugs very rapid—just like the uptake of oxygen. Gaseous drugs will also mostly be eliminated by the lungs, and similarly promptly—just like the exhalation of carbon dioxide.[8] Rapid uptake and elimination is advantageous in the application of inhalation anesthetics such as ether, N_2O, and isoflurane, since it allows for accurate control and swift adjustments of the depth of anesthesia.

Nongaseous drugs can be applied by inhalation only as aerosols. The aerosol droplets are deposited in the bronchi but do not reach the alveoli; therefore, aerosols are a form of topical application to the bronchi and are used for treating afflictions of the latter. A common example are glucocorticoids, dispensed with nebulizers in the treatment of asthma.

3.3.5 Other forms of drug application

Dermal application of drugs has two cases: (1) topical application for the treatment of skin disease, and (2) dermal application for systemic use. Topical

[8]For the most part. However, some inhalation anesthetics are quite lipophilic and therefore accumulate in fat tissue, from where they slowly distribute back into the blood before being exhaled; this causes the prolonged odor of patients after surgery.

treatment of skin diseases allows for high local drug concentrations and minimal side effects on the rest of the body. Dermal application for systemic use is slow and inefficient with most drugs, since the skin is quite well isolated, with layers of compacted cells, keratin and ceramide, and uptake in this case is typically slow and inefficient. Notable exceptions are very hydrophobic compounds such as organic solvents and solutes they may carry. Nerve gases like diisopropylfluorophosphate (see Figure 6.26) have been made hydrophobic for the very purpose of skin penetration. The typically slow rate of uptake across the skin can be used to sustain prolonged, slow delivery, for example in the application of nicotine to weaning smokers.

Mucosal drug application exploits the fact that, compared to the skin, the epithelial barrier is much thinner. Moreover, the veins underlying the mucous membranes in the two preferred places, the nose and the rectum, are not drained toward the liver. Therefore, the first-pass effect is circumvented.

Cocaine or antidiuretic hormone (ADH) are commonly applied through the nose, as was tobacco in the olden days. ADH is a peptide, which tells us that even peptides can make it across this mucous membrane. Nitroglycerin is applied sublingually, that is, below the tongue. Acetaminophen is often applied rectally, particularly in children. Rectal application will increase the bioavailability of this drug relative to oral uptake, because the first-pass effect is absent.[9]

Depending on the drug and the location, uptake after mucosal application can be quite fast—nitroglycerin or cocaine take effect within a few minutes.

3.4 DRUG DISTRIBUTION

After their uptake into the systemic circulation, drugs are distributed between different compartments. Considering the capillary walls and the cell membranes as major barriers to distribution results in four major compartments: (1) The *intravascular volume* is the space inside the blood vessels; it contains both the plasma volume and the intracellular volume of the blood cells with ~4% and with ~3% of the total body volume, respectively. (2) The *interstitial volume* is the extracellular space outside of the blood vessels and is 2–3 times larger than the intravascular volume. (3) The aqueous *intracellular volume* amounts to approximately 40% of the total body volume. (4) Body fat occupies another share of the intracellular volume; it is an important reservoir for lipophilic drugs. This volume is more variable than the other ones, but values in the range of 5–15% are not uncommon.

The four compartments are illustrated in Figure 3.11. You may notice that their combined volume falls short of 100%. This is because some parts of the body, such as bone minerals and muscle protein, are not accessible to any solutes.

The ionic composition of the fluid that fills the interstitial space closely resembles that of blood plasma; this reflects the fact that the capillary endothe-

[9]More precisely, diminished—the venous blood is not drained toward the liver at its very end of the rectum, but a few centimeters above it is.

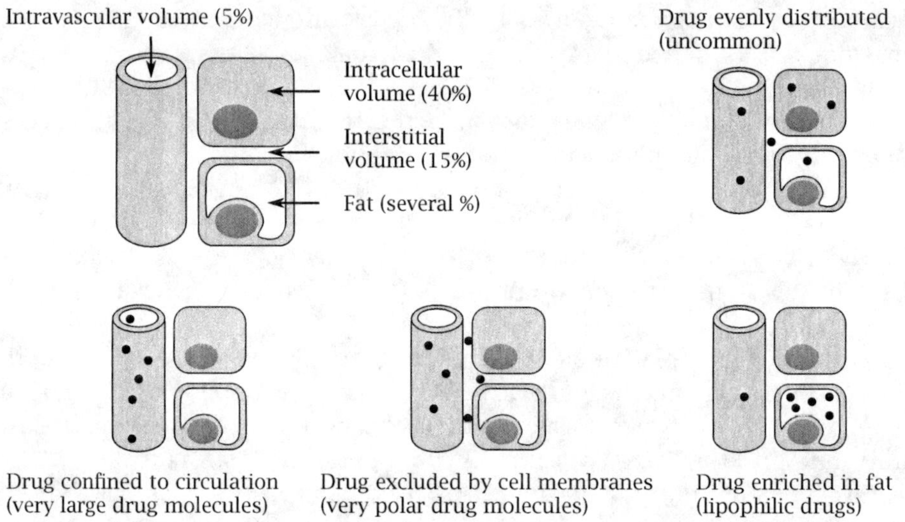

Figure 3.11 Major compartments of drug distribution. Large drug molecules will remain confined to the circulation and therefore occur in the plasma at high concentrations. Drugs that permeate capillaries and cell membranes easily will show lower plasma concentrations, particularly when they are lipophilic and become sequestered inside fat cells.

lium is freely permeable to small solutes. However, it has a considerably lower protein concentration, since most plasma proteins substantially exceed the size limit imposed by the capillary endothelium and are therefore confined to the intravascular space.[10] This restriction also applies to tPA and factor VIII, our two model drugs for intravenous application. The reason why this does not cause a problem is that both interact with the blood coagulation system, meaning that their target sites are also in the circulation.

Few drugs will distribute evenly among these compartments. Factors that will cause the distribution to be uneven include (1) the exclusion of membrane-impermeant drugs from the intracellular volume, (2) the accumulation of lipophilic drugs in fat tissue, and (3) the propensity of many drugs for binding to plasma proteins, which are much more abundant in the blood than in the interstitial fluid.

3.4.1 The volume of distribution

An uneven distribution between the intravascular space and the various extravascular spaces implies that we cannot use the plasma concentration of a drug directly to measure its total amount in the body. To account for un-

[10] The endothelial size barrier breaks down in tissues affected by inflammation or tumor invasion. This enables immune effector proteins from the blood plasma to seep into the disease focus, and it is the basis of several experimental approaches to the selective delivery of anti-tumor drugs using macromolecules or liposomes as carriers (Section 14.3.1). Increased leakiness is also observed at the blood–brain barrier in meningitis; this promotes distribution of antibiotics into the liquor space. It may be absent, however, in localized infections (brain abscesses), which therefore can prove less amenable to antibiotic treatment.

even distribution, a coefficient termed the *volume of distribution* (V_d) has been invented. It is defined simply as

$$V_d = \frac{n_{\text{drug}}}{[\text{drug}]_{\text{plasma}} \times \text{body weight}} \tag{3.1}$$

where n_{drug} is the number of drug molecules in the system. The volume of distribution can be measured experimentally by determining the plasma concentration at steady state. A steady state can be established for example by continuous infusion, which at some point will establish equilibrium with elimination. The V_d is not a real volume; it just happens to be measured in units of liters per kilogram, hence the fancy name. For large molecules that are prevented from leaving the bloodstream or from partitioning into cells, V_d will be small and similar to the plasma volume. In contrast, V_d will be very high, sometimes up to 10 L/kg, for lipophilic drugs that accumulate in the fat tissue.

It should go without saying the volume of distribution for any given drug varies between individuals. Nevertheless, average values are useful for example to estimate the amount of a drug present in the system from the plasma concentration.

3.4.2 Protein binding

All of the major compartments a drug distributes between contain proteins, and drugs may to varying extents bind to those proteins. The most abundant protein in the blood plasma is albumin, which occurs at a concentration of approximately 60 g/L. Albumin acts as a carrier for free fatty acids, which by way of hydrophobic interaction bind to multiple binding sites on each albumin molecule.[11] These binding sites can also bind drug molecules, particularly hydrophobic ones, which therefore are often bound to an extent of more than 90% of their total concentration in the blood plasma.

Since the blood plasma has a much higher protein concentration than the interstitial space, protein binding will favor retention of a drug in the intravascular volume, at least in the short term. Protein binding is usually rapidly reversible, so that the bound fraction of the drug is not lost; it can yet dissociate, equilibrate with the intracelluar compartment, which is also rich in protein, and bind to its target subsequently. However, one important consequence of plasma protein binding is that it will interfere with glomerular filtration of the drug in the kidneys, which is an important step in drug excretion (see below).

3.4.3 Kinetics of drug distribution

You will have noted that, in the compartment model of distribution described above, all organs have been lumped together. However, depending on the drug and target site in question, one or several organs may have to be considered separately. This is often the case for the central nervous system, which is protected by the blood–brain barrier. The initial phase of drug distribution is also

[11] This prevents the amphiphilic, detergent-like fatty acid molecules released by fat cells from partitioning into and potentially damaging the cell membranes of blood and endothelial cells.

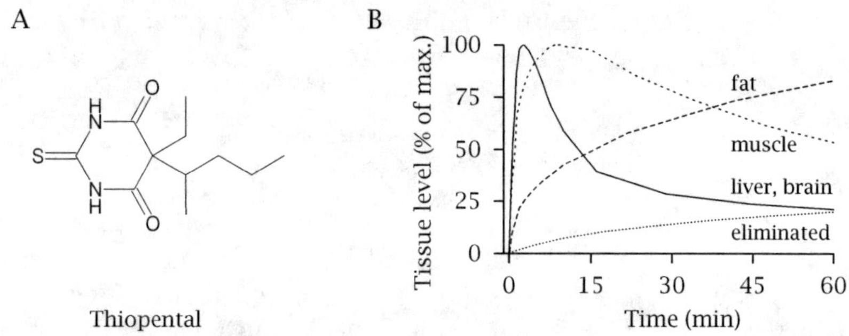

Figure 3.12 Kinetics of thiopental distribution. **A:** Structure of thiopental. **B:** Time course of thiopental levels in various tissues after intravenous application. After a very short initial phase of accumulation in the lung (not shown), thiopental redistributes to other strongly perfused organs such as the brain and the liver, then to skeletal muscle (which is not very strongly perfused in a resting patient) and finally to fat tissue, which is least strongly perfused but accumulates the lipophilic drug. (Figure redrawn from original data in Ref. [10].)

strongly affected by the varying rates of perfusion of different organs. This is illustrated in Figure 3.12 for the drug thiopental, a barbiturate that is used by intravenous injection to induce narcosis of short duration for minor surgical procedures. Thiopental is a very lipophilic molecule that readily penetrates the brain by diffusion across the lipid bilayers in the blood-brain barrier. Very shortly after injection, the concentration in the brain peaks, and for a few minutes the level is high enough to induce narcosis. The high initial concentration in the brain is related to the fact that the brain receives a very large fraction of the cardiac output (~20%). However, after a short time, the drug leaves the brain again and accumulates in lean tissues such as skeletal muscle, from where it later redistributes to the body fat. To a lipophilic molecule, fat provides the most accommodating environment; however, since it is only weakly perfused, substance exchange works more slowly than with the other tissues.

In the particular case of thiopental, the effect of the drug is not terminated by its elimination, as is usually the case, but solely by its redistribution from the brain to inert tissue reservoirs. The ultimate elimination of thiopental is far slower than its redistribution; it takes days to complete and involves hepatic metabolism, followed by urinary excretion.

3.5 DRUG ELIMINATION VIA THE KIDNEYS

A sizable fraction of all drugs is eliminated entirely or to a significant extent through the kidneys. As a rule of thumb, drugs can undergo renal elimination if they are hydrophilic, whereas hydrophobic drug molecules often undergo metabolism to more hydrophilic derivatives in the liver before elimination. To understand renal drug elimination, we first need to consider the basics of kidney function.

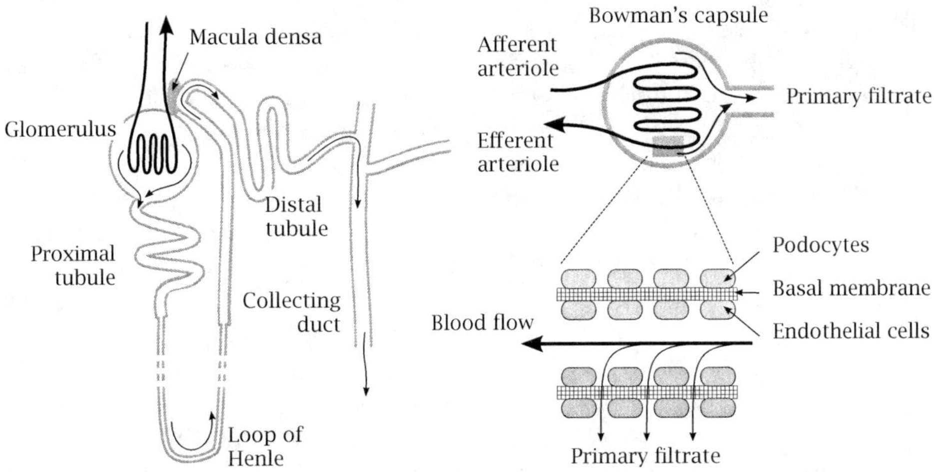

Figure 3.13 The nephron. **A:** A nephron consists of a glomerulus and a tubular part that can be divided into the proximal tubule, the loop of Henle, and the distal tubule; the latter joins a collecting duct that drains several nephrons. The primary filtrate that is formed in the glomerulus is extensively post-processed by active and passive transport across the epithelium that lines the tubular section. **B:** Glomerular filtration is driven by the hydrostatic pressure gradient across the wall of the arteriole; the exclusion size limit is imposed by the basal membrane.

3.5.1 Functional anatomy of the kidney

The flow of blood plasma through the kidneys amounts to approximately 0.6 liters (0.15 gallons) per minute. Since ~40% of the blood volume consists of red cells, the overall blood flow is about 1 liter per minute. Both kidneys together have a mass of only about 300 grams. Per gram of tissue, the kidneys have the highest rate of blood flow among all organs, including the heart and the brain. The strong perfusion is a key aspect of the very high rate of elimination that the kidneys achieve with suitable solutes.

The kidney tissue is organized into a large number of similar functional units, each of which performs all stages of urine processing. A single unit is called a *nephron* (Figure 3.13). It has a very unique structure and traverses both the cortex and the medulla—that is, the outer and the inner tissue layers of the kidney. In the cortex, we find the glomerulus as well as the proximal tubule and the distal tubule of the nephron. The intervening part between the proximal and distal tubules descends into the medulla and is called the *loop of Henle*. Each nephron "distills" urine from the blood plasma in three stages:

1. Filtration of blood plasma occurs in the glomerulus. The filtrate is funneled into the proximal tubule and passes through all the tubular sections of the nephron.
2. Metabolites and salts are reabsorbed into the blood by specific, active transport across the epithelium within the tubular sections. Water is reabsorbed by osmosis.
3. Secretion from the blood plasma into the nascent urine also occurs by active transport in the tubules.

The filtration apparatus in the glomerulus consists of a flexuous stretch of small arteries, or *arterioles*, the walls of which act as a sieve. The inner layer is covered by endothelial cells with multiple slits or fenestrations. An outer, likewise discontinuous layer is formed by so-called podocytes ("foot cells"). Sandwiched between the two cell layers is the basal membrane. The latter is continuous and spans the slits that cut through the endothelium and the podocyte layer. Therefore, the basal membrane determines the selectivity of the filter. As in the capillaries of the general circulation, the molecular weight cutoff is in the range of 10 kDa, so that most protein molecules will be retained.[12] The overall flow rate of filtration can be adjusted by the endothelial cells and podocytes through widening or narrowing the slits; however, this will not change the molecular selectivity of the filter.

In ordinary capillaries of other tissues, the pressure gradient across the capillary wall is mostly balanced by the colloid-osmotic effect of intravascular albumin, so there is little net filtration across the basal membrane. In contrast, the transmural pressure in the glomerular arterioles is much higher, and it diverts a full 20% of the plasma volume flow across the basal membrane. This amounts to approximately 170 L (44 gal) of primary filtrate every day. With respect to smaller solutes such as glucose, salts and amino acids, the composition of the primary filtrate is virtually the same as that of blood plasma. It would of course be unsustainable to lose water and metabolites at such a staggering rate.[13] You will therefore be relieved to learn that the bulk of both water and solutes undergo reabsorption already at the very next stage of processing in the proximal tubule. This applies in particular to glucose and amino acids, which are recovered mostly through sodium cotransport. Chloride and bicarbonate ions are likewise recovered to a large extent in the proximal tubule.

Additional water is reclaimed within the loop of Henle, where an unusually high salt concentration in the surrounding interstitial fluid within the kidney medulla makes it possible to concentrate the urine to approximately four times the osmolality of blood plasma. Fine tuning of urine volume and ion concentrations occurs in the distal tubule and the collecting ducts. Protons are secreted or recovered through exchange for potassium; therefore, the urine pH may vary between 5 and 9, in keeping with the prevailing metabolic situation.

Drugs in the circulation may be affected by all stages of urine processing. If they are sufficiently small and not protein-bound, they will undergo filtration. As discussed above, all the major classes of transporter proteins responsible for solute reuptake and secretion—ABC, OAT, and OCT—contain members with fairly broad substrate specificity. In addition, hydrophobic solutes may undergo reuptake without specific transport, simply by diffusion across the

[12]The appearance of protein in the urine—called *proteinuria*—is a key symptom of *glomerulonephritis*, an autoimmune disease that afflicts the glomeruli.

[13]What is the use, then, of producing so much filtrate and expelling so many useful solutes in the first place? I like to think of it as shifting the burden of proof—after filtration, each solute has to prove its right to remain in the system by qualifying for reuptake. Unknown solutes that don't fit the substrate specificity of any active transporter will be flushed out by default.

lipid membranes of the epithelial cells. The rate of renal elimination of most drugs is therefore at least somewhat impacted by secretion and reuptake.

3.5.2 Kinetics of renal drug elimination

As a rule, the rates of all three processes in renal elimination—filtration, secretion, and reuptake—will be proportional to the concentration of the solute in the blood plasma.[14] This means that the overall rate of elimination should also be proportional to the plasma concentration

$$\frac{dn_\mathrm{D}}{dt} = k \times [\mathrm{D}]_\mathrm{plasma} \tag{3.2}$$

where D is the drug and $\frac{dn_\mathrm{D}}{dt}$ is the number of drug molecules that are being eliminated per time interval.

To measure $\frac{dn_\mathrm{D}}{dt}$, we simply collect the urine over a defined period of time, typically 24 hours, and determine the volume and drug concentration:[15]

$$\frac{dn_\mathrm{D}}{dt} = [\mathrm{D}]_\mathrm{urine} \times \text{flow rate}_\mathrm{urine} = [\mathrm{D}]_\mathrm{urine} \times \frac{V_\mathrm{urine}}{24\,\mathrm{h}} \tag{3.3}$$

We combine equations 3.2 and 3.3 and solve for k:

$$k = \text{flow rate}_\mathrm{urine} \times \frac{[\mathrm{D}]_\mathrm{urine}}{[\mathrm{D}]_\mathrm{plasma}} \tag{3.4}$$

The term on the right-hand side of Equation 3.4 is called the *renal clearance* of D. The experimentally determined clearance will measure the actual rate constant of renal elimination if that process is indeed of first order. Measuring the clearance of drugs or model compounds with known amenability to glomerular filtration or tubular secretion can be used to assess the kidney function, which will help to adjust the dosages of drugs with small therapeutic indices.

3.5.3 Determination and use of the glomerular filtration rate

Inulin is a metabolically inert polysaccharide of approximately 6 kDa. It is considered to be (1) not bound to plasma proteins and therefore fully amenable to filtration in the glomerulus and (2) not subject to secretion or reuptake in the tubules.[16] Therefore, the concentration of inulin in the primary filtrate should be the same as in the blood plasma, and all inulin molecules that are filtrated will also show up in the final urine. As the initial volume of the filtrate

[14] The rate of reuptake should be proportional to the concentration in the nascent urine, which in turn depends on the rates of filtration and secretion, and therefore like the latter should also be proportional to the plasma concentration.

[15] This assumes that the plasma concentration and rate constant are invariable over time. If they are not, collecting for 24 hours will at least balance circadian variations.

[16] Other substances that may be more convenient to quantify can serve the same purpose, for example, iothalamate [1].

is trimmed down due to tubular water reuptake, the concentration of inulin in the remaining urine increases reciprocally:

$$\frac{[\text{inulin}]_{\text{urine}}}{[\text{inulin}]_{\text{filtrate}}} = \frac{\text{flow rate}_{\text{filtrate}}}{\text{flow rate}_{\text{urine}}} \tag{3.5}$$

Since inulin filtration is quantitative, its concentration in the filtrate equals its plasma concentration. We can substitute accordingly and solve for the filtrate volume flow:

$$\text{flow rate}_{\text{filtrate}} = \text{flow rate}_{\text{urine}} \times \frac{[\text{inulin}]_{\text{urine}}}{[\text{inulin}]_{\text{plasma}}} \tag{3.6}$$

Equation 3.6 tells us that the inulin clearance directly measures the volume flow of the glomerular filtrate, or the glomerular filtration rate (GFR). The GFR will affect the elimination of many drugs. It can be significantly reduced in elderly patients or in those with kidney diseases. With drugs that have a small therapeutic index, GFR measurements are used to adjust the dosage regimen in order to keep the plasma concentrations in the therapeutic range.

In clinical practice, the GFR is measured not with inulin but with creatinine, which is formed at a constant rate in the degradation of creatine in muscle. The plasma concentration of creatinine is kept constant by the equilibrum of biosynthesis and excretion, so that the creatinine clearance can be determined from a single measurement of the plasma concentration and of the total amount excreted in 24 hours.

As a further simplification, we can assume that the rate of creatinine synthesis, if normalized for body weight, is equal between individuals. Since the plasma creatinine level is in a dynamic equilibrium, the rate of creatinine urine excretion should equal that standard rate of synthesis, and therefore we can do away with collecting the urine altogether and estimate the glomerular filtration rate from the plasma creatinine concentration alone. While the accuracy of this simplified method is limited, it is still useful to detect major deviations of the glomerular filtration rate. This is often sufficient to make necessary adjustments to initial drug dosage regimens; the further course of therapy can then be adjusted using direct measurements of the plasma concentration of the drug itself.

3.5.4 The effectiveness of tubular secretion

In the kidney tissue, blood vessels and kidney tubules run alongside one another, which minimizes the distance for solute exchange between blood and nascent urine by way of secretion and reuptake. Exactly how effective tubular secretion can be is strikingly illustrated by the model compound *p*-aminohippuric acid (*p*-AH). The concentration of *p*-AH in the kidney veins is $\leq 10\%$ of that in the kidney arteries, meaning that *p*-AH is nearly quantitatively transferred from the blood to the urine during a single passage through the kidneys! As we have seen, glomerular filtration alone could remove only some 20%, so *p*-AH must be eliminated mostly by tubular secretion. The concentration of *p*-AH in the kidney arteries is the same as in the general circulation. Therefore, if we neglect the small residual amount in the kidney veins, we can write:

$$[p\text{-AH}]_{\text{plasma}} \times \text{flow rate}_{\text{plasma}} = [p\text{-AH}]_{\text{urine}} \times \text{flow rate}_{\text{urine}} \tag{3.7}$$

If we solve for the flow rate$_{\text{plasma}}$ and compare the result to Equation 3.4, we see that it is given simply by the *p*-AH clearance. Therefore, the *p*-AH clearance measures the maximum possible rate of renal drug elimination, which also corresponds to the renal plasma flow. While this would seem to be an interesting parameter in pharmacotherapy, the *p*-AH clearance is not routinely determined in clinical practice.

3.5.5 Non-equilibrium elimination kinetics

In the foregoing, we have always assumed that the drug or model compound be present at constant plasma concentration, which would require its application at constant rate, for example by continuous intravenous infusion. More commonly, however, drugs are applied discontinuously. To deal with this case, we will assume that at time zero we have a known number n_0 molecules of drug (D) in the system, and that distribution has reached equilibrium; these conditions will often be fulfilled a short time after intravenous injection of a known amount of the drug. As before, we assume the rate of elimination to be proportional to the plasma concentration. We can then write, for the number n of drug molecules remaining in the system:[17]

$$\frac{dn}{dt} = -k \times [\text{D}]_{\text{plasma}} \tag{3.8}$$

Equation 3.1 relates the number of drug molucules to the plasma concentration and can be restated as

$$n = [\text{D}]_{\text{plasma}} \times V_{\text{d}} \times \text{body weight} \tag{3.9}$$

Combining the two equations, integration, and solving for $[D]_{\text{plasma}}$ gives us:

$$[\text{D}]_{\text{plasma}} = [\text{D}]_{\text{plasma},0} \times e^{-\frac{k}{V_{\text{d}} \times \text{body weight}} t} \tag{3.10}$$

We can express the same relationship using the *half-life* of the plasma concentration:

$$t_{1/2} = \ln 2 \times \frac{V_{\text{d}} \times \text{body weight}}{k} \tag{3.11}$$

[17]The presence of the minus sign here but not in Equation 3.2 reflects the fact that n here denotes the number of drug molecules remaining in the system, whereas in Equation 3.2 it denotes those being excreted.

If we express the clearance as k over body weight, as is commonly done, this simplifies to

$$t_{1/2} = \ln 2 \times \frac{V_d}{CL} \tag{3.12}$$

Note that, in the foregoing, the only assumption made about elimination was that it follow first-order kinetics. This commonly holds for both renal and metabolic elimination,[18] and therefore the concepts of half-life and of clearance can be applied to metabolic elimination as well. If a drug is eliminated both in the kidneys and in the liver, the clearances will be additive, and the half-life can be obtained from the sum using Equation 3.12.

3.5.6 The rate of renal drug elimination is highly variable

How fast can renal elimination possibly work? Tubular secretion will work only with small molecules, which will typically not be confined to the circulation but at least distribute into the interstitial space. This results in a volume of distribution of V_d = 0.2L/kg. If the drug is quantitatively extracted from the blood plasma perfusing the kidneys, like *p*-aminohippurate, its clearance will be 0.6L/(min×70 kg). Using Equation 3.12, this works out to a half-life of approximately 15 min.

This model calculation constitutes a limiting case, and with most real drugs, the half-life of renal elimination is significantly longer. Still, it illustrates just how powerful the renal apparatus really is, and it tells us that even with drugs that are resistant to metabolic inactivation the very speed of renal elimination may limit their usefulness for therapy. As an example, we may consider penicillin G (Figure 11.8), which is cleared from the circulation with a half-life of about one hour by the same mechanism as *p*-AH. In the early days of penicillin therapy, when the drug still was very scarce, this rapid elimination was a major problem. The urine of patients receiving penicillin therapy was actually collected, and the secreted penicillin recovered. This problem was overcome by the development of probenecid (Figure 3.7), which inhibits the excretion of penicillin across the apical membrane of the tubule epithelia and very pronouncedly prolongs the retention of penicillin in the body.[19]

Another example of a drug that is swiftly eliminated is cimetidine. Its clearance is smaller than that of *p*-AH but exceeds that of of inulin, which indicates that it, too, must be actively secreted. At the other end of the scale, we find drugs like phenobarbital or chloroquine, which in their non-metabolized forms undergo negligible renal elimination. This is a common finding with drugs that show the following characteristics: 1) high volume of distribution—if V_d is high, this means that the plasma concentration is low, and accordingly very few

[18] A notable exception is the metabolism of alcohol, which follows zero-order kinetics, which means that its elimination rate is independent of the remaining concentration.

[19] Penicillin is no longer scarce, and so nowadays it is simply applied in dosages large enough to compensate for its rapid elimination. However, penicillin does not distribute well into the brain. In the treatment of brain abscesses, probenecid can be used to raise the plasma concentration, and thereby achieve sufficient levels within the brain. Probenecid is also used as a uricosuric drug in the therapy of gout (see Section 10.2.2).

drug molecules will enter the kidneys to begin with; 2) high degree of protein binding—glomerular filtration will operate only on the unbound fraction in the plasma; 3) lipophilic nature—this is correlated with the former two traits, and it also facilitates reuptake by diffusion across the cell membranes in the tubular epithelial cells.[20]

If a drug is lipophilic but ionizable, the effectiveness of renal elimination will depend on the urine pH. A good example is amphetamine (Figure 6.22), which has an amino group that at low pH values will be mostly protonated; at higher pH values, the unprotonated fraction will be high enough for significant reuptake by nonionic diffusion. This can be therapeutically exploited to accelerate the elimination of drugs; since the kidneys are responsible for adjusting the blood plasma pH, they will secrete alkaline or acidic urine in response to the infusion of sodium bicarbonate or ammonium chloride, respectively.[21]

If the poison or drug is membrane-permeant and neither acidic nor basic, the only way to accelerate renal elimination is to increase the urine volume. In this case, the rate of elimination will simply be proportional to the volume flow of urine. This strategy is called *forced diuresis*.[22]

While the hydrophobicity of a drug does have an effect on the rate of its elimination in the kidneys, the correlation between the two parameters is actually quite feeble. Specific structural features that do not change the global physicochemical properties can have an overriding effect. For example, the two drugs quinine and quinidine are diastereomers of each other, yet quinidine undergoes tubular secretion with a fourfold higher rate, presumably through stereoselective active transport [7]. Interaction with drug-metabolizing enzymes, and even plasma protein binding, may be similarly affected.

3.6 QUANTITATIVE EXPERIMENTAL METHODS IN PHARMACOKINETICS

The traditional approach to pharmacokinetics relied on assaying drug concentrations in urine and feces and in readily accessible body compartments such as blood and cerebrospinal fluid. Together with the determination of laboratory parameters that measure kidney and liver function, such measurements are usually sufficient for drug monitoring in clinical therapy. However, more sophisticated methods are used for investigating the pharmacokinetics of drugs during development. In particular, imaging methods that have been developed in recent decades allow the experimenter to follow the distribution of drugs between different organs and their elimination in greatly increased resolution of space and time. Two key imaging methods are magnetic resonance imaging

[20] For example, chloroquine has a logP of 3.7, a distribution volume of 196 L/kg, and a half-life of 408 h [4]. The slow elimination of chloroquine translates into long dosage intervals—when it was in use for malaria prophylaxis, weekly application was sufficient.

[21] Alkalinization would be used, for example, with barbiturates; acidification would be used in metamphetamine intoxication.

[22] Another, more effective but also more involved method for the accelerated elimination of hydrophobic drugs such as barbiturates is *hemoperfusion*. Here, blood is diverted from a large artery (typically in the thigh), passed over a hydrophobic solid-phase absorber, and fed back into the corresponding vein.

(MRI) and positron emission tomography (PET). MRI and PET have complementary strengths and limitations that are related to their underlying physical principles.

PET is based on the use of drug molecules that are labeled with a radioactive isotope that emits a positron. In an environment replete with electrons, the positron will immediately undergo positron–electron annihilation. This produces two γ particles of equal and characteristic energy that fly off in opposite directions and can therefore be discriminated from background radiation using a coincidence detector. Serial images are constructed from data accumulated by rotating the detector around the patient or experimental animal in successive transversal planes, much as in X-ray–based computer tomography scans.[23]

Like other radioactivity-based methods, PET is very sensitive, which makes it possible not just to track the mass flow of the drug analogs but even to visualize the receptor saturation in specific tissues. Its main limitation lies in the physical half-life of the positron-emitting isotope, which may be too short to allow observation over the entire pharmacokinetically relevant timespan. A short half-life also poses significant challenges for the synthesis and purification of the drug analog. This is the case for ^{11}C and ^{15}O, which have half-lives of just a few minutes. Isotopes with longer half-lives are ^{18}F (109 minutes) and ^{124}I (4.2 days). Of course, a longer half-life is equivalent to a lower specific activity and thus to lower sensitivity.

MRI uses stable isotopes that have an uneven number of nucleons. The most common such isotope is simply regular hydrogen (^{1}H), and conventional diagnostic MRI imaging is based on the variation between tissues in their content of water. Obviously, because of the large amount of hydrogen present in the tissues, drugs cannot be detected using their hydrogen signal. The highest sensitivity can be achieved with fluorine (^{19}F), which can often be substituted for hydrogen in drug molecules without causing a major functional change.

An advantage of MRI is that the observation time is not limited by any inherent instability of the probe. However, it is considerably less sensitive than PET. In research on small experimental animals, this may be partially compensated for by using very strong magnetic fields and by dosages that are higher than those used for therapy. Very large dosages will leave the pharmacokinetic parameters unchanged only to the extent that the usual assumption of first-order kinetics still applies, which may be questionable.

The experimental detail available from these advanced methods calls for similarly advanced mathematical models that are a far cry from the simplistic compartmentalization scheme discussed above (Section 3.4), and that will take into account the different rates of blood perfusion and other properties of individual organs. Such models are important in drug development but less so in clinical routine.

[23]Advanced detectors allow the calculation of the point of decay directly from the difference in the time of flight of the two γ particles. Since the latter travel at the speed of light, this requires arrival times to be measured on a scale of picoseconds.

STUDY QUESTIONS

3.1 In Chapter 1, we briefly mentioned oligonucleotides and their pharmacokinetic limitations. Let us consider a hypothetic oligonucleotide with a length of 15 bases. (1) What routes of drug application should be most useful? (2) What compartments should it be able to distribute into? Which barriers should it be unable to cross? (3) How would it be eliminated? (4) Can you find experimental studies that confirm or contradict your predictions?

3.2 Lysergic acid diethylamide (LSD) is a hallucinogenic drug used by some to dispel suburban boredom. Reportedly, the hallucinations can be either very pleasant or very frightening ("bad trip").

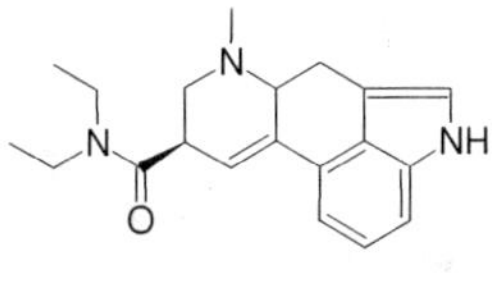

Lysergic acid diethylamide (LSD)

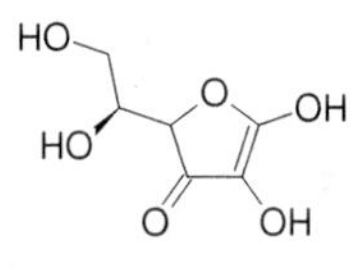

Ascorbic acid (vitamin C)

In case of a bad trip, a recipe from the underground advises to use large doses of ascorbic acid (vitamin C) to shorten its duration. Can you suggest a mechanism for how this might work?

3.3 The plots below show some correlations of pharmacokinetic parameters for approximately 250 clinically used drugs listed in Ref. [4].

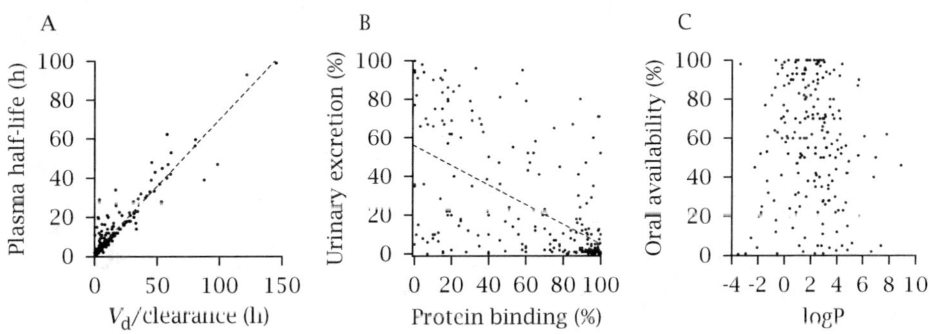

A: Correlation between half-life and ratio of distribution volume and clearance. Some drugs with very long lifetimes have been removed from this sample. **B**: Correlation of renal elimination and protein binding. **C**: Correlation of logP (the logarithm of the octanol-water partition coefficient; obtained from emolecules.com) and oral availability. What do you make of these data? Are the correlations, or the lack thereof, expected or unexpected?

REFERENCES

[1] R. Agarwal, "Chromatographic estimation of iothalamate and p-aminohippuric acid to measure glomerular filtration rate and effective renal plasma flow in humans," *J. Chromatogr. B Biomed. Sci. Appl.* 705 (1998), 3–9.

[2] S. G. Aller et al., "Structure of P-glycoprotein reveals a molecular basis for poly-specific drug binding," *Science* 323 (2009), 1718–1722.

[3] S. Bihorel et al., "Modulation of the brain distribution of imatinib and its metabolites in mice by valspodar, zosuquidar and elacridar," *Pharm. Res.* 24 (2007), 1720–1728.

[4] L. L. Brunton, J. S. Lazo, and K. L. Parker, eds., *Goodman and Gilman's The pharmacological basis of therapeutics,* 11th ed., McGraw Hill, 2005, 1794–1888.

[5] T. Cheng et al., "Computation of octanol-water partition coefficients by guiding an additive model with knowledge," *J. Chem. Inf. Model.* 47 (2007), 2140–2148.

[6] R. J. P. Dawson and K. P. Locher, "Structure of the multidrug ABC transporter Sav1866 from Staphylococcus aureus in complex with AMP-PNP," *FEBS Lett.* 581 (2007), 935–938.

[7] D. A. Notterman et al., "Stereoselective renal tubular secretion of quinidine and quinine," *Clin. Pharmacol. Ther.* 40 (1986), 511-517.

[8] D. Ribatti et al., "Development of the blood-brain barrier: a historical point of view," *Anat. Rec. B New Anat.* 289 (2006), 3–8.

[9] U. Tröger et al., "Tetraparesis associated with colchicine is probably due to inhibition by verapamil of the P-glycoprotein efflux pump in the blood-brain barrier," *Br. Med. J.* 331 (2005), p. 613.

[10] D. R. Wada et al., "Computer simulation of the effects of alterations in blood flows and body composition on thiopental pharmacokinetics in humans," *Anesthesiology* 87 (1997), 884–899.

CHAPTER 4

DRUG METABOLISM

As discussed in Chapter 3, the rates of excretion in the urine vary greatly between different drug molecules. On average, the fraction excreted unchanged in the urine is less than 50% [2]. The remainder undergoes some kind of *biotransformation*, which is a fancy name for metabolism. Most of this metabolism is accomplished by enzyme systems that have evolved for the disposal of *xenobiotics*, that is, natural compounds that are poisonous or at least serve no useful purpose in human metabolism, and that must be prevented from accumulating in the body. These enzymes also metabolize many synthetic chemicals and drugs. Major classes of drug-metabolizing enzymes are (1) cytochrome P450, (2) UDP-glucuronosyltransferases, (3) glutathione-*S*-transferases, (4) sulfotransferases, and (5) *N*-acetyltransferases. Several enzymes whose primary substrates are not xenobiotics but physiological metabolites also contribute to drug metabolism, for example, monoamine oxidase, xanthine oxidase, and nitric oxide synthase.

Each of the major enzyme classes contains several to several dozen different forms. This multitude arises in part from multiple independent genes and in part from alternate splicing. Enzymes within one family share the same overall structure and reaction mechanism, yet they vary in their substrate specificity, even though specificities often overlap among various family members.

Substrate specificities may also overlap between enzyme classes. For example, a phenolic hydroxyl group in a given drug molecule may be a substrate for

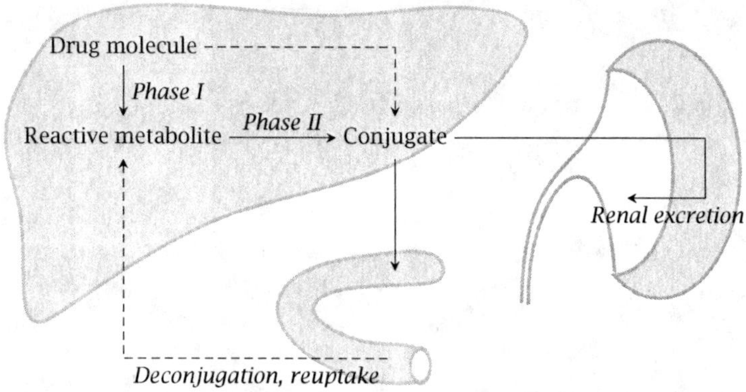

Figure 4.1 The role of metabolism in drug elimination. Most drug biotransformation reactions occur in the liver. With many drugs, hydroxylation by cytochrome P450 or other *phase I reactions* precede conjugation with glucuronic acid, glutathione or other functional groups (collectively referred to as *phase II reactions*). The conjugated drugs may be exported into the bloodstream and be eliminated in the kidneys, or they may be excreted into the bile and then reach the intestine. In the large intestine, they may undergo deconjugation by bacterial enzymes and reuptake. Reabsorbed drugs will be drained to the liver and undergo another round of conjugation and biliary excretion; this is referred to as *enterohepatic cycling*.

sulfotransferases, glucuronosyltransferases, and acetyltransferases. Through alternate and successive transformations, a single drug may give rise to a large number of metabolites, which may vary with respect to pharmacological activity, toxicity, and rate of excretion.

The ability of the sizable yet limited number of enzymes and membrane transporters to cope with a virtually boundless number of drugs and poisons, without taking an excessive toll on physiological metabolites, is really quite remarkable. Many drugs are metabolized by multiple enzymes. On the other hand, if a drug is a substrate for a single enzyme or membrane transporter only, allelic gene variations of that enzyme and transporter may profoundly affect its rate of elimination. The study of such allelic variations belongs to the field of pharmacogenetics.

4.1 OVERVIEW: POSSIBLE FUNCTIONAL OUTCOMES OF DRUG METABOLISM

By and large, metabolism of drugs results in their inactivation and accelerated excretion (Figure 4.1). However, other outcomes are possible.

1. Metabolites may retain pharmacological activity. Sometimes such an *active metabolite* is more stable than the original drug. Some active metabolites have become drugs in their own right. Fexofenadine, a histamine H_1 receptor antagonist derived from terfenadine (Figure 4.2), has even supplanted its parent compound, because the latter turned out to cause side effects on cardiac excitation when the metabolic conversion to fexofenadine was impeded by drugs that inhibit cytochrome P450 enzymes [6].

2. Metabolism may be *required* to confer pharmacological activity to a drug. A common case is activation by hydrolysis of resorption esters, that is, drugs in

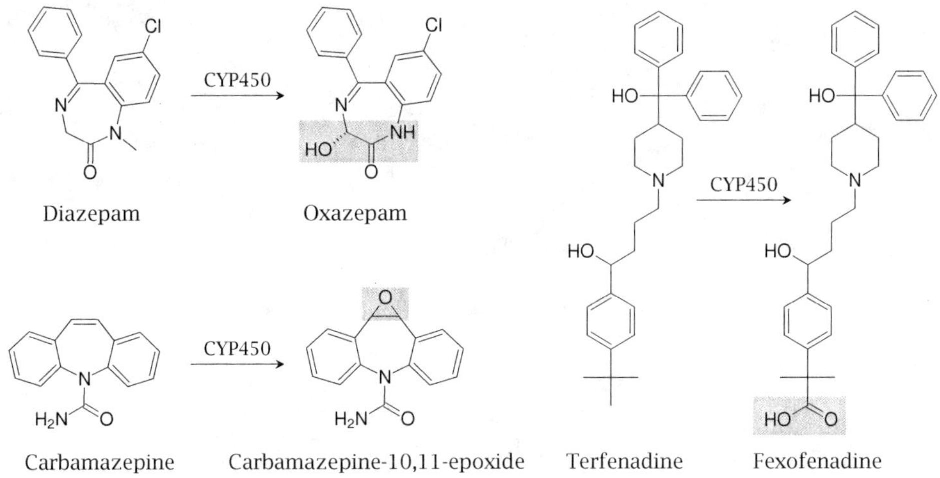

Figure 4.2 Examples of active drug metabolites. All metabolites shown here are formed by cytochrome P450 enzymes. Oxazepam and fexofenadine have become drugs in their own right.

which some polar or ionizable groups are masked in order to facilitate uptake across the intestinal epithelium. Drugs that require metabolic activation are referred to as *prodrugs.*

3. Biotransformation may inactivate a drug molecule but in a way that does not accelerate elimination. For example, acetylation of aliphatic amino groups will prevent their protonation and therefore tend to reduce polarity, which may impede excretion in the kidneys.

4. Drug metabolism may *cause* toxicity. An example is the formation of reactive epoxides from initially inert aromatic compounds, and the oxidation of acetaminophen, whose metabolite sequesters glutathione (see Section 4.5.3).

While enzymes capable of metabolizing drugs are found in many tissues, the highest activity is found in the intestinal epithelium and in the liver, in keeping with the major route of uptake of xenobiotics.[1] Accordingly, orally applied drugs may be extensively metabolized even before they reach the systemic circulation, during their first passage through the liver; this constitutes the so-called first-pass effect. Since the liver is very strongly perfused, hepatic metabolism also affects drugs that have reached the systemic circulation, either by surviving the first-pass effect or through parenteral application.

Apart from enzymes in the body tissues, those of bacteria in the large intestine contribute to drug metabolism. The milieu in the large intestine is anaerobic, and accordingly drugs may undergo reductive reactions. Another common modification is the cleavage of glucuronic acid conjugates, which may lead to reabsorption of the drug and to enterohepatic cycling.

[1] An example of a drug that undergoes metabolism outside the liver is the β-lactam antibiotic imipenem (see Figure 11.8). It is cleaved rapidly by an esterase found in the kidney and has to be protected by the simultaneous application of cilastatin, an inhibitor of that esterase.

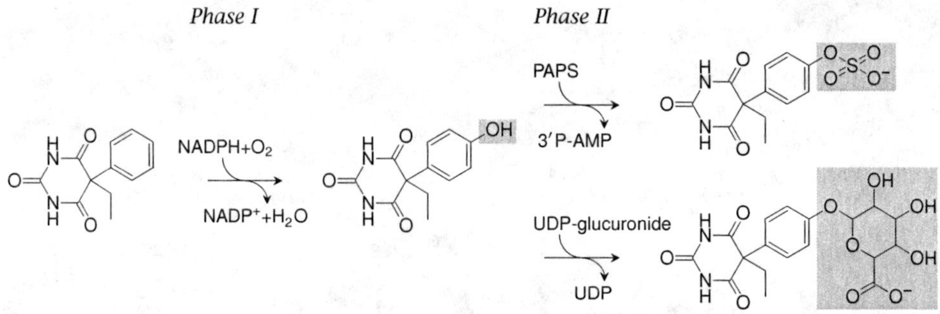

Figure 4.3 Phases of phenobarbital metabolism. In phase I, phenobarbital is hydroxylated by cytochrome P450. In phase II, the hydroxyl group is conjugated with glucuronic acid, which is catalyzed by UDP-glucuronosyltransferase, or alternatively with sulfate by sulfotransferase. PAPS is 3′-adenosine-5′-phosphosulfate.

4.2 PHASE I AND PHASE II REACTIONS

Traditionally, reactions in drug metabolism are classified as phase I and phase II reactions. The basic idea of this distinction can be illustrated using the drug phenobarbital as an example (Figure 4.3). In its unmodified form, phenobarbital is quite hydrophobic and is eliminated only at a very low rate. To render it more suitable for excretion, it must first be transformed to a more hydrophilic metabolite. An effective means of increasing hydrophilicity of a drug molecule consists in its conjugation with glucuronic acid by UDP-glucuronosyltransferase. However, phenobarbital does not have any good functional group that could serve as a target for conjugation. Such a target site can be introduced by hydroxylation, which is mediated by cytochrome P450. This preparatory hydroxylation is considered a phase I reaction, and the subsequent conjugation with glucuronic acid is considered a phase II reaction.

Apart from increasing a drug's polarity, glucuronidation will usually inactivate the molecule. In addition, it will render the drug suitable for excretion by the kidneys or into the bile. Several membrane transporters in both the kidneys and the liver, both of the SLC and the ABC families, recognize glucuronides as substrates and ensure efficient elimination.[2]

Phase I reactions, which often involve the forcible oxidation of an inert drug molecule, tend to be rate-limiting. Glucuronidation and other phase II reactions occur between reactive groups on the substrates and activated cosubstrates and therefore are more facile. In some cases, for example, with glutathione and reactive epoxides, conjugations may even proceed without enzymatic catalysis.

The distinction between phase I and phase II reactions captures an important pattern in drug metabolism, yet there is no point in trying to apply it universally. Some drugs may undergo more than two successive reactions, whereas others may require no phase I modification at all and instead undergo conjugation directly. The latter case is illustrated by morphine. This drug has two free

[2]The glucuronic acid conjugates are called *glucuronides*, not *glucuronates*, because the bond created is a glycosidic bond, not an ester bond. The carboxylate group of glucuronic acid remains free and ionizable and contributes to the overall hydrophilicity.

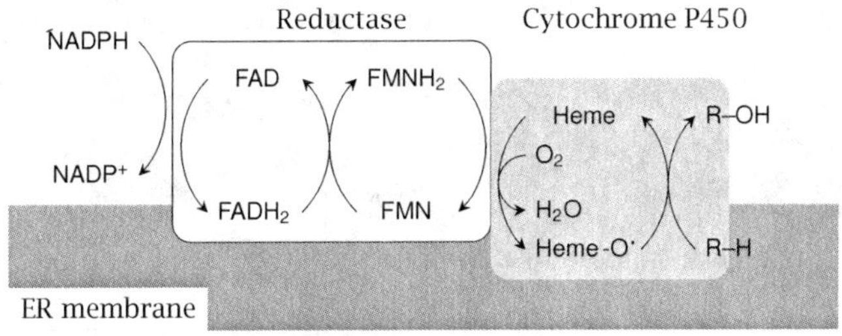

Figure 4.4 Mode of action of cytochrome P450 enzymes. Most isoforms are associated with the membrane of the endoplasmic reticulum (ER), which facilitates processing of lipophilic substrates that initially may reside in the apolar phase of the membrane. Cytochrome P450 reductase acquires two electrons from NADPH and delivers them successively to heme in the active site of cytochrome P450, where they reduce one of the two atoms of molecular oxygen to water. The second oxygen atom reacts with the substrate, which may result in hydroxylation (as shown here) or a number of other oxidative modifications.

hydroxyl groups, to either or both of which a UDP-glucuronosyltransferase in the liver ER will attach a glucuronic acid moiety. The glucuronide will again be excreted both in the urine and the bile.

4.3 CYTOCHROME P450

Enzymes of the cytochrome P450 family are responsible for most phase I reactions. Cytochrome P450 enzymes are very widespread in nature and occur in both eukaryotic and prokaryotic cells. In eukaryotic cells, they mostly reside in the membrane of the smooth endoplasmic reticulum, but some are found in the mitochondria. They vary in substrate specificity, and, where they bind the same substrates, they may modify different atoms within them. The exact kind of modification performed appears to be controlled more by the nature of the substrate and its orientation in the active site than by any intrinsic properties of the enzyme. In total, 57 different cytochrome P450 enzymes have been identified in humans; among these, about a dozen participate significantly in drug metabolism. The levels of enzyme activity present in the cell vary substantially between individual enzymes. The single most abundant enzyme, CYP3A4 (Figure 4.5), participates in 50% of all metabolic reactions that affect drugs in clinical use. Other isoforms with prominent roles in drug metabolism are 1A2, 2C9, and 2D6.

A cytochrome P450 enzyme works in conjunction with cytochrome P450 reductase, which supplies it with electrons from NADPH, using FAD and FMN in the electron transfer process (Figure 4.4). The two electrons are sequentially delivered to the heme cofactor in the active center of the cytochrome, which in turn transfers them to one of the two oxygen atoms of O_2 to yield water. The other oxygen atom is retained at the heme in a highly reactive state and can attack many organic molecules, even ones that are intrinsically quite inert.

With phenobarbital as well as many other drugs, oxidation by cytochrome P450 yields a phenolic hydroxyl group. However, the reaction may have various

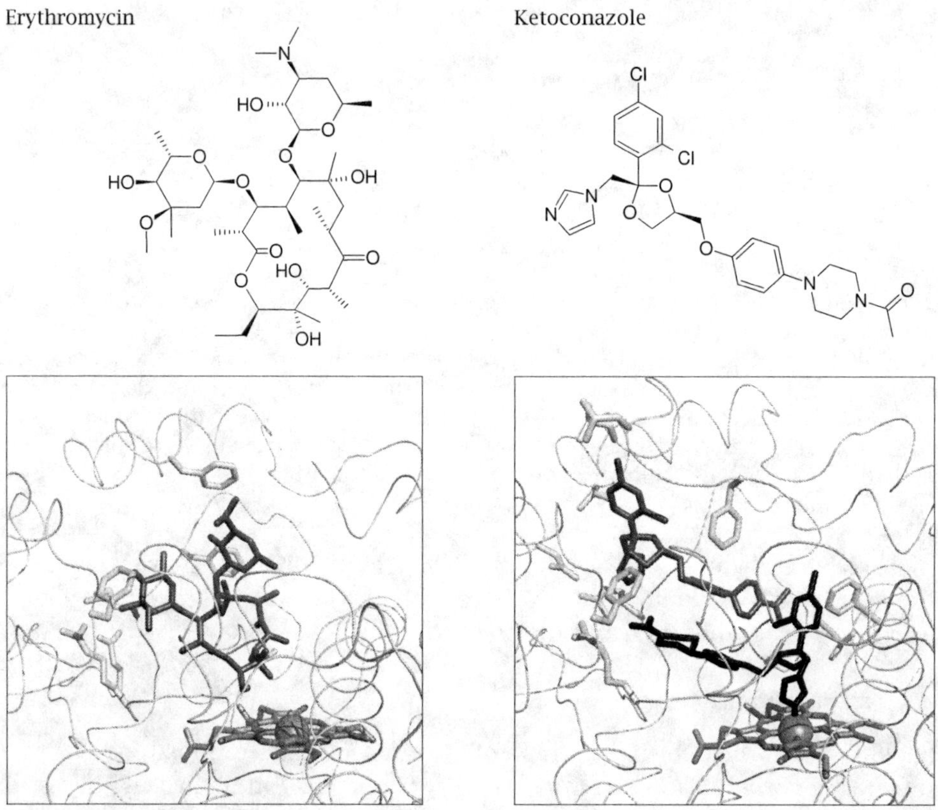

Figure 4.5 Promiscuous substrate binding by cytochrome P450 3A4 [3]. The enzyme was crystallized in the presence of erythromycin or of ketoconazole, respectively. Two molecules of ketoconazole are bound in the active site; the orientation of the upper one is similar to the structure shown above. The two drugs interact with different active site amino acid residues (white sticks). The two structures are shown from the same angle, and differences in the trace of the enzyme's polypeptide backbone reflect conformational changes induced by the drugs. (Figure rendered from 2j0d.pdb and 2v0m.pdb [3]. A colored version of this figure is shown in the insert in the middle of this book.)

other outcomes, which include (1) hydroxylation of aliphatic carbon and of amines, (2) formation of epoxides, (3) sulfoxide formation, (4) dealkylation, (5) ring aromatization, and (6) oxidation of hydroquinones to quinones. Some of these reactions are shown in Figure 4.6. Cytochrome P450 reaction products may be harmful and must then be scavenged by subsequent metabolism. For example, epoxides may react with nucleic acids and cause mutations. Epoxide groups may be converted to vicinal hydroxyl groups by epoxide hydroxylase or may alternatively undergo conjugation with glutathione.

The activity of CYP3A4 and several other cytochrome P450 enzymes can be increased by transcriptional induction. This may cause drug interactions, since one drug may induce an enzyme that also metabolizes another, unrelated drug (Section 4.6). On the other hand, cytochrome P450 enzymes may be inhibited by

$$\mathrm{R{-}H \xrightarrow{[O]} R{-}OH} \qquad \text{A}$$

$$\mathrm{RCH_2{-}OH \xrightarrow{[O]} RCH{=}O + H_2O}$$

$$\mathrm{RCH{=}O \xrightarrow{[O]} RCOOH}$$

$$\mathrm{R_2N{-}H \xrightarrow{[O]} R_2N{-}OH} \qquad \text{B}$$

$$\mathrm{R_3N \xrightarrow{[O]} R_3N{=}O}$$

$$\mathrm{R_2S \xrightarrow{[O]} R_2S{=}O}$$

$$\mathrm{RO{-}CH_2R \xrightarrow{[O]} ROH + O{=}CHR} \qquad \text{C}$$

$$\mathrm{R_2N{-}CH_2R \xrightarrow{[O]} R_2NH + O{=}CHR}$$

$$\mathrm{R{-}HC{=}CH{-}R \xrightarrow{[O]} R{-}HC\overset{O}{\triangle}CH{-}R} \qquad \text{D}$$

Figure 4.6 Some oxidative reactions catalyzed by cytochrome P450. **A**: Hydroxylation of aliphatic and aromatic carbon, and oxidation of alcohols and aldehydes; **B**: Hydroxylation and oxidation of amines and sulfides; **C**: Dealkylation of amines and ethers; **D**: Epoxide formation.

therapeutic drugs or by food constituents. For example, ketoconazole, an antifungal drug, is a competitive inhibitor of fungal cytochrome P450 enzymes that are required in ergosterol synthesis. As a side effect, it also binds and inhibits human CYP3A4 (see Figure 4.5). Grapefruit juice contains furanocoumarins that are first converted to reactive intermediates by CYP3A4 and then form covalent adducts with the enzyme. The inactivation of CYP4A3 will slow down the inactivation of drugs that are applied concurrently.

Inhibition of cytochrome P450 can also be applied for therapeutic purposes. A practical application of this principle is the combination of two different inhibitors of HIV protease. The drug ritonavir inhibits CYP3A4 and is used mostly for the sake of improving absorption of the second drug, lopinavir.

Enzymes other than cytochrome P450 that are involved in oxidative drug metabolism include the flavin-containing monooxygenases and monoamine oxidases. The former resemble cytochrome P450 enzymes in being membrane-associated and accepting electrons from NADPH for the activation of molecular oxygen. Their range of reactions appears to be narrower than that of cytochrome P450 enzymes, and flavin-containing monooxygenases are not known to be subject to induction by drugs. Monoamine oxidases mainly oxidize endogenous catecholamines and related drugs but may also contribute to the oxidation of other amines.

4.4 REDUCTIVE DRUG METABOLISM

Enzymatic reduction is another class of reactions that often constitute phase I transformations. The range of drugs that are substrates for reduction is not as wide as that for oxidation yet still fairly diverse. Functional groups that are subject to reduction include quinones, aromatic nitrates, other nitrogen-containing groups, sulfoxides, and alkyl halides. The following enzymes participate in reductive drug metabolism:

1. Cytochrome P450 reductase. While the main function of this protein is to supply electrons to cytochrome P450, it can also transfer electrons to other substrates, including drugs.
2. Cytochrome P450 enzymes themselves, which may transfer electrons to substrates other than oxygen.
3. Cytochrome b5 proteins. These hemoproteins transfer one electron at a time and participate in the reduction of nitrogen compounds, alkyl halides, and quinones.
4. NADH:quinone oxidoreductase, or diaphorase, which mediates the two-electron reduction of quinones to hydroquinones. Another variety of diaphorase uses NADPH or reduced nicotinamide riboside (NRH) as cosubstrates. Notwithstanding the name, the substrates reduced by diaphorase are not limited to quinones.
5. Thioredoxin reductase, a flavoprotein that accepts two electrons from NADPH and uses them to reduce thioredoxin, which in turn reduces protein disulfide bonds. Thioredoxin reductase may also reduce other substrates such as sulfoxides.

Reduced drug molecules usually have to undergo conjugation for final detoxification and excretion. If conjugation does not occur, they may be reoxidized, which results in redox cycling (see Section 4.4.3). Intermediates or products of reductive drug metabolism may be reactive toward cellular macromolecules and therefore toxic. This is exploited therapeutically in the activation of the antitumor prodrug CB 1954 (Figure 4.12) and of metronidazole, which undergoes reduction selectively in anaerobic organisms and can be used against these (see Section 11.4.1).

4.4.1 Reduction of nitroaromatic and other nitrogen compounds

Reduction plays a prominent role in the metabolism of aromatic nitrogen compounds. Reduction of a nitro group may go all the way to the corresponding amine, with nitroso and hydroxylamino groups occurring as intermediates. This requires six electrons overall and involves both one- and two-electron reductive steps. Single-electron transfers will give rise to radicals, which may react with cellular macromolecules, or with molecular oxygen to form superoxide radicals.

Nitrogen reduction may also start at the level of nitroso groups, hydroxylamino groups, amidoximes and azo groups. All reduced forms of nitrogen can potentially be oxidized again, which gives rise to futile redox cycling and oxidative stress due to formation of reactive oxygen species. Such toxicity is suppressed by trapping the reductively formed amino or hydroxylamino groups in subsequent conjugation reactions.

4.4.2 Quinone reduction

Another prominent class of substrates for reductive metabolism are quinones, which are reduced to hydroquinones. The latter may easily be oxidized again, often without requiring enzymatic catalysis, which results in redox cycling. One-electron reduction of quinones produces semiquinones, which may engage in radical reactions. An additional mechanism of quinone toxicity is the formation of adducts with cellular nucleophiles. This is exemplified by the metabolism of acetaminophen (see Figure 4.9). For effective detoxification, hydroquinones must be conjugated and excreted.

4.4.3 Redox cycling

We have seen that the pathways and enzymes for both oxidative and reductive drug metabolism are of fairly low specificity. This can give rise to redox cycles, in which a drug molecule is first oxidized and then reduced again. When this happens, each drug molecule acts effectively as a catalyst for the depletion of reduction equivalents and often also for the generation of reactive oxygen species.

The most important scavenger of reactive oxygen species is reduced glutathione. In healthy individuals, the cellular reserves for regenerating reduced glutathione are not easily exhausted. However, the capacity to reduce glutathione is diminished in a condition called *favism*, which is caused by mutations of glucose-6-phosphate dehydrogenase. This enzyme catalyzes the first step in the hexose monophosphate shunt, the major pathway for the regeneration of NADPH, which in turn is required for the reduction of glutathione. The defect mainly affects red blood cells. It becomes manifest after ingestion of isouramil and related compounds found in broad beans (*Vicia faba*) or after application of drugs such as primaquine. Redox cycling depletes reduced glutathione (Figure 4.7). Reactive oxygen species that are no longer scavenged then inflict oxidative damage to the red cell membranes by way of lipid peroxidation, which ultimately causes destruction of the cells (hemolysis).

4.5 CONJUGATION REACTIONS

Conjugation reactions involve cosubstrates that contain smaller or larger functional groups in an activated form. These are attached to the drug molecule, often at functional sites that were created in a preceding phase I reaction. These activated functional groups include (1) glucuronic acid, (2) glutathione, (3) the acetyl group, (4) sulfate, (5) various amino acids, and (6) the methyl group. Conjugation can have various effects on a drug or drug metabolite:

1. It may abolish pharmacological activity. This is more likely to occur with large and polar moieties.

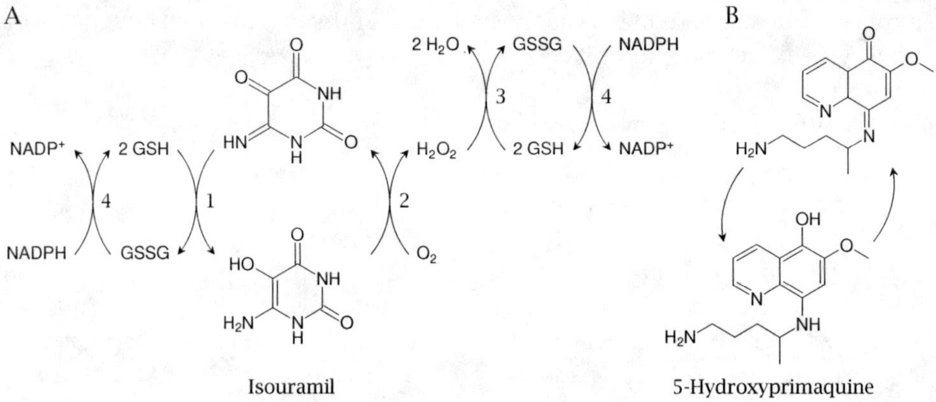

Figure 4.7 Redox cycling of isouramil and of 5-hydroxyprimaquine. **A:** Isouramil, which occurs in the broad bean, undergoes redox cycling at the expense of reduced glutathione (GSH), which it converts to the disulfide (GSSG). Reactions (1) and (2) occur spontaneously. Reaction (3) is catalyzed by glutathione peroxidase and reaction (4) by glutathione reductase. **B:** The 5-hydroxy derivative of the malaria drug primaquine, which is formed by cytochrome P450 metabolism, can engage in a similar redox cycle. Both cycles can exhaust the capacity for regeneration of NADPH and hence glutathione in the red blood cells of patients with hereditary glucose-6-phosphate dehydrogenase deficiency.

2. It may mask functional groups on the drug molecule that would otherwise cause harm by reaction with cellular macromolecules. On the other hand, in a few cases, conjugation may actually increase reactivity.

3. The increased polarity of the conjugate, and the existence of transport proteins with specificity for conjugates will facilitate excretion in the kidney or liver.

The cosubstrates used in drug conjugation are not specific for these reactions but have other roles in metabolism. UDP-glucuronic acid and 3′-phosphoadenosine-5′-phosphosulfate (PAPS) provide building blocks for the synthesis of mucopolysaccharides. *S*-Adenosylmethionine donates methyl groups in the synthesis of phosphatidylcholine from phosphatidylethanolamine, whereas acetyl-CoA has a key role in energy metabolism.

4.5.1 Glucuronidation

The most common points of glucuronic acid attachment are phenolic or alcoholic hydroxyl groups on the substrate molecule. Other possible target sites include carboxylic acids, amines, hydroxylamines, and thiol groups. This versatility explains why glucuronidation is the most common type of drug conjugation.

Conjugation with glucuronic acid is accomplished by several UDP-glucuronosyltransferases, which use UDP-glucuronic acid as the cosubstrate. UDP-glucuronosyltransferases are integral membrane proteins. They are located in the endoplasmic reticulum, that is, in the same location as the cytochrome P450 enzymes. The two enzymes often cooperate in the metabolism of hydrophobic or amphiphilic substrates, which are enriched within the lipid bilayer.

Glucuronidation occurs at the luminal side of the ER membrane, which means that transport of UDP-glucuronic acid from the cytosol and of the glucuronide product to the cytosol is required; the details of transport across the ER membrane are not fully understood. Once in the cytosol, glucuronides are efficiently extruded from the cell, mostly by transporters of the ABC family. If the conjugate is secreted into the blood, final elimination will typically occur by tubular secretion in the kidney. Liver cells may also directly secrete the conjugates into the bile.[3] Similarly, in the small intestine, conjugates may be directly secreted from the epithelial cells into the lumen.

Bacteria in the large intestine may enzymatically cleave glucuronides in order to then utilize the glucuronic acid; one such bacterium is *Escherichia coli.* The deconjugated drug or primary metabolite may then be taken up again from the intestine and reach the liver, where it may be conjugated and secreted once more. This so-called *enterohepatic cycling* may considerably delay drug elimination. An example is the drug digitoxin (Figure 6.9), which is used in the treatment of heart disease. The half-life of digitoxin is extended to several days by enterohepatic cycling. It may nevertheless be preferred over digoxin, which is structurally similar yet renally eliminated, in those patients who have impaired kidney function and therefore may eliminate digoxin at unpredictable rates.

4.5.2 Sulfation

Sulfotransferases occur both as soluble proteins in the cytosol and bound to intracellular membranes. The ones that are involved in drug metabolism belong to the soluble class; the membrane-associated ones are located in the Golgi apparatus and are concerned with the sulfation of glycosaminoglycans and proteins. Both classes are structurally similar and require the cosubstrate 3′-phosphoadenosine-5′-phosphosulfate (PAPS). Attachment of a sulfate group will render the drug substrate more hydrophilic and, usually, less active; there are, however, examples of drug activation by sulfation. Some sulfate conjugates exhibit enhanced reactivity toward cellular nucleophiles because sulfate constitutes an effective leaving group.[4] Like other reactive intermediates, these are likely to react with glutathione, which will protect the cell from most potential damage.

The substrate specificities of UDP-glucuronosyltransferases and sulfotransferases overlap, so it is not uncommon to find both types of conjugates with a

[3]Excretion into the bile is the norm for the diglucuronide of the endogenous metabolite bilirubin, which is formed in the degradation of heme. Bilirubin diglucuronide is modified by bacteria in the large intestine; its conversion products give the feces their characteristic color. If biliary secretion is blocked, as may be the case in hepatitis or when a tumor compresses the bile duct, the feces turn pale. Bilirubin diglucuronide accumulates in the body, and the patient turns yellow.

[4]Mutagenesis with dimethylsulfate, which has been used in genetic research, is based on this principle. Dimethylsulfate is a bifunctional reagent that introduces crosslinks into DNA molecules.

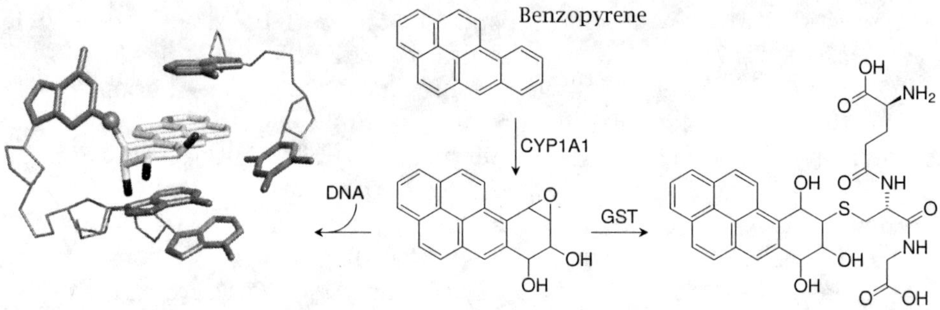

Figure 4.8 Metabolism of benzopyrene. Epoxidation is performed by cytochrome P450, in particular CYP1A1. Some of the metabolite molecules are hydrolyzed by epoxide hydrolase or conjugated by glutathione-*S*-transferase (GST). Those that remain may react covalently with DNA when the reactive metabolite intercalates between the stacked base pairs, giving rise to mutations and potentially cancer. (Structure of the DNA adduct rendered from 2ia6.pdb [1].)

single drug (see Figure 4.3). Due to the limited supply of PAPS, the capacity of sulfation is lower than that of glucuronidation, which means that at high drug concentrations the glucuronides tend to predominate.

4.5.3 Glutathione conjugation

Glutathione is a tripeptide that contains one cysteine residue, the sulfhydryl or thiol group of which is a very strong nucleophile and engages in a multitude of redox and conjugation reactions. It is of crucial importance in the detoxification of reactive oxygen species that form as the byproducts of oxygen transport and biochemistry. Its concentration in the cytosol is in the millimolar range, reaching 5–10 mM in metabolically active organs such as the liver, and it is kept mostly in the reduced form at the expense of NADPH by glutathione reductase. Because of its high concentration and reactivity, reaction rates with many electrophiles are significant even without enzymatic activation. Conversion of glutathione to its reactive form, the thiolate anion, is facilitated by glutathione-*S*-transferases. These enzymes occur in soluble and membrane-associated forms and in high abundance, amounting to several percent of the cellular protein content.

Glutathione conjugation is important in the detoxification of reactive metabolites, which mostly stem from cytochrome P450 metabolism. Examples of such toxic phase I metabolites are the epoxides formed from polycyclic aromatic hydrocarbons such as benzopyrene (Figure 4.8) and quinones such as *N*-acetyl-p-benzoquinone imine (NAPQI), which is formed from acetaminophen (Figure 4.9).

While conjugation will consume a stoichiometric amount of glutathione, this consumption will typically not significantly deplete the very high intracellular levels of glutathione. NAPQI is an exception to this rule. If acetaminophen is overdosed, formation of NAPQI can exceed the available glutathione, and the excess can then react with protein thiol groups, which may result in liver

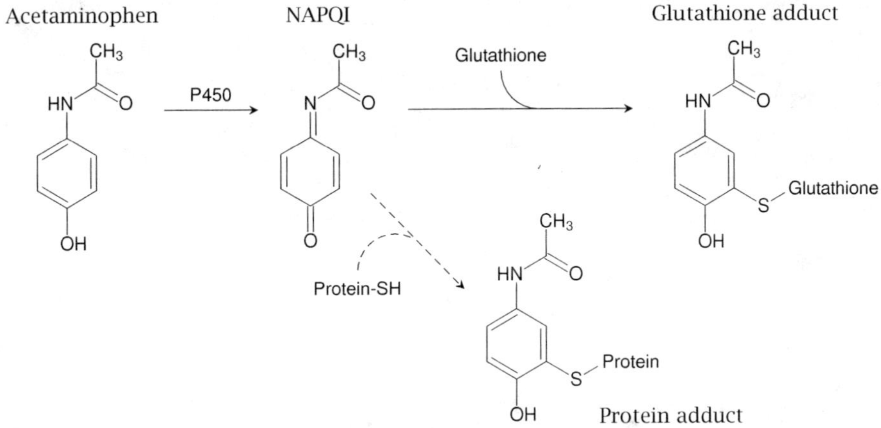

Figure 4.9 Metabolism of acetaminophen. Oxidation by CYP3A4, CYP2E1, or other cytochrome P450 isoforms yields *N*-acetyl-*p*-benzoquinone imine (NAPQI). The latter is conjugated with glutathione. When acetaminophen is overdosed, NAPQI may deplete cellular glutathione. The excess will then react with protein thiols, leading to liver cell damage and tissue necrosis.

cell destruction. Induction of cytochrome P450 by ethanol or other drugs may increase NAPQI formation and therefore acetaminophen toxicity.

The levels of glutathione and of glutathione-*S*-transferases may vary between different cell types. Increased expression of glutathione-*S*-transferase is selected for in tumor cells under treatment with alkylating agents such as cyclophosphamide, which are used in tumor therapy. An elegant strategy has been devised to turn this resistance mechanism on its head. The prodrug canfosfamide is *activated* rather than inactivated by glutathione-*S*-transferase P1-1 [10]. Proton abstraction by the enzyme triggers a β-elimination in the prodrug, which releases an alkylating agent (see Figure 4.10). The drug is currently undergoing clinical testing.

4.5.4 Acetylation

The *N*-acetyltransferases NAT1 and NAT2 are important in the metabolism of drug molecules that are aromatic amines or hydrazines. The activated acetyl groups are donated by the ubiquitous cosubstrate acetyl-CoA.

Aromatic amines and hydrazides are quite reactive and may undergo different routes of metabolism. If they are not *N*-acetylated, they may be converted by cytochrome P450 to hydroxylamines. The latter may be further oxidized to nitroso derivatives, which are chemically reactive and may engage in redox cycling. Hydroxylamines may also undergo *O*-acetylation or sulfation. The resulting metabolites are often reactive toward nucleophilic groups on proteins and nucleic acids. If they react with DNA, this can lead to mutations and cancer. Reaction with proteins may cause the immune system to mount a reaction to the modified proteins, which may result in autoimmune disorders such

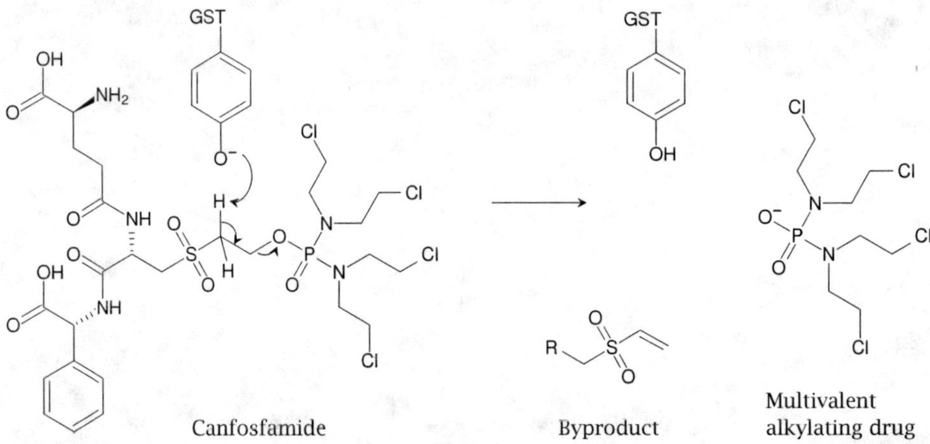

Figure 4.10 Activation of the antitumor prodrug canfosfamide by glutathione transferase (GST) P1-1 [10]. The reaction is initiated by a tyrosine residue in the enzyme's active site. The released, active drug contains two *N*-mustard moieties and behaves as a DNA-alkyating agent (see Section 12.4.5). Canfosfamide targets tumor cells that over-express glutathione-*S*-transferase, usually as the result of selection under treatment with other cytotoxic drugs.

as lupus-like syndrome, agranulocytosis, or toxic epidermal necrolysis (Lyell's syndrome).[5]

To the extent that *N*-acetylation of aromatic amines and hydrazines prevents these more toxic routes of metabolism, it should be beneficial. On the other hand, once hydroxylamines have been formed, *N*-acetyltransferases may increase their reactivity and toxicity by *O*-acetylating them to form acetoxy derivatives.[6] On balance, however, acetylation tends to reduce drug toxicity. This is evident from the effects of allelic variants of *N*-acetyltransferase 2 (NAT2). This variation was first observed with the tuberculostatic drug isoniazid (INH). With a random sample of patients, the rates of INH acetylation yield a bimodal histogram (Figure 4.11). The phenotypic groups represented by the two peaks are referred to as *slow acetylators* and *fast acetylators*, respectively, and they correspond to different alleles of NAT2. Slow acetylators are more susceptible to toxic effects of INH and of other drugs that undergo acetylation, such as procainamide, hydralazine, or dapsone [9]. They are also more likely to develop cancer of the urinary bladder and other organs when exposed to aromatic amines. If acetylation is slow, this should allow a greater fraction of amines to be converted to hydroxylamines first. Subsequent acetylation would then yield a correspondingly greater fraction of reactive acetoxy compounds.

The creation of toxic compounds through the interplay of reduction and acetylation is illustrated by the metabolic activation of the experimental cancer

[5]Coupling to carrier proteins is a commonly used technique for raising antibodies against small molecules, or *haptens* that on their own would fail to elicit an immune response.

[6]In some cases, *N*-hydroxylation by cytochrome P450 may occur even after *N*-acetylation, whereafter the acetyl group may be transferred to the hydroxyl group; this will again form a reactive acetoxy-group.

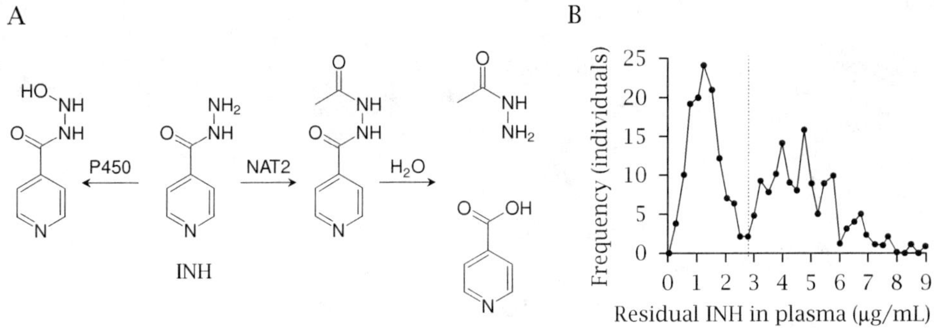

Figure 4.11 Hepatic metabolism of isoniazid (INH). **A**: The most important route of metabolic conversion is acetylation by *N*-acetyltransferase NAT2. The acetylated product may hydrolyze to release acetylhydrazide, which is a reactive metabolite. Nevertheless, *N*-acetylation reduces toxicity, since more toxic metabolites would otherwise arise through *N*-hydroxylation by cytochrome P450. **B**: Bimodal distribution of acetylation rates. The INH plasma concentration remaining 6 h after application of a single dosage of INH is inversely related to the rate of acetylation. The left and right peaks represent fast and slow acetylators, respectively. (Figure redrawn from original data in Ref. [5].)

drug CB 1954 (Figure 4.12). An aromatic nitro group is first reduced to the hydroxylamine by diaphorase. Reduction is followed by acetylation, which is mediated by NAT2. This yields a bifunctional reagent, which may introduce crosslinks into DNA and thereby cause cytotoxicity. Therefore, in this particular case, phase I and phase II reactions conspire to *create* toxicity, rather than prevent it.

4.5.5 Methylation

Methylation uses *S*-adenosylmethionine as a cosubstrate. Hydroxyl groups of drug molecules may be methylated by the enzyme catechol-*O*-methyltransferase (COMT), whose physiological function is the degradation of catecholamines. N-methylation may be catalyzed by phenylethanolamine-*N*-methyltransferase, which functions in the synthesis of epinephrine from norepinephrine in the adrenal glands (see Section 6.8.2).

Enzymatic *S*-methylation is the key reaction in the metabolism of mercaptoguanine and azathioprine, which are antimetabolites of purine synthesis and are used in the treatment of tumors. The enzyme responsible, thiopurine-*S*-methyltransferase, is subject to allelic variation. Clearance of the drugs will be delayed in patients with a low enzyme activity; this can cause severe toxicity if the dosage is not adjusted accordingly.

While glucuronidation, sulfation and glutathione conjugation increase the polarity of their substrates, this is not the case with acetylation and methylation. Therefore, these modifications may not accelerate elimination by renal or hepatic excretion. They may, however, still abolish the drug's ability to bind to its target. In this context, it is interesting to note that many of the bacte-

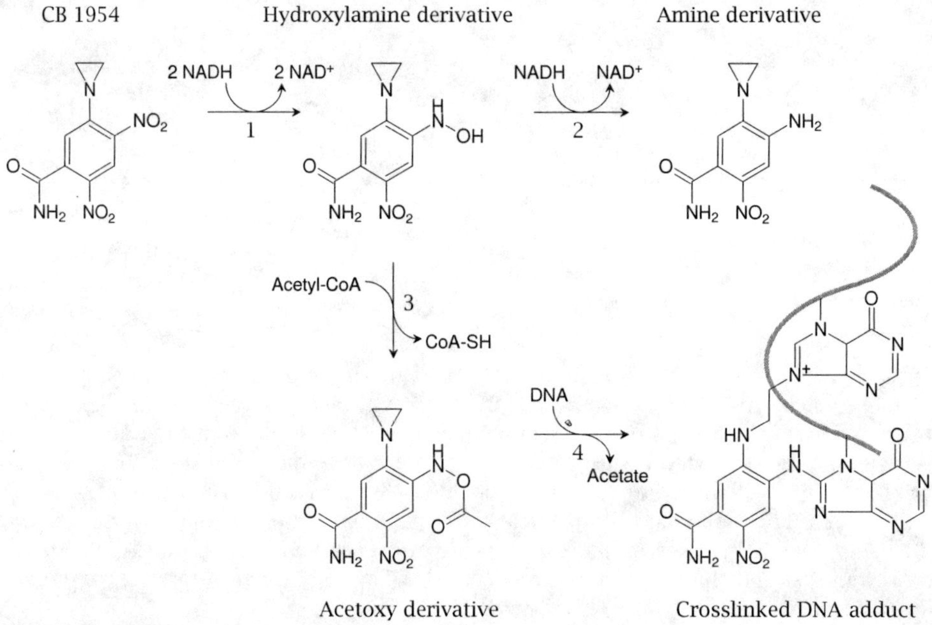

Figure 4.12 Reduction and conjugation of the cytotoxic drug 5-(aziridin-1-yl)-2,4-dinitrobenzamide (CB 1954). Reduction by diaphorase (1) yields the 4-hydroxylamine, which can either be further reduced to the amine (2) or undergo *O*-acetylation (3) by NAT2 [7,8]. The acetoxy derivative is a bifunctional reagent that can induce crosslinks in DNA molecules (4) and thus cause cytotoxicity.

rial enzymes that inactivate antibiotics and cause resistance are methyl- and acetyltransferases (see Section 13.1.2).

4.5.6 Amino acid conjugation

Like acetylation, amino acid conjugation results in the formation of amide bonds. This always involves the amino group of the amino acid. Substrates therefore are restricted to carboxylic acids; examples are salicylic acid and benzoic acid. Amino acids used for conjugation include glycine, glutamine, and taurine. The conjugation occurs in two steps. In the first step, the substrate organic acid is activated to its coenzyme A thioester. In the second step, the coenzyme A is displaced by the amino group of the amino acid, which results in the formation of an amide bond.

Apart from xenobiotics, amino acid conjugation applies also to bile acids, which to a significant extent occur in the bile as glycine and taurine conjugates. As an example, the conjugation of phenylacetate with glutamine is shown in Figure 4.13. This conjugation is part of an alternative pathway for phenylalanine degradation. In an interesting twist, glutamine conjugation of exogenous phenylacetate can be used to achieve elimination of excess nitrogen in enzyme defects of the urea cycle (see Section 10.1.3).

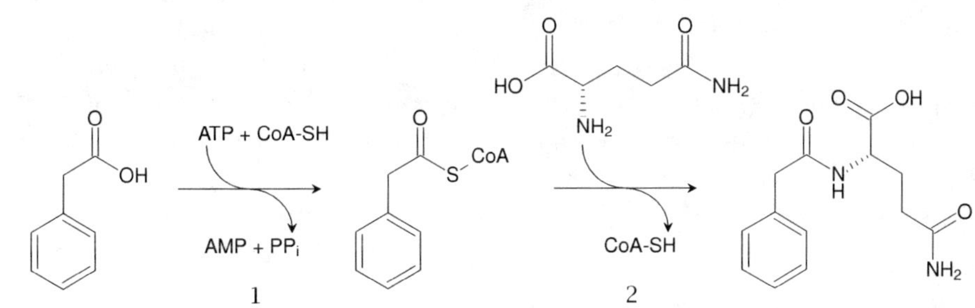

Figure 4.13 Conugation of phenylacetate with glutamine. Enzymes: (1) phenylacetate:CoA ligase, (2) glutamine-*N*-phenylacetyltransferase.

4.5.7 Conjugation and membrane transport

Hydrophobic drug molecules will be rendered considerably more hydrophilic after conjugation with glutathione, sulfate, or glucuronic acid. The conjugates will therefore be much less capable of passive diffusion across lipid bilayers. Active membrane transport of conjugates is mediated by various transporters of the ABC and OAT classes that accept glutathione conjugates and glucuronides as substrates. In kidney, gut, and liver, these transporters mediate excretion of conjugates into the nascent urine, intestinal lumen, and bile, respectively. At the blood-brain barrier, they ensure exclusion of conjugates from the central nervous system. The combination of reduced passive transport and increased active transport means that the body gains control over the movements of the conjugates. So, in a sense, conjugation tags drug molecules for elimination. Excretion of conjugates by active transport is sometimes referred to as *phase III* of drug metabolism.

4.6 ENZYME INDUCTION

The classic textbook example of a metabolic substrate that induces the transcription of its cognate enzymes is lactose, which greatly increases the rate of transcription from the *lac* promoter in *Escherichia coli*; this upregulates of both the metabolizing enzyme, β-galactosidase, and a specific transport protein for uptake, lactose permease. With xenobiotics and mammalian cells, the molecular mechanism of induction is a little different, yet the result—increased activity of corresponding enzymes and transporters—is the same.

Induction of enzymes that metabolize xenobiotics was first observed with polycyclic aromatic hydrocarbons. Benzopyrene and related compounds bind to the aryl hydrocarbon receptor (AHR). Prior to ligand binding, this receptor protein resides in the cytosol. The receptor-ligand complex is transported to the nucleus and binds to a related protein, aryl receptor nuclear translocator (ARNT). The protein heterodimer binds to a DNA consensus motif with the sequence 5′-GCGCT-3′, which occurs in multiple locations on the genome and is termed the *xenobiotic-responsive element* (XRE). Binding results in an increased rate of transcription of genes nearby.

Attention was initially focused on the transcriptional induction of cytochrome P450 enzymes. AHR induces the subtype CYP1A1, which performs oxidative metabolism of benzopyrene. However, it has since become clear that reductive enzymes and phase II enzymes such as UDP-glucuronosyltransferases, glutathione-*S*-transferases, and *N*-acetyltransferases are induced as well. Transcriptional induction also encompasses reductive enzymes as well as P-glycoprotein and other transport proteins.

Other prominent receptors that mediate transcriptional induction of drug-metabolizing enzymes are the pregnane X receptor (PXR), the constitutive androstane receptor (CAR), and the so-called peroxisome proliferator-activated receptors (PPAR), which all belong to the family of nuclear hormone receptors (see Section 7.1.1). They bind overlapping yet different sets of drugs or environmental chemicals and in turn induce different members of the various xenobiotic enzyme families.

The pregnane X receptor is activated by a particularly large number of drugs. It induces CYP3A4, which is one reason for the prominent role of this enzyme in phase I metabolism. Strong activators of PXR among clinically important drugs are phenytoin and phenobarbital, which are used to treat epilepsy, and rifampicin, which is used to treat tuberculosis. Most drugs that induce CYP3A4 are also substrates of it, as are synthetic estrogens and progestins used for contraception. This leads to a variety of clinically important drug interactions. For example, oral contraception will no longer work reliably under treatment with rifampicin or phenytoin, and dosages of phenytoin will have to be increased during concomitant treatment with rifampicin.

4.7 METABOLISM-RELATED TOXICITY AS A THERAPEUTIC TARGET

Benzopyrene and related compounds, formed during combustion of tobacco, are metabolically activated by cytochrome P450 to epoxides (Figure 4.8) and are the most common cause of lung cancer. Similarly, the introduction of epoxy groups into aflatoxins, a group of aromatic compounds produced by the fungus *Aspergillus flavus*, may cause liver cancer. Considering that the liver and the small intestine have the highest activities of cytochrome P450, it is remarkable that both organs have comparatively low rates of primary cancer.[7] This suggests that phase II enzymes, which are also highly abundant in the liver and the small intestine, are very effective at protecting both organs from the carcinogenic phase I metabolites.[8] However, formation or distribution of phase

[7]Intestinal cancer almost always originates in the large intestine, and the liver is far more often afflicted by metastatic cancer than by tumors that originate within the liver itself. Moreover, primary liver cancer is most often related to virus infections, not to chemical toxicity.

[8]A second line of defense against toxic epoxides and other DNA-modifying agents are DNA repair enzymes; deficiencies of these enzymes are associated with increased incidence of tumors. However, repair enzymes do not catch all instances of DNA modification, and sometimes damage to the DNA is caused by the repair process itself.

I metabolites is not limited to these organs, and other tissues may not have the same capacity for their detoxification.

A plausible strategy to prevent carcinogenesis through metabolic toxicity therefore is to increase the capacity of phase II metabolism. The drug oltipraz activates the transcription factor Nrf2, which increases the expression of several phase II enzymes, including glutathione-*S*-transferase. Its use for the prevention of cancer caused by the toxic metabolites of aromatic hydrocarbons and aflatoxins is currently under study.

STUDY QUESTIONS

4.1 As we have seen in Section 2.5, drugs differ with respect to their therapeutic indices. Between drug L with a large therapeutic index and drug S with a small therapeutic index, which one would be more prone to interference by enzyme induction by a third drug T?

4.2 Hepatic metabolism of the antiepileptic drug valproic acid can give rise to 4-enevalproic acid and 2,4-dienevalproic acid (see figure) as well as other desaturated metabolites. These metabolites cause liver toxicity.

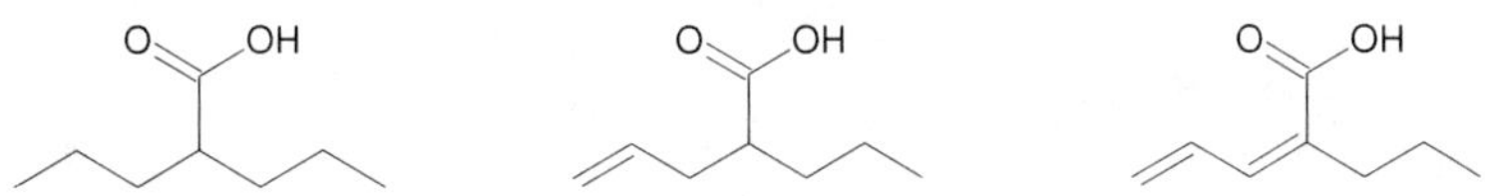

Valproic acid (VPA) 4-ene VPA 2,4-diene-VPA

Which type of enzyme do you consider most likely to be responsible for this conversion, and what is the most likely mechanism of toxicity?

4.3 In Section 10.1, it is discussed that hereditary enzyme defects often become manifest only after birth, while in the uterus the fetus is protected from the effects of accumulating toxic metabolites due to equilibration with the maternal circulation.

Will this also apply to glucose-6-phosphate dehydrogenase deficiency (see Section 4.4.3), that is, will a fetus with this enzyme defect be protected if its mother has intact glucose-6-phosphate dehydrogenase activity?

REFERENCES

[1] J. Bauer et al., "A structural gap in Dpo4 supports mutagenic bypass of a major benzo[a]pyrene dG adduct in DNA through template misalignment," *Proc. Natl. Acad. Sci. U. S. A.* 104 (2007), 14905–14910.

[2] L. L. Brunton, J. S. Lazo, and K. L. Parker, eds., *Goodman and Gilman's The pharmacological basis of therapeutics,* 11th ed., McGraw Hill, 2005, 1794–1888.

[3] M. Ekroos and T. Sjögren, "Structural basis for ligand promiscuity in cytochrome P450 3A4," *Proc. Natl. Acad. Sci. U. S. A.* 103 (2006), 13682–13687.

[4] A. Fura, "Role of pharmacologically active metabolites in drug discovery and development," *Drug Discov. Today* 11 (2006), 133–142.

[5] H. W. Goedde, E. Schoepf, and D. Fleischmann, "Studies on pharmacogenetics. I. The enzymatic acetylation of isonicotinic acid hydrazide (INH)," *Biochem. Pharmacol.* 13 (1964), 1671–1675.

[6] P. K. Honig et al., "Terfenadine-ketoconazole interaction. Pharmacokinetic and electrocardiographic consequences," *JAMA.* 269 (1993), 1513–1518.

[7] R. J. Knox et al., "Bioactivation of CB 1954: reaction of the active 4-hydroxylamino derivative with thioesters to form the ultimate DNA-DNA interstrand crosslinking species," *Biochem. Pharmacol.* 42 (1991), 1691–1697.

[8] D. J. Mitchell and R. F. Minchin, "E. coli nitroreductase/CB1954 gene-directed enzyme prodrug therapy: role of arylamine N-acetlytransferase 2," *Cancer Gene Ther.* 15 (2008), 758–764.

[9] M. V. Relling, "Polymorphic drug metabolism," *Clin. Pharm.* 8 (1989), 852–863.

[10] A. Satyam et al., "Design, synthesis, and evaluation of latent alkylating agents activated by glutathione S-transferase," *J. Med. Chem.* 39 (1996), 1736–1747.

CHAPTER 5

G PROTEIN-COUPLED RECEPTORS

G protein-coupled receptors (GPCRs) represent the largest protein family among the receptors for hormones and neurotransmitters in the human body. There are more than 200 different GPCRs for which the ligands are known. Nearly as many "orphan" receptors, that is ones without any known ligands, have been found in the human genome. In addition, there are several hundred sensory GPCRs that mediate our senses of taste and smell.

The large number of GPCRs translates into a very prominent role in physiological regulation. From mood to circulation, from blood clotting to inflammation and gastric acidity, G protein-coupled receptors are involved in the control of any physiological function one can think of. Accordingly, they are very important drug targets; approximately 50% of all drugs in clinical use today act on GPCRs, and this percentage may grow even higher in the future. Because of this, it is appropriate to consider the structure and function of G protein-coupled receptors in some detail. Some example drugs and their receptors are listed in Table 5.1.

5.1 OVERVIEW

G protein-coupled receptors reside in the plasma membrane, facing both the intra- and the extracellular space. Most receptors are activated by the binding of

Table 5.1 Examples of drugs that act on G protein-coupled receptors. Some drugs act on other receptors in addition to those listed.

Drug	Receptor	Drug action	Clinical use	See Section
Tolazoline	α-Adrenergic	Inhibitor	Antihypertensive	6.8.2
Salbutamol	β_1-Adrenergic	Partial agonist	Bronchodilatation	6.8.2
Atropine	Muscarinic	Inhibitor	Pupil dilation	6.8.3
Haloperidol	Dopamine D_2	Inhibitor	Antipsychotic	6.8.2
Morphine	Opioid	Agonist	Painkiller	6.8.5
Losartane	Angiotensin	Inhibitor	Antihypertensive	1.3.2
Desmopressin	Vasopressin V_2	Agonist	Antidiuretic	7.2.1
Clopidogrel	P2Y (ADP)	Inhibitor	Platelet inhibition	10.4.4
Cinacalcet	Calcium-sensing	Agonist	Parathyroid gland inhibition	7.5.2

diffusible ligand molecules from the extracellular side. Upon binding to its cognate ligand, the entire receptor molecule undergoes a conformational change, which causes its intracellular surface to bind and activate a *heterotrimeric G protein*, which in turn will activate or inhibit some downstream intracellular signalling cascade. The receptor may return to its inactivated state by dissociation of the ligand; it may also undergo inactivation through phosphorylation by specific *GPCR kinases.* Phosphorylated receptors may be removed from the membrane to intracellular compartments.

Among the physiological ligands of G protein-coupled receptors, polypeptides are more numerous than small molecules such as dopamine, histamine, or acetylcholine. However, the larger share of the currently available drugs target receptors for small molecules. Many of these drugs were derived from the corresponding physiological mediators, and frequently their design predates any detailed knowledge of the molecular features of drug–receptor interaction (see Section 1.3).

By comparison, the number of currently available synthetic drugs that target peptide receptors is relatively small. For reasons related to molecular size and stability in vivo (see Chapter 3), the usefulness of peptides as drugs is limited. For the design of a synthetic drug molecule that targets a peptide's receptor, one wants to know the key features of the peptide that directly and avidly interact with the complementary ones on the receptor molecule. Ideally, such knowledge would be obtained from a high-resolution structure of the receptor–ligand complex. It is, however, more difficult to obtain crystal structures of integral membrane proteins than for water-soluble ones, and thus far the structure of only two GPCRs—that of rhodopsin, and more recently the β-adrenergic receptor [5]—have been experimentally determined. Therefore, the hypothetical structures of other GPCRs have to be pieced together from those of receptor fragments and by homology-based molecular modeling, using the rhodopsin structure as a template. While this situation poses challenges to drug development, synthetic small molecule ligands for several peptide receptors have already been introduced into clinical practice, and novel drugs that target additional peptide receptors are being developed.

5.2 GPCR STRUCTURE

The key structural feature conserved among all GPCRs is expressed in the synonymous names *heptahelical receptors* or 7-TM receptors. All GPCRs have exactly seven helical transmembrane domains, which are connected by mostly relatively short loop segments. The N-terminus always faces the extracellular space, and the C-terminus resides in the cytosol. The helices are packed against each other in a roughly circular bundle (Figure 5.1). With quite a few receptors for small ligands, the binding sites are located close to the center of this bundle. In this way, the ligand interacts with amino acid residues on several helices and can induce movements of these helices relative to one another.

At least in the rhodopsin molecule and related receptors, there is a short eighth helix within the C-terminal extension. It is oriented at a right angle to the transmembrane helices and is packed against the cytoplasmic membrane. This helix has a regulatory role in receptor activation (see Section 5.6.1).

5.3 STRUCTURAL GPCR FAMILIES

While there is no significant degree of sequence conservation that applies to the entire class of G protein-coupled receptors, some key features are conserved within each of the three major families. One such feature is the distribution of proline residues. Prolines break the continuity of α helices and thus have a strong impact on the overall conformation and conformational freedom of protein molecules. It is therefore interesting to note that each GPCR receptor family has its own distinctive set of conserved prolines; this suggests that the mechanics of activation are conserved within each family but differ between them (see Section 5.4.2).

5.3.1 The rhodopsin-like family (family A)

This family, which is named after its most prominent member, is the largest one. It is characterized by the following structural features: (1) The C-terminal extension contains a cysteine residue, which is anchored to the membrane through a posttranslationally attached palmitoyl group, thereby creating an additional membrane-attached loop following the last transmembrane helix. (2) The third interhelical loop, which is located at the cytoplasmic surface, contains a conserved aspartate/glutamate-arginine-tyrosine (DRY) motif. This motif plays an important role in receptor activation. (3) Transmembrane helix 6 and several others contain single proline residues in conserved positions.

Most receptors in this family have only short N-termini. Exceptions include some peptide hormone receptors as well as the protease-activated receptors (see Section 5.4.3).

5.3.2 The glucagon receptor-like family (family B)

Receptors in this family mostly bind peptides, the binding sites of which reside in their extended N-termini. The family contains its own set of conserved

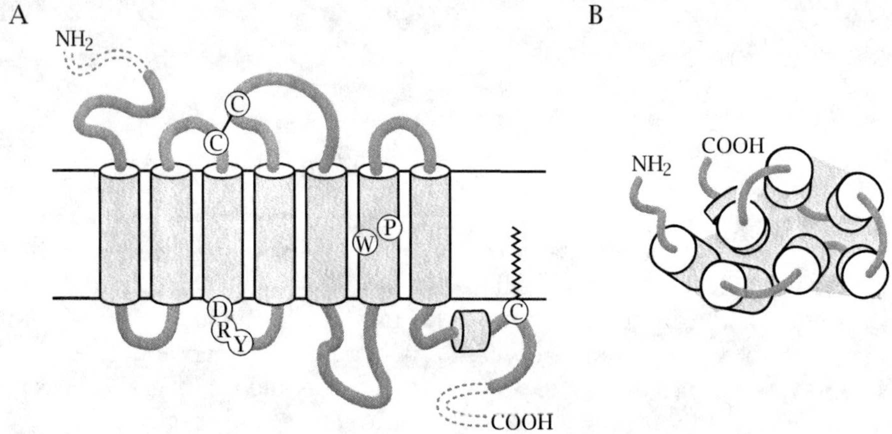

Figure 5.1 G protein-coupled receptor structure. **A:** Snake diagram. Several conserved features found in the rhodopsin family are indicated, such as the DRY motif at the cytoplasmic end of helix 3, the proline and tryptophan residues in helix 6, and the short C-terminal helix 8. **B:** Top view, showing the approximately circular arrangement of the helices in the membrane plane. With many receptors, the crevice in the center contains the ligand binding site.

proline residues and lacks the features listed above for family A, in particular the DRY motif. The family also includes the unusual LNB receptors (see Section 5.4.3).

5.3.3 Metabotropic neurotransmitter receptors (family C)

The name of this family serves to distinguish them from the other major type of neurotransmitter receptors, the ligand-gated channels. The metabotropic glutamate receptors fall into this group. However, not all GPCRs that bind neurotransmitters belong to this family—the dopamine and serotonin receptors for example are members of family A—nor do all these receptors bind neurotransmitters.

Receptors in this family have long N-termini, fewer conserved prolines, and a relatively short third intracellular loop that carries a conserved NEAK motif. Like the aforementioned DRY motif, the NEAK motif contains two opposite charges, and it may serve a similar role in receptor activation.

5.4 ACTIVATION OF GPCRS

Most experimental data on the molecular aspects of GPCR activation have been obtained with rhodopsin and other members of the rhodopsin-like family. At least within this group, the mechanisms of activation appear to be quite similar.

5.4.1 Ligand binding sites

Much of the evidence on the locations of ligand binding sites has been gathered by site-directed mutagenesis and by construction of *chimeric* receptors. With these techniques, it is possible to correlate loss or preservation of affinity

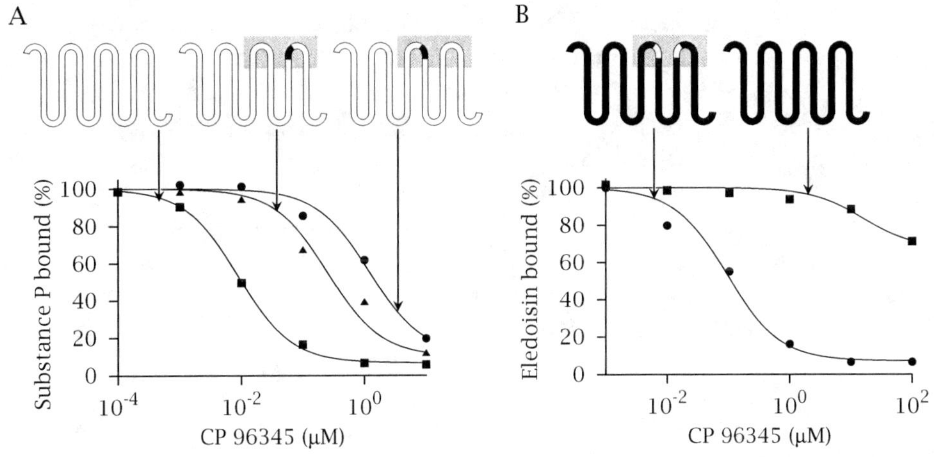

Figure 5.2 Locating the binding site for a synthetic receptor antagonist using receptor chimeras. **A**: The peptide named Substance P is displaced from its receptor by the drug CP 96345. Replacement of two short sequence segments by the corresponding ones from the homologous but CP 96345-insensitive NK_3 receptor significantly reduces the affinity for the inhibitor. **B**: The reciprocal exchange renders the NK_3 receptor sensitive to inhibition by CP 96345. (Figure redrawn from original data in Ref. [9].)

for a ligand with individual amino acid residues or with contiguous sequence stretches. An example experiment is depicted in Figure 5.2.

We have previously likened the activation of a receptor by its ligand to the actuation of an enzyme by an allosteric effector (see Section 2.2.5). Many enzymes have more than one allosteric binding site, which implies that the activating conformational transition can be induced from different locations within the molecule. One might therefore expect that with G protein-coupled receptors, the locations of binding sites may be diverse, too. This is borne out in various ways: (1) With rhodopsin-like receptors that have short N-termini, ligands bind directly to the transmembrane helices or to the short extracellular loops that connect them. In contrast, with those receptors that have large N-terminal domains—mostly from families B and C—ligands apparently bind to the N-termini only. (2) With some receptors, synthetic agonists or antagonists have been found that bind residues not engaged by the physiological ligand. Moreover, antibodies have been characterized that activate or inhibit GPCRs, typically binding in locations other than the binding sites of the physiological ligands.

A synthetic ligand that binds outside the regular binding site may bind simultaneously with the physiological ligand and therefore may behave like a truly allosteric effector. An interesting example is the noncompetitive inhibition of human metabotropic glutamate receptors by the experimental drug 7-(hydroxyimino)cyclopropan[b]-chromen-1a-carboxylic acid ethylester (CPCCOEt) [11]. While the physiological ligand (glutamate) binds in the large N-terminal domain of that receptor, the antagonist binds to a separate site on transmembrane helix 7 [17].

GPCR-binding antibodies offer an interesting approach to the development of subtype-specific receptor antagonists. For example, muscarinic M_2 receptors, which are involved in certain types of cardiac arrhythmias, can be inhibited by specific antibody fragments [21]. On the other hand, in Graves' disease, autoantibodies activate the thyroid-stimulating hormone (TSH) receptor, which causes excessive release of thyroid hormones (see Section 7.3). In some cases, it has been observed that only native, bivalent antibody molecules elicit receptor activation, while their monovalent F_{ab} fragments do not. This is most likely related to receptor oligomerization (see section 5.5). Considering that antibodies are much larger than typical GPCR ligands, most antibodies that activate or inhibit GPCRs likely recognize epitopes located outside the regular ligand-binding pockets.

Notwithstanding the variability of binding sites in principle, tried and true binding motifs may occur similarly in receptors for various ligands. For example, the mediators epinephrine, dopamine, histamine, and serotonin, and acetylcholine (see Section 6.8.2) all possess an amine group, which engages an aspartic acid residue in the third transmembrane that is conserved among all their respective receptors.

5.4.2 Conformational changes associated with activation

The conformational changes that underly GPCR activation appear to be similar between rhodopsin and the homologous receptors that have been studied in detail, such as the β_2-adrenergic receptor. This conformational switch is characterized by the following key features:

1. When no ligand is bound, the receptor is locked down in its inactive state by several strategically placed, non-covalent intramolecular bonds. Activation requires these bonds to be broken.
2. Transition to the active state involves a major movement of helix 6 relative to the other helices.
3. Some GPCRs can adopt different active conformations in response to the binding of different ligands that result in preferential activation of different G proteins.

There is a large body of evidence pertinent to each of these effects, obtained through a host of biochemical and biophysical techniques; we will consider a few example experiments. Many of these studies have used rhodopsin as a model. Rhodopsin is not activated by ligand binding; instead, the absorption of a photon causes a cis-trans isomerization in its covalently bound prosthetic group retinal. The effect of this steric change on the receptor is considered to be equivalent to the event of ligand binding in other receptors.[1]

Movement of helix 6. The conformational dynamics of rhodopsin and other G protein-coupled receptors have been studied by site-directed mutagenesis in conjunction with various spectroscopic methods. An intrinsic spectroscopic probe that occurs in most proteins is the amino acid tryptophan. By individually

[1] Another unique feature of rhodopsin is its location in an intracellular membrane compartment, the so-called *disk membranes*, which are stacked inside the rod and cone cells of the retina.

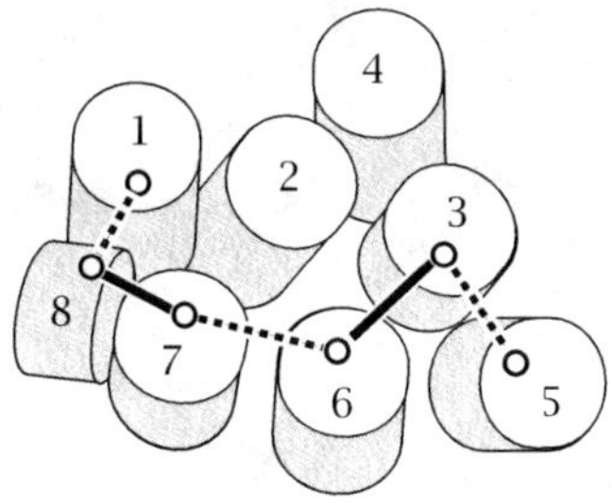

Figure 5.3 Analysis of conformational changes required for the activation of rhodopsin with engineered disulfide bonds. The bonds connect various helices at the cytoplasmic surface. Bonds drawn in solid lines prevent activation, those drawn in dashed lines do not [12,7].

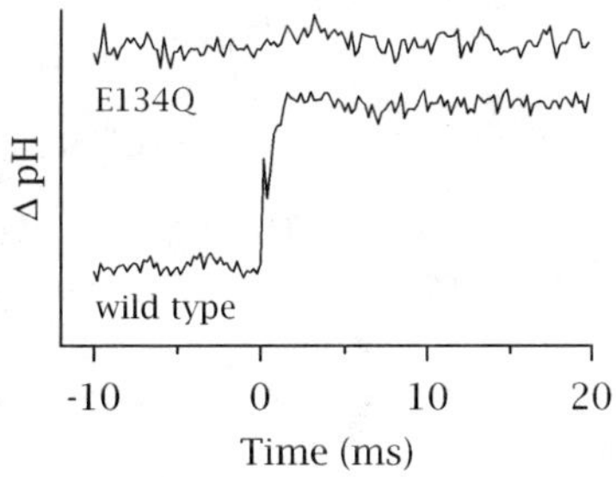

Figure 5.4 Protonation of residue E134 rhodopsin in response to light stimulation. Wild-type rhodopsin causes a swift change in the pH of an unbuffered solution upon photostimulation at t=0; the mutant E134Q does not. (Figure redrawn from original data in Ref. [1].)

replacing each tryptophan residue within rhodopsin, it was determined that activation has the most pronounced effect on residue W265, which is located in helix 6, suggesting a conformational change affecting this helix (see Section 5.8.1).

Extrinsic spectroscopic probes are most conveniently introduced into proteins using mutant cysteine residues (see Section 5.8.2). Cysteine mutants labeled with nitroxide radicals, in particular, have been used extensively with single and double cysteine mutants of rhodopsin [12]. The spin of the unpaired electron of a nitroxide radical causes it to behave as a little magnet, the movements and interactions of which can be detected by electron paramagnetic resonance spectroscopy (EPR).

If two mutant cysteine residues are sufficiently close to one another in the folded structure of the protein, they will form a disulfide bond. Pairs of mutant cysteines may form disulfide bonds, which will then constrain the conformational flexibility of the protein. When helix 6 is tied to helix 3 by such an engineered disulfide bond, rhodopsin can no longer be activated (Figure 5.3), which confirms the functional significance of the mobility of helix 6.

The movement of helix 6 appears to swivel around its conserved proline residue. Proline residues play a special role in defining the conformational flexibility of protein molecules. The side chain of proline is linked to its α amino group, which therefore does not carry a hydrogen and cannot participate in the hydrogen bonding pattern that underlies the formation of α helices. A proline residue thus introduces a kink and some measure of flexibility into an otherwise regular and rigid helix. In keeping with the assumed role of the conserved proline in helix 6, mutations in its vicinity that affect the local mobility also shift the equilibrium of receptor activation [24]. As stated before, the positions of proline residues are conserved within GPCR families but vary between them; this suggests that the conformational transitions that lead to activation may differ as well.

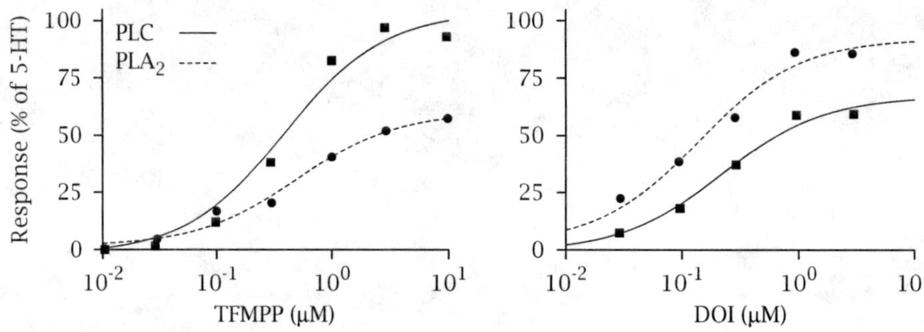

Figure 5.5 Agonist-specific coupling of 5-HT_2 serotonin receptors. The receptor binds two different G proteins. $G\alpha_q$ activates phospolipase C (PLC), which releases inositol-triphosphate (solid lines), whereas $G\alpha_{12}$ activates phospolipase A_2 (PLA_2), which releases arachidonic acid (dashed lines). 3-Trifluoromethylphenyl-piperazine (TFMPP) activates PLC more strongly than PLA_2, whereas 1-(2,5-dimethoxy-4-iodophenyl)-2-aminopropane (DOI) shows the opposite preference. (Figure redrawn from original data in Ref. [2].)

Ground state-stabilizing interactions. With most receptors and in the absence of ligand, the inactive ground state is much more prevalent, and therefore more stable, than the activated state. The inactive state is stabilized by several intramolecular bonds, which need to be broken in order for activation to occur.

One such interaction involves the DRY motif, which is located at the intracellular end of helix 3 and occurs in rhodopsin and many other receptors in the rhodopsin family. The first two amino acids in this motif—aspartate (D) or glutamate (E), and arginine (R)—carry opposite charges, which in the ground state form an ionic bond. With rhodopsin, it has been directly demonstrated that receptor activation induces protonation of the glutamate residue (Figure 5.4). Protonation will break the ionic bond with arginine, which thereby becomes free to engage in other interactions that are conducive to activation.

Mutating the conserved glutamate to glutamine prevents protonation and also formation of the bond between D/E and R in the first place; accordingly, receptors with this mutation exhibit increased unstimulated activity.

Multitude of conformational states induced by different ligands. The literature contains several reports on *agonist-specific coupling* or *stimulus trafficking* of G protein-coupled receptors. An example is shown in Figure 5.5. As discussed in Section 2.2.6, agonist-specific coupling implies the existence of more than one active conformational state of the receptor in question. Direct evidence of agonist-dependent conformational heterogeneity has been obtained with fluorescently tagged derivatives of the β-adrenergic receptor [14].

5.4.3 GPCR activation without diffusible ligands

Some G protein-coupled receptors may be activated without the binding of any diffusible ligands. These include (1) constitutively active receptors, (2) the so-called LNB-7TM receptors, and (3) protease-activated receptors.

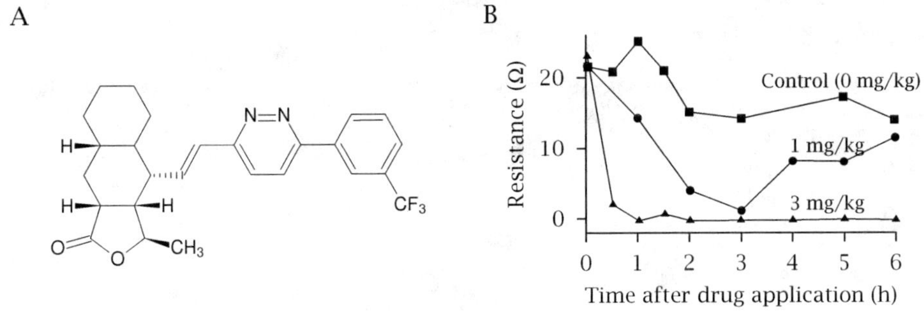

Figure 5.6 Inhibition of the protease-activated receptor PAR-1. **A**: Structure of the synthetic inhibitor used in the illustrated experiment. **B**: Inhibition of thrombocyte aggregation in experimental animals. Plasma samples were obtained at various times after application of the inhibitor and spiked with a peptide PAR-1 agonist. Thrombocyte aggregation is measured through the electrical resistance between electrodes immersed into the sample; as the thrombocytes aggregate on the electrode surfaces, the resistance increases. (Figure redrawn from original data in Ref. [4].)

Constitutive activity. Receptors may acquire mutations that shift the intrinsic equilibrium between the inactive and active states, such that the active fraction becomes significant in the absence of ligand. Such receptors are referred to as *constitutively active mutants.* However, constitutive activity also occurs with some wild-type receptors, such as the cannabinoid receptors, which regulate synaptic transmission in the central nervous system.

LNB-7TM receptors. These possess long N-termini, and the amino acid sequences of their membrane-embedded heptahelical portions place them into the structural family B.

The long N termini contain a variety of repeated domains related to growth factors and adhesion molecules. These N-termini bind to other receptors on adjacent cells; therefore, LNB-7TM molecules are really receptors and ligands at the same time. They are involved in the regulation of cell growth and differentiation [25] and are of great interest to fundamental research, but their functions are not yet sufficiently well understood to make them very relevant to current pharmacology.

Protease-activated receptors. These receptors, which belong to the rhodopsin family, have no physiological diffusible ligands, but instead are switched on when proteases such as thrombin or factor Xa cleave a peptide fragment from the extracellularly located N-terminus of the receptor molecule. This exposes a peptide motif at the truncated N-terminus that then acts as an intramolecular ligand to activate the receptor. Activation is irreversible, since the ligand remains tethered to the receptor and cannot be diluted out after reversible dissociation from the binding site.

It is probably not a coincidence that protease-activated receptors prominently occur on cells such as thrombocytes and neutrophile granulocytes that are themselves expended in a single, irreversible round of activation; however, they are also found on longer-lived cells. Thrombocytes and granulocytes are

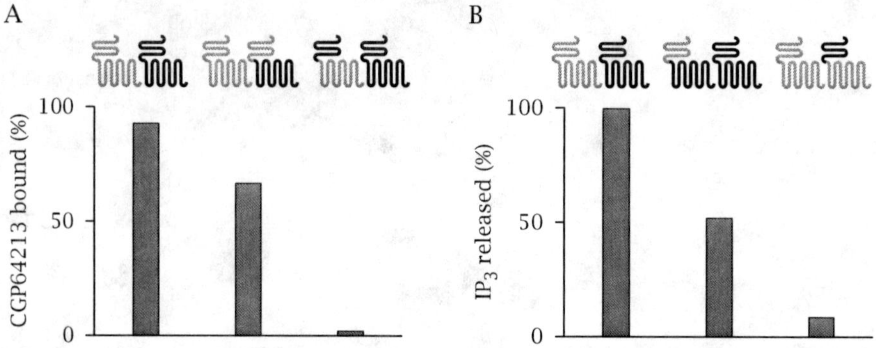

Figure 5.7 Functional specialization in $GABA_B$ receptor heterodimers. Cells were transfected with both $GABA_B$-1 receptor (gray) and $GABA_B$-2 receptor (black), or with combinations of either with chimeric receptors. **A:** The $GABA_B$-2 N-terminus is dispensable for ligand binding. **B:** G protein activation, as measured by subsequent IP_3 release, remains partially functional with receptor dimers lacking the $GABA_B$-1 transmembrane domain. (Figure redrawn from original data in Ref. [8].)

key elements in blood coagulation and inflammation, and drugs that inhibit leukocyte and thrombocyte activity are among the most commonly prescribed drugs in current use. Therefore, protease-activated receptors have attracted considerable interest as drug targets. Apart from their unusual mode of ligand generation, these receptors function in the same way as more conventional GPCRs, and they are therefore susceptible to the effects of synthetic soluble agonists and antagonists that target the binding site of the intramolecular peptide ligand. Figure 5.6 illustrates an inhibitor that prevents thrombocyte aggregation, which is a major contributing factor in the development of thrombosis.

Intriguingly, protease-activated receptors have also been implicated in functions of the central nervous system such as learning and addiction; therefore, it seems advisable to design inhibitors for clinical use such that they are excluded by the blood brain barrier.

5.5 GPCR DIMERIZATION AND OLIGOMERIZATION

The original concept of GPCR action involved single receptor molecules interacting with individual G proteins. However, it is now clear that very many GPCRs are indeed dimers or even larger oligomers. While some receptors are functional in both monomeric and oligomeric states, others are functional exclusively as oligomers. For example, the $GABA_B$ receptor will not even be expressed on the cell surface as a monomer, and its dimeric state is an essential prerequisite of its activity (Figure 5.7).

GPCR oligomers may be *homomers*, meaning that they form from like subunits, or *heteromers*, which consist of different types of subunits. With the $GABA_B$ receptor and most other heteromers that have been characterized so far, the different subunits are fairly closely related and bind the same ligands. However, there are some intriguing examples of heteromers forming from sub-

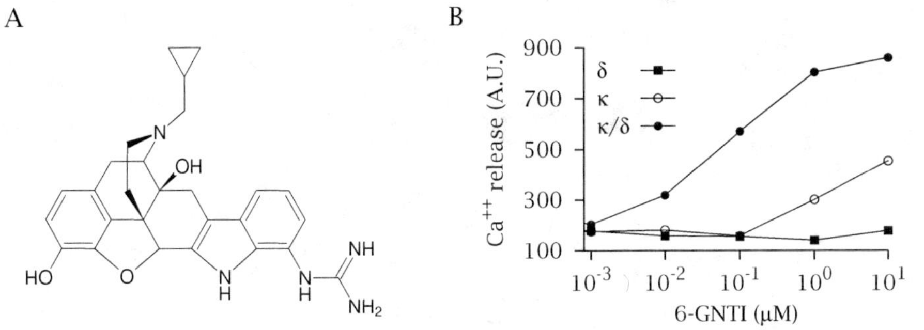

Figure 5.8 Selective activation of $\kappa\delta$-opioid receptor heteromers by 6-guanidinonaltrindole (6-GNTI). **A:** Structure of 6-guanidinonaltrindole. **B:** Dose-response curves of homomeric and heteromeric κ and δ receptors expressed in cell culture, measured through the release of Ca^{2+} downstream of phospholipase C. (Figure redrawn from original data in Ref. [27].)

units as diverse as β-adrenergic and leukotriene receptors, suggesting a whole new dimension of complexity in GPCR signaling.

GPCR oligomerization may have significant consequences for drug action. Firstly, oligomeric receptors may show cooperative behavior (see Section 2.2.6). Secondly, receptor heteromers may differ from their constituent subunits in their ligand specificity. Thirdly, receptor heteromers themselves are potential drug targets that may offer a higher degree of tissue selectivity than their corresponding monomeric or homomeric receptors.

When multiple receptor preparations exhibit different affinities for and responses to a set of ligands, this would traditionally be considered evidence of multiple receptor subtypes. It now turns out that some of these operationally defined subtypes may indeed be heteromers of known receptors.

An example of novel ligand specificity caused by heteromer formation is provided by opioid receptors. Heterodimers of κ- and δ-opioid receptors do not respond to various drugs that stimulate homogeneous κ or δ receptors. On the other hand, they do respond to the novel agonist 6-guanidinonaltrindole [13]. Since this agonist activates the homogeneous receptors only weakly, it allows for selective stimulation of the heteromeric ones (Figure 5.8). The κ/δ receptor heteromers are found in the spinal cord but apparently not the brain, which illustrates the potential gains in therapeutic selectivity associated with targeting receptor heteromers.

While the observation of heterodimer specificity of a monovalent agonist in the study cited was fortuitous, there have been efforts to target receptor heteromers in a rational manner by combining two different ligands with specificity for the constituent subunits in a single bifunctional molecule. Similarly, homomers can be targeted selectively by combining two identical ligand molecules. Since on the receptor dimer the two binding sites will be located at some distance from each other, the two pharmacophores will have to be joined using a flexible spacer. In a study on bivalent agonists of muscarinic acetylcholine receptors, a bell-shaped relationship between spacer length and the extent of

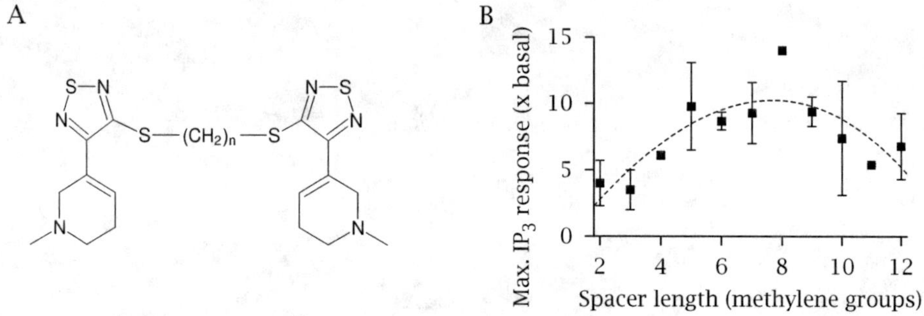

Figure 5.9 Bivalent agonists of muscarinic acetylcholine receptors. **A:** General structure of a series of the bivalent derivatives of the agonist 1-methyl-1,2,5,6-tetrahydropyridyl-1,2,5-thiadiazole. The number of methylene groups in the central spacer was varied systematically. **B:** Maximal inositoltriphosphate release in response to bivalent agonists of varying spacer length. (Figure redrawn from original data in Ref. [22].)

receptor activation was observed (Figure 5.9), so that at least in this case it was possible to change the efficacy of the dimeric drug through variation of a single, simple structural parameter.[2]

5.6 G PROTEINS

A GPCR communicates its activated state to the cell through one or more heterotrimeric GTP-binding proteins, or G proteins for short. These proteins reside at the inner surface of the cytoplasmic membrane, to which they are attached through posttranslational lipid modifications. Each G protein combines one subunit each of the α, β and γ types. Gα contains the binding site for GTP as well as the GTP hydrolase activity that converts the bound GTP molecule to GDP. The interaction with GTP controls the reversible dissociation of Gα from the β and γ subunits; the latter two remain associated with each other throughout the entire cycle of activation (see Figure 5.10).

GTP binds to Gα when the heterotrimer is associated with an activated receptor (Figure 5.10). GTP binding triggers a conformational switch that causes Gα to dissociate; both Gα and G$\beta\gamma$ thus become free to interact with their respective effector molecules. The conformational switch reverts when GTP is hydrolyzed, which is carried out by Gα itself but is accelerated by its interaction with the effector and with further regulatory proteins. The GDP-bound Gα subunit will swiftly reassociate with the G$\beta\gamma$ dimer; therefore, the lifetime of the activated state of the G protein corresponds to that of the associated GTP molecule.

5.6.1 Interaction of G proteins with GPCRs

G proteins may bind to nonactivated GPCRs with low affinity. Whether this actually leads to significant precoupling before ligand binding in vivo is subject

[2]The authors of the study cited did not interpret their findings in terms of receptor oligomerization; however, oligomerization of muscarinic receptors has been demonstrated [18].

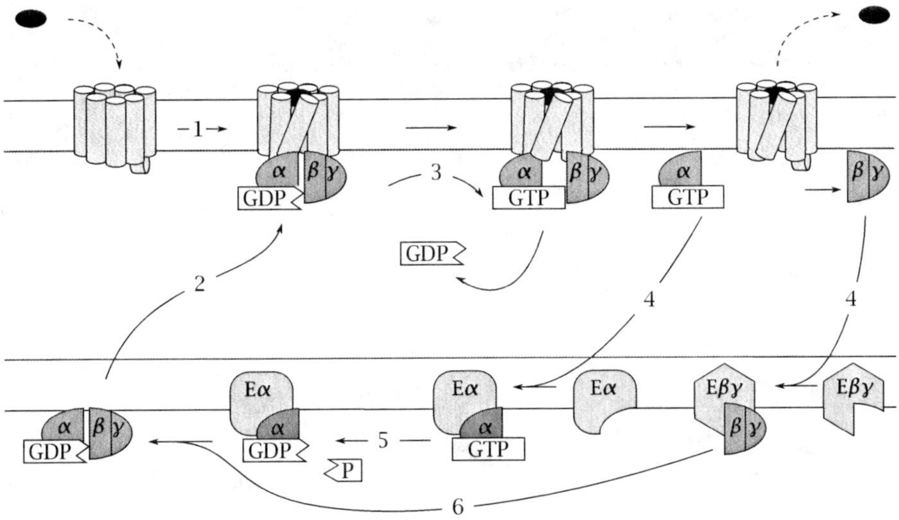

Figure 5.10 The G protein cycle. After the receptor has been activated by ligand binding (1), the heterotrimeric G protein binds to its cytoplasmic surface (2), whereupon GTP replaces GDP on Gα (3). Gα and Gβγ then dissociate and bind to their respective effector proteins (4). Gα cleaves its associated GTP (5), then leaves its effector and associates with the same or another Gβγ dimer (6) to await another round of activation.

to ongoing debate, and it may well depend on the individual combination of receptor and G protein. Activated receptors expose a high-affinity binding site, which involves the third loop, that is, the one containing the DRY motif in rhodopsin-like receptors. It is assumed that the outward movement of helix 6 creates a crevice that accommodates the C-terminal end of the Gα subunit.

Additional sites both on the G protein, including the Gβγ dimer, and on the receptor participate in the interaction. In studies on the leukotriene B_4 receptor, a special role has been ascribed to the short C-terminal helix 8. Activation of the G protein causes a decrease in this receptor's affinity for its ligand. Ablation of helix 8 does not interfere with G protein activation, but it prevents the decrease in ligand affinity, indicating that helix 8 imposes a negative feedback on receptor activation [20].

5.6.2 G protein subtypes and their functions

There are multiple subtypes for each of the three components of heterotrimeric G proteins. The largest number of functionally different variants exists for the Gα subunit, which therefore is the most important determinant of the entire heterotrimer's specificity.

Gα subunits can be arranged into several classes according to homology and to their major effector proteins (see Table 5.2). Each class contains several members, whose names, however, do not always clearly indicate this membership. While the $G\alpha_s$, $G\alpha_i$ and $G\alpha_q$ families all have clear-cut cases of GPCRs that act mainly through them to cause their pharmacologically relevant intracellular

Table 5.2 Major types of G proteins and examples of affiliated receptors.

Class	Effectors and effects	Related GPCRs (examples)
$G\alpha_s$	Stimulation of adenylate cyclase (various types)	β-Adrenergic, 5-HT_4, 5-HT_6, 5-HT_7, D_1, D_5; ACTH
$G\alpha_{i/o}$	Inhibition of adenylate cyclase; activation of extracellular signal-regulated kinase (ERK)	α_2-Adrenergic, 5-HT_1, D_2, D_3, D_4, $GABA_B$
$G\alpha_{q/11}$	Stimulation of phospholipase Cβ (various subtypes)	α-Adrenergic, 5-HT_2, H_1, $GABA_B$
$G\alpha_{12/13}$	Indirect activation of RhoA GTPase and of phospholipase A_2	5-HT_4, AT_1, protease-activated receptors

effects, the $G\alpha_{12/13}$ family is less well defined with respect to receptor and effector affiliation.

$G\beta\gamma$ dimers interact with a set of effectors that includes some $G\alpha$ effectors such as adenylate cyclase and phospholipase C (Figure 5.11). They also activate potassium channels of the inward rectifier type and some voltage-gated calcium channels (see Chapter 6). These effects may be synergistic or antagonistic to those evoked by the simultaneously activated $G\alpha$ subunits. In addition, $G\beta\gamma$ dimers stimulate GPCR kinases, which exert a negative feedback on GPCR activation. Interaction of $G\beta\gamma$ with receptor tyrosine kinases such as the epithelial growth factor (EGF) receptor is one of several mechanisms by which GPCR activation may influence transcriptional regulation.

Since the GPCRs are much more numerous than G proteins, the signals arriving at several different receptors will converge on the same G proteins. Notably, this convergence is not restricted by GPCR families; for example, the β-adrenergic receptors from family A and the glucagon receptor from family B both couple to $G\alpha_s$. On the other hand, one receptor will often couple to more than one G protein, sometimes in an agonist-dependent way (see Figure 5.5).

5.6.3 Experimental study of G protein activation

A widely used biochemical technique to measure GPCR and G protein activation relies on the detection of GTP binding to $G\alpha$. This requires physical separation of the G protein-bound GTP from the unbound remainder. If plain GTP were used for this assay, the built-in GTPase activity of the $G\alpha$ subunit would cleave most of the bound GTP during the time required for this separation, and therefore interfere with the measurement. This problem is overcome by using GTPγ^{35}S, which is the same as GTP but with one of the oxygen atoms of the terminal phosphate group replaced by radioactive sulfur. This sulfur atom renders the compound resistant to cleavage by $G\alpha$, and at the same time its radioactivity facilitates quantification of the bound nucleotide. An example application of this assay is shown in Figure 2.5C on page 23.

The GTPγ^{35}S is not selective for any specific $G\alpha$ subunit; to make it so, it is necessary to purify individual $G\alpha$ subunits and separately incubate them with

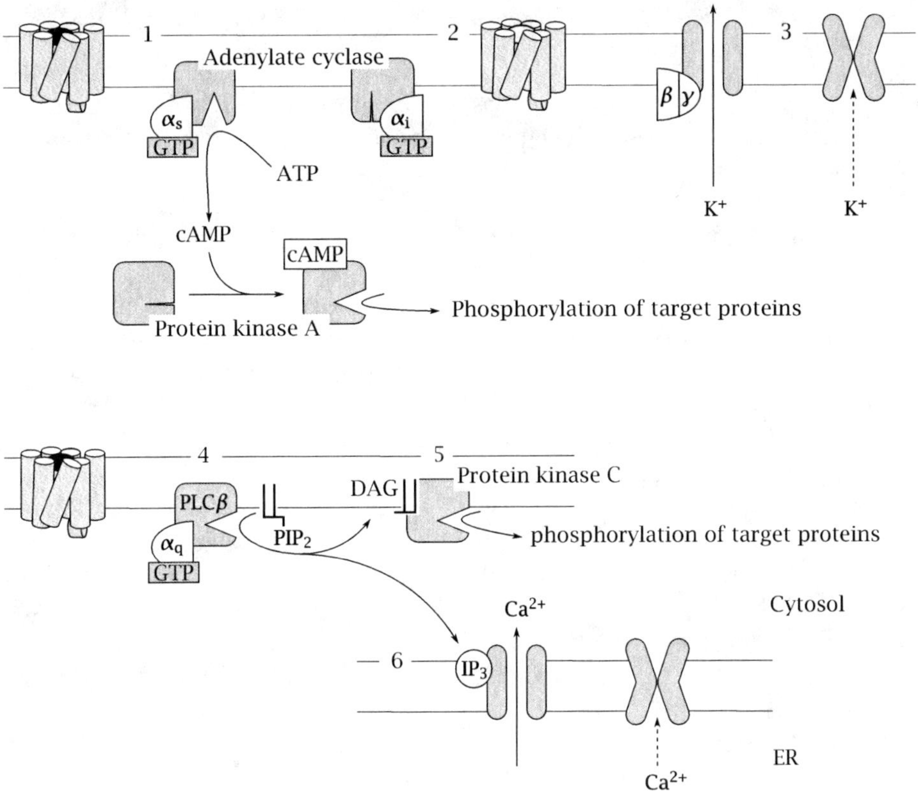

Figure 5.11 Some G protein effector mechanisms. Gα_s stimulates adenylate cyclase (1), which releases cAMP, an allosteric activator of protein kinase A. Gα_i has the opposite effect (2). G$\beta\gamma$ opens K^+ channels of the Kir type (3). Gα_q stimulates phospholipase Cβ (4), which cleaves phosphatidylinositol-bis-phosphate (PIP$_2$) into two secondary messengers: diacylglycerol (DAG), which remains in the cytoplasmic membrane and activates protein kinase C (5), and inositoltriphosphate (IP$_3$), which activates a cognate calcium channel in the ER membrane (6).

the GPCR of interest. Other techniques permit the study of G protein activation in whole cells. Most members of the G$\alpha_{i/o}$ family are susceptible to inhibition by pertussis toxin, a protein toxin secreted by the bacterium *Bordetella pertussis*, the causative agent of whooping cough. This toxin contains an enzyme that attaches a ADP-ribose moiety to G$\alpha_{i/o}$ and thereby inactivates it. Inhibition of a cellular response by pertussis toxin therefore suggests that the response is mediated by a member of the G$\alpha_{i/o}$ family. Selective suppression of individual G proteins is also possible with DNA- or RNA-based knockout techniques.

Fluorescence imaging techniques allow one to observe the interaction of G proteins with their receptors, or of Gα with G$\beta\gamma$ subunits, without specific functional assays. An example study is discussed in Section 5.8.3.

5.7 GPCR PHOSPHORYLATION, ENDOCYTOSIS, AND G PROTEIN-INDEPENDENT SIGNALING

Most GPCRs bind their ligands reversibly, and they may simply revert to the inactive conformation when the ligand dissociates. An additional mechanism of inactivation that mostly affects ligand-bound GPCRs is phosphorylation. GPCR kinases are often activated downstream of the signal transmitted by the receptor itself, so that their effect upon the receptor constitutes negative feedback mechanism.

The target residues of phosphorylation are located in the third loop, and the modification may immediately prevent the further interaction with its cognate G protein or, sometimes, switch the specificity to another G protein. An example is the phosphorylation-dependent switch of the β-adrenergic receptor from $G\alpha_s$ to $G\alpha_i$, which reverses the effect upon the downstream effector adenylate cyclase from stimulation to inhibition (see Table 5.2).

Receptor phosphorylation is involved in *tachyphylaxis*, the rapid reduction of the functional response of a cell or organ to receptor activation. This is illustrated in Figure 5.12, which shows the effect of the β-adrenergic receptor agonist isoproterenol on the strength of contraction of mouse heart muscle, measured as the rate of the rise in intraventricular pressure. The strength of contraction declines rapidly in the face of sustained levels of isoproterenol. Genetic knockout of GPCR kinase 2 significantly diminishes this decline.

GPCR phosphorylation also sets the stage for the binding of β-arrestins to the receptor, which will completely prevent any further productive interaction with G proteins. At the same time, receptor-bound β-arrestins interact with several other proteins, which results in the movement of the receptor to clathrin-coated pits and from there to intracellular compartments through endocytosis. The endocytosed receptor may shed the β-arrestin, undergo dephosphorylation, and return to the plasma membrane; alternatively, it may be completely degraded. One of the factors that decide the fate of the receptor is the affinity of β-arrestin binding, which varies with the receptor and is determined mostly by its C-terminus. β-Adrenergic receptors, for example, bind β-arrestin more loosely and have a greater chance to be reactivated than angiotensin receptors and protease-activated receptors, which are more likely to undergo degradation.

Intriguingly, the binding of β-arrestin to a GPCR also may result in *G protein-independent signaling.* An example is the activation of a protein kinase named extracellular signal-regulated kinase (ERK) by the β_2-adrenergic receptor. The activation is biphasic, such that an early, intense stimulation is followed by a late, weaker one. The early phase, but not the late phase, is prevented by H-89, an inhibitor of protein kinase A, which is activated through $G\alpha_s$. Inhibiting the expression of β-arrestin using a small interfering RNA (see Section 13.2.1) *reduces* ERK activation (Figure 5.13). G protein-mediated signalling should be enhanced, if anything, by diminished β-arrestin activity; therefore, these findings suggest that β-arrestin itself mediates the G protein-independent activation of ERK.

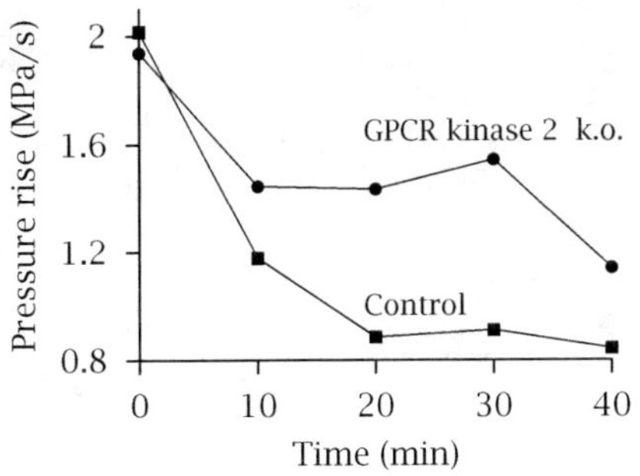

Figure 5.12 Knockout of GPCR kinase 2 reduces tachyphylaxis of the myocardial response to β-adrenergic receptor stimulation. The agonist isoproterenol was applied continuously through the course of the experiment. (Figure redrawn from original data in Ref. [19].)

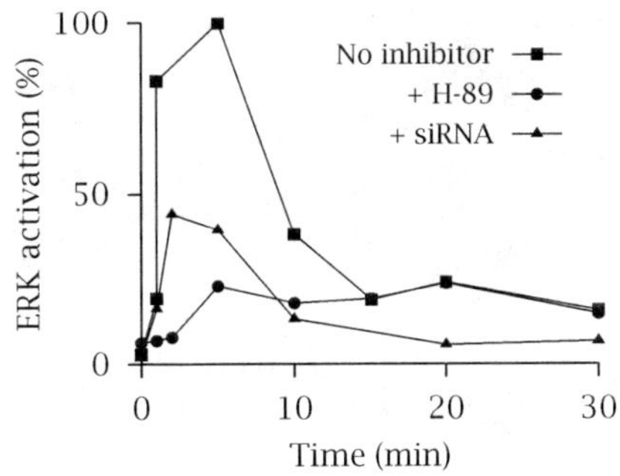

Figure 5.13 Activation of extracellular signal-regulated kinase (ERK) by β-adrenergic receptors in response to isoproterenol, in presence of the adenylate cyclase inhibitor H-89 and of siRNA specific for β-arrestin. See text for details. (Redrawn from data in Ref. [23].)

5.8 APPENDIX

This section provides additional background and data on some of the experimental studies of G protein-coupled receptors that were discussed above.

5.8.1 Tryptophan absorbance and fluorescence

The side chains of aromatic amino acids in proteins absorb ultraviolet light, and they also emit some of the absorbed energy through fluorescence. These absorption and fluorescence signals can be used to study protein structure and function. Among the three aromatic amino acids, tryptophan absorbs and fluoresces most strongly. It also is less abundant than phenylalanine and tyrosine. Therefore, the contributions of individual tryptophan residues to a protein's overall signal are usually most readily discernible.

Figure 5.14A illustrates typical absorption and fluorescence spectra of a tryptophan residue. Both spectra are affected by changes to the environment of the tryptophan residue such as overall polarity and formation or disruption of hydrogen bonds. For example, the dashed curves in the figure depict spectral shifts that may result from an increased exposure of a tryptophan residue to solvent. If such a spectral shift occurs when the protein changes conformation, it indicates that the tryptophan residue in question is affected by this change.

A spectral change can be represented as a *difference spectrum*, which is obtained by subtracting experimental spectra collected under two different conditions. Figure 5.14B shows difference spectra for two individual tryptophan residues in the rhodopsin molecule. These were obtained from absorption spectra of rhodopsin mutants before and after the light-induced conformational change [16]. The difference spectrum for W35 is essentially flat, suggesting that W35 is not affected by this conformational change. In contrast, a significant change is visible in the spectrum of residue W265. The spectral change resembles that of the model compound N-acetyltryptophanamide (NATA) when it is

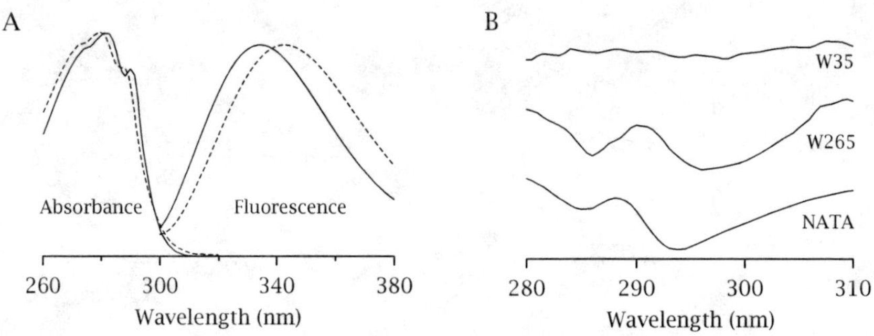

Figure 5.14 Tryptophan absorption and fluorescence emission spectra. **A:** Representative tryptophan absorption and fluorescence emission spectra in apolar environment (solid lines) and polar environment (dashed lines). **B:** Absorption difference spectra of tryptophan residues 35 and 265 in rhodopsin, and of the model compound N-acetyl-tryptophanamide (NATA). See text for details. (Figure redrawn from original data in Ref. [16].)

transferred from water to the aprotic solvent dimethylsulfoxide (DMSO). This similarity suggests that W265 experiences a change in hydrogen bonding. In any case, the observed spectral change indicates that W265, which is located in helix 6 of the rhodopsin molecule, participates directly in the conformational rearrangement that underlies rhodopsin activation.

5.8.2 Cysteine mutagenesis and site-directed labeling

Among all standard amino acids, only cysteine carries a sulfhydryl group, which has unique chemical reactivity. Cysteine residues can be substituted for other amino acids at strategic locations within a protein molecule. The reactivity of the mutant cysteines can then be used to probe the function of the protein in various ways:

1. The accessibility of the mutant cysteines to water-soluble sulfhydryl-specific reagents can be determined. If a cysteine does not react, it is either buried inside the protein itself or within a membrane; this principle is often used to examine the structure of integral membrane proteins. Changes in the reactivity of a cysteine residue can detect its participation in changes of conformation.

2. The mutant cysteines can be selectively labeled with sulfhydryl-reactive spectroscopic probes. The spectral signals of the probes can then be watched as the protein adopts different functional and conformational states.

Useful spectroscopic probes include nitroxide radicals, as well as a variety of fluorescent dyes. Some fluorescent dyes behave similar to tryptophan, in that their fluorescence emission spectra are sensitive to the polarity of their environment. This environmental sensitivity can be used to detect conformational transitions, interaction with membrane lipids and other effects.

Several coupling chemistries that are commonly used to modify cysteine residues are depicted in Figure 5.15. Pairwise introduction and modification of cysteines gives additional possibilities:

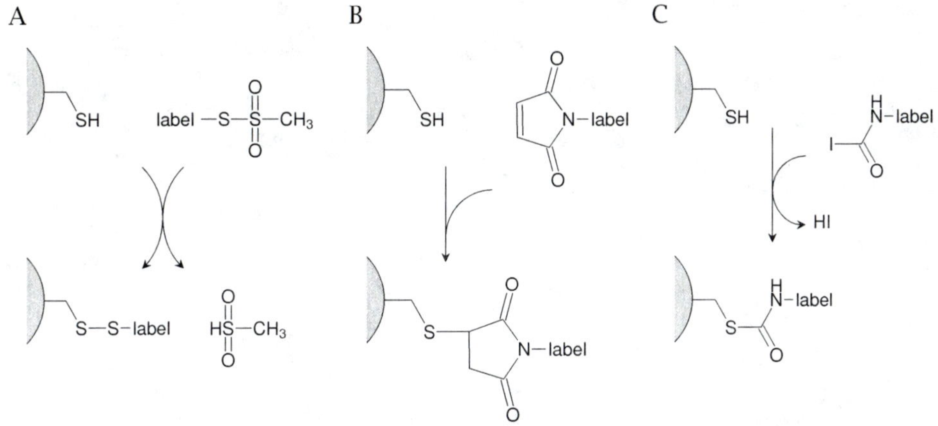

Figure 5.15 Reactive groups for thiol-specific modification of cysteine residues. **A**: Methylthiolsulfonate. **B**: Maleimide. **C**: Iodoacetamide. In each reaction scheme, the *label* can be a fluorescent or a paramagnetic probe, a crosslinker, or whatever else one needs to introduce into the protein. Note that methylthiolsulfonate modification gives a disulfide, which can be reduced again under mild conditions with free thiols. The other two modifications yield thioethers and are irreversible.

1) If the two cysteine residues are sufficiently close to one another, they will react to form a disulfide bond. Such a bond will constrain the local conformational flexibility; freedom can be restored with thiol-reducing agents such as dithiothreitol. The effect of the disulfide bond on protein activity will provide insight into the functional significance of local conformational mobility (see Figure 5.3).

2) If two nitroxide labels are introduced into the same protein molecule, their magnetic fields will interact with each other across distances of up to approximately 2.5 nm. This can be detected with EPR spectroscopy and used to measure distances and observe conformational changes. Combinations of fluorescent dyes can be used for fluorescence energy transfer in a similar manner.

5.8.3 Fluorescence resonance energy transfer

Fluorescence resonance energy transfer (FRET) is a physical effect that causes excitation energy to migrate from one fluorescent molecule, the donor, to another one, the acceptor [6]. This only happens if the emission spectrum of the donor overlaps the excitation spectrum of the acceptor. Nevertheless, FRET does *not* involve emission of an actual photon. Rather, it occurs by direct coupling of the two molecular dipoles. The *Förster radius* is the distance between the two fluorophores at which exactly 50% of all photons absorbed by the donor migrate to the acceptor. While its exact value depends on the two fluorophores in question, it is generally in the range of a few nanometers, which is similar to the dimensions of proteins and other biological macromolecules [26]. FRET is therefore a useful tool to study the behavior and interactions of macromolecules.

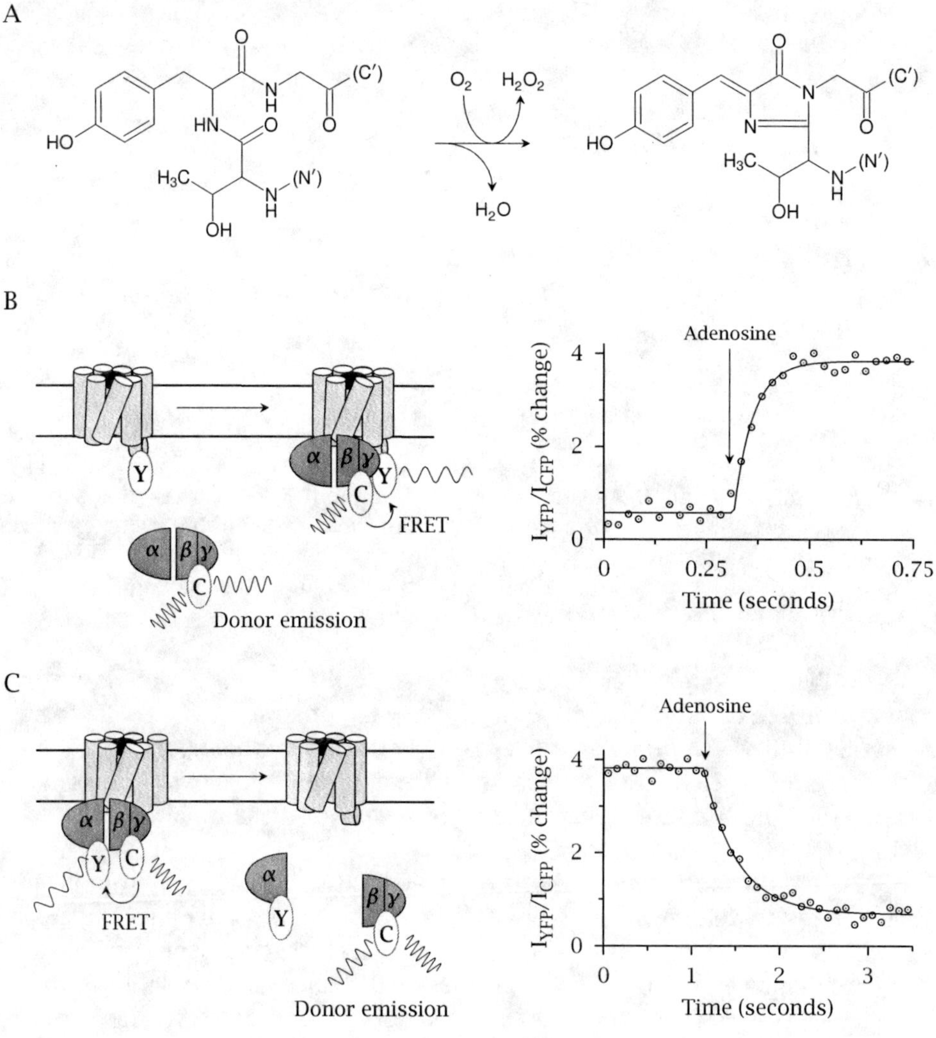

Figure 5.16 Fluorescence resonance energy transfer (FRET) between intrinsically fluorescent proteins—application to G protein-coupled receptors. **A:** Autocatalytic formation of the fluorophore in the *Aequoria* green-fluorescent protein. In the mutant cyan-fluorescent protein (CFP), the tyrosine residue is replaced by tryptophan. The yellow-fluorescent protein (YFP) contains a mutant tyrosine residue that is not part of the fluorophore but engages in π-stacking interactions with the GFP fluorophore. **B:** Application of FRET to the binding of G proteins to an activated GPCR. Cyan-fluorescent protein (C) was translationally fused to the γ subunit of G_S, and the yellow-fluorescent protein (Y) to the A_{2A} adenosine receptor. Addition of adenosine triggers binding of G_S to the receptor, and the ensuing FRET from CFP to YFP is detected as in increase in the ratio of YFP emission to CFP emission. **C:** In this experiment, the α subunit and the γ subunit were labeled with YFP and CFP, respectively. Dissociation of the G protein is evident as a decrease in FRET. Note the difference in the timescales between B and C: Dissociation of the G protein is notably slower than its binding to the activated receptor. (Figure redrawn from original data in Ref. [10].)

In order to use FRET, the proteins of interest must be fluorescently labeled. In living cells, this is most readily accomplished using the green fluorescent protein (GFP) of *Aequoria* jellyfish. Even more useful than GFP itself are engineered GFP mutants that give brighter fluorescence and also have shifted absorption and emission spectra, and which can be employed in combination as donor–acceptor couples in FRET experiments.

The magic feature of GFP and its variants is the autocatalytic formation of a fluorophore from three amino acid residues, which does not require any additional proteins or cofactors and therefore works in any cell. A protein of interest can be labeled simply through translational fusion with one of the fluorescent proteins.

Figure 5.16 illustrates the application of FRET between two GFP derivatives to the interaction of G proteins with GPCRs. A notable finding of this study was that the dissociation of the G protein heterotrimer is the kinetically limiting step in GPCR-mediated signaling [10].

STUDY QUESTIONS

5.1 The plot below shows competition experiments with H_4 histamine receptors from different species ("monkey" represents Cynomolgus monkey; data from Ref. [15]).

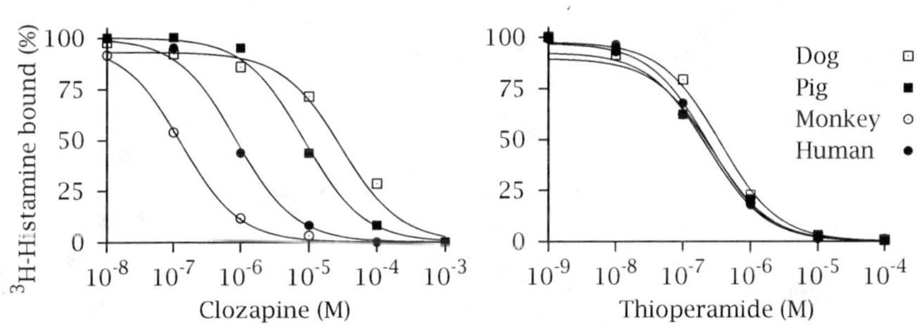

(1) Propose an explanation for the observed differences between clozapine and thioperamide. (2) What do these observations suggest for the validity of testing drugs in animal experiments?

5.2 The plot below shows the effect of an experimental drug, CGP7930, on $GABA_B$ receptors, alone and in combination with GABA [3]. What appears to be the mode of action of CGP7930? What does it tell us about the activity of GABA itself?

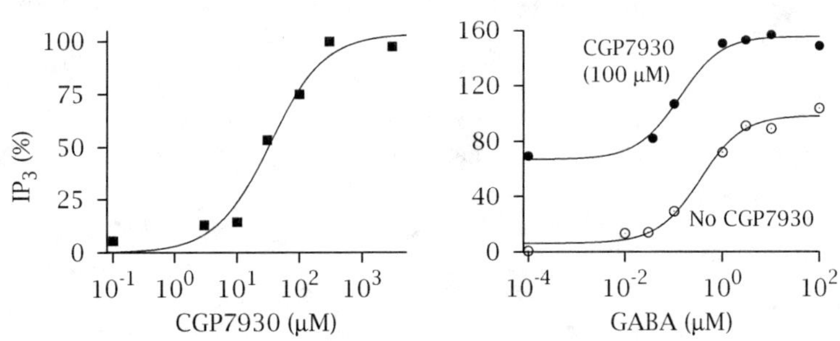

5.3 Figure 5.16B displays the FRET that occurs when an adenosine receptor associates with its cognate G protein (G_S), and Figure 5.16C shows the reduction of FRET that occurs when the same G protein dissociates upon interaction with the same receptor.

The absorption spectra of the donor (CFP) and the acceptor (YFP) are well separated. Therefore, most of the acceptor intensity should arise through FRET from the donor, as opposed to direct excitation by the laser beam used to excite the donor. One would then expect that binding of donor to acceptor, or their dissociation, should have a strong effect on the acceptor fluorescence.

Nevertheless, in both experiments, the change in the ratio of donor and acceptor fluorescence is quite small (close to 4%). How can this be explained?

REFERENCES

[1] S. Arnis et al., "A conserved carboxylic acid group mediates light-dependent proton uptake and signaling by rhodopsin," *J. Biol. Chem.* 269 (1994), 23879–23881.

[2] K. A. Berg et al., "Effector pathway-dependent relative efficacy at serotonin type 2A and 2C receptors: evidence for agonist-directed trafficking of receptor stimulus," *Mol. Pharmacol.* 54 (1998), 94–104.

[3] V. Binet et al., "The heptahelical domain of GABA(B2) is activated directly by CGP7930, a positive allosteric modulator of the GABA(B) receptor," *J. Biol. Chem.* 279 (2004), 29085–29091.

[4] S. Chackalamannil et al., "Discovery of potent orally active thrombin receptor (protease activated receptor 1) antagonists as novel antithrombotic agents," *J. Med. Chem.* 48 (2005), 5884–5887.

[5] V. Cherezov et al., "High-resolution crystal structure of an engineered human beta2-adrenergic G protein-coupled receptor," *Science* 318 (2007), 1258–1265.

[6] T. F
"orster, "Intermolecular energy migration and fluorescence," *Ann. Phys.* 2 (1948), 55–75.

[7] O. Fritze et al., "Role of the conserved NPxxY(x)5,6F motif in the rhodopsin ground state and during activation," *Proc. Natl. Acad. Sci. U. S. A.* 100 (2003), 2290–2295.

[8] T. Galvez et al., "Allosteric interactions between GB1 and GB2 subunits are required for optimal GABA(B) receptor function," *EMBO J.* 20 (2001), 2152–2159.

[9] U. Gether et al., "Different binding epitopes on the NK1 receptor for substance P and non-peptide antagonist," *Nature* 362 (1993), 345–348.

[10] P. Hein et al., "Gs activation is time-limiting in initiating receptor-mediated signaling," *J. Biol. Chem.* 281 (2006), 33345-33351.

[11] E. Hermans, S. R. Nahorski, and R. A. Challiss, "Reversible and non-competitive antagonist profile of CPCCOEt at the human type 1α metabotropic glutamate receptor," *Neuropharmacology* 37 (1998), 1645–1647.

[12] W. L. Hubbell et al., "Rhodopsin structure, dynamics, and activation: a perspective from crystallography, site-directed spin labeling, sulfhydryl reactivity, and disulfide cross-linking," *Adv. Protein Chem.* 63 (2003), 243-290.

[13] B. A. Jordan and L. A. Devi, "G-protein-coupled receptor heterodimerization modulates receptor function," *Nature* 399 (1999), 697–700.

[14] B. K. Kobilka and X. Deupi, "Conformational complexity of G-protein-coupled receptors," *Trends Pharmacol. Sci.* 28 (2007), 397–406.

[15] H. D. Lim et al., "Molecular determinants of ligand binding to H4R species variants," *Mol Pharmacol* 77 (2010), 734–743.

[16] S. W. Lin and T. P. Sakmar, "Specific tryptophan UV-absorbance changes are probes of the transition of rhodopsin to its active state," *Biochemistry* 35 (1996), 11149–11159.

[17] S. Litschig et al., "CPCCOEt, a noncompetitive metabotropic glutamate receptor 1 antagonist, inhibits receptor signaling without affecting glutamate binding," *Mol. Pharmacol.* 55 (1999), 453–461.

[18] A. W. Ma et al., "Recovery of oligomers and cooperativity when monomers of the M2 muscarinic cholinergic receptor are reconstituted into phospholipid vesicles," *Biochemistry* 46 (2007), 7907–7927.

[19] S. J. Matkovich et al., "Cardiac-specific ablation of G-protein receptor kinase 2 redefines its roles in heart development and β-adrenergic signaling," *Circ. Res.* 99 (2006), 996–1003.

[20] T. Okuno et al., "Leukotriene B4 receptor and the function of its helix 8," *J. Biol. Chem.* 280 (2005), 32049–32052.

[21] J. Peter et al., "Modulation of the M2 muscarinic acetylcholine receptor activity with monoclonal anti-M2 receptor antibody fragments," *J. Biol. Chem.* 279 (2004), 55697–55706.

[22] W. G. Rajeswaran et al., "Design, synthesis, and biological characterization of bivalent 1-methyl-1,2,5,6-tetrahydropyridyl-1,2,5-thiadiazole derivatives as selective muscarinic agonists," *J. Med. Chem.* 44 (2001), 4563–4576.

[23] S. K. Shenoy et al., "β-Arrestin-dependent, G protein-independent ERK1/2 activation by the β2 adrenergic receptor," *J. Biol. Chem.* 281 (2006), 1261–1273.

[24] L. Shi et al., "Beta2 adrenergic receptor activation. Modulation of the proline kink in transmembrane 6 by a rotamer toggle switch," *J. Biol. Chem.* 277 (2002), 40989–40996.

[25] Y. Shima et al., "Opposing roles in neurite growth control by two seven-pass transmembrane cadherins," *Nat. Neurosci.* 10 (2007), 963–969.

[26] L. Stryer and R. P. Haugland, "Energy transfer: a spectroscopic ruler," *Proc Natl Acad Sci U S A* 58 (1967), 719–726.

[27] M. Waldhoer et al., "A heterodimer-selective agonist shows in vivo relevance of G protein-coupled receptor dimers," *Proc. Natl. Acad. Sci. U. S. A.* 102 (2005), 9050–9055.

CHAPTER 6

PHARMACOLOGY OF CELL EXCITATION

Cell excitation provides the most rapid means of functional regulation of cells and of communication between them. Excitable cells include nerve cells, striated muscle cells in the heart and the skeletal muscles, smooth muscle cells in blood vessels and interior organs, and some endocrine gland cells (Figure 6.1). They are key in many of the physiological processes that we want to control by drug therapy. Examples of such therapeutic applications are (1) the blockade of nerve conduction for local anesthesia, (2) the moderation of nerve cell excitability in the treatment of epilepsy, (3) the reduction of smooth muscle cell tone in blood vessels to reduce blood pressure, and (4) the stabilization of mood in the treatment of bipolar disorder.

The fundamental prerequisite of cell excitation is the electrical potential across the cytoplasmic membrane. This potential exists in all living cells.[1] In non-excitable cells, the membrane potential is constant, and the cell interior is always electrically negative relative to the outside. This also applies to excitable cells that are not currently excited and thus are at their *resting potential*. However, excitable cells experience changes of the membrane potential, such that the orientation of the membrane potential is transiently reversed. These

[1]Membrane potentials also exist across intracellular membranes. For example, the potential across the inner mitochondrial membrane is the major driving force of ATP synthesis. However, intracellular potentials don't have a prominent role in cell excitation.

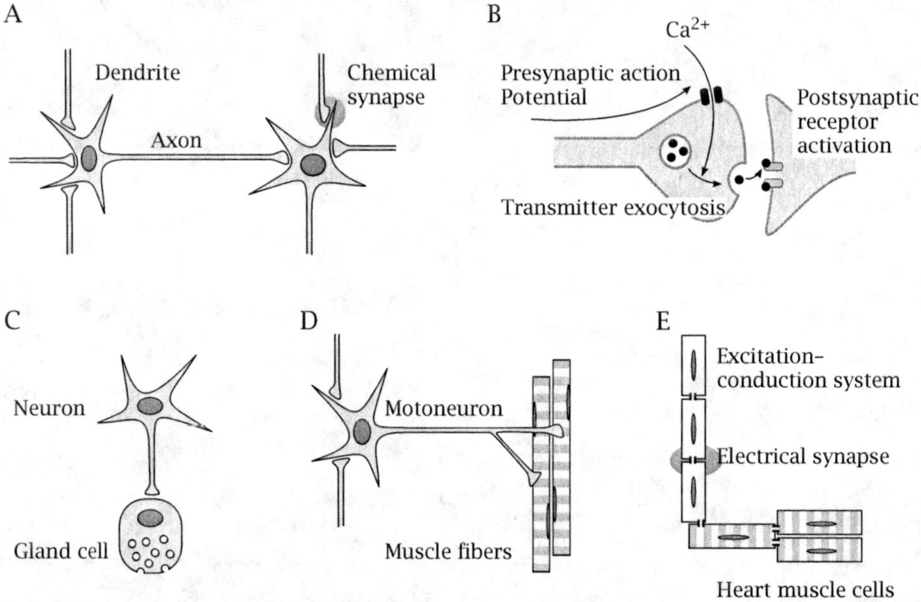

Figure 6.1 Types of excitable cells. **A:** Nerve cells communicate through chemical synapses, which are formed between axon terminals of presynaptic cells and dendrites of postsynaptic cells. **B:** In a chemical synapse, a presynaptic action potential induces the release of transmitter from the axon terminal into the synaptic cleft. Binding of transmitter to the postsynaptic receptors influences the excitation of the postsynaptic cell. **C:** In some endocrine glands, the release of hormones from gland cells into the circulation is controlled by nerve cells through chemical synapses. **D:** Skeletal muscle cells are controlled by α-motoneurons, which originate in the brain stem and the spinal cord. **E:** The heart possesses two types of excitable cells. Those within the excitation-conduction system spontaneously produce the rhythm, which they then propagate to the regular "worker" muscle cells. Communication occurs through electrical synapses, in which ions can flow directly through cytoplasm bridges or *gap junctions* between cells.

reversals are called *action potentials.* Action potentials spread along the membrane of the cell, with speeds that vary greatly but reach more than 100 m/s in the fast-conducting, long fibers that connect the brain and the spinal cord with distant parts of the body.

Excitable cells communicate with one another through synapses. In a synapse, the cell membranes of the two participating cells are closely aligned. The space between them, which is only a few nanometers wide, is the synaptic cleft. When an action potential arrives at the presynaptic side of a synapse, it triggers the release of a transmitter substance by the cell. The released transmitter diffuses across the synaptic cleft toward the postsynaptic cell membrane, where it binds to cognate receptors (Figure 6.1B). Depending on the functional type of the receptor, this will promote or inhibit the formation of a new action potential in the postsynaptic cell.

While the basic principles of function are the same for nerve and muscle cells, the details of organization and regulation vary considerably. Skeletal mus-

Table 6.1 Intra- and extracellular concentrations of major ions. The resulting equilibrium potentials (E_0) are calculated using the Nernst equation (see Section 6.9.1).

Ion	Cytosolic	Extracellular	E_0 at 37°C
K^+	150 mM	6 mM	- 86 mV
Na^+	15 mM	150 mM	+ 62 mV
Ca^{2+}	100 nM	1.2 mM	+ 126 mV
Cl^-	9 mM	150 mM	- 70 mV

cle is simplest. A skeletal muscle cell has no spontaneous activity of its own, and it is entirely controlled by a single nerve cell, usually an α-motoneuron. Synapses between motoneurons and muscle cells are often called *motor endplates.* In a motor endplate, each presynaptic action potential triggers a postsynaptic one. The frequency of postsynaptic action potentials controls the contractile force of the muscle cell.

In contrast to skeletal muscle, the heart can sustain its own activity. This is due to specialized pacemaker cells that produce action potentials spontaneously at regular intervals and so generate the heart rhythm. While the rhythm set by the pacemaker cells can be modulated by excitatory and inhibitory signals from the autonomic nervous system and from circulating hormones, the heart will continue to beat in the absence of any such input.

Neurons in the central nervous system and in peripheral ganglia and plexus[2] likewise receive excitatory and inhibitory input. In contrast to muscle cells, however, neurons often interact with hundreds or thousands of cells upstream and downstream. From the symphony of input it receives through its upstream or *afferent* synapses, a neuron determines its own response, which it encodes in the frequency of action potentials that it will fire and send downstream to its *efferent* synapses.

Most neurons have intricate, tree-like shapes with many branches and twigs to form synapses with all these other cells. The afferent or receiving branches of a neuron are called *dendrites.* A nerve cell's main efferent structure is the *axon.* It likewise may branch out to form numerous presynaptic terminals.

Many of the proteins that conduct action potentials along nerve or muscle cells and relay them across synapses are targets for therapeutic drugs. These proteins also are very commonly targeted by natural poisons, some of which have found application as drugs or have served as models for drug development.

In the remainder of this chapter, we will cover quite a range of drugs and drug targets. We will first focus on those proteins that sustain and propagate the action potential on a single cell, and then move on to the drug targets involved in synaptic transmission.

[2]Ganglia (singular: ganglion) and plexus (singular: plexus) are groups of neurons that occur outside the central nervous system and perform some local or regional signal processing.

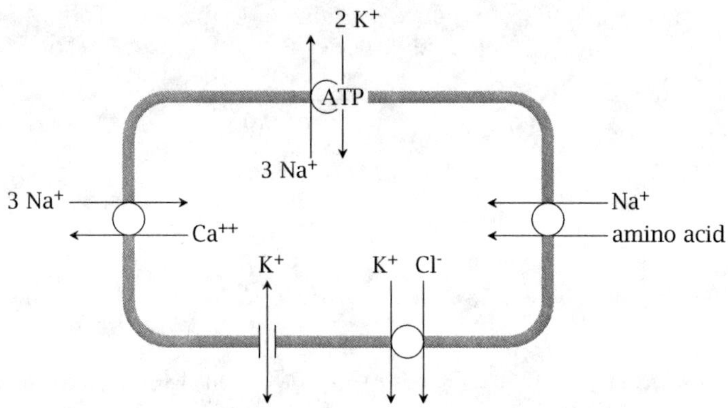

Figure 6.2 Major ion transport processes across the cytoplasmic membrane. Na^+/K^+-ATPase keeps up the concentration gradients for the two major cations. The high extracellular sodium concentration drives calcium antiport as well as symport of amino acids and other metabolites. Potassium symport drives the export of chloride. Potassium leak channels establish a negative-inside diffusion potential.

6.1 IONS, PUMPS, AND CHANNELS

Both resting potentials and action potentials depend on the existence of ion concentration gradients across the membrane in question. The major ion species involved are K^+, Na^+, Ca^{2+}, and Cl^- (see Table 6.1). Ion gradients result from the activities of three types of membrane protein:

1. Ion pumps use ATP to transport ions against their concentration gradients. Directly or indirectly, they are responsible for maintaining all ion gradients. The most abundant ion pump is Na^+/K^+-ATPase (Figure 6.3). The function of ion pumps is referred to as *primary active transport.*

2. Ion exchangers and cotransporters link different ion gradients together, so that gradients can be established for ions that may have no or not enough specific pumps. This function is called *secondary active transport.* In symport, both solutes are transported in the same direction, whereas in antiport one solute is exchanged for the other. Examples are the Na^+/Ca^{2+} exchanger and the K^+/Cl^- cotransporter in the cytoplasmic membrane.

3. Ion channels. These proteins facilitate the passive diffusion of ions down their electrochemical gradients. This is known as *passive transport.*

Active transporters establish and maintain the ion gradients, whereas ion channels tap into these gradients to control the resting and action potentials. Both resting potentials and action potentials are diffusion potentials. The causation of diffusion potentials by ion channels is explained in Section 6.9.1.

Ion channels often are specific for a single ion species, but some are promiscuous. Most ion channels can open and close. They may do so in response to a ligand binding to them, or to changes in the surrounding electrical field. In addition to these *ligand-gated* and *voltage-gated* channels, there are *leak* channels, which are continually open. In particular, K^+ leak channels cause the K^+ ion gradient to dominate the resting membrane potential.

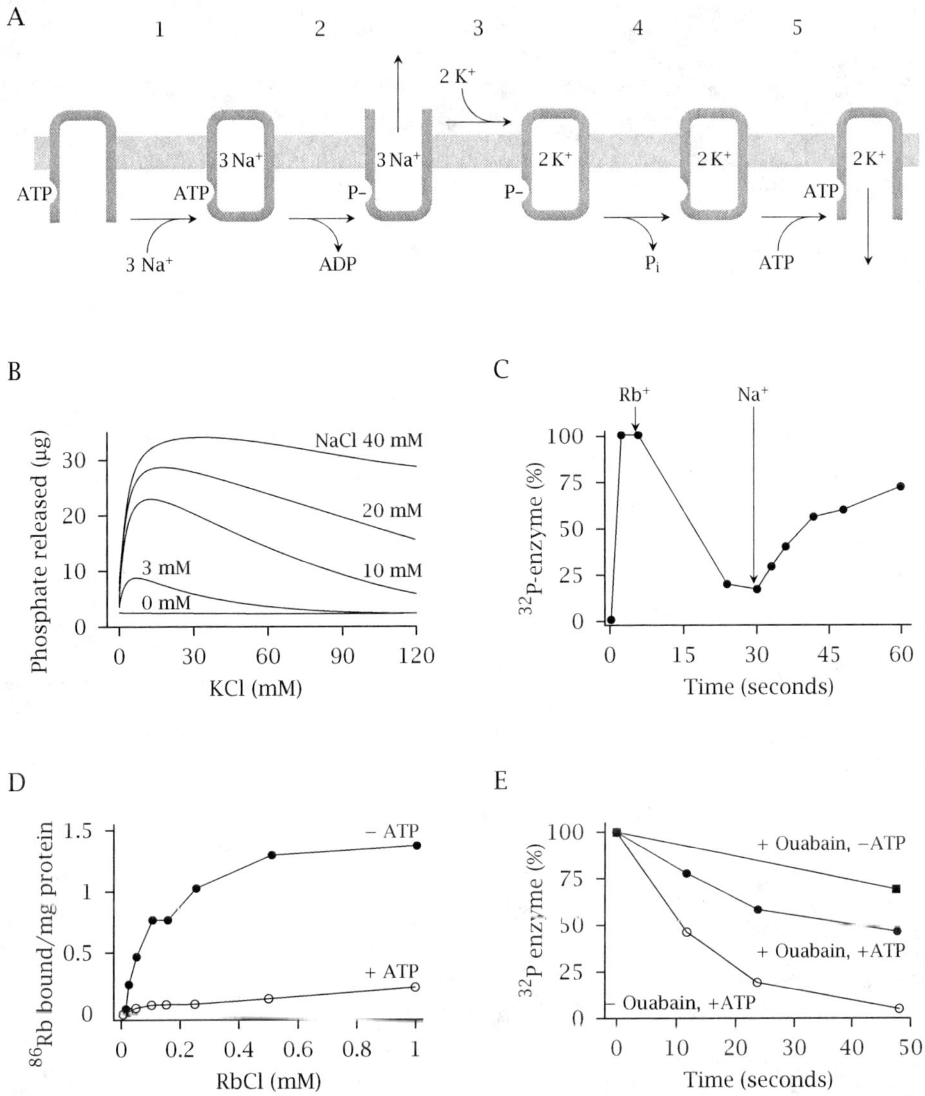

Figure 6.3 Function of Na^+/K^+-ATPase – overview and experimental evidence. **A:** The catalytic cycle. ATP and Na^+ bind from the intracellular side. Steps 1–5 are explained in the text (Section 6.2). **B:** ATPase activity as a function of KCl and NaCl concentrations [25]. KCl alone does not elicit any activity. Maximal activity is observed at approximately equal concentrations of KCl and NaCl. **C:** Effects of Rb^+ and of Na^+ on the phosphorylation state of the enzyme [19]. At time zero, only Na^+ is present, and phosphorylated enzyme accumulates. Rb^+, which mimics K^+, dephosphorylates the enzyme; Na^+ added in excess over Rb^+ phosphorylates it again. **D:** ATP is required to release Rb^+ from tight binding to the enzyme. Tight binding is interpreted as occlusion of the ion within the enzyme [6]. **E:** Inhibition of Na^+/K^+-ATPase by ouabain [23]. At time zero, an excess of nonradioactive ATP is added to a preparation of enzyme that has been phosphorylated with radioactive ^{32}P-ATP. As the enzyme turns over, the radioactive phosphate is replaced with the nonradioactive. (Panels B–E redrawn from the references cited.)

The flux of ions through channels and secondary transporters works relentlessly to dissipate the ion gradients. This necessitates the continuous activity of the ion pumps, which consumes a sizable fraction of every cell's metabolic energy. The major ion transport processes that are responsible for maintaining the ion gradients and the resting potential across the cytoplasmic membrane are summarized in Figure 6.2.

6.2 ATP-DRIVEN ACTIVE ION TRANSPORT

The most abundant ATP-driven ion transporter is Na^+/K^+-ATPase. In many cells, this enzyme is responsible for the greater share of all ATP consumed. Figure 6.3A illustrates how ATP hydrolysis is coupled to ion transport: (1) the ATP-bound form of the enzyme is open to the cytosol and accepts three Na^+ ions, which become occluded inside the enzyme; (2) the occluded enzyme transfers the terminal phosphate of ATP to itself and releases ADP, concomitantly with eversion and release of Na^+ to the extracellular space; (3) two K^+ ions bind from the outside and become occluded; (4) phosphate is hydrolyzed; (5) ATP binds and opens the enzyme to the intracellular side. This releases K^+ and restores the enzyme to its initial state.

The intracellular levels of Na^+ and Ca^{2+} are linked to one another through antiport (Figure 6.2). Therefore, inhibition of Na^+/K^+-ATPase will indirectly increase the intracellular concentration of calcium, which in muscle cells promotes contraction. The Na^+/K^+-ATPase inhibitors ouabain, digoxin, and digitoxin can be used to treat cardiac insufficiency, that is, weak heart muscle function. Small amounts of ouabain and several other similar compounds, some of which have not been obtained in pure form or structurally characterized [12,17], are also synthesized in the adrenal glands and may function as a separate class of steroid hormones.

There are several other ATP-driven ion pumps that are structurally homologous to Na^+/K^+-ATPase and function in a similar way. One of them pumps Ca^{2+} ions out of the cell. Another one pumps protons out of the parietal cells of the stomach mucous membrane. This pump has no role in cell excitability; instead, it produces gastric acid. Its pharmacological inhibition with omeprazole is used in the treatment of ulcers and gastritis to prevent the induction or aggravation of tissue damage by gastric acid.

6.3 VOLTAGE-GATED CHANNELS AND THE ACTION POTENTIAL

While transporters and leak channels control the resting potential, the action potential is dominated by the gated channels. Broadly speaking, ligand-gated channels occur in synapses and initiate the action potential, whereas its propagation across the expanse of the cell surface is controlled by voltage-gated channels.

At the resting potential, most voltage-gated channels remain closed. They only open when the cell membrane is depolarized, that is, when the cell interior becomes electrically positive, or at least substantially less negative relative to the exterior.

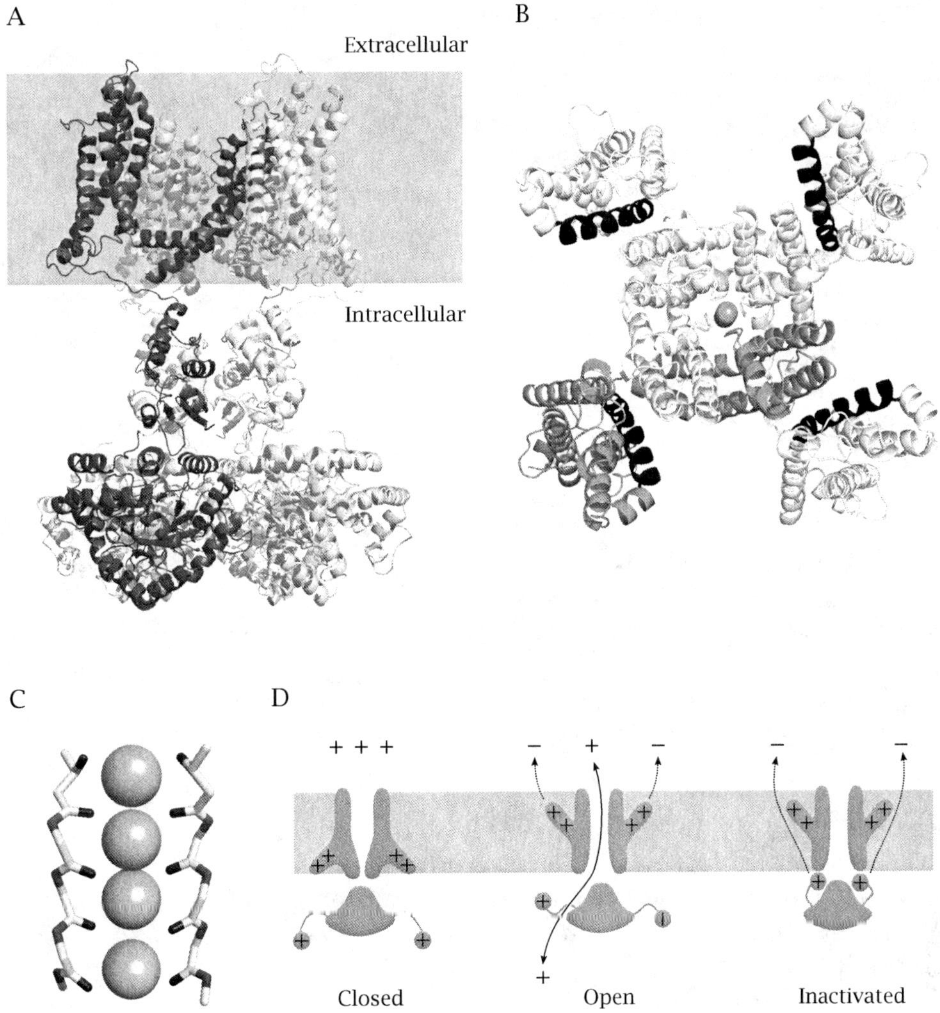

Figure 6.4 Structure and function of a voltage-gated potassium channel. **A:** Overall structure. The gray rectangle delineates the membrane-embedded portion. One of four identical subunits is rendered in dark gray. The N-terminal inactivation domain that is located in the cytosolic portion (see D) is missing from this structure. **B:** View from the extracellular side. The membrane-embedded portion consists of an inner layer, which contains the selectivity filter (see C) and an outer layer that mediates gating (see D). An arginine-rich helix that occurs in each subunit and forms the voltage sensor of the opening gate is rendered in black. **C:** Structure of the K^+ selectivity filter. Backbone oxygen atoms tightly enclose the dehydrated K^+ ions. Each of the four subunits contributes one backbone segment, but for clarity only two segments are shown. **D:** The voltage-dependent gating mechanism. When the membrane potential is reversed, the membrane-embedded voltage sensor moves outward, which loosens up the inner layer and opens the channel. A separate inactivation gate, which is flexibly attached to the cytosolic portion of the channel and also carries a positive net charge, then responds and moves to plug the channel. (Structures in A–C rendered from 2r9r.pdb [15]. A colored version of this figure is shown in the insert in the middle of this book.)

The voltage-sensitive switches built into these channel proteins consist of domains that are both mobile and charged, and thus will move in response to changes in the surrounding electric field. Within each channel, there are two separate switches. The first one opens the channel in response to depolarization, whereas the second one *inactivates* the channel, in the face of continued depolarization. The channel thus is open for only a short period of time.

Most excitable cells contain large numbers of voltage-gated channel molecules for both sodium and potassium. Heart and smooth muscle cells additionally contain calcium channels in abundance. Each of these classes of channels contains several subtypes, which are associated with different cell types and tissues. These subtypes can to some degree be selectively addressed by clinical and experimental drugs.

The voltage-gated channels for Na^+, K^+, and Ca^{2+} are structurally homologous. The structure of a voltage-gated K^+ channel (K_V) is shown in Figure 6.4. The channel consists of four subunits. One part of the channel is embedded in the membrane, whereas another domain protrudes into the cytosol. One of the membrane-embedded helices forms the voltage sensor for opening the channel. The sensor helix carries several positively charged arginine residues. This is an unusual feature for transmembrane helices, which typically contain very few charges or none at all. The second sensor, which is responsible for channel inactivation, is flexibly attached to the intracellular domain. Upon membrane depolarization, it moves to plug the intracellular vestibule of the channel.

6.3.1 Sodium and potassium channels propagate the action potential

Figure 6.5 illustrates how the opening of sodium channels becomes a self-sustaining wave of depolarization that propagates along the entire expanse of the membrane. This wave constitutes the action potential. While Na^+ and K^+ channels are structurally similar and respond to membrane depolarization in much the same way, the functional consequences of their responses are opposed to one another:

1. Because of the opposite concentration gradients, open Na^+ channels will cause membrane depolarization, that is, a more positive-inside membrane potential, while open K^+ channels will cause repolarization and even hyperpolarization, that is, a more negative-inside potential.
2. The Na^+ channels respond faster than the K^+ channels and therefore dominate the early upswing phase of the action potential, whereas the K^+ channels control the late phase, in which the membrane potential is restored back to normal. The overall duration of the action potential is often around 1–2 ms.

Although the simplified illustration in Figure 6.5 suggests otherwise, the membrane potential does not actually have to be fully reversed for the sodium channels to open. As soon as the local potential reaches the threshold at approximately −55 mV, sodium channels will begin to open, and the action potential will be triggered. This threshold is a property of the channel molecules themselves, and it differs between different types of voltage-gated channels.

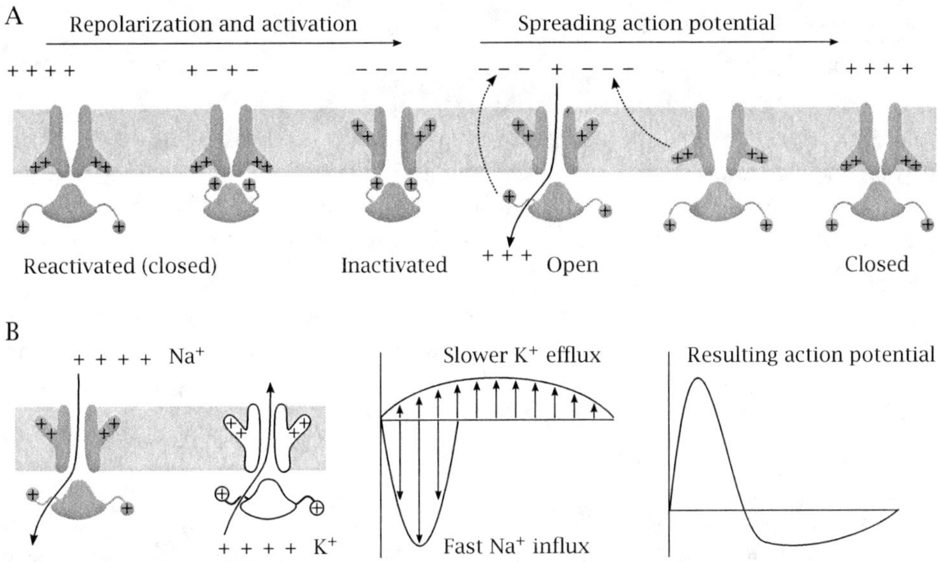

Figure 6.5 Propagation of action potentials by voltage-gated channels. **A:** Voltage-gated channels have two separately operated gates, which are voltage-sensitive conformational switches. They exist in three functional states. The closed state gives way to the open state when an approaching wave of depolarization pulls open the first gate. With Na_V channels, this leads to Na^+ influx, which in turn enlarges the depolarized membrane area. The second gate then closes, and the channel inactivates in spite of the depolarized state of the membrane. After the membrane becomes repolarized, both gates revert to their initial conformations. **B:** Interplay of Na_V and K_V channels in shaping the action potential. Both channels open in response to depolarization, causing Na^+ influx and K^+ efflux. The Na^+ channels open and inactivate faster and therefore dominate the early phase of the action potential. The slower response of the K^+ channels dominates the later phase and causes a transient hyperpolarization of the membrane. (A colored version of this figure is shown in the insert in the middle of this book.)

6.3.2 Functions of voltage-gated calcium channels

The concentrations of Na^+, K^+ and Cl^- ions both inside and outside the cell are quite high and so are not significantly changed by the flow of ions through open channels. In contrast, the concentration of Ca^{2+} in the cytosol is very low (see Table 6.1). As a consequence, it significantly increases when Ca_V channels open. This increase represents a biochemical signal, which is picked up inside the cell by calcium-binding proteins such as calmodulin and troponin. Therefore, calcium channels have a key role in translating changes in the membrane potential to changes in intracellular function. In muscle cells, the Ca^{2+} influx triggers contraction; this is explained in more detail in Section 6.9.4. In presynaptic nerve terminals cells, Ca^{2+} influx causes transmitter exocytosis.

In specialized pacemaker cells both in the brain and in the heart, Ca_V channels are also responsible for the generation of spontaneous periodic action potentials.

6.4 CHANNELS CONTROLLED BY INTRACELLULAR LIGANDS

Among ligand-gated channels, we can broadly distinguish channels that are controlled by extracellular ligands from those regulated by intracellular ligands. Extracellularly binding ligands are mostly neurotransmitters. The corresponding synaptic receptor channels or *ionotropic receptors* are discussed in Section 6.7.2. Intracellular ligands may be second messengers that relay extracellular signals, often also in synaptic transmission, or they may function in cellular autoregulation.

Cyclic nucleotide-gated (CNG) channels. Structurally, CNG channels have some similarity to K_V channels, with four subunits that each contain six transmembrane helices. Each subunit also contains one cyclic nucleotide binding site. Binding of cAMP or cGMP opens the channel for Na^+, K^+, and Ca^{2+}. On balance, this will cause partial membrane depolarization and also raise the intracellular calcium level.

The levels of cAMP and cGMP are affected by activation of many GPCRs and also by other signaling pathways such as insulin receptors and nitric oxide synthase. CNG channels therefore can translate extracellular chemical signals into changes of the membrane potential.

Some CNG channels are controlled both by the membrane potential and by cyclic nucleotide binding, thus defying the distinction between ligand-gated and voltage-gated channels. Such hyperpolarization-activated CNG channels, or HCN channels for short, participate in the generation of the rhythm in cardiac pacemaker cells. Mutations affecting these channels can cause cardiac arrhythmias.

Inward rectifier potassium (K_{ir}) channels. K_{ir} channels are also tetrameric but have a simpler structure than the voltage-gated channels, with only two transmembrane helices in every subunit.[3] When open, K_{ir} channels hyperpolarize the cell. One notable group in this family are the G protein-coupled inward rectifiers, or GIRK channels, which are activated by the $\beta\gamma$ dimers of some G proteins. Another important K_{ir} channel is the K_{ATP} channel. It is associated with the sulfonylurea receptor, which can bind two molecules of ATP and then proceeds to close the K_{ATP} channel.

In the β cells of the pancreatic islets, the sulfonylurea receptor is involved in the control of insulin release. In muscle cells, the K_{ATP} channel appears to serve a protective function—an exhausted cell that is running low on ATP will open the K_{ATP} channels. This will hyperpolarize the membrane and cause the cell to sit out some rounds of excitation in order to catch its breath.

6.5 TRANSIENT RECEPTOR POTENTIAL (TRP) CHANNELS

Transient receptor potential or TRP channels are cation channels that can be activated by very diverse stimuli, including mechanical force, heat, changes

[3]The name of this family of channels refers to the fact that they conduct K^+ ions more readily into the cell than out of it. This is brought about by intracellular ions such as Mg^{2+} and spermidine, which enter and plug the channel lumen from inside when the membrane potential is reversed.

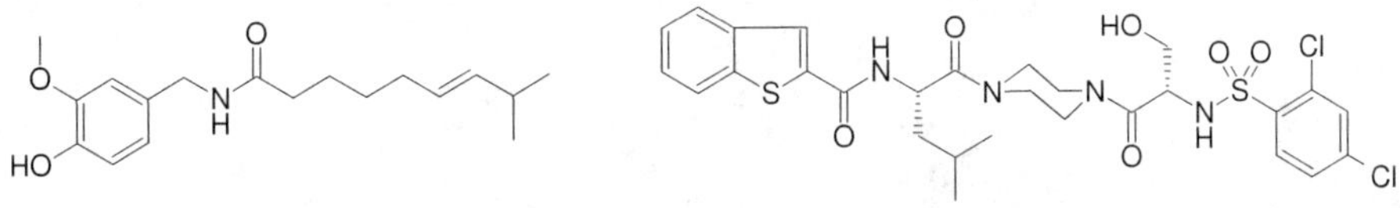

Capsaicin GSK1016790A

Figure 6.6 Agonists of transient receptor potential (TRP) channels. Capsaicin, which is contained in chili peppers, activates the TRPV1 channel. The experimental drug GSK1016790A [30] activates the mechanosensitive TRPV4 channel, which acts as a blood pressure sensor.

in pH, and various intra- and extracellular ligands. Individual channels can respond to more than one kind of stimulus. For example, the TRPV1 channel is activated by both heat and capsaicin (Figure 6.6), the active ingredient of hot pepper. While TRP channels are permeable to various cations, their functionally most significant conductivity is for Ca^{2+}.

TRP channels are also important in the perception of pain and other sensory qualities. Mechanosensitive TRP channels in the skin mediate the sense of touch, and similar channels in the hair cells of the inner ear detect sound waves. Mechanosensitive channels in the blood vessels mediate feedback regulation of blood pressure. An experimental agonist (see Figure 6.6) of the mechanosensitive TRPV4 channel, which is found in vascular endothelia, strongly reduces blood pressure in animals [30]. This effect is mediated by the influx of calcium and the subsequent activation of nitric oxide synthase (see Section 8.2).

6.6 VOLTAGE-GATED CHANNELS OF NERVE CELLS AS DRUG TARGETS

The channels and other ion transport proteins that we have covered so far control the excitability of single cells and allow us to understand some important mechanisms of drug action. We will consider these now before moving on to synaptic transmission.

6.6.1 Sodium channel blockers in local anesthesia

Blocking voltage-gated sodium channels is a straightforward way to suppress the propagation of action potentials. Total, systemic inhibition of Na_V channels will paralyze all nervous activity. This is the mode of action of lethal neurotoxins such as tetrodotoxin, which is found in pufferfish (Figure 6.7). Therapeutic Na_V channel inhibition must therefore be limited. In local anesthesia, Na_V channel blockade is restricted through local drug application.

The first Na_V channel-blocking drug to be used for local anesthesia was cocaine.[4] Because of its psychotropic and addictive effects (see Section 6.8.2),

[4]The discovery of this drug's anesthetic activity was facilitated by its liberal recreational use in Vienna in the late 19th century. Karl Köller, who is credited with this invention, recounts [20]: "On a certain occasion, another colleague, Dr. Engel, shared some cocaine with me, on the tip of his pocket-knife, and observed: 'It really numbs the tongue!' To which I replied: 'Yes, this has been noticed by everyone who has eaten it.' At that moment I realized that I was carrying in my pocket the local anesthetic I had been searching a few years ago."

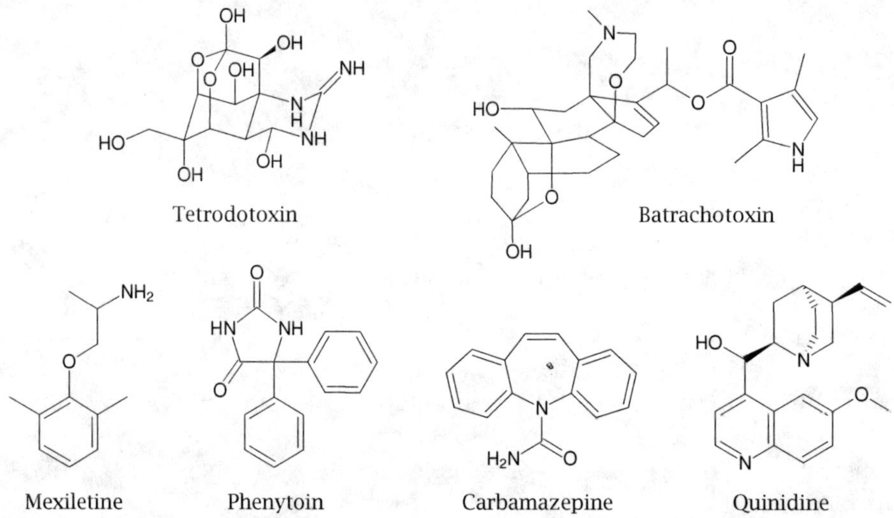

Figure 6.7 Drugs and poisons that act on voltage-gated sodium (Na_V) channels. Batrachotoxin interacts with the channel from within the lipid bilayer and activates it. It has no therapeutic use but is used in experimental studies on Na_V channels. All other molecules are Na_V channel blockers. Tetrodotoxin is the poison found in pufferfish. Mexiletine is a more metabolically stable analog of lidocaine; it is used in some forms of cardiac arrhythmia, as is quinidine. Phenytoin and carbamazepine are used to treat epilepsy and some other neurological diseases.

cocaine has been superseded as a local anesthetic by structurally similar compounds such as lidocaine (Figure 6.8). Local anesthetics affect sensory and motor nerve fibers alike, which causes the unpleasant facial paralysis that accompanies anesthesia in dental procedures.[5]

The channel block caused by lidocaine can be observed with single channel molecules. The drug reduces the conductivity of the open state, which is referred to as a *fast block*. In addition, it also slows down the reactivation of the channel, which is observed as a *slow block*. Interestingly, the two types of block can be assigned to separate moieties of the lidocaine molecule (Figure 6.8). The block by lidocaine is *use-dependent*, which means that the channels must be in the open state before the drug can access its binding site. Use dependence is observed with many channel-blocking drugs.

6.6.2 Voltage-gated channels and epilepsy

Epilepsy has many variations, but a common motif is a pathological increase in the spontaneous activity and excitability of groups of neurons. Several forms of epilepsy can be attributed to mutant ion channels. Pharmacological modulation of ion channels is widely used in these cases, as well as in other forms of epilepsy.

[5]While some peripheral nerves are purely sensory or motoric, the branches of the trigeminal nerve that are blocked in dental procedures contain both motor and sensory nerve fibers.

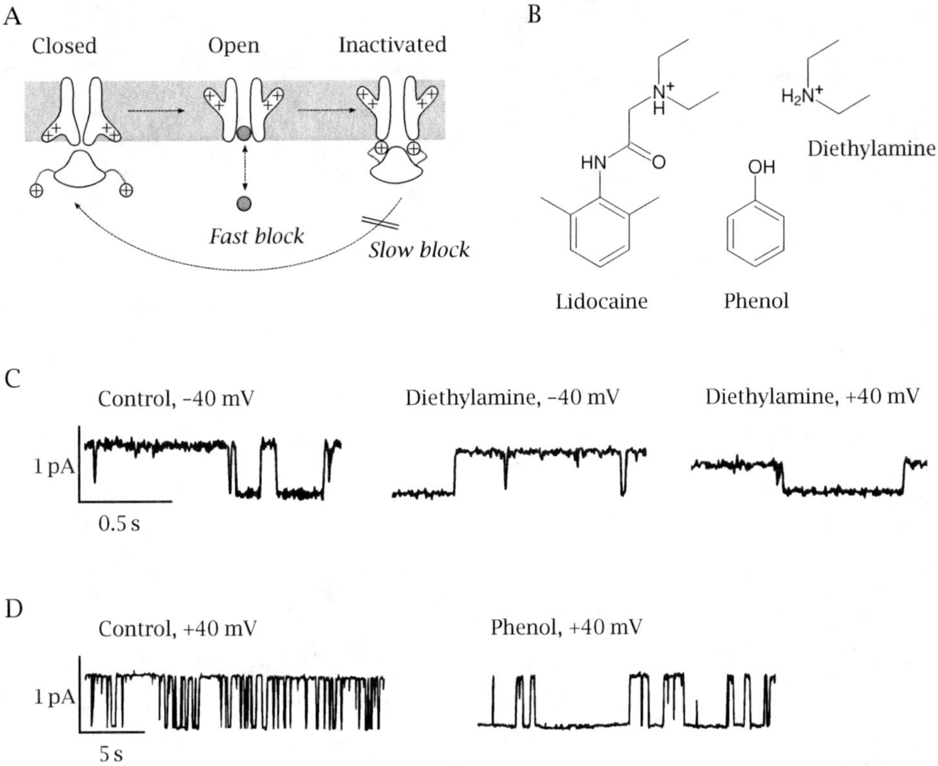

Figure 6.8 Fast block and slow block of Na_V channels. **A:** Fast block occurs when a drug reversibly binds within the channel lumen and obstructs it. Slow block is observed when a drug binds to the inactivated state of the channel and delays its reactivation. **B:** Structure of the sodium channel blocker lidocaine, and of diethylamine and phenol, which resemble two different moieties of the lidocaine molecule. **C:** Na_V channel conductance in the presence of diethylamine. The channels were activated with batrachotoxin. Diethylamine binds and dissociates faster than can be resolved by the instrument, resulting in an apparent decrease in the conductance of the open state. The extent of this decrease depends on the membrane potential. **D:** Na_V channel conductance in the presence of phenol (note the the timescale is different from that in C). Phenol causes extended intervals of channel inactivity but does not change the conductivity of the open state. (Panels C and D adapted, with permission, from [31].)

Inhibition of Na_V channels reduces neuronal excitability. Examples of Na_V inhibitors used in the treatment of epilepsy are phenytoin and carbamazepine. Unlike local anesthetics, they only cause slow channel block, that is, they delay channel reactivation. This results in a less pronounced inhibition of the channel. Both phenytoin and carbamazepine are also potent inducers of cytochrome P450 enzymes and thereby may interfere with the activity of other drugs (see Section 4.6).

While the activity of most nerve cells depends on upstream input, some groups of nerve cells in the brain generate intrinsic, spontaneous activity. This occurs in a manner similar to that of the pacemaker cells in the heart and involves two different types of Ca_V channels, the T-type and the L-type channels

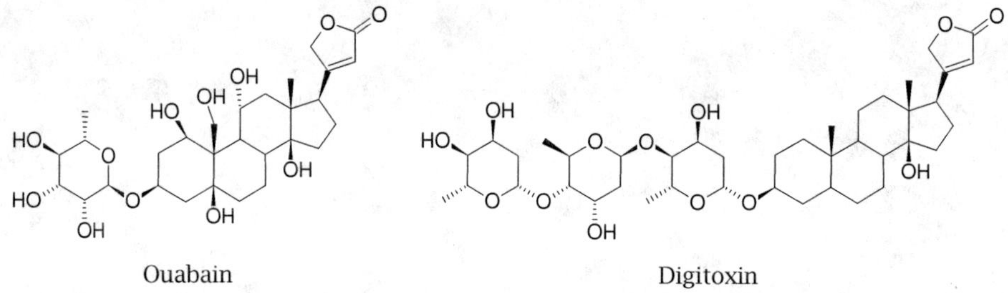

Figure 6.9 Structures of the Na^+/K^+-ATPase inhibitors ouabain and digitoxin.

(see Section 6.9.3). Mutations in T-type channel genes can cause a specific form of epilepsy known as *generalized absences.* This condition is treated with ethosuximide, which inhibits T-type channels [5].

In contrast to Na_V and Ca_V channels, K_V channels mediate membrane hyperpolarization, so that drugs that promote the open state of these channels will reduce excitability. This therapeutic approach has only recently been put into practice. The drug retigabine (Figure 6.11) promotes opening of KCNQ or type 7 K_V channels, which occur in the CNS. It is currently in clinical trials.

6.6.3 Ion channels and transporters in cardiovascular disease

Drugs that modulate cardiac ion channels and transporters can be used in different pathological conditions. In cardiac insufficiency, the contractile strength of the heart muscle is weakened. In the various forms of arrhythmia, the heartbeat is too fast, too slow, or irregular, which interferes with proper heart function.

The most common cause of cardiac insufficiency is *ischemic* heart disease. Here, arteriosclerosis compromises the perfusion of the heart muscle, which leads to tissue damage (see Section 10.4). Cardiac insufficiency is often made worse by arterial hypertension, which forces the heart to pump against increased resistance.

Cardiac arrhythmias also often arise due to ischemic heart disease. Certain forms of arrhythmias are due to mutations of ion channels.

***Digitalis* glycosides.** In cardiac insufficiency, the contractile strength of the heart muscle cells can be increased with ouabain and the *Digitalis* glycosides digoxin and digitoxin (Figure 6.9). As discussed in Section 6.2, these drugs inhibit Na^+/K^+-ATPase. The increase of the intracellular Na^+ concentration reduces the effectiveness of Ca^{2+} export from the cytosol via Na^+ antiport and therefore raises the cytosolic calcium concentration. Since calcium controls the interaction of actin and myosin, the functional effect is an increased strength of heart muscle contraction.

Additional signaling mechanisms have been described for *Digitalis* glycosides applied as drugs as well as for endogenous *Digitalis*-like steroids [2], but a clear picture of their physiological significance has not yet emerged.

Calcium channel blockers. L-type and T-type calcium channels cooperate in generating the heart rhythm within the cells of the cardiac excitation–conduction system. L-type channels also mediate excitation–conduction coupling in the regular heart muscle cells, which we will refer to in this chapter as the *worker* cells. The roles of Ca_V and other ion channels in heart excitability are explained in more detail in the appendix to this chapter (Section 6.9.3).

L-type channel blockers such as verapamil and nifedipine diminish the strength of the heart muscle contraction. They also slow down the heartbeat, and in some situations restore a regular rhythm to an irregularly beating heart. All of these effects improve the metabolic situation of heart tissue that is trying to get by on limited oxygen supply.

L-type channel blockers also induce relaxation of vascular smooth muscle cells, thereby lowering flow resistance and blood pressure. This reduces the workload and further improves the metabolic situation of a diseased heart. Note, however, that the effect of calcium channel blockers on heart contractility is opposite to that of *Digitalis* glycosides. In advanced cardiac insufficiency, this can be counterproductive. To avoid the reduction of contractile strength, we can look to inhibitors of T-type channels rather than L channels, since they will act preferentially on the excitation–conduction system and spare the regular "worker" muscle cells.

The drug mibefradil inhibits both T and L channels. Its effect on L channels has been ascribed to an active metabolite, and an experimental derivative that does not produce this metabolite and reportedly is selective for T-type channels has been synthesized [9].

Figure 6.10 shows the structures of some representative calcium channel-blocking drug molecules.

K_{ir} channel agonists. Cardiac G protein-coupled inward rectifier (GIRK) channels are activated downstream of muscarinic acetylcholine receptors. Muscarinic antagonists such as atropine or ipratropium bromide (Section 6.8.3) reduce the K^+ permeability, which will promote membrane depolarization and can be useful for accelerating an abnormally slow heartbeat.

K_{ir} channels are also found in association with the sulfonylurea receptor. When activated, the sulfonylurea receptor causes the channel to close. The receptor antagonist diazoxide opens the channel and therefore dampens excitability. The same effect also occurs in smooth muscle cells in the vasculature, where reduced contractility relaxes the blood vessels and lowers the blood pressure. Minoxidil sulfate (Figure 6.11), which is formed by enzymatic sulfation of the prodrug minoxidil, has the same mode of action.[6] Agonists of the sulfonylurea receptor such as tolbutamide are used in type 2 diabetes, where they increase the secretion of insulin in pancreatic islet β cells (see Section 10.3.3). Diazoxide and minoxidil, as a side effect, reduce insulin secretion and therefore can worsen glucose control in overtly or latently diabetic patients. The drug iptakalim has been reported to open vascular K_{ir} channel subtypes but close the ones in β cells [16], which would offer an attractive therapeutic option for the fairly common combination of type 2 diabetes and hypertension.

[6] In addition, minoxidil also stimulates hair growth. This side effect seems to appeal to some, but not all patients; research into the causes of this variation in preference continues.

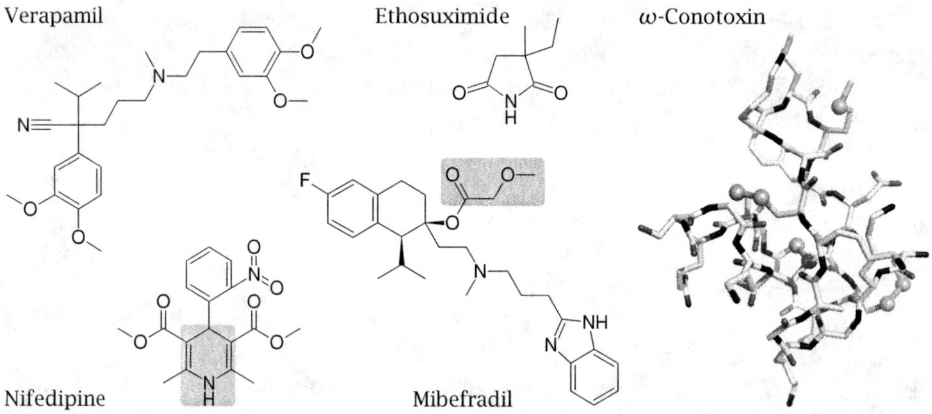

Figure 6.10 Inhibitors of voltage-gated calcium channels. Nifedipine and verapamil both inhibit L-type Ca_V channels. These channels are also called *dihydropyridine receptors* (DHPRs), after the dihydropyridine group that occurs in some inhibitors; it is highlighted in the structure of nifedipine. Mibefradil inhibits both L-type and T-type channels. Its effect on DHPR channels has been ascribed to an active metabolite, in which the highlighted substituent is hydrolytically removed. Replacement of this group with a cyclopropyl moiety produces a selective T-type channel blocker [9]. Ethosuximide selectively blocks T type channels. The cone snail ω-conotoxins selectively block N type channels. The figure shows ω-conotoxin CVID [1]. Sulfurs are rendered as spheres to highlight disulfide bonds.

Figure 6.11 shows the molecular structures of some representative potassium channel blockers and openers.

Drugs that act on other cardiac ion channels. All channels that shape the action potential in either the excitation–conduction system or the worker muscle cells are potentially involved in arrhythmias and relevant as drug targets. The malfunctioning cells that produce aberrant rhythms can either be located in the excitation–conduction system, or they can be among the "worker" muscle cells. In the latter case, the excitability of the workers can be selectively reduced using inhibitors of sodium channels, since the latter have no prominent role in the excitation–conduction system. Na_V channel blockers so used include mexiletine, lidocaine and quinidine.

Among the various types of K_V channels, the hERG channel has a prominent role in the termination of action potentials in the heart.[7] This channel is interesting mainly not as a drug target but as an *antitarget*. It is inhibited by numerous drugs from different classes, and such inhibition can cause potentially severe arrhythmias. Therefore, the effect on hERG channels is now routinely examined during the development of new drugs.

Hyperpolarization-activated cyclic nucleotide-gated (HCN) channels participate in generating the rhythm and have a net depolarizing effect on the membrane potential. The drug ivabradine blocks HCN channels, which slows down

[7]The acronym stands for "human ether-à-go-go-related gene". Like several other cryptic and vaguely outrageous monikers, this one has been invented by *Drosophila* geneticists.

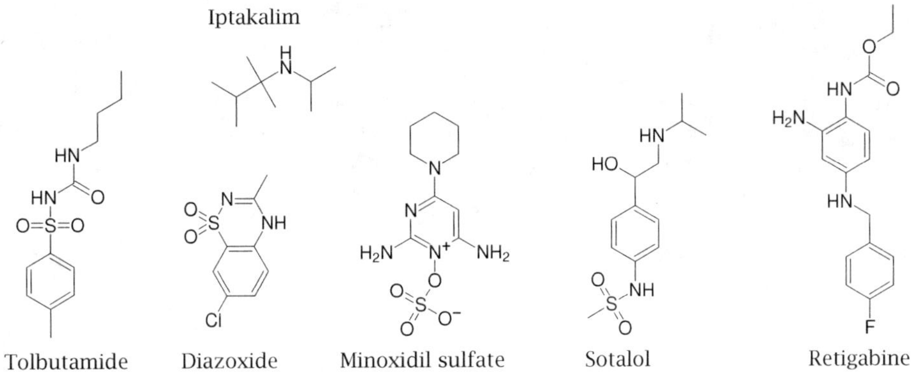

Figure 6.11 Drugs that interact with potassium channels. Tolbutamide is a sulfonylurea receptor agonist and closes the K_{ir} channel associated with that receptor, which raises insulin secretion from pancreatic β cells. Diazoxide and minoxidil sulfate are sulfonylurea receptor antagonists. They open K_{ir} channels on vascular smooth muscle cells, inducing relaxation. Iptakalim reportedly opens K_{ir} channels in the vasculature but closes those in β cells [16]. Sotalol inhibits K_V channels, which delays repolarization and prolongs the action potential. It also blocks β-adrenergic receptors. Retigabine activates KCNQ channels, which belong to the K_V family and occur in neurons. It reduces neuronal excitability and is used to treat epilepsy.

the heartbeat and is useful in some forms of arrhythmia. Antagonists of β-adrenenergic receptors reduce HCN channel activity by lowering the intracellular level of cAMP. This is one of several effector mechanisms of these drugs.

6.7 SYNAPTIC TRANSMISSION

When an action potential arrives at the presynaptic nerve ending, the influx of calcium through N-type Ca_V channels triggers the exocytosis of neurotransmitter vesicles. The released transmitter activates postsynaptic receptors. Depending on the type of the receptor, this will either promote or inhibit excitation of the postsynaptic cell.

6.7.1 Excitatory and inhibitory postsynaptic potentials

In the motor endplate, each action potential in the α-motoneuron will trigger an all-out action potential in the postsynaptic muscle cell. This makes sense, since the muscle cell is controlled by a single neuron. In contrast to muscle cells, however, most nerve cells communicate with numerous upstream cells, and firing an all-out action potential in response to excitatory input from any one of them would not be sensible. Instead, in excitatory synapses between nerve cells, a single presynaptic action potential will cause a postsynaptic depolarization that remains localized and below the firing level; this is called an *excitatory postsynaptic potential* or EPSP. The firing level may be reached if the presynaptic cells fires repeatedly before the earlier EPSPs have dissipated, so that the depolarization occurs in a cumulative fashion; this effect is called *temporal summation.* Alternatively, the firing level may be reached through spatial

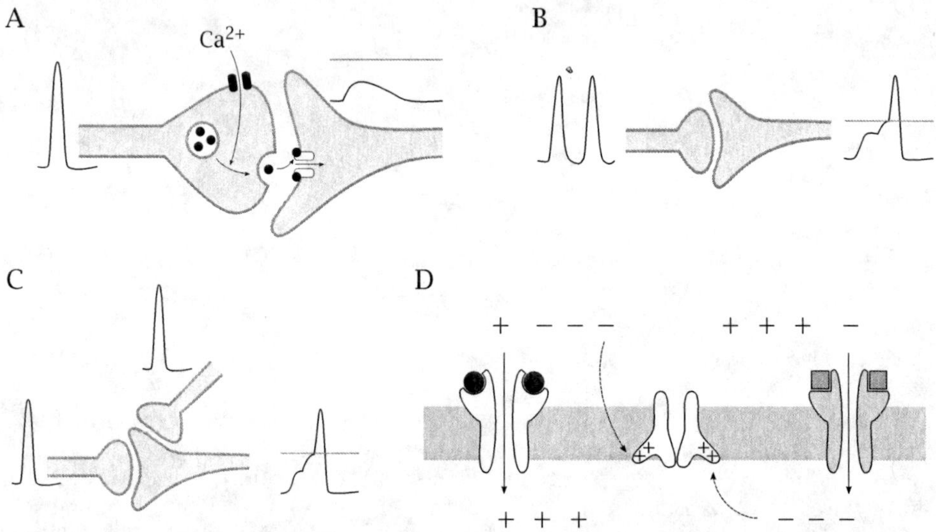

Figure 6.12 Summation of postsynaptic potentials. **A:** In excitatory synapses between nerve cells, a single presynaptic action potential causes a localized partial depolarization, the excitatory postsynaptic potential (EPSP). A single EPSP will remain below the firing level and not trigger a postsynaptic action potential. **B:** An action potential may form if a single synapse fires repeatedly before the EPSP dissipates; this is known as *temporal summation.* **C:** Alternatively, in *spatial summation*, the EPSPs caused by neighboring synapses firing at the same time may jointly trigger an action potential. **D:** Inhibitory synapses, which often employ chloride channels, cause inhibitory postsynaptic potentials (IPSPs), which interfere with the activation of voltage-gated channels by EPSPs in the vicinity.

summation when several vicinal nerve terminals fire simultaneously (Figure 6.12).

The activity of excitatory synapses is balanced by inhibitory ones. Major transmitter receptors in inhibitory synapses are the chloride channels that are activated by glycine or GABA. When activated, they cause a localized inhibitory postsynaptic potential or IPSP, which will detract from EPSPs that occur in the vicinity. The fundamental role of these inhibitory synapses is evident from the dramatic consequences, such as tetanus and epileptic seizures, that result from their failure.

6.7.2 Neurotransmitter receptors

Neurotransmitter receptors fall into two major categories: G protein-coupled receptors (GPCRs) and ionotropic receptors. GPCRs do not conduct ions directly but instead control various intracellular signaling pathways (see Chapter 5). Cognate GPCRs exist for a large number of neurotransmitters, including amino acids, catecholamines, and peptides. Most transmitters have multiple GPCR variants that couple to different intracellular cascades. As outlined in Section 6.4, some of these pathways produce intracellular messengers that in turn control ligand-gated channels, which then indirectly affect the membrane potential.

Ionotropic receptors are ligand-gated channels that are activated directly by their cognate neurotransmitters via extracellular binding sites. Depending on their ion selectivity, they either depolarize or hyperpolarize the membrane. Ionotropic receptors belong to several major structural families: (1) the nicotinic acetylcholine receptor and related receptors in the cys-loop family, (2) the ionotropic glutamate receptors, and (3) the purine P2X receptors that bind ADP or ATP.

The subunits of all these ionotropic receptors are encoded by multiple genes, and they assemble in various combinations and proportions. Specific subunit combinations often occur in specific physiological roles and locations, and they sometimes can be targeted selectively by suitable drugs.

With the exception of glycine, all transmitters that activate ionotropic receptors also have cognate receptors in the GPCR family. Often, the two receptor types occur within the same synapses.

6.7.3 The cys-loop receptor family

This family of ionotropic receptors includes the nicotinic acetylcholine receptor (NAR) and the 5-HT_3 serotonin receptor, which are excitatory cation channels, as well as the $GABA_A$ and glycine receptors, which are chloride channels and therefore inhibitory. Among these receptors, the NAR has been most widely studied and serves as a model for the structure and function for the entire family. These studies showcase some interesting biochemical methods, and we will consider them in some detail.

Structure of the NAR. The preferment of the NAR for biochemical studies is due to a very practical reason: availability. The receptor can be isolated from the electric organs of electric eel or electric ray like *Torpedo californica*, which use strong electric discharges to incapacitate their prey or drive away predators. These discharges are triggered through cholinergic synapses, in which the receptor is highly abundant. Vesicles obtained from postsynaptic membranes through homogenization and centrifugation contain the receptor in remarkable density and purity [3].

The structure of NAR in postsynaptic membrane vesicles has been studied in detail by Unwin and coworkers using electron crystallography. In this technique, a beam of electrons—and not X rays, which are normally used in protein crystallography—is diffracted by an ensemble of protein molecules that are arranged in a regular two-dimensional lattice, such as that formed spontaneously by the NAR in the aforementioned fish electric organ vesicles. From diffraction patterns obtained at different angles of the incident electron beam, a three-dimensional electron density map can be constructed. This is illustrated in Figure 6.13. The resolution of this map is lower than that typically attainable in X-ray crystallography, but it still can guide the construction of a detailed model of the molecular structure.

Remarkably, electron crystallography has provided structural models of NAR in both the open and in the closed states. Considering that the lifetime of the open state is only a few milliseconds, how could one possibly observe it in a crystal?

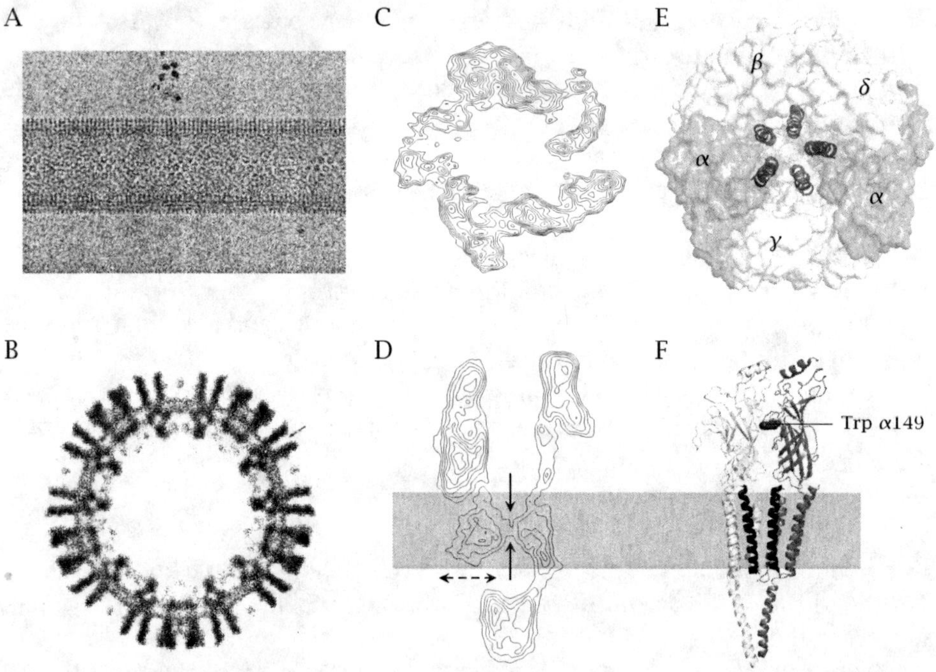

Figure 6.13 Structure of the nicotinic acetylcholine receptor, as elucidated by electron crystallography. **A,B:** Electron micrographs of tubular synaptic membrane particles isolated from *Torpedo* electric organs. The cross section in **B** shows the major structural features of the receptor. The formation of dense, regularly packed arrays as seen in **A** permits the construction of higher resolution contour maps. **C,D:** Contour maps obtained by electron crystallography. In the side view, a constriction within the bilayer is clearly visible. The receptor has an intracellular domain with lateral openings. **E:** High resolution structural model, top view. The receptor consists of five subunits, each of which contributes one helix to the gate located at the level of the lipid bilayer. **F:** Side view of the structural model. Only one of the α chains and the adjacent γ chain are shown. The tryptophan residue at position 149 of the α chain is part of one of the two acetylcholine binding sites, which are located at the α–γ and the α–δ interfaces, respectively. (**A–D** courtesy of Nigel Unwin. **E** and **F** rendered from 2bg9.pdb [28]. A colored version of this figure is shown in the insert in the middle of this book.)

The solution to this problem is as simple as it is ingenious. The NAR receptor molecules in the postsynaptic membranes, despite being arrayed in two-dimensional crystals, are nevertheless alive; that is, they can bind acetylcholine and then change to the open state. Therefore, instead of crystallizing the open state, which would be impossible, Unwin took those preexisting crystals and converted them to the open state [27]. The challenge was to open all receptor molecules synchronously, and then arrest them in the open state during its very short lifetime. This he accomplished with the crafty apparatus illustrated in Figure 6.14.

Figure 6.13 shows raw electron micrographs and the derived contour maps and molecular models of the NAR, in top and side views. In the top view, the identities of the individual subunits are also indicated; these have been

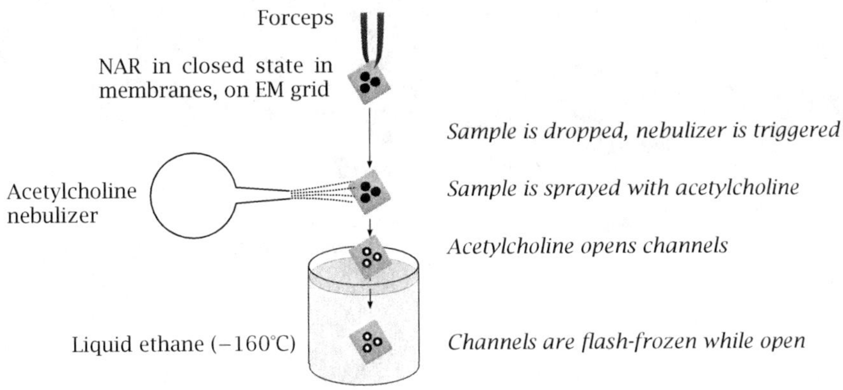

Figure 6.14 Experimental setup for trapping the nicotinic acetylcholine receptor in the open state for electron crystallography studies. The membrane vesicles containing the receptors, in 2D crystal arrangement and in the closed state, are mounted on an EM sample support and held in a forceps. When the sample is dropped, it passes through a stream of vaporized acetylcholine. Binding of acetylcholine opens the channels, which immediately thereafter plunge into cold, liquid ethane and are shock-frozen in the open state. All further sample processing must occur at low temperatures so as to preserve the open state for observation.

determined using samples labeled with antibody fragments specific for each subunit. The subunit composition of the NAR varies between different tissues. The $\alpha_2\beta\gamma\delta$ pattern shown here occurs in motor endplates and in fish electric organs, both of which are derived from the mesoderm during embryonic development. Neuronal NAR, including the one found in the autonomic ganglia, is $\alpha_3\beta_2$ or $\alpha_2\beta_3$. This structural difference allows for the selective action of several agonists and antagonists on muscle or the ganglia, respectively (see Section 6.8.3).

In the side view, a significant part of the receptor protrudes from the membrane into the extracellular space. The central lumen is fairly wide in this extracellular domain but narrower at the level of the lipid bilayer. This bottleneck functions as the gate that opens in response to acetylcholine binding. It is lined by a ring of five α helices, one from each of the receptor's five subunits. In the functionally closed state, the lumen between these helices is appreciable but smaller than the radius of hydrated ions, which therefore cannot pass across. Binding of acetylcholine causes some of these helices to twist, which breaks the bonds between them and opens the gate.

As with voltage-gated channels, the gate itself does not impose any selectivity for specific ions. Instead, ion selectivity is associated with a separate functional domain. Mutagenesis experiments with the homologous 5-HT_3 serotonin receptor have mapped the selectivity filter to the lateral openings in the cytoplasmic domain of the receptor [13]. However, the selectivity is still not very strict, and the NAR has appreciable conductivity for Na^+, K^+, and Ca^{2+}.

Location of the NAR acetylcholine binding site. The acetylcholine binding site has been characterized by photoaffinity labeling. In this method, the receptor is incubated with a chemically reactive analog of the ligand. The receptor

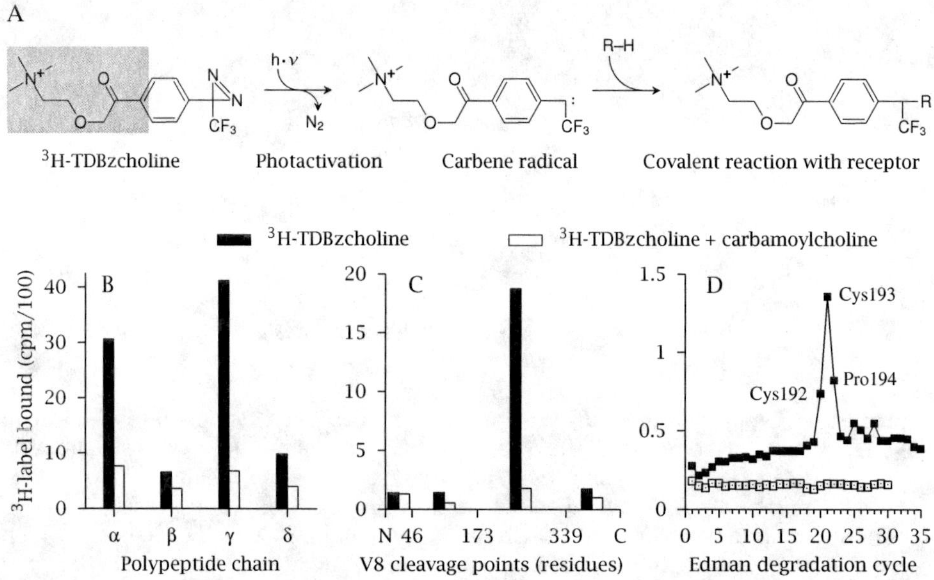

Figure 6.15 Mapping of the acetylcholine binding site in the nicotinic acetylcholine receptor by affinity labeling. **A:** Structure and reactivity of the photoactivatable and radiolabeled ligand ^{3}H-TDBzcholine. The acetylcholine moiety within the molecule is highlighted. Activation with UV light releases nitrogen and produces a highly reactive carbene radical, which immediately reacts with amino acid residues in the vicinity. **B–D:** Analysis of ^{3}H-label incorporation into NAR. Samples represented by black symbols were prepared with ^{3}H-TDBzcholine only. Control samples additionally contained unlabeled carbamoylcholine in excess to dislodge ^{3}H-TDBzcholine from the binding site. **B:** Incorporation of ^{3}H-TDBzcholine into the four different subunit types. Most of the carbamoylcholine-displaceable label is bound to the α and the γ chains. **C:** Incorporation of the label into proteolytic fragments of the α chain. The purified α chain was digested with staphylococcal protease V8, and fragments were separated by HPLC and identified by N-terminal sequencing. Most of the label is found in a 20-kDa fragment of the chain that spans residues 174–339. **D:** Edman degradation of the labeled 20 kDa fragment. Residues 192–194 contain the most incorporated label, indicating that they are part of the acetylcholine binding site. (Figure redrawn from original data in Ref. [4].)

protein is then fragmented, and the amino acid residues that have covalently reacted with the ligand analog are identified. Numerous photoaffinity labeling studies have been performed with the NAR; Figure 6.15 illustrates one of them [4]. In this study, both the α and γ chains are labeled intensely, indicating that the binding site is located at the interface of the two chains. Similar studies with different affinity labels indicate a second binding site between the second α chain and the δ chain.

Since the acetylcholine binding site is located between subunits, differences in the subunit composition between NARs in different cell types imply that the binding sites should be different, too. Indeed, NAR agonists and antagonists are often specific for certain receptor subtypes.

The nature of the NAR–acetylcholine interaction. Acetylcholine has a positive charge, which is shared by virtually all NAR agonists and antagonists, most commonly in the form of tertiary or quaternary amino groups. What role does this positive charge play in binding to the receptor? The acetylcholine binding site at the α–γ subunit interface is not rich in negative charges. It does, however, contain multiple tryptophan side chains, suggesting a role for cation–π interactions. In this type of molecular interaction, the fixed charge of a cation is accommodated by the mobile π electron cloud of an aromatic side chain.

The importance of cation–π interactions in acetylcholine binding has been experimentally demonstrated. The codons of tryptophan residues located in the binding site in the cloned α chain were mutated to *amber* stop codons (TAA). The mutant messenger RNAs were obtained by in vitro transcription and injected into frog oocytes, together with suppressor-tRNAs that had been coupled to fluorinated tryptophan residues. This resulted in the expression of receptor molecules carrying fluorinated tryptophan residues at those strategic positions.

Fluorine is strongly electronegative and therefore withdraws π electron density from the aromatic ring of tryptophan, which will attenuate cation–π interactions and should thereby reduce the affinity of the receptor for acetylcholine. This was indeed observed [33].

Receptor desensitization. When motor endplates are continuously exposed to acetylcholine or cholinergic agonists, their response will decline within seconds (Figure 6.16A). This is due to NAR desensitization, which can be understood in terms of a model of channel function resembling the one we have seen before with voltage-gated channels. The receptor can cycle between three functional states. Upon acetylcholine binding, the resting state converts to the open state. The open state then converts to the inactivated state, with the ligand still bound. The receptor reactivates only after acetylcholine dissociates (Figure 6.16B).

In the absence of acetylcholine, the resting state is the most thermodynamically stable one, so most receptors will be in this state. Ligand binding favors both the open state and the inactivated state over the resting state. Opening is faster than inactivation, but the inactivated state is more stable. Therefore, at saturating ligand concentration, most receptor molecules will initially open but then accumulate in the inactivated state.

The GABA$_A$ receptor. This receptor is a chloride channel and therefore causes inhibitory postsynaptic potentials. It has a key role in inhibitory transmission in the brain. Blockade of this receptor with antagonists such as pentylenetetrazole and bicuculline causes epileptic seizures. Such drugs have been used in the past as stimulants but are now used only for experimental studies. Inhibition of the GABA$_A$ receptor can also occur as a side effect of other drugs, for example, very high dosages of penicillin, which again can result in seizures.

GABA$_A$ receptor agonists such as benzodiazepines and barbiturates are useful in the treatment of epilepsy but also have numerous other applications, such as sedation and narcosis. The GABA$_A$ receptor is also the target of gaseous and soluble general narcotics. Different families of drugs bind to separate sites on the receptor molecule. These different binding sites seem to vary

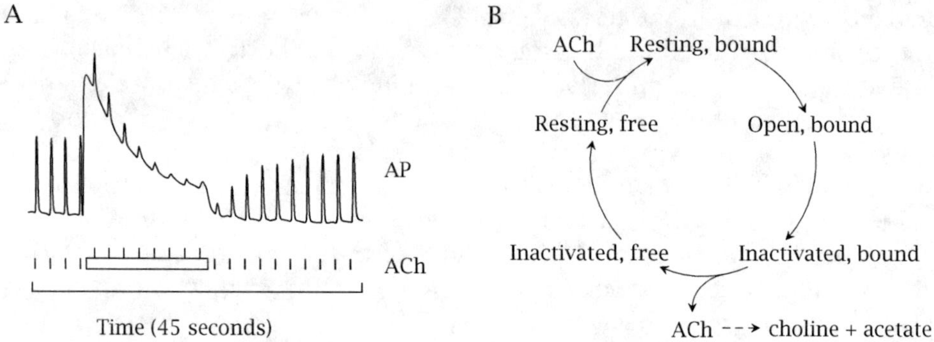

Figure 6.16 Desensitization of the nicotinic acetylcholine receptor. **A:** Experimental observation. Acetylcholine (ACh) was electrophoretically applied to frog motor endplates, and the resulting postsynaptic action potentials (AP) were recorded. Repetitive short ACh stimuli evoke action potentials of uniform amplitude. When continuous application starts, a strong depolarization occurs, which then declines even in the face of continued ACh application. At the same time, the response to superimposed short ACh pulses also declines. After continuous ACh application is stopped, the response to short ACh pulses recovers over a period of several seconds. (Drawn after an original figure in Ref. [11].) **B:** Kinetic scheme for acetylcholine binding, receptor activation, and desensitization. When ACh binds, the receptor initially opens but then inactivates. Opening is faster than inactivation, but the inactive conformation is more stable, so that after sufficient time most receptors accumulate in this state. They revert to the resting, responsive state only after ACh dissociates and is cleaved to choline and acetate by cholinesterase.

in the maximal extent of activation that can be achieved through them; for example, benzodiazepines achieve a lower degree of receptor activation, and are therefore safer for use as tranquilizers or hypnotics than barbiturates.

Specific subtypes of the $GABA_A$ receptors are also the targets of ethanol and of its experimental competitive antagonist Ro15-4513 [29], which is a benzodiazepine derivative. While Ro15-4513 blocks the typical behavioral effects of ethanol, it fails to prevent the lethal effects that arise at higher ethanol dosages [18], suggesting that at high concentrations ethanol affects a broader spectrum of $GABA_A$ receptors and possibly other receptors as well. Similarly, the effects of narcotic gases are only partially inhibited by $GABA_A$-specific receptor antagonists [32].

Glycine receptors. The second major inhibitory ionotropic channel is the glycine receptor, which is also a chloride channel. On α-motoneurons, the inhibitory action of glycine receptors is essential to balance the excitatory action of glutamate receptors.

5-HT$_3$ receptors. The 5-HT$_3$ receptor is the only ionotropic receptor for serotonin; all other serotonin receptors belong to the GPCR family. Like the NAR, the 5-HT$_3$ receptor is a cation channel. In both the intestinal tract and the CNS, the 5-HT$_3$ receptor mediates emetic stimuli, and specific 5-HT$_3$ inhibitors such as ondansetron have proven very successful in suppressing the severe emesis that often accompanies antitumor chemotherapy.

6.7.4 Ionotropic glutamate receptors

Glutamate is the most abundant neurotransmitter in the brain, with both GPCRs and ionotropic receptors. Ionotropic glutamate receptors have four subunits each and form a distinct structural family. They are classified into three major groups according to their response to the synthetic glutamate analogs NMDA, AMPA, and kainate (see Figure 15.1). All of these pass Na^+ ions and therefore trigger excitatory postsynaptic potentials. NMDA receptors and some AMPA receptors also pass Ca^{2+} ions, which triggers additional signals in the postsynaptic cell. The receptor types overlap in their distribution, and many synapses contain combinations of various types.

The function of AMPA and kainate receptors in evoking excitatory postsynaptic potentials is quite straightforward. In contrast, the NMDA receptor has some unique properties that give it a more complex role. Binding of glutamate alone is not sufficient to activate this receptor. Instead, a second site, located on a specific receptor subunit, binds glycine or D-serine and also needs to be occupied in order to activate the receptor. In addition, the receptor responds only if the postsynaptic membrane is already partially depolarized by EPSPs that were triggered via other receptors, for example AMPA receptors. This means that the NMDA receptor detects sustained stimulation with two different transmitters. Through Ca^{2+} influx, it mediates effects in the postsynaptic cell that are referred to as *long-term potentiation* and *long-term depression*, respectively, and are important in memory and learning.

6.7.5 Purine P2X receptors

These receptors are activated by ATP. They form a distinct structural family with trimeric subunit stoichiometry. P2X receptors are cation channels and have a role in pain perception and conduction. Within the CNS, the extracellular ATP that stimulates the receptor is released by exocytosis, like other transmitters. In the periphery, the ATP may be released by cell damage, which will then activate pain receptors. P2X receptor antagonists are still at the experimental stage.

6.7.6 Mechanisms of drug action in synapses

In a chemical synapse, the information that in both participating cells is propagated as an electrical signal is transiently recoded as a chemical signal. Therefore, synapses are a logical target for therapeutic intervention with chemicals. The various steps in synaptic transmission that can be targeted by drug action are summarized in Figure 6.17:

1. Direct agonists and antagonists bind to the postsynaptic receptors and activate or block them, respectively.

2. In many synapses, there are both post- and presynaptic receptors for the transmitter. The presynaptic receptors are usually inhibitory, so that the transmitter exerts a negative feedback on its own release. Where the presynaptic receptors are of a different type than the postsynaptic ones, agonists

selective for the presynaptic receptor will ultimately reduce the stimulation of postsynaptic receptors.

3. The inactivation of released transmitter can be inhibited. Acetylcholine and some peptide transmitters are inactivated by hydrolytic enzymes, and inhibitors for these enzymes exist. The other major mechanism of inactivation is cellular reuptake into the presynaptic neuron or into surrounding glia cells. Inhibition of reuptake is a common mode of drug action with dopamine, norepinephrine and serotonin.

4. Within the presynaptic neuron, synthesis, breakdown, or vesicular accumulation of the transmitters may be inhibited.

5. Negative feedback from the postsynaptic cell on the presynaptic cell is referred to as *retrograde signaling*. The most important retrograde mediators are the endocannabinoids anandamide and 2-arachidonylglycerol, which stimulate the CB_1 receptor. This receptor is also stimulated by *Cannabis* alkaloids (Section 9.8).

6. False transmitters inhibit synaptic transmission primarily by displacing the regular transmitter from vesicular storage. They are released in response to presynaptic action potentials like the regular ones; they fail to stimulate the postsynaptic receptor and may in addition mediate presynaptic negative feedback. Examples are α-methyldopa and guanethidine, which displace norepinephrine.

Drugs that act on targets other than neurotransmitter receptors are sometimes referred to as *indirect agonists* or *antagonists*, respectively. In the following sections, we will see examples of several indirect mechanisms. However, it should be noted that direct agonists and antagonists are more commonly used in practice. The reason for this is that neurotransmitter receptors are more differentiated and diverse than most other targets and therefore allow more specific action.

6.8 PHARMACOLOGY OF INDIVIDUAL TRANSMITTERS

Individual neurotransmitters typically have multiple types of receptors. These may occur in different locations within the nervous system, and sometimes also on other cell types. These different receptor types often are associated with different physiological functions, and at the same time they often can be selectively addressed. For example, histamine occurs both in the brain, where it mediates arousal, and in the stomach, where it mediates the secretion of hydrochloric acid. The major histamine receptors in the brain are of the H_1 type, whereas H_2 receptors are found in the stomach. Therefore, selective H_2 antagonists can be used to block gastric acid secretion without side effects on the CNS.

The blood–brain barrier affords another means for drug selectivity. In our example, H_1 receptors are found not only in synapses in the brain but also on vascular endothelia and other cells in the periphery, where they mediate type I hypersensitivity reactions such as hay fever. Early H_1 receptor antagonists were effective in suppressing these allergic symptoms but also caused seda-

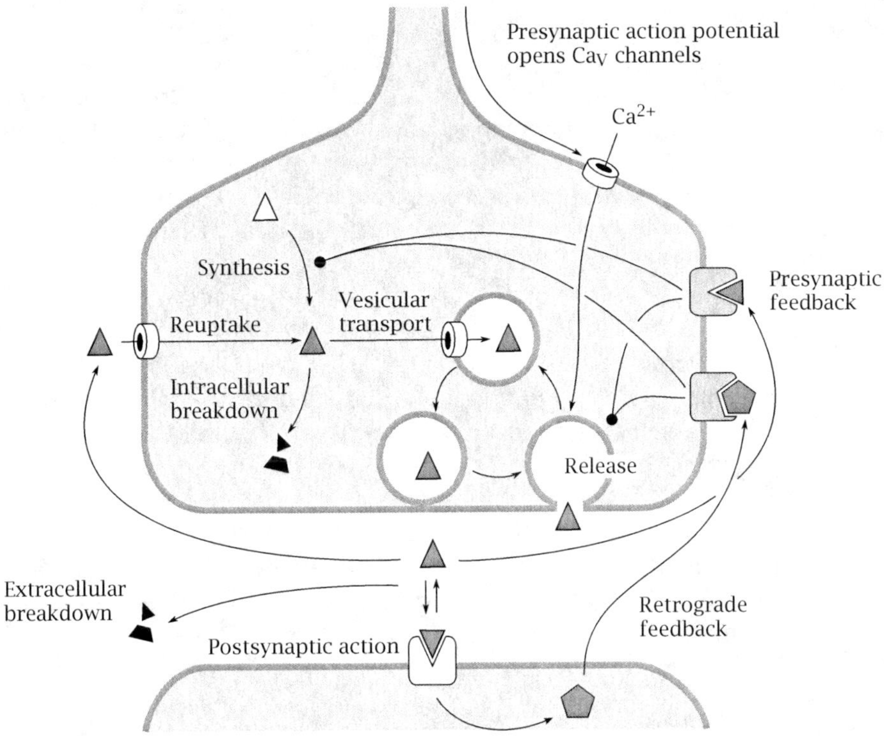

Figure 6.17 Function of a chemical synapse. Exocytosis of transmitter is triggered by N-type Ca_V channels. Released transmitter binds and activates postsynaptic receptors. In response, the postsynaptic cell may release mediators that exert retrograde negative feedback on the presynaptic cell. The transmitter itself can cause negative presynaptic feedback, through receptors that may differ from those on the postsynaptic membrane. The transmitter is either broken down extracellularly or scavenged by presynaptic reuptake. Transmitter vesicles are recycled and loaded with transmitter through active transport. The pool of transmitter in the presynaptic cell is controlled by the rates of synthesis, degradation, and reuptake. All of these processes can be influenced by drug therapy.

tion, which was due to their effect on the H_1 receptors in the CNS. Modern H_1 blockers do not cross the blood–brain barrier, thus avoiding sedation.

6.8.1 Amino acids

Glutamate is the most important excitatory transmitter in the brain, and its derivative γ-aminobutyric acid (GABA) is the major inhibitory one. Both transmitters have receptors in the ionotropic channel and the GPCR families. Glycine is another inhibitory transmitter with a cognate ionotropic receptor. In addition, glycine also acts as a cotransmitter on the NMDA-type glutamate receptor, as does D-serine.

Following their release from presynaptic cells, amino acid transmitters are inactivated by reuptake into presynaptic cells or into glia cells, where they are metabolically degraded.

Glutamate. The fundamental role of glutamate receptors in the brain has motivated numerous drug development efforts, in the context of virtually every conceivable neurological or psychiatric disorder. However, the yield of clinically successful drugs is not commensurate with these efforts [7]. The NMDA receptor antagonists ketamine and phencyclidine induce psychotic symptoms, and at higher dosages cause *dissociative anesthesia*, a state of mind in which awareness is reduced and pain perception subdued; these drugs can be used in minor surgery. Cycloserine, an antibiotic drug that is used in the treatment of tuberculosis, is also a partial agonist at the glycine/D-serine site of the NMDA receptor. It reportedly provides some benefit in schizophrenia. Drugs with the same mode of action but higher potency and efficacy are under investigation.

Inhibition of kainate and AMPA receptors is of interest for the treatment of *excitotoxicity*. This term refers to the compounding role of glutamate release in hypoxic damage to nerve cells. Disruption of the oxygen supply stalls ATP production and Na^+/K^+-ATPase. This leads to membrane depolarization and triggers action potentials, which then cause glutamate release and more cell depolarization. The result is a vicious cycle that accelerates the exhaustion of the nerve cells and triggers apoptosis through excessive calcium influx. While therapeutic inhibition of excitotoxicity with glutamate receptor antagonists appears sound in principle, it remains to be seen whether it can be as successful in clinical practice as it may be in carefully scheduled and timed animal experiments.

Metabotropic glutamate receptors form a distinct structural group among the GPCRs, with binding sites located on large N-terminal domains (Section 5.3.3). They are often located presynaptically, where they exert negative feedback on transmitter release. Agonists for specific receptor subtypes are in the experimental stage, mostly for a variety of psychiatric disorders or symptoms.

Several agonists and antagonists of glutamate receptors are shown in Figure 15.1. Cycloserine is shown in Figure 11.7.

GABA. GABA (γ-aminobutyric acid) is synthesized by decarboxylation of glutamate. Three receptor types have been distinguished. $GABA_A$ and $GABA_C$ are ligand-gated channels; among these, only the $GABA_A$ channel is currently important as a drug target (see Section 6.7.3). $GABA_B$ receptors are GPCRs. They often couple to $G\alpha_{i/o}$ proteins, causing inhibition of adenylate cyclase and consequently of CNG channels, and activation of K^+ (GIRK) channels. They are found on both pre- and postsynaptic membranes. The $GABA_B$ agonist baclofen is used to treat spasticity, that is, overshooting muscle tone, in various neurological disorders.

Some GABA receptor agonists and antagonists, as well as the GABA reuptake inhibitor tiagabine, are shown in Figure 6.18. The structures of the $GABA_A$ agonists diazepam and phenobarbital are shown in Figures 4.2 and 4.3, respectively.

Glycine. Like the $GABA_A$ receptor, the glycine receptor is a Cl^- channel, and its activation hyperpolarizes the postsynaptic cell, subduing its activity. Glycine receptors exert inhibitory control over the α-motoneurons in the brain stem

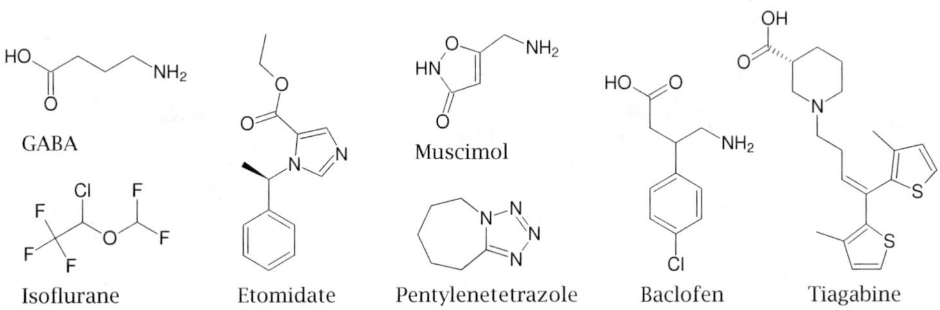

Figure 6.18 Drugs interacting with GABA receptors. The transmitter itself is shown for comparison. Isoflurane is an inhalation anesthetic, and etomidate is an intravenously applied anesthetic. Pentylenetetrazol is a $GABA_A$ antagonist that is used experimentally to induce seizures. Muscimol is an ingredient of fly agaric and a $GABA_A$ and $GABA_C$ agonist. Baclofen is a $GABA_B$ agonist that is used in the treatment of spasticity. Tiagabine is an inhibitor of presynaptic GABA reuptake, which is used in the treatment of epilepsy.

and spinal cord. Glycinergic inhibition is an integral and essential part of motor control, as is evident from the dramatic effects of its disruption in tetanus. This disease is caused by wound infections with the anaerobic bacterium *Clostridium tetani*, which produces tetanus toxin. This protein toxin is taken up by motoneurons and through retrograde axonal transport travels all the way up to the spinal cord, where it translocates across the synaptic cleft into presynaptic inhibitory glycinergic neurons. There, it becomes active as a protease and cleaves synaptobrevin, a protein molecule that is essential for the exocytosis of transmitter vesicles [22]. Deprived of glycinergic inhibition, the motoneurons begin to fire in an all-out fashion, causing incessant maximal stimulation of the skeletal muscles, to the point where tendons may snap and bones may break.[8]

Symptoms that resemble tetanus are triggered by strychnine, a competitive antagonist of the glycine receptor (Figure 6.19). This well-known poison has in the past been used as a stimulant and in the treatment of barbiturate intoxications. This latter application has no sound basis in pharmacology, never worked very well, and has been superseded by standard intensive care procedures such as hemoperfusion (see footnote on page 63) and artificial respiration.

Two different glycine reuptake transporters have been characterized. The type I glycine transporter occurs in the forebrain, where glycinergic synapses occur as part of the reward circuit (see Section 6.8.2). Inhibitors of the type I transporterssuch as Org 29535 (Figure 6.19) are under investigation in the treatment of psychosis and addiction.

[8]Tetanus is life-threatening because the permanent muscle contraction interferes with breathing. This permanent contraction can be broken with cholinergic antagonists (muscle relaxants; see Section 6.8.3), and the patients can then be kept alive through artificial respiration. The effect of the toxin, and the requirement for treatment, lasts for several weeks. Preventive vaccination with a chemically inactivated preparation of tetanus toxin induces antibodies that bind and neutralize the toxin.

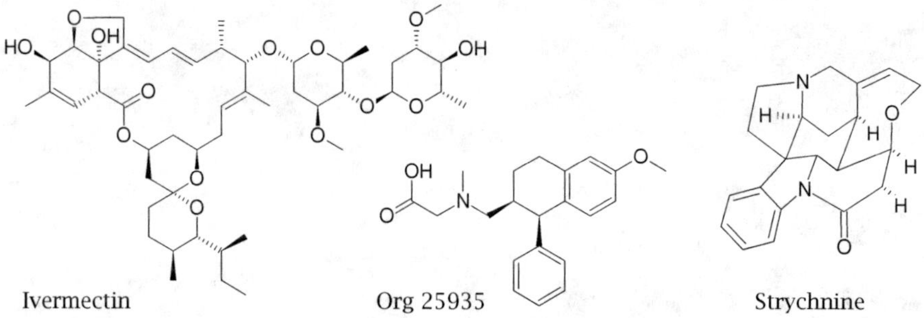

Ivermectin Org 25935 Strychnine

Figure 6.19 Drugs that interact with glutamatergic and glycinergic chloride channels. Ivermectin is an allosteric agonist of the glycine receptor [24]. Its main target is a glutamate receptor/chloride channel that occurs in non-vertebrates including some human parasites, but not in humans; it is used in the treatment of parasite infections. Strychnine is an antagonist of the glycine receptor. Org 29535 is an inhibitor of glycine reuptake that is being investigated for the treatment of psychosis and addiction.

6.8.2 Monoamines

Monoamine transmitters are biosynthetically derived from amino acids. Tyrosine is hydroxylated to L-DOPA, which gives rise to dopamine, norepinephrine and epinephrine. Serotonin and melatonin are derived from tryptophan, whereas histamine is derived from histidine. These transmitters occur in both the central nervous system and the periphery, and some of them act both as transmitters (that is, locally between adjacent cells) and as hormones, which are distributed systemically or regionally through the circulation. All monoamines act through multiple types of cognate GPCRs. Serotonin additionally has an ionotropic receptor in the cys-loop family (Section 6.7.3).

In the CNS, the monoamines have modulatory roles, facilitating or inhibiting functions and behaviors that intrinsically employ glutamate or GABA as the major cogwheels of operation. Accordingly, the neurons that release monoamine transmitters are far less numerous, and their cell bodies found in fewer locations than those of glutamatergic neurons. These cells do, however, project their axons across much of the CNS, so that a myriad of other neurons are exposed to their modulatory influence. Interestingly, drugs that influence the activity of monoamine transmitters have found more widespread and more compelling applications than those that directly interfere with glutamate.

The various monoamines are loaded into synaptic vesicles by the same vesicular monoamine transporter (VMAT). Similarly, two forms of monoamine oxidase (MAO) handle the degradation of all monoamines. Inhibitors of VMAT or MAO therefore have poor selectivity and are no longer widely used in modern pharmacotherapy. Presynaptic reuptake of monoamines is more specific, with separate transporters for serotonin, dopamine, and norepinephrine, and selective inhibitors of these are currently in clinical use.

Many drugs that act on monoamine receptors and transporters have powerful and immediate effects on vigilance and mood, and some of them are popular as drugs of abuse. The same receptors and transporters are also targeted by

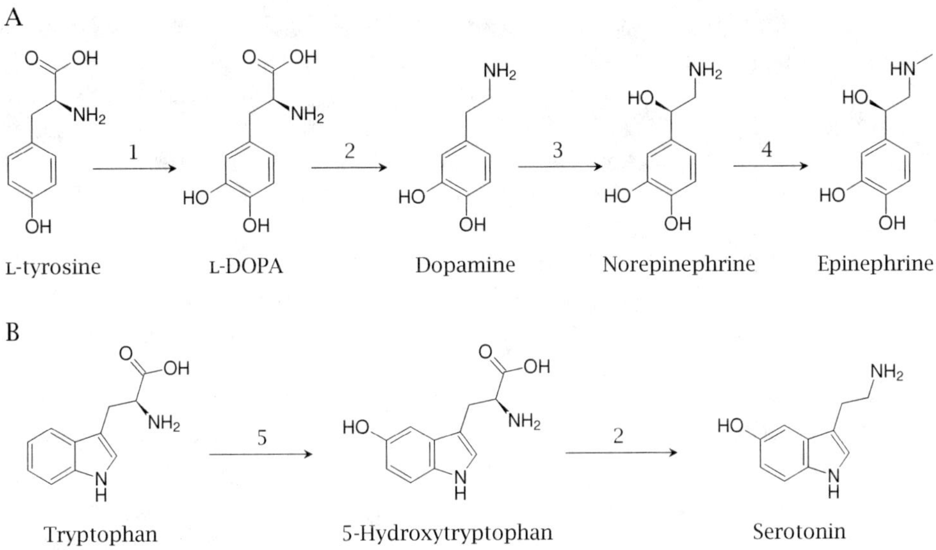

Figure 6.20 Biosynthesis of the catecholamines (**A**) and of serotonin (**B**). Enzymes: (1) Tyrosine hydroxylase, (2) L-aromatic amino acid decarboxylase or DOPA decarboxylase, (3) Dopamine β-hydroxylase, (4) Norepinephrine methyltransferase, (5) Tryptophan hydroxylase.

clinically used antidepressant drugs. In contrast to the instant stimulation and gratification that is the object of drug abuse, however, the antidepressant effect takes days, if not weeks, of continued treatment to assert itself. It is assumed to occur downstream of monoamine receptor stimulation, but its molecular mechanisms are currently not clear.

Dopamine. Dopamine occurs mostly in the central nervous system. The degeneration of dopaminergic neurons in a specific group of neurons in the brain stem, the *substantia nigra*, causes Parkinson's disease, which is characterized by the dysregulation and inhibition of voluntary movement. It can be treated by the replacement of dopamine in the form of its prodrug L-DOPA (see Section 14.2.1).

Dopaminergic neurons in the midbrain and forebrain play key roles in the control of behavior, emotion, and learning. They occur in the *reward circuit*, a functional system that ties together various centers in the brain and employs several other neurotransmitters in addition to dopamine, including endorphines and endocannabinoids. The reward circuit reinforces acquired behavior and is important in learning, but also in addiction. Several drugs of addiction, including alcohol, nicotine, cannabis, and opioids, act on receptors in the reward circuit. The "rewarding" and addictive effects of these drugs can often be inhibited by the application of dopamine receptor antagonists, indicating that the corresponding receptors function upstream of dopamine within the reward circuit.

Several dopamine receptor subtypes are important drug targets of antipsychotic drugs. Many such drugs, for example, haloperidol, are antagonists at the

D_2 receptor subtype. The "atypical" antipsychotic drug clozapine has high affinity only for the D_4 receptor subtype, and also for several serotonin receptors. Its lack of D_2 receptor inhibition coincides with the absence of Parkinson-like side effects that plague the use of haloperidol and similar drugs. While clozapine is now widely and successfully used in antipsychotic therapy, the development of drugs that selectively target dopamine receptor subtypes D_3 or D_4 without acting on serotonin receptors has not resulted in clinically effective antipsychotics, indicating a major role for serotonin receptors in the effect of clozapine.

Dopamine receptors in the pituitary control the release of prolactin, a peptide hormone that in turn controls lactation. The dopamine receptor agonist bromocriptine, a semisynthetic ergot alkaloid derivative, is used to treat excessive release of prolactin that may result from hormonal dysregulation or tumorous proliferation of prolactin-producing cells in the pituitary. It is also used in an ancillary role in the treatment of Parkinson's disease.

Serotonin. Serotonin is derived from tryptophan via tryptophan hydroxylase and aromatic amino acid decarboxylase (Figure 6.20). It has twelve receptor subtypes of the GPCR type and one ligand-gated channel, the 5-HT_3 receptor (Section 6.7.3). Antagonists of various GPCR subtypes are used in the treatment of migraine and of anxiety and other psychiatric conditions. Serotonin-selective inhibitors of presynaptic reuptake are used in the treatment of depression. 5-HT_3 receptor antagonists such as odansetron have largely done away with the severe nausea that used to plague patients undergoing cancer chemotherapy with cytotoxic drugs (see Chapter 12).

Several drugs that act on dopaminergic and serotoninergic synapses are shown in Figure 6.21.

Norepinephrine and epinephrine. Norepinephrine and epinephrine, which are biosynthetically derived from dopamine (see Figure 6.20), occur in both the CNS and the periphery. They are found not only in synapses but also as circulating hormones, and their receptors are not confined to excitable cells, although heart muscle cells and smooth muscle cells in the vasculature are important targets in the periphery. Circulating norepinephrine and epinephrine are released by the medulla (the central part) of the adrenal glands. Norepinephrine and epinephrine are the major mediators in the sympathetic part of the autonomic nervous system (see Section 6.9.5).

Norepinephrine and epinephrine have overlapping receptor specificities. α-Adrenergic receptors respond to both norepinephrine and epinephrine, whereas β-adrenergic receptors respond mainly to epinephrine. Both α and β receptors have several subtypes, and agonists and antagonists specific for these subtypes have distinct clinical uses.

Adrenergic receptors have a key role in cardiovascular regulation, and a great many drugs have been developed that interfere with this function, most commonly to control blood pressure and heart rate. In clinical practice, the most important ones are straightforward GPCR agonists and antagonists, but many other modes of action have been realized as well. The structures of several drugs that interact with adrenergic synapses are shown in Figure 6.22.

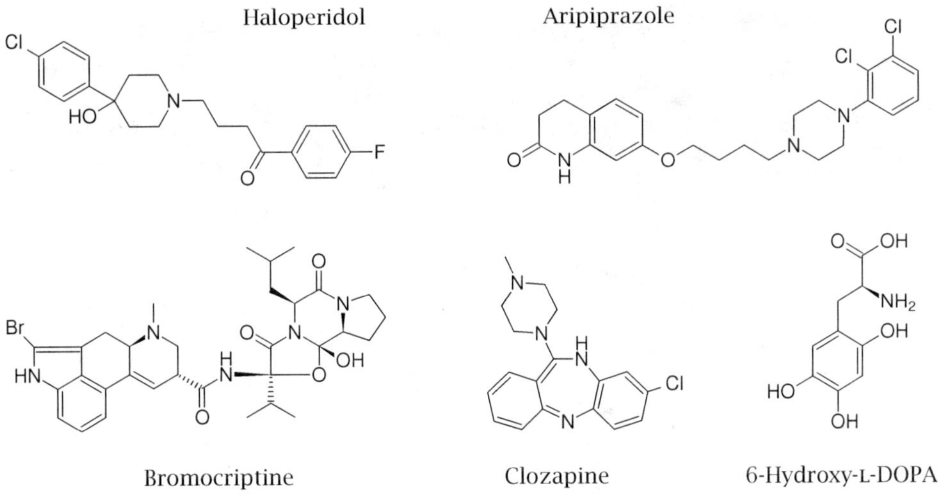

Figure 6.21 Drugs that interact with dopaminergic and serotoninergic synapses. Haloperidol is a first-generation antipsychotic and a D_2 dopamine receptor antagonist. Bromocriptine is an agonist at the same receptor and is used to suppress prolactin secretion and in Parkinson's disease. Clozapine is a second-generation antipsychotic with relatively low affinity for D_2 receptors but higher antagonistic activity on D_3 and D_4 dopamine receptors as well as 5-HT_{2A} serotonin receptors. Aripiprazole is another second-generation antipsychotic that is similar to clozapine but is a partial agonist at the D_2 receptor. 6-Hydroxy-L-DOPA is transported into dopaminergic neurons by reuptake and damages the cells through inhibition of the respiratory chain. It is not used clinically but is used in research to induce a Parkinson-like state in experimental animals.

α-Adrenergic receptor agonists and antagonists. α_1-Adrenergic receptors occur prominently on smooth muscle cells in the vasculature, on which they induce contraction via the $G\alpha_q$/PLC pathway and calcium, thus increasing flow resistance and blood pressure. Accordingly, α_1 agonists can be used to treat hypotension, and antagonists to treat hypertension. While most α_1 receptor antagonists bind noncovalently, the covalent inhibitor with phenoxybenzamine and its use in pheochromocytoma have been covered in Section 2.2.4.

α_2 Receptors occur both pre- and postsynaptically in the CNS and the periphery. The presynaptic receptors mediate feedback inhibition of norepinephrine release via $G\alpha_i$, thereby indirectly lowering blood pressure, and α_2 agonists can be used to treat hypertension. Postsynaptic α_2 receptors in the CNS are targeted by agonists in the treatment of neuropathic pain, that is, pain originating from some disturbance within the nervous system itself.

β-Adrenergic receptor agonists and antagonists. β_1-Adrenergic receptors occur prominently in the heart, where they activate adenylate cyclase and thereby increase the heart rate and strength of contraction. Agonists are occasionally used to support these functions, but more commonly antagonists (β blockers) are used to subdue them, in order to protect a diseased heart from excessive

Isoproterenol (β ↑) Terbutaline (β_2 ↑) Xylometazoline (α ↑) Clonidine (α_2 ↑)

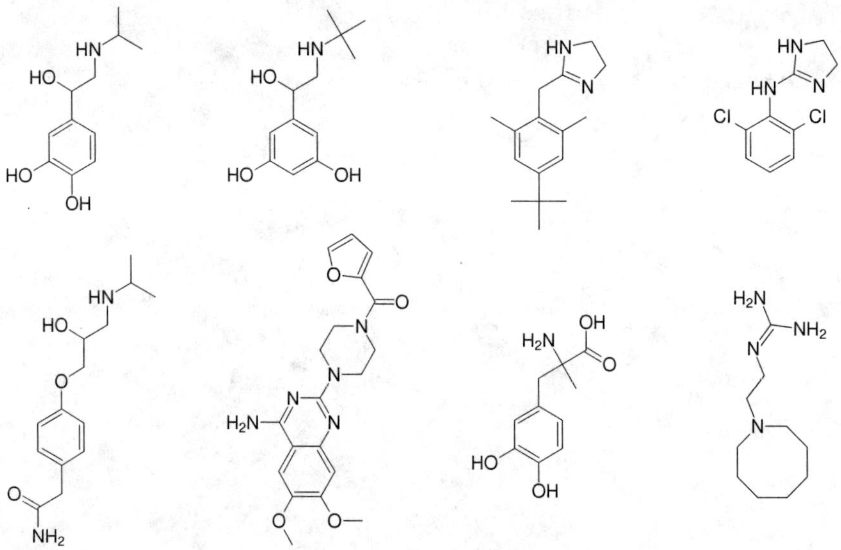

Atenolol (β_1 ↓) Prazosine (α_1 ↓) α-Methyl-DOPA Guanethidine

Figure 6.22 Receptor agonists or antagonists and false transmitters at adrenergic synapses. Up- and downward arrows indicate the activities of direct receptor agonists and antagonists, respectively. α-Methyl-DOPA and guanethidine act as false transmitters in noradrenergic synapses; the former penetrates the blood–brain barrier, whereas the latter does not.

load. Application of β blockers is one of very few therapeutic strategies with proven long-term benefits in patients that have suffered from myocardial infarction.

Activation of β_2 receptors on smooth muscle cells in the bronchi and the uterus induces relaxation, and receptor agonists are used to relieve bronchial constriction in asthma and to suppress premature labor. For patients who require β blockers for their heart disease but also suffer from bronchial obstruction, β_1-selective antagonists offer some benefit over nonselective β blockers.

False transmitters. This therapeutic approach to antihypertensive therapy is rather ingenious but no longer widely used. An example is the drug α-methyl-DOPA, which makes its way to the CNS and then undergoes decarboxylation in the same way as L-DOPA does (see Figure 6.20). The resulting metabolite, α-methyldopamine, is then further converted to α-methylnorepinephrine, which accumulates in synaptic vesicles of adrenergic synapses, displacing the normal transmitter. Upon release, it fails to stimulate the postsynaptic α_1 receptors but does activate presynaptic α_2 receptors, inducing feedback inhibition of transmitter release.

The drug guanethidine works by the same mechanism (except for the metabolic conversions) but does not effectively penetrate the CNS and is limited to adrenergic neurons in the periphery.

Inhibitors of monoamine transport. The vesicular monoamine transporter, VMAT, is inhibited by reserpine (Figure 6.23). This inhibition affects all monoamine transmitters. Reserpine was previously used in the treatment of hypertension and of mania, but the broad scope of its action renders it prone to side effects, including depression, and it is no longer in use.

The transporters for presynaptic reuptake are more specific for individual monoamine transmitters, and inhibitors that inhibit those for norepinephrine or serotonin alone or in combination are now commonly used as antidepressants. Cocaine acts by the same mechanism but inhibits the reuptake of norepinephrine, dopamine, and serotonin all at once. All of these actions contribute to its psychotropic and addictive effects.

Amphetamine and related drugs affect both the presynaptic reuptake of monoamine transmitters and their storage in the presynaptic vesicles. They cause leakage of the transmitters from the vesicles back into the cytosol; the released transmitters then undergo retrograde transport from the presynaptic cell into the synaptic cleft by the reuptake transporters. Amphetamine is structurally similar to monoamines but more hydrophobic, and both specific inhibition of the transport proteins and nonionic diffusion across the vesicle membranes, followed by displacement of transmitters from binding sites within the vesicles, are believed to contribute to its overall effect [26].

Inhibitors of monoamine degradation. The enzymes monoamine oxidase (MAO) and catecholamine-*O*-methyltransferase (COMT) cooperate in the degradation of catecholamines (Figure 6.24A). MAO occurs inside neurons and in the gastrointestinal tract and the liver. There are two variants. Norepinephrinergic neurons contain MAO_A, whereas MAO_B occurs in neurons that use serotonin or histamine as transmitters. The fact that the MAO_B-selective inhibitor selegiline is beneficial in Parkinson's disease suggests a role of MAO_B in the degradation of dopamine, too.

MAO inhibitors were traditionally used as antidepressants, but they have been largely superseded by reuptake inhibitors or receptor antagonists. Most of these drugs are covalent inhibitors; an example is tranylcypromine (Figure 6.24C).

Inhibition of MAO in the intestinal tract and the liver interferes with the degradation of aromatic amines such as tyramine or phenylethylamine, which are found in fermented foods and beverages. Such amines can penetrate nerve cells by nonionic diffusion and then act in a manner similar to amphetamine, releasing stored monoamine transmitters and inducing dangerous spikes in heart rate and blood pressure. This effect is known as the "cheese reaction", after just one type of fermented food that can induce it.[9] Noncovalent MAO inhibitors are less prone to this complication, but they also seem to be less effective as antidepressants. They have so far not found widespread application.

[9]Many microbes decarboxylate amino acids to amines in order to buffer the acids that they also accumulate through the fermentation of carbohydrates. Decarboxylation of tyrosine and phenylalanine will yield tyramine and phenylethylamine, respectively.

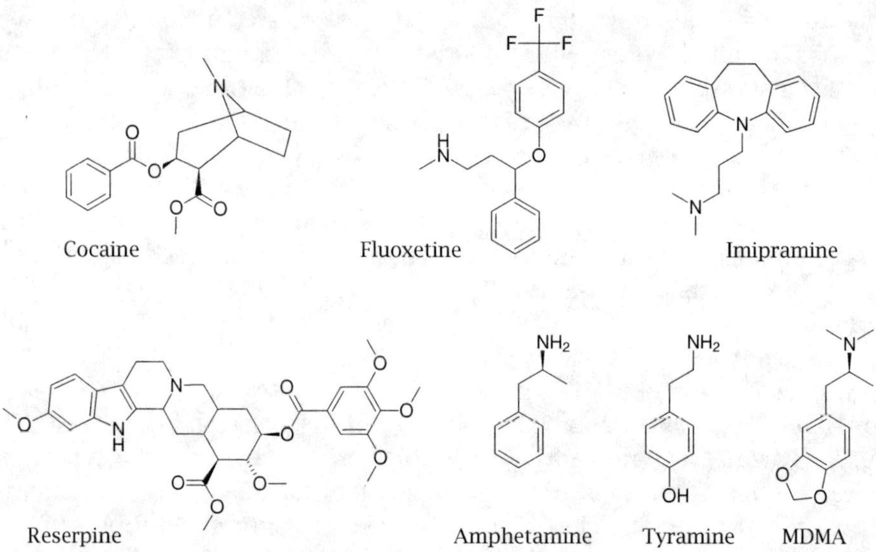

Figure 6.23 Drugs that act on the membrane transport of monoamine transmitters. Cocaine inhibits the presynaptic reuptake of dopamine, serotonin, and norepinephrine. Imipramine and fluoxetine are serotonin-selective reuptake inhibitors (SSRIs) and are used as antidepressants. Imipramine is enzymatically demethylated to desipramine, which itself is also used as an antidepressant drug and blocks the reuptake of norepinephrine. Reserpine blocks the vesicular monoamine transporter (VMAT) and prevents the accumulation of all monoamine transmitters in synaptic vesicles. Amphetamine and tyramine release dopamine and norepinephrine from synaptic vesicles and then promote their retrograde transport from the cytosol into the synaptic cleft. *N*-Methyl-3,4-methylenedioxyamphetamine (MDMA, "Ecstasy") does the same to serotonin.

COMT is specific for the degradation of catecholamines. It methylates the 3′-hydroxyl group of the ring (see Figure 6.24A) using *S*-adenosylmethionine as a cosubstrate. Inhibitors of COMT such as entacapone will prolong the lifetime of catecholamines, including dopamine, and can be used in conjunction with L-DOPA in the treatment of Parkinson's disease.

Histamine. Four different receptors of the GPCR type, H_1–H_4, have been characterized. Histamine receptors occur on neurons in the central nervous system but also on cells of other types in the periphery. The H_1 receptors in the brain are mostly relevant to drug therapy because of the drowsiness that occurs as a side effect in the treatment of allergies with H_1 antagonists. This side effect has been turned into an application, with some time-honored anti-allergic drugs now being sold as hypnotics.

H_3 receptors in the brain mediate presynaptic feedback in histaminergic synapses. In addition, they also inhibit release of other monoamines and of acetylcholine. H_3 antagonists have been tried clinically on patients with Alzheimer's disease, obesity, and diverse other potentially lucrative indications [21].

In the periphery, H_1 receptors occur on endothelial cells in the circulation and respond to histamine released from mast cells, which are a special type of white blood cell. This release is triggered by antigens binding to IgE antibodies

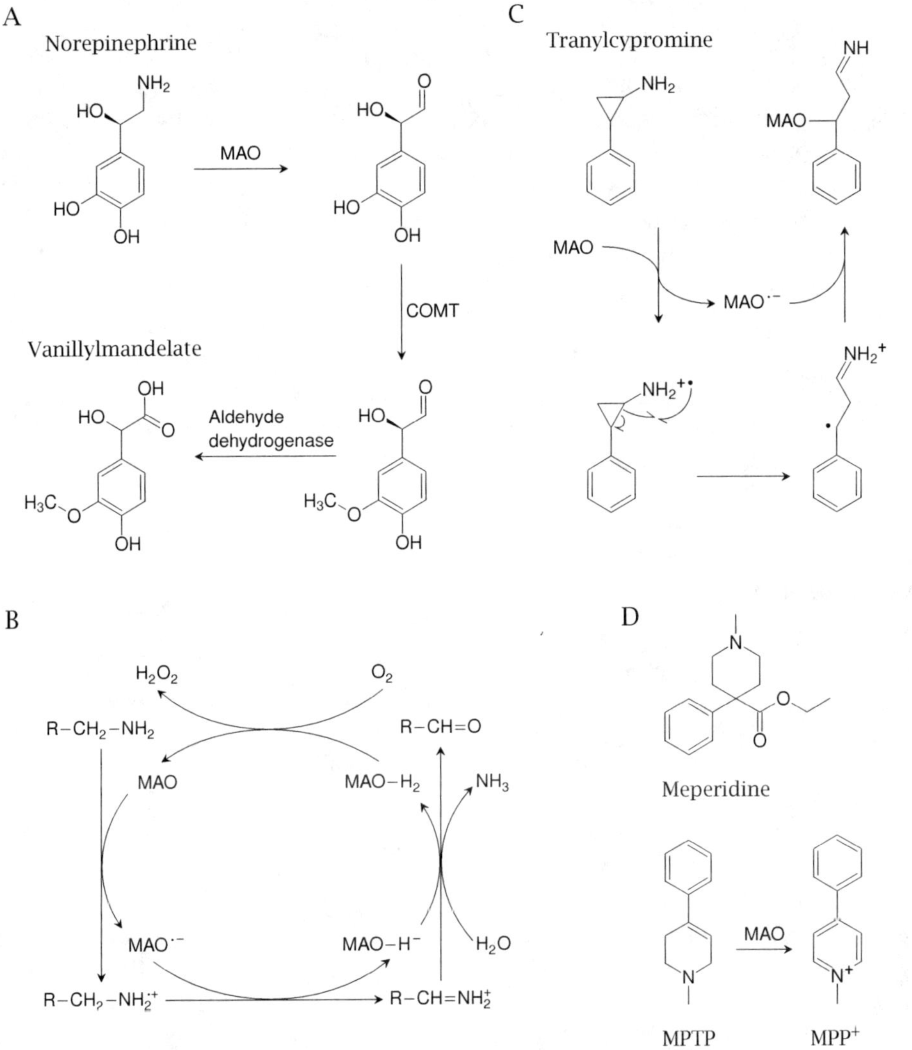

Figure 6.24 Monoamine oxidase (MAO) and catechol-*O*-methyltransferase (COMT) in the degradation of catecholamines. **A:** Degradation of norepinephrine (overview). MAO and COMT act in any order. The aldehyde group can either be oxidized to the acid as shown, or reduced to the alcohol by aldehyde reductase. The degradation of epinephrine and dopamine is similar. **B:** Mechanism of MAO. The reaction begins with the abstraction of a single electron from the substrate, which turns both the enzyme and the substrate into radicals. **C:** Mechanism-based inhibition of MAO by tranylcypromine. The amino group is attached to a reactive cyclopropyl ring, which upon the initial electron withdrawal opens to produce another radical intermediate. The latter then undergoes radical recombination with the enzyme. **D:** MAO-induced toxicity of MPTP (N-methyl-4-phenyltetrahydropyridine). MPTP occurred as a side product in the illicit synthesis of meperidine, a synthetic opioid. It is converted by MAO to *N*-methyl-4-phenylpyridine (MPP^+), which is accumulated inside dopaminergic neurons via the dopamine reuptake transporter [10]. It kills these neurons through inhibition of the respiratory chain. MAO inhibitors protect against this toxicity [8].

that are located on the surface of the mast cells. As a consequence of histamine release, blood vessels relax and capillaries become leaky, wheals rise up and itchiness sets in. This mechanism, which is referred to as *type I hypersensitivity*, operates in hay fever and similar allergies. While in most cases type I allergies produce more discomfort than real damage, in severe reactions the vascular leakiness and plasma sequestration can cause acute circulatory collapse and shock. This is more likely when the allergen is parenterally applied, such as in bee sting or penicillin allergies.[10]

H_2 receptors occur on parietal cells in the gastric mucosa, where they mediate the secretion of gastric acid. H_2 receptor blockers are used in the treatment of gastric and duodenal ulcers, which are not caused but are compounded by acidity, and of gastroesophageal reflux, which when unchecked not only is painful but also promotes development of esophageal cancer.

Figure 1.4 shows the structures of the H_1 antagonist cyclizine and the H_2 antagonist cimetidine.

6.8.3 Acetylcholine

Acetylcholine, the first neurotransmitter to be discovered, has multiple roles in the periphery and the central nervous system, and this diversity translates into multiple therapeutic applications of drugs that act on cholinergic synapses.

Acetylcholine has two major types of receptors, named nicotinic and muscarinic receptors after their respective prototypical agonists. Both are important drug targets. Some agonists and antagonists for both receptor types are shown in Figure 6.25.

Nicotinic acetylcholine receptors occur in the motor endplates, as well as in the autonomic ganglia of both the sympathetic and the parasympathetic parts of the autonomic nervous system. Muscarinic receptors occur on most of the effector cells controlled by the parasympathetic system (see Section 6.9.5).

Acetylcholine differs from the other transmitters discussed thus far by its mode of inactivation. It does not undergo presynaptic reuptake, at least not directly; instead, it is inactivated by cholinesterase, which is located at the outer surface of the postsynaptic cells. The choline produced by cholinesterase undergoes presynaptic reuptake and in the cell is converted back to acetylcholine by choline acetyltransferase. Inhibitors of cholinesterase are used as drugs and poisons.

Nicotinic receptor agonists and antagonists. The most widely used nicotinic receptor agonist is, of course, nicotine itself. Its pleasurable stimulatory effect is mediated by nicotinic acetylcholine receptors in the brain. Nicotine consumption apparently lowers the incidence of Alzheimer's disease; this benefit is reproducible in mouse models of the disease [14].[11]

[10] H_1 blockers only work *before* exposure to the allergen. If an acute situation has already arisen, histamine antagonists come too late; instead, the circulation must be supported with epinephrine and other measures.

[11] The former German chancellor Helmut Schmidt, who is now well into his nineties, remains a chain-smoking fixture on TV and continues to impress with poise and ready wit. Other notable smokers who retained young minds in old age were Winston Churchill and Deng Xiaoping.

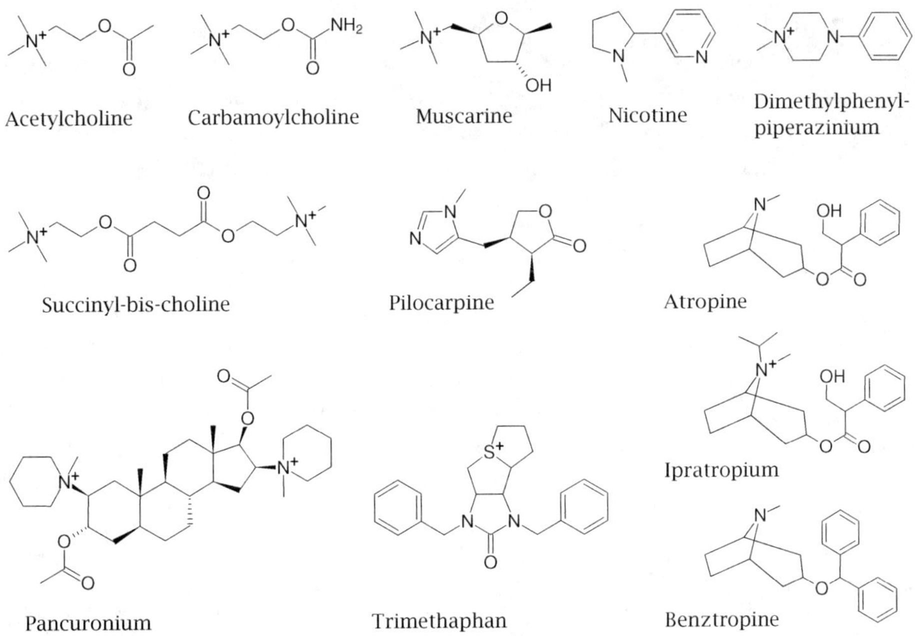

Figure 6.25 Structures of acetylcholine and of cholinergic receptor agonists and antagonists. Carbamoylcholine is active on both muscarinic and nicotinic receptors, but more so on the former. Its cleavage by cholinesterase causes carbamoylation and transient covalent inactivation of the enzyme. Dimethylphenylpiperazinium is a nicotinic agonist that is excluded by the blood–brain barrier. Succinyl-bis-choline is a nicotinic agonist that induces "depolarizing block" in motor endplates and is used as a muscle relaxant. Pilocarpine is a muscarinic agonist that is used in ophthalmology. Pancuronium and trimethaphan are nicotinic antagonists that are used as a muscle relaxant and as a ganglion-blocking agent, respectively. Atropine is a muscarinic antagonist. Ipratropium is an atropine analog with lower, and benztropine one with greater, penetration of the blood–brain barrier.

Nicotinic receptor antagonists are widely used in combination with narcotic gases in systemic narcosis. The purpose of systemic narcosis is twofold: (1) suppression of consciousness and, with it, pain perception; and (2) suppression of movements, including reflex movements triggered at the level of the spinal cord in response to painful stimuli. Unconsciousness can be achieved with dosages of gaseous narcotics lower than those required for suppression of spinal cord activity. Nicotinic receptor antagonists disconnect the muscle cells from the motoneurons and therefore remove the need to suppress spinal cord activity. This permits a reduction in the dosage of gaseous narcotics, and therefore of the side effects associated with the latter. NAR inhibitors such as *d*-tubocurarine and pancuronium that act preferentially on motor endplates, as opposed to autonomic ganglia, are referred to as *muscle relaxants.*

Neuromuscular blockade can also be achieved with the NAR agonist succinyl-bis-choline. This drug causes *depolarizing block,* which involves, but is not fully explained by, the receptor desensitization observed under continuous stimulation (see Section 6.7.3).

Blockade of the neural NAR in autonomic ganglia with so-called *ganglion blockers* such as trimethaphan was used in the past to treat hypertension. This approach is fraught with significant side effects, however, since it affects both the sympathetic and the parasympathetic parts of the autonomic nervous system (see Section 6.9.5). It has been largely abandoned in favor of more specific therapies.

Muscarinic receptor agonists and antagonists. The muscarinic acetylcholine receptors, named after the agonist muscarine, are GPCRs. They are found both in the brain and in the periphery on cells that are controlled by the peripheral parasympathetic system, such as smooth or heart muscle cells and gland cells (see Section 6.9.5). Muscarinic receptors occur in various subtypes, but most practically applied drugs don't discriminate between these.

In motor control within the CNS, muscarinic receptors are antagonistic to dopamine receptors, and antimuscarinic drugs such as benztropine are used in an auxiliary role in the treatment of Parkinson's disease. At high dosages, muscarinic antagonists may cause confusion and hallucinations.[12] Muscarinic receptors outside the CNS mediate most of the effects of the parasympathetic nervous system. When targeting muscarinic receptors in the periphery, side effects on the CNS can be avoided through the use of antagonists that are excluded by the blood–brain barrier, such as ipratropium bromide, an atropine derivative with a quaternary amine (Figure 6.25). These drugs are useful in the treatment of asthma and for persuading a sluggish sinoatrial node to generate a faster and more regular heart rhythm.

Muscarinic agonists are used after abdominal surgery to accelerate resumption of intestinal and bladder activity. Local application to the eye is used in the treatment of glaucoma, a condition in which the pressure within the eyeball is pathologically increased. They promote contraction of the ciliary muscle, which indirectly relieves the compression of *Schlemm's canal*, a tiny conduit through which excess fluid is drained from the eyeball.

Muscarinic agonists that reach the brain produce a mildly stimulating effect. *Areca* nuts, which contain arecoline, and the leaves of *Pilocarpus* shrubs, which contain pilocarpine, are consumed in some parts of the world as stimulants.

Cholinesterase inhibitors. Inhibitors of acetylcholine esterase reduce the rate of acetylcholine inactivation and therefore increase the level of cholinergic activity, without discriminating between nicotinic and muscarinic synapses. Complete inhibition of cholinesterase is fatal, and these inhibitors are used as insecticides and as nerve agents in chemical warfare.[13] An interesting medical application is *myasthenia gravis*, an autoimmune disease in which the nicotinic acetylcholine receptors in the motor endplates are inactivated by autoantibodies. This reduces the response of muscle cells to the transmitter released by

[12]Muscarine, like muscimol, is found in fly agaric (*Amanita muscaria*). Of the two, muscimol is the more abundant and important poison in this mushroom.

[13]In World War II, both sides developed highly potent cholinesterase inhibitors and amassed them in enormous amounts, but ultimately shied away from their use. They were again manufactured in Iraq under Saddam Hussein and used both in Iraq's war with Iran and in attacks on the domestic Kurdish population.

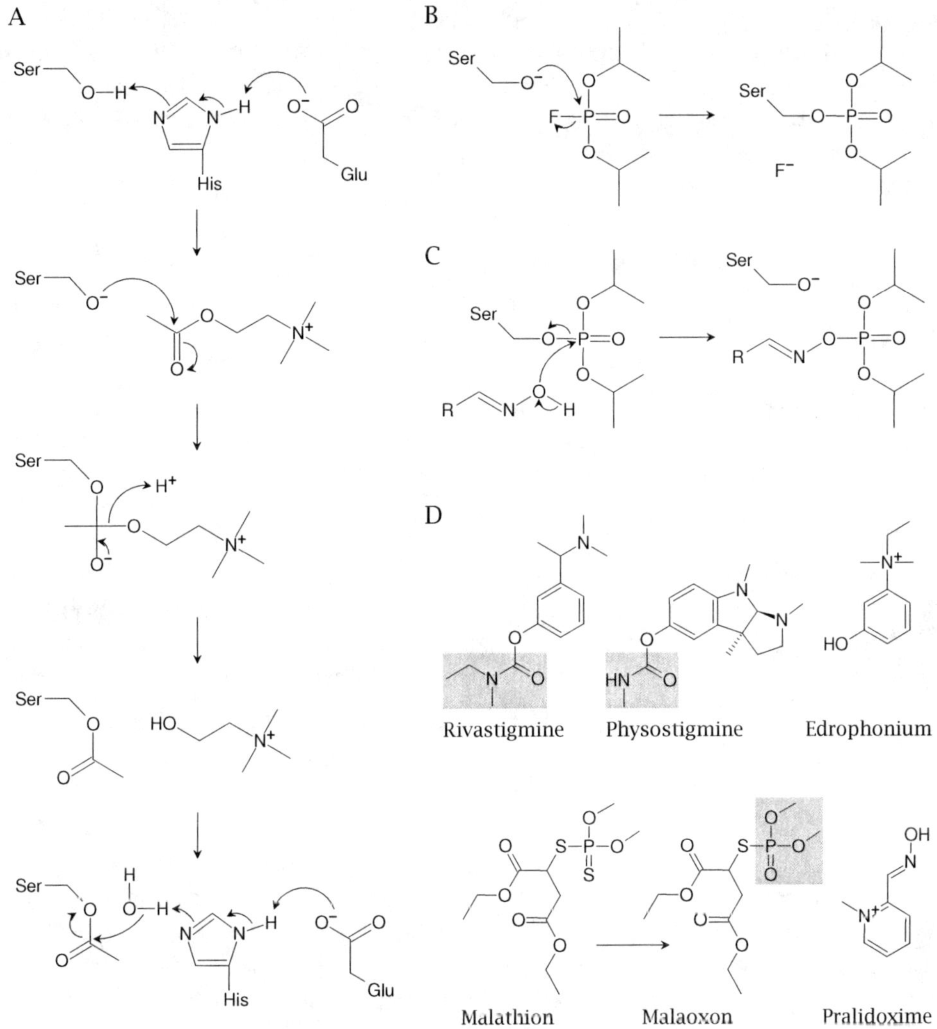

Figure 6.26 Catalytic mechanism and inhibitors of cholinesterase. **A:** The catalytic mechanism of cholinesterase resembles that of serine proteases. A glutamate and a histidine residue deprotonate the catalytic serine, which then performs nucleophilic attack on the substrate. The tetrahedral transition state gives way to an acylated form of the enzyme, which is then hydrolyzed. **B:** Diisopropylfluorophosphate (DFP) inactivates cholinesterase covalently. DFP and other organophosphates structurally resemble the tetrahedral transition state and therefore bind the enzyme very avidly. **C:** Covalently inactivated cholinesterase cannot reactivate itself but can be reactivated by oximes such as pralidoxime. **D:** Structures of cholinesterase inhibitors and of the reactivating drug pralidoxime. Edrophonium is a noncovalent inhibitor, whereas rivastigmine and physostigmine are covalent ones. The groups that end up bound to the active site serine are highlighted. All three drugs are used clinically. Malathion is an insecticide; it is metabolically activated to malaoxon. Like DFP, malaoxon converts the enzyme to an alkylphosphate. Malathion is inactivated in human plasma through enzymatic cleavage of the ester bonds in the leaving group, and therefore has relatively low toxicity for humans.

the α-motoneurons, resulting in weakness or even total disruption of voluntary movement. Partial inhibition of cholinesterase extends the lifetime of acetylcholine in the synapse and amplifies the response generated from the remaining postsynaptic receptors.[14]

The mechanism of cholinesterase and the structures of several inhibitors are shown in Figure 6.26.

6.8.4 Adenosine and adenine nucleotides

ATP is stored in synaptic vesicles, often along with catecholamines. Upon release, it can bind to postsynaptic receptors directly, or it can undergo dephosphorylation to adenosine, which then binds to adenosine receptors. Another related transmitter is adenine dinucleotide (Ap4A), in which two adenosine molecules are connected by a tetraphosphate bridge. The receptors for these transmitters are referred to as *purine receptors*, but some of them respond also to uridine and cytidine nucleotides. The P2X purine receptors are ionotropic. They occur in sensory neurons in the spinal cord and are of interest to pharmacological research because of their role in pain perception.

The other purine receptors are GPCRs. Among them, the A_{2A} subtype responds to adenosine and mediates anxiolytic and sedative effects. Caffeine and related methylxanthines are antagonists at this receptor. The related ADP receptors, which are of interest mostly for their role in thrombocyte activation, are discussed in Section 10.4.4.

6.8.5 Peptides

Peptide transmitters are numerous, but their physiological roles are often incompletely understood. Peripherally applied peptides are excluded by the blood–brain barrier and are susceptible to proteolytic cleavage. Practical agonists and antagonists would have to be small organic molecules, which are currently still unavailable for most peptide receptors. This shortage of specific drugs hinders not only practical pharmacology but also the characterization of the physiological roles of peptide transmitters. Several peptides appear in totally different functional roles in the CNS and the periphery; for example, vasopressin modulates social behavior in the CNS, whereas in the periphery it serves as a hormone and regulates water retention in the kidneys (Section 7.2.1). Such multiple functions would have to be taken into account in the development of receptor agonists or antagonists.

Endorphins and enkephalins. One major exception from the dearth of synthetic analogs are endorphins and enkephalins, the endogenous ligands of opioid receptors. The poppy alkaloid morphine is a strong agonist by itself and has served as a starting point for the synthesis of other agonists and antagonists.

[14]In the clinically similar Lambert–Eaton syndrome, autoantibodies against presynaptic Ca_V channels interfere with acetylcholine release. Cholinesterase inhibitors can be of use in this disease as well.

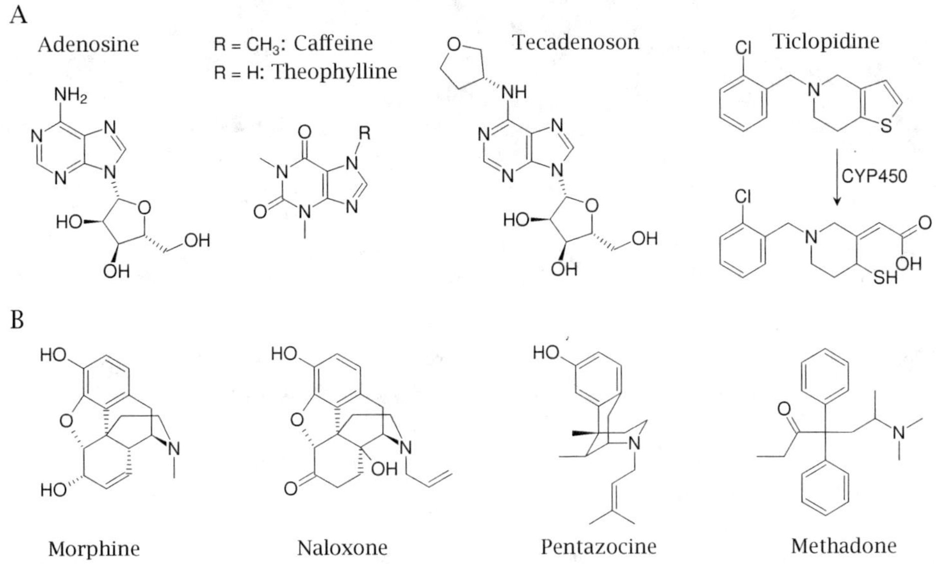

Figure 6.27 Drugs that interact with purine receptors (**A**) and with opioid receptors (**B**). Adenosine is shown for comparison. Caffeine and theophylline are antagonists at A_{2A} adenosine receptors. Theophylline is effective in the treatment of asthma. Tecadenoson is an agonist at adenosine receptors that is used in the treatment of some forms of cardiac arrhythmia. Ticlopidine is an ADP receptor antagonist that inhibits thrombocyte aggregation. It requires metabolic activation by cytochrome P450. Morphine is an agonist, and naloxone an antagonist at μ-, κ- and δ-opioid receptors. Pentazocine acts as a partial agonist primarily at μ-opioid receptors; it is a strong analgesic. Methadone is a full μ-receptor agonist and is used as a substitute for heroin.

All opioid receptors are GPCRs and exert analgesic effects at several stages of pain perception and transmission, and opioid receptor agonists suppress pain more strongly than do any other non-narcotic agents. Opioid receptors also are part of the reward circuit (see Section 6.8.2), which is responsible for the addictive effect of opioids. The three major receptor subtypes are named μ, δ, and κ. There are numerous natural and synthetic agonists, partial agonists, and antagonists for opioid receptors, some but not all of which are structurally similar to morphine (Figure 6.27B). A very high level of opioid receptor activation inhibits respiration, which causes death in accidental or intentional drug overdoses in addicts. While receptor antagonists may be of use in the treatment of an incipient intoxication, they would come too late in manifest opioid-induced apnea, which instead requires immediate intubation and mechanical respiration.

Morphine receptor subtypes may occur in various states of association, and receptor heterodimers may exhibit distinct pharmacological properties (see Figure 5.8).

6.9 APPENDIX

This section expands on some of the physiological foundations of the material covered in this chapter.

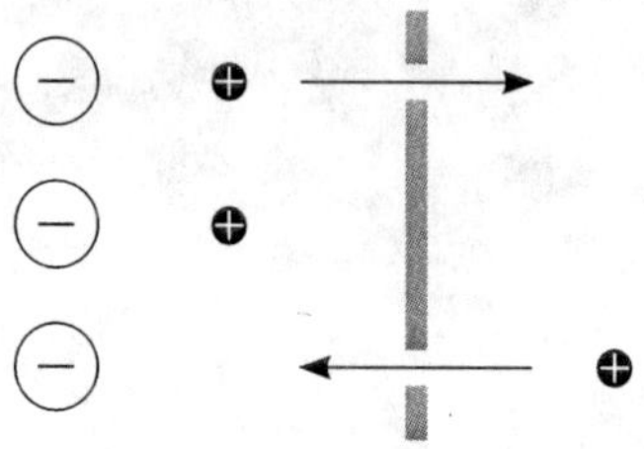

Figure 6.28 The driving forces that generate diffusion potentials across membranes. The membrane must be selectively permeable for some ion species but not its counterion, and for the permeative ion there must be a concentration gradient across the membrane. The permeative ion will diffuse across the membrane until the electrostatic potential created by the increasing charge imbalance becomes equally strong as the entropy associated with the concentration gradient.

6.9.1 The physics of diffusion potentials across membranes

Both the resting potential and the action potential are *diffusion potentials.* A diffusion potential forms if (1) the membrane is selectively permeable for a given ion species but not for its counterion, and (2) the ion concentrations are different on the two sides of the membrane. Then, the permeant ions—in Figure 6.28, the cations—are subject to two opposite driving forces. Entropy drives them down their concentration gradient. However, as ions cross the membrane, they become separated from their counterions, which creates an electrostatic potential that pulls them back home. When a certain number of ions have crossed, the potential becomes as strong as the concentration gradient, and no more net flux of ions occurs. The potential at this point, the *equilibrium potential,* is given by the Nernst equation:

$$\Delta E = \frac{RT}{zF} \ln \frac{[\text{cation}]_{\text{outside}}}{[\text{cation}]_{\text{inside}}} \tag{6.1}$$

In this equation, R, z, and F are constants, and in the human body T is virtually constant also. Therefore, the equilibrium potential is essentially determined by the ion concentration gradient alone.

The Nernst equation deals only with a single diffusible ion species, but in a cell there are several. The two most prevalent cations, Na^+ and K^+, have roughly opposite gradients and equilibrium potentials (see Table 6.1). This situation is accounted for by the Goldman equation. The Goldman equation introduces one additional parameter, the permeability P, which has a different value for each ion present. The contribution of each ion gradient to the overall membrane potential is weighted for this ion's permeability. Taking into account the three most prevalent ions, K^+, Na^+, and Cl^-, the Goldman equation reads as

$$\Delta E = \frac{RT}{zF} \ln \frac{P_{\text{K}}\,[\text{K}^+]_{\text{outside}} + P_{\text{Na}}\,[\text{Na}^+]_{\text{outside}} + P_{\text{Cl}}\,[\text{Cl}^-]_{\text{inside}}}{P_{\text{K}}\,[\text{K}^+]_{\text{inside}} + P_{\text{Na}}\,[\text{Na}^+]_{\text{inside}} + P_{\text{Cl}}\,[\text{Cl}^-]_{\text{outside}}} \tag{6.2}$$

The opposite concentration gradients for K^+ and Na^+ are of similar magnitudes, but the permeability for K^+ is higher, and so the actual membrane resting potential is closer to the K^+ equilibrium potential than to that of Na^+. Also note that, because of its negative charge, the terms for chloride are reversed relative to those for K^+ and Na^+, such that the intracellular Cl^- level is in the numerator, not the denominator. The high extracellular Cl^- concentration therefore promotes a negative-inside potential, just like the high intracellular K^+ concentration does.

What is the molecular basis of different permeabilities for different ions? This is where the ion channels come in. K^+ leak channels cause this ion's high permeability and establish its domination of the resting potential. However, as other channels open and close, the changing permeabilities will shift the weight from one ion to the other. Voltage-gated sodium channels open swiftly and forcefully; this will raise the permeability for Na^+ to above that of K^+, so that now the Na^+ concentration gradient gets to dominate the membrane potential. On most excitable cells, this is what triggers and sustains the action potential.

6.9.2 Electrophysiological characterization of single channel molecules

Integral membrane proteins such as membrane channels can be mobilized from membranes with detergents and purified. When such purified channel molecules are brought into contact with an artificial lipid bilayer, they tend to spontaneously insert into it. This can be used for functional studies. A bilayer is suspended between two buffer reservoirs, each of which contains an electrode (Figure 6.29A). A small amount of purified channel protein is introduced into one reservoir; the right amount to achieve membrane insertion of just one channel molecule is determined by trial and error.

The controlling electronics that are connected to the electrodes can maintain a fixed voltage (V) across the membrane and record the resulting current (A) as the channel opens and closes. A typical voltage-gated channel will be closed, and the current will be zero as long as it is held at a voltage resembling the resting potential. When the voltage is raised, the channel opens, and the current jumps up to a value that reflects the channel's conductivity. Channels often oscillate between the open and closed state while activated.

The same experimental setup can also be applied to ligand-gated channels. Agonists or antagonists can be can be applied to the reservoir adjoining the extra- or intracellular face of the channel, and their effects can be observed.

The patch clamp technique allows the study of one or a few channel molecules within intact cells or cell membranes (Figure 6.29B). A glass capillary that is connected to an electrode is lowered onto the cell surface. When it touches the membrane, it spontaneously forms an electrical seal, so that all current through the electrode now has to go through the membrane patch underneath. In *whole-cell mode*, the other electrode is immersed in the extracellular buffer. The current between the two electrodes passes the cell membrane twice, but it is limited by the point of highest resistance, which is the small membrane

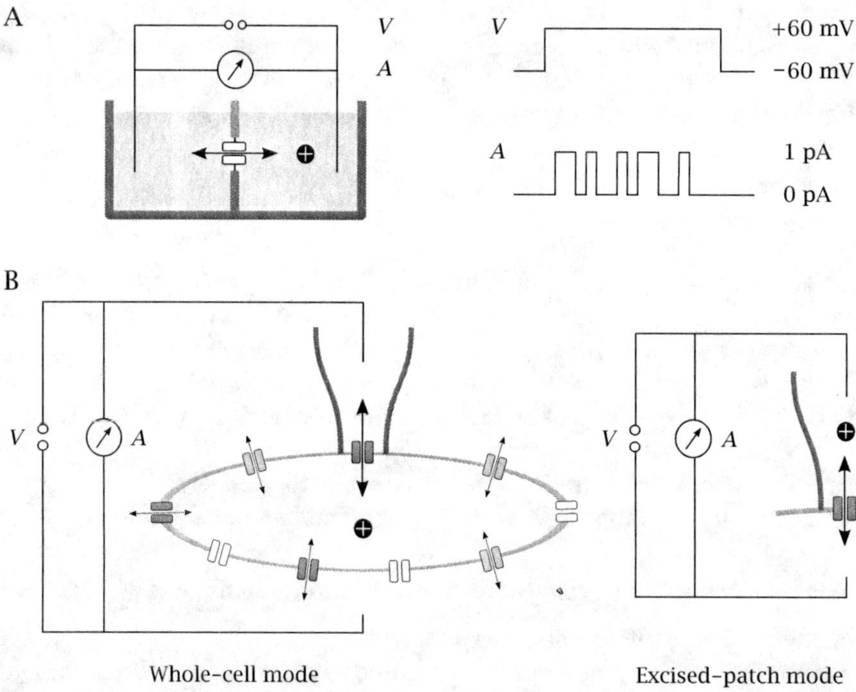

Figure 6.29 Electrophysiological methods for studying single channel molecules. **A:** Channel molecules can be purified and inserted into an artificial lipid bilayer. Electrodes are inserted into the buffer reservoirs on both sides of the bilayer. Voltage can be applied, and the opening and closing of the channel in response to the change in the potential across the membrane can be observed by monitoring the ensuing current, which often changes in discrete jumps of constant magnitude that is proportional to the conductivity of the open channel. **B:** In the patch clamp technique, the channel is observed in its natural environment within the cell membrane. Isolated observation of a single channel is accomplished by placing a glass micropipette on top of it, such that it forms a seal with the cell membrane. All current through the pipette must then go through the channel as well.

patch underneath the capillary, and specifically the channel or channels that this membrane patch contains.

In *excised-patch mode*, the cell-adherent capillary is withdrawn, and the attached membrane patch is torn from the cell. This allows the experimenter to freely vary the composition of the solutions on both the extracellular and the intracellular side of the membrane.

6.9.3 Ion channels and the heart rhythm

The electrophysiology of the heart differs considerably from that of nerve or skeletal muscle cells. While a single contraction of a skeletal muscle fiber is sustained by a burst of repetitive action potentials of very short duration, the heart generates only a single action potential for each contraction. This action potential lasts several hundred milliseconds and is sustained by L-type Ca_V channels that inactivate much more slowly than other voltage-gated channels.

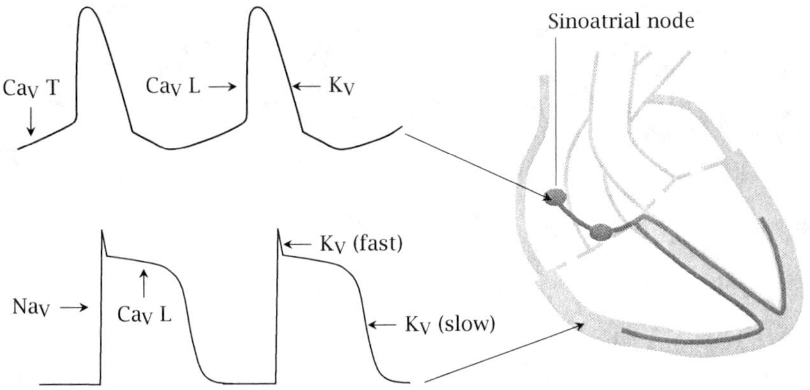

Figure 6.30 Voltage-gated channels and action potentials in the heart. Action potentials are generated in the sinoatrial node. T type Ca$_V$ channels open slowly at the resting potential, causing a gradual depolarization. Once depolarization reaches the firing level of the faster DHPR or L type Ca$_V$ channels, these channels open and trigger an action potential. The action potential travels down the fibers of the excitation-conduction system (dark gray) to reach the muscle cells (light gray). In the latter, the action potential is initiated by Na$_V$ channels and sustained for the duration of the contraction through Ca$_V$ channels. Repolarization is mediated by K$_V$ channels and L channel inactivation.

The rhythm of the heartbeat is generated in the sinoatrial node, which is the topmost part of the heart's *excitation-conduction system.* From there, the action potentials spread along the other parts of this system—the atrioventricular node, the bundle of His, and the Purkinje fibers—towards the regular worker muscle cells. Transmission from cell to cell occurs through electrical synapses or gap junctions.

The roles of the various types of voltage-gated channels, and the resulting action potentials, are illustrated in Figure 6.30. The rhythm arises from the cooperation of two different types of Ca$_V$ channels in the sinus node. The T-type channel acts as the timer. It has a lower activation threshold than other voltage-gated channels, and therefore undergoes slow spontaneous activation from the resting potential. Once the slow depolarization carried by the T-type channels reaches the firing level of the L-type channels, these respond with a fast and forceful depolarization that sets off the action potential. Repolarization is mediated by L-type channel inactivation and by K$_V$ channel opening.

Worker muscle cells do not have T-type channels and therefore normally have no spontaneous activity. They do, however, have Na$_V$ channels that cause a quick initial rise and spike of the action potential. After the Na$_V$ channels inactivate, the action potential is sustained at a slightly lower level by slowly inactivating L-type channels. As in the excitation-conduction system, inactivation is mediated by K$_V$ channels.

Drugs that act on the T-type Ca$_V$ channels will selectively affect the excitation-conduction system, whereas drugs that affect Na$_V$ channels make it possible to selectively address the worker muscle cells. Drugs that target L-type Ca$_V$ channels will affect both cell types. All three types of drugs have specific applications in the treatment of various types of cardiac arrhythmias.

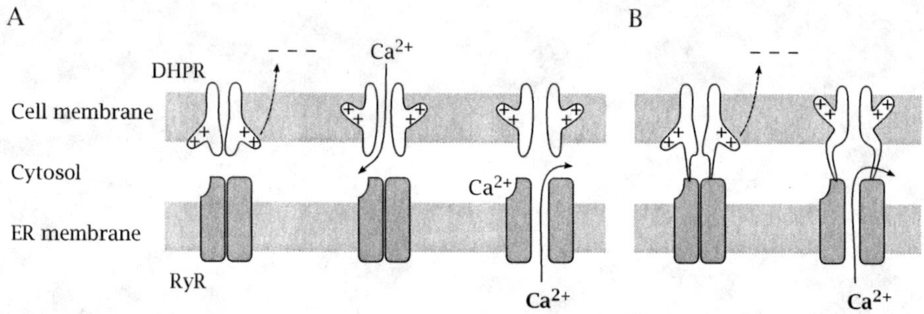

Figure 6.31 Interaction of the dihydropyridine receptor and the ryanodine receptor Ca^{2+} channels in heart muscle and skeletal muscle. **A:** In the heart, activation of the ryanodine receptor (RyR) requires influx of Ca^{2+} through the dihydropyridine receptor (DHPR), which opens during an action potential. **B:** In skeletal muscle, the two channels are hooked up directly to each other. The conformational change that occurs in the DHPR upon depolarization is transmitted to the RyR and opens it; no actual influx of extracellular Ca^{2+} is necessary.

6.9.4 Excitation–contraction coupling in muscle cells

In all types of muscle cells, voltage-gated calcium channels play a key role in excitation–contraction coupling, that is in translating action potentials to the contraction of the cell. However, there are some differences in the details of the coupling mechanism between the various muscle tissues that create opportunities for selective drug action.

Heart and skeletal muscle. In striated muscle, a key aspect of excitation-contraction coupling is the amplification of the initial calcium signal inside the cell through the release of additional calcium from intracellular stores. Apart from the extracellular space, a relatively high calcium concentration also exists in the endoplasmic reticulum and the mitochondria. Active transporters and ligand-gated channels regulate the flux into and out of these intracellular calcium stores. Among these, the ryanodine receptor (RyR), a calcium-activated calcium channel in the membrane of the endoplasmic reticulum, plays a prominent role in excitation–conduction coupling.

When L-type Ca_V channels in the cytoplasmic membrane open in response to a propagating action potential, calcium flows in and activates the RyR, which releases a secondary wave of calcium from the endoplasmic reticulum. This amplification is necessary to saturate troponin, a calcium-binding protein that controls the interaction of actin and myosin, which in turn causes muscle contraction.

The L-type Ca_V channel in the cytoplasmic membrane is also named *dihydropyridine receptor* (DHPR) after a group of inhibitory drugs. The DHPRs on skeletal and heart muscle are very similar, except for one crucial detail. In both tissues, the cytoplasmic membrane and the ER membrane, and with them the DHPR and the RyR, are in immediate proximity.

In skeletal muscle, the DHPR is directly connected to the ryanodine receptor (Figure 6.31). Here, when an action potential sweeps the cytoplasmic membrane,

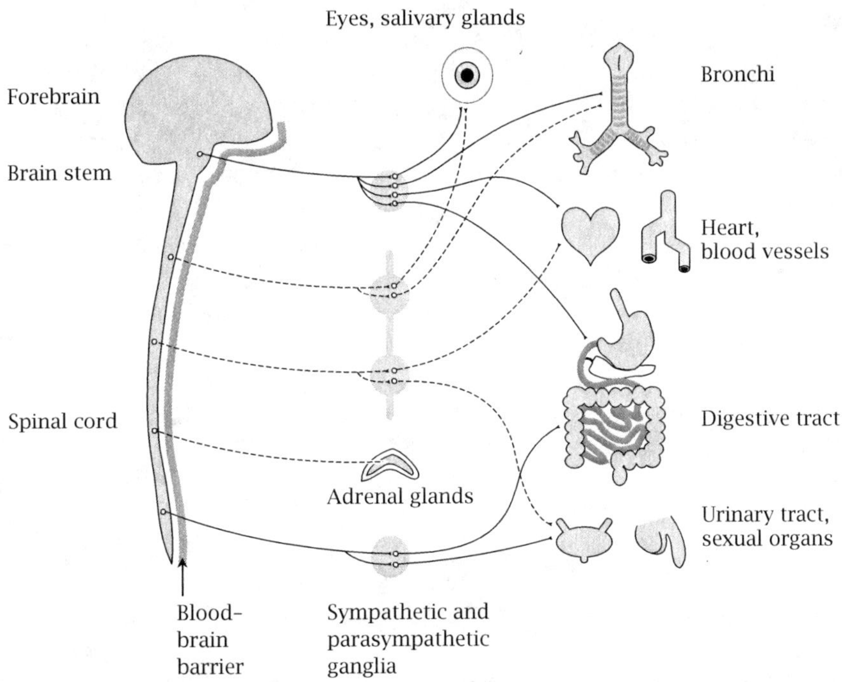

Figure 6.32 Organization of the autonomic nervous system. The autonomic nervous system consists of parasympathetic parts (solid lines) and sympathetic parts (dashed lines). Each connection between the central part of the autonomic nervous system and a peripheral organ contains two neurons connected in series. The cell body of a first neuron is located inside the CNS, but its efferent synapses are located inside a sympathetic or parasympathetic ganglion, outside the blood-brain barrier. The cell body of a second neuron is located in this ganglion.

the conformational change that is triggered within the DHPR is communicated directly to the RyR. Therefore, the DHPR acts as a voltage sensor for the RyR, which thus does not have to wait for the influx of extracellular Ca^{2+} but can directly respond to the action potential.

In contrast, in the heart, the DHPR and the RyR are not directly connected, so that the opening of the RyR depends on an actual influx of extracellular Ca^{2+} through the DHPR channel. DHPR channel blockers therefore have a much stronger effect on the heart than on skeletal muscle, even if they bind to the channels on both cell types.

Smooth muscle. In smooth muscle cells, contraction is activated by a protein phosphorylation cascade rather than by troponin, and release of large amounts of calcium from intracellular stores is not required. Nevertheless, as in heart muscle, the initial signal depends on the influx of calcium through DHPR channels. Therefore, DHPR antagonists will reduce contractile strength in both heart muscle and vascular smooth muscle. Both effects will reduce blood pressure (see Section 6.6.3).

6.9.5 The autonomic nervous system

The autonomic nervous system (Figure 6.32) comprises the parts of the central and peripheral nervous systems that deal with the unconscious control and homeostasis of organ function. It is conventionally subdivided into the *sympathetic* and the *parasympathetic* systems, which in most organs have opposite regulatory effects. The sympathetic system is activated during physical exertion and increases heart rate and blood pressure, relaxes smooth muscles in the bronchi to improve ventilation, and increases the blood flow to skeletal muscle at the expense of the intestinal organs. The parasympathetic system has the opposite effects, favoring digestion and regeneration over exertion. As an exception from this antagonistic pattern, the two systems cooperate in controlling the function of the sexual organs.

Signals from the central part of the autonomic nervous system to a peripheral organ are relayed by two neurons connected in series. The cell body of a first neuron is located inside the CNS, but its efferent synapses are located inside a sympathetic or parasympathetic ganglion, outside of the blood brain barrier. The cell body of a second neuron is located in this ganglion, and its axon runs within peripheral nerves or along blood vessels toward the peripheral organ.

The synapses in both sympathetic and parasympathetic ganglia contain nicotinic acetylcholine receptors. Antagonists of these receptors are referred to as *ganglion blockers* (Section 6.8.3). The receptors in the secondary parasympathetic synapses are muscarinic acetylcholine receptors, whereas those in the secondary sympathetic synapses are mostly α-adrenergic receptors. This distinction allows for more selective drug action. Examples are the use of α receptor blockers in hypertension, muscarinic antagonists in asthma, and muscarinic agonists to activate bladder or intestinal function.

STUDY QUESTIONS

✎ **6.1** Find out about the workings of fish electric organs. How are they capable of generating electric discharges of several hundred volts or more?

✎ **6.2** Ionotropic receptors go through a functional cycle that involves resting, active and inactivated states (Figure 6.16B). This cycle can only be unidirectional if it receives an input of energy—otherwise, it would constitute a perpetual motion machine. What is the source of this energy?

✎ **6.3** The molecular mode of action of tetanus toxin was discussed in Section 6.8.1. A closely related (yet more popular) toxin is botulinum toxin. Remarkably, while tetanus induces excess muscular activity, botulinum toxin does the exact opposite, that is it induces muscle paralysis. Find out how this difference comes about.

✎ **6.4** Some ion channels, such as the major voltage-gated channels, pass ions in their dehydrated form, whereas others such as the NAR pass hydrated ions. Ion channels also differ in their degree of ion selectivity. Is there a connection between ion hydration and channel selectivity?

REFERENCES

[1] D. J. Adams et al., "Omega-conotoxin CVID inhibits a pharmacologically distinct voltage-sensitive calcium channel associated with transmitter release from preganglionic nerve terminals," *J. Biol. Chem.* 278 (2003), 4057–4062.

[2] A. Y. Bagrov and J. I. Shapiro, "Endogenous digitalis: pathophysiologic roles and therapeutic applications," *Nat. Clin. Pract. Nephrol.* 4 (2008), 378–392.

[3] A. Brisson and P. N. Unwin, "Tubular crystals of acetylcholine receptor," *J. Cell Biol.* 99 (1984), 1202–1211.

[4] D. C. Chiara et al., "Identification of amino acids in the nicotinic acetylcholine receptor agonist binding site and ion channel photolabeled by 4-[(3-trifluoromethyl)-3H-diazirin-3-yl]benzoylcholine, a novel photoaffinity antagonist," *Biochemistry* 42 (2003), 271–283.

[5] D. A. Coulter, J. R. Huguenard, and D. A. Prince, "Characterization of ethosuximide reduction of low-threshold calcium current in thalamic neurons," *Ann. Neurol.* 25 (1989), 582–593.

[6] B. Forbush, "Rapid release of ^{42}K and ^{86}Rb from an occluded state of the Na,K-pump in the presence of ATP or ADP," *J. Biol. Chem.* 262 (1987), 11104–11115.

[7] A. C. Foster and J. A. Kemp, "Glutamate- and GABA-based CNS therapeutics," *Curr. Opin. Pharmacol.* 6 (2006), 7–17.

[8] R. E. Heikkila et al., "Protection against the dopaminergic neurotoxicity of 1-methyl-4-phenyl-1,2,5,6-tetrahydropyridine by monoamine oxidase inhibitors," *Nature* 311 (1984), 467–469.

[9] L. Huang et al., "NNC 55-0396 [(1S,2S)-2-(2-(N-[(3-benzimidazol-2-yl)propyl]-N-methylamino)ethyl)-6-fluoro-1,2,3,4-tetrahydro-1-isopropyl-2-naphtyl cyclopropanecarboxylate dihydrochloride]: a new selective inhibitor of T-type calcium channels," *J. Pharmacol. Exp. Ther.* 309 (2004), 193–199.

[10] J. A. Javitch et al., "Parkinsonism-inducing neurotoxin, N-methyl-4-phenyl-1,2,3,6 -tetrahydropyridine: uptake of the metabolite N-methyl-4-phenylpyridine by dopamine neurons explains selective toxicity," *Proc. Natl. Acad. Sci. U. S. A.* 82 (1985), 2173–2177.

[11] B. Katz and S. Thesleff, "A study of the desensitization produced by acetylcholine at the motor end-plate," *J. Physiol.* 138 (1957), 63–80.

[12] A. Kawamura et al., "On the structure of endogenous ouabain," *Proc. Natl. Acad. Sci. U. S. A.* 96 (1999), 6654–6659.

[13] S. P. Kelley et al., "A cytoplasmic region determines single-channel conductance in 5 HT3 receptors," *Nature* 424 (2003), 321–324.

[14] Q. Liu et al., "Dissecting the signaling pathway of nicotine-mediated neuroprotection in a mouse Alzheimer disease model," *FASEB J.* 21 (2007), 61–73.

[15] S. B. Long, E. B. Campbell, and R. Mackinnon, "Crystal structure of a mammalian voltage-dependent Shaker family K^+ channel," *Science* 309 (2005), 897–903.

[16] N. Misaki et al., "Iptakalim, a vascular ATP-sensitive potassium (KATP) channel opener, closes rat pancreatic β-cell KATP channels and increases insulin release," *J. Pharmacol. Exp. Ther.* 322 (2007), 871–878.

[17] M. G. Nicholls et al., "Ouabain, a circulating hormone secreted by the adrenals, is pivotal in cardiovascular disease. Fact or fantasy?," *J. Hypertens.* 27 (2009), 3–8.

[18] A. Poling, H. Schlinger, and E. Blakely, "Failure of the partial inverse benzodiazepine agonist Ro15-4513 to block the lethal effects of ethanol in rats," *Pharmacol. Biochem. Behav.* 31 (1988), 945–947.

[19] R. L. Post, C. Hegyvary, and S. Kume, "Activation by adenosine triphosphate in the phosphorylation kinetics of sodium and potassium ion transport adenosine triphosphatase," *J. Biol. Chem.* 247 (1972), 6530–6540.

[20] A. dos Reis, "Sigmund Freud (1856-1939) and Karl Köller (1857-1944) and the discovery of local anesthesia," *Rev. Bras. Anestesiol.* 59 (2009), 244–257.

[21] K. Sander, T. Kottke, and H. Stark, "Histamine H3 receptor antagonists go to clinics," *Biol. Pharm. Bull.* 31 (2008), 2163–2181.

[22] G. Schiavo et al., "Tetanus and botulinum-B neurotoxins block neurotransmitter release by proteolytic cleavage of synaptobrevin," *Nature* 359 (1992), 832–835.

[23] A. K. Sen, T. Tobin, and R. L. Post, "A cycle for ouabain inhibition of sodium- and potassium-dependent adenosine triphosphatase," *J. Biol. Chem.* 244 (1969), 6596–6604.

[24] Q. Shan, J. L. Haddrill, and J. W. Lynch, "Ivermectin, an unconventional agonist of the glycine receptor chloride channel," *J. Biol. Chem.* 276 (2001), 12556–12564.

[25] J. C. Skou, "The influence of some cations on an adenosine triphosphatase from peripheral nerves," *Biochim. Biophys. Acta* 23 (1957), 394–401.

[26] D. Sulzer et al., "Mechanisms of neurotransmitter release by amphetamines: a review," *Prog. Neurobiol.* 75 (2005), 406–433.

[27] N. Unwin, "Acetylcholine receptor channel imaged in the open state," *Nature* 373 (1995), 37–43.

[28] N. Unwin, "Refined structure of the nicotinic acetylcholine receptor at 4A resolution," *J. Mol. Biol.* 346 (2005), 967–989.

[29] M. Wallner, H. J. Hanchar, and R. W. Olsen, "Ethanol enhances $\alpha_4\beta_3\delta$ and $\alpha_6\beta_3\delta$ γ-aminobutyric acid type A receptors at low concentrations known to affect humans," *Proc. Natl. Acad. Sci. U. S. A.* 100 (2003), 15218–15223.

[30] R. N. Willette et al., "Systemic activation of the transient receptor potential vanilloid subtype 4 channel causes endothelial failure and circulatory collapse: Part 2," *J. Pharmacol. Exp. Ther.* 326 (2008), 443–452.

[31] G. W. Zamponi and R. J. French, "Dissecting lidocaine action: diethylamide and phenol mimic separate modes of lidocaine block of sodium channels from heart and skeletal muscle," *Biophys. J.* 65 (1993), 2335–2347.

[32] Y. Zhang et al., "Neither GABA(A) nor strychnine-sensitive glycine receptors are the sole mediators of MAC for isoflurane," *Anesth. Analg.* 92 (2001), 123–127.

[33] W. Zhong et al., "From ab initio quantum mechanics to molecular neurobiology: a cation-pi binding site in the nicotinic receptor," *Proc. Natl. Acad. Sci. U. S. A.* 95 (1998), 12088–12093.

CHAPTER 7

HORMONES

Messenger molecules vary profoundly with respect to their scope of action. The effect of neurotransmitters such as glutamate and acetylcholine is limited to a single synapse. In contrast, glucocorticoid hormones, which are secreted by the adrenal glands into the blood, reach and affect virtually every cell in the body. Other hormone molecules fall between these two limiting cases. Their scope of action may be limited by regional distribution and by organ-specific expression of their receptors. Some messenger molecules, for example, vasopressin or norepinephrine, serve as both hormones and transmitters.

Diseases that involve hormone-producing, or *endocrine* glands are manifold, but their pathophysiology mostly boils down to either excessive or wanting hormone secretion. A lack of hormone may result from a developmental deficiency and become evident during childhood. In adults, gland tissue may be destroyed by tumors, autoimmune diseases, or other causes. Treatment consists in substitution of the lacking hormone, often for the patient's lifetime.

Excessive hormone activity may be due again to autoimmune diseases or to hormone-producing tumors, which may be benign or malignant. Aside from surgical treatment, gland tumors may be amenable to symptomatic or causal drug therapy.

7.1 HORMONE RECEPTORS

A large number of hormone receptors are G protein-coupled receptors, which are discussed as a group in Chapter 5. Receptor tyrosine kinases are exemplified by the insulin receptor (see Section 10.3.1). The growth hormone receptor is not a kinase itself but is associated with the so-called Janus kinase that triggers a cascade of protein phosphorylation in much the same way as the insulin receptor. Steroid hormones and thyroid hormones act primarily through nuclear hormone receptors.

7.1.1 Nuclear hormone receptors

These receptors are a homologous family of transcriptional regulator proteins whose ligands include hormones, xenobiotics, vitamins, and lipid metabolites (Table 7.1). These receptors bind to specific short DNA motifs, referred to as *response elements*, in the vicinity of gene promoters. DNA binding occurs in the absence or presence of the cognate hormone or other ligand. However, the effect on transcription differs between the two cases. Without ligand, the DNA-bound receptors recruit inhibitory proteins that suppress transcription from the adjacent promoters. In contrast, when the ligand is bound to the receptor, the latter recruits auxiliary proteins such as histone acetylase that collectively facilitate access of RNA polymerase to the promoters, and transcription increases.

Activated nuclear hormone receptors are dimers, either homodimers of the same receptor, or heterodimers of two receptors with different ligand specificity. Each receptor monomer contains a DNA-binding domain and a ligand-binding domain, so that two ligand molecules and two DNA half-sites are bound by a receptor dimer.

The length of a DNA half-site bound by one receptor monomer is 6 base pairs. These half-sites vary little between the different receptors. The mutual specificity of DNA response elements and receptor dimers arises mostly from the relative spacing and orientation of the two half-sites, and of the corresponding DNA-binding domains in the receptor dimers. The two hexanucleotide half-sites can be separated by one or several intervening base pairs. If we consider that one intervening base pair introduces not only a translational offset of 0.34 nm along the DNA molecule but also a rotational offset by 36°, it becomes plausible that receptor dimers can readily discern the proper spacing of the half-sites. Furthermore, the two DNA hexamers can be arranged in a parallel or antiparallel fashion. This variability is illustrated in Figure 7.1, which shows the DNA-binding domain of thyroid hormone receptor (TRβ) as a homodimer and also as a heterodimer with retinoid X receptor (RXR). The DNA half-sites recognized by TR and RXR are the same, but they occur with different orientations and spacing in the two motifs recognized by the two different dimers.

7.2 THE HYPOTHALAMUS AND THE PITUITARY GLAND

The hypothalamus and the pituitary gland jointly control much of the activity of the entire endocrine system. The hypothalamus is located at the bottom

Table 7.1 Nuclear hormone receptors and their ligands.

Ligand types	Ligands	Receptors	Receptor subtypes
Hormones	Thyroxine, triiodothyronine	Thyroid hormone receptors	α, β_1, β_2
	Cortisone, cortisol	Glucocorticoid receptor	
	Aldosterone	Mineralocorticoid receptor	
	Testosterone, dehydroepiandrosterone	Androgen receptor	
	Estradiol	Estrogen receptor	α, β
	Progesterone	Progesterone receptor	
Vitamin derivatives	1,25-Dihydroxy-cholecalciferol	vitamin D receptor	
	All-*trans*-retinoic acid	Retinoic acid receptor	α, β, γ
	9-*cis*-Retinoic acid	Retinoid X receptor	α, β, γ
Intermediates of lipid metabolism	Fatty acids	Peroxisome proliferator-activated receptors	α, γ, δ
	Oxysterols	Liver X receptor	α, β
	Bile acids	Bile acid receptor	
Drugs, xenobiotics	Carbamazepine, rifampicin, etc.	pregnane X receptor	
	Phenobarbital	Constitutive androstane receptor	

of the forebrain. The pituitary (or hypophyseal) gland is directly connected to it but extends out of the brain and of the blood–brain barrier. It consists of two distinct parts, the anterior lobe and the posterior lobe, which arise from separate origins during embryonic development. The posterior lobe merely contains the nerve endings of neurons whose cell bodies are located within the hypothalamus. These nerve endings release their mediators—the peptide hormones oxytocin and vasopressin—not into synaptic clefts but directly into the blood. This phenomenon is known as *neurosecretion* and nicely blurs the line between neural and endocrine signaling.

The anterior lobe contains regular endocrine gland cells that produce and secrete various peptide hormones. The rate of hormone release by the anterior lobe is regulated by the hypothalamus through local secretion of releasing factors that again are mostly peptide molecules. The secretion of these hypothalamic factors is *pulsatile*, occurring in short bursts that vary in both frequency

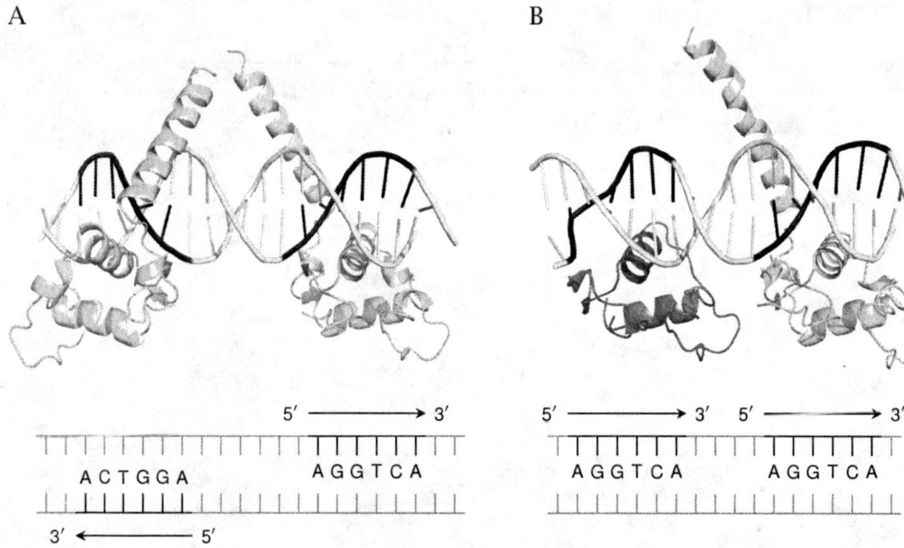

Figure 7.1 Recognition of DNA hormone response elements by thyroid receptor (TRβ) and retinoid X receptor (RXRα). The consensus sequence that is recognized by both receptors is AGGTCA. **A:** The TRβ homodimer recognizes two instances of the consensus sequence that occur on opposite strands and point away from one another (inverted palindrome). **B:** The TRβ–RXRα heterodimer recognizes two instances of the consensus sequence that are located on the same strand (direct repeats). In both A and B, only the DNA-binding domains of the receptor molecules are shown; the remaining domains would attach to the tips of the upward pointing helices. (A) rendered from 3m9e.pdb [7], (B) from 2nll.pdb [20].

and amplitude. These variations may follow some circadian rhythm, or they may be tied to stages of growth and development or of the menstrual cycle.

The hypophyseal hormones affect various peripheral organs (Table 7.2). Some of the target organs are again endocrine glands. Specifically, the thyroid gland, the adrenal cortex, and the hormone-producing cells in the ovaries and testicles are controlled by hormones of the anterior pituitary. The hormones produced by those peripheral glands exert negative feedback on the hypothalamus and also directly on the pituitary. For example, thyroid hormones inhibit the secretion of both hypothalamic thyrotropin releasing hormone (TRH) and of the pituitary hormone thyrotropin (thyroid stimulating hormone, TSH). The same pattern of feedback regulation also applies to the sexual steroid hormones, which are produced in the ovaries and testes, and to the glucocorticoids, which are produced in the cortex of the adrenal glands (Figure 7.2).

The cells in the anterior pituitary are specialized for the production of individual hormones. Tumors generally arise from single cells, and therefore tumors of the anterior pituitary will typically overproduce individual hormones. On the other hand, destructive processes such as disrupted perfusion or infiltration by tumors that originate outside the pituitary tend to destroy the entire anterior lobe, and occasionally the posterior lobe as well, thereby disrupting the production of many hormones at once.

Table 7.2 Hormones of the anterior lobe of the pituitary gland. Control by hypothalamic hormones is stimulatory unless stated otherwise.

Pituitary hormone	Hypothalamic control	Target organ	Physiological effect
Thyroid-stimulating hormone (TSH)	Thyrotropin-releasing hormone (TRH)	Thyroid gland	Proliferation, secretion of triiodothyronine (T_3) and tetraiodothyronine (T_4)
Corticotropin (ACTH)	Corticotropin-releasing hormone (CRH)	Adrenal gland (cortex)	Secretion of cortisol, dehydroepiandrosterone (DHEA)
Follicle-stimulating hormone (FSH)	Gonadotropin-releasing hormone (GnRH)	Ovaries	Growth of follicle, secretion of estrogens
		Uterus	Growth of mucous membrane (decidua)
		Testes	Formation of sperm cells (spermatogenesis)
Luteinizing hormone (LH)	GnRH	Ovaries	Ovulation, secretion of progestins
		Uterus	Maintenance of mucous membrane (decidua)
		Testes	Secretion of testosterone
Growth hormone	Stimulated by growth hormone-releasing hormone (GHRH), inhibited by somatostatin	Many	Tissue proliferation; metabolic regulation; release of insulin-like growth factor (IGF-1)
Prolactin	Stimulated by TRH, inhibited by dopamine	Mammary glands	Proliferation, lactation

7.2.1 Vasopressin and oxytocin

Vasopressin and oxytocin, the two hormones released by the posterior pituitary, are nonapeptides with similar structures (Figure 7.3). Both are produced by proteolysis of precursor peptides and have an amidated C-terminus that arises by oxidative cleavage of an extra glycine residue. Both peptides act on GPCRs that also share extensive homology but regulate quite different physiological processes. The receptors activate the intracellular phospholipase C/inositoltriphosphate pathway via G_q but also other G proteins and downstream effectors.

Vasopressin, also called *antidiuretic hormone* (ADH), has a key role in the regulation of blood pressure and of fluid balance. Its release is stimulated by high osmolality of the blood plasma and the extracellular fluid. Vasopressin receptors occur in two types. Activation of the V_1 receptor causes vasoconstriction and raises the blood pressure. The V_2 receptor controls the activity of

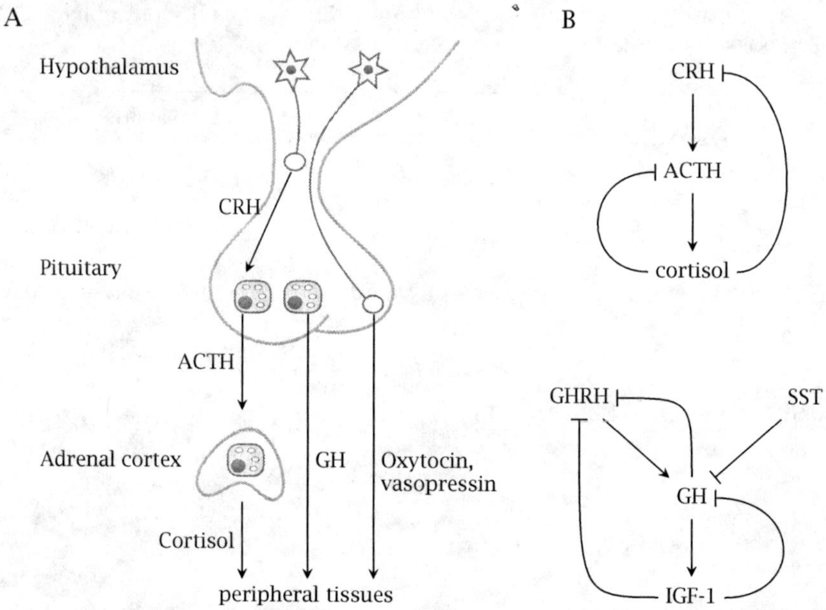

Figure 7.2 Regulation of hormone secretion by the *hypothalamic-pituitary axis* (examples). **A:** The pituitary is located adjacent to the hypothalamus. The posterior pituitary contains the axon endings of hypothalamic nerve cells, which secrete oxytocin and vasopressin directly into the circulation. The anterior pituitary contains gland cells that are activated by specific hypothalamic hormones, for example, corticotropin-releasing hormone (CRH). The cells of the anterior pituitary release peptide hormones that stimulate peripheral glands like adrenocorticotropic hormone (ACTH) or other tissues like growth hormone (GH). **B:** Regulatory patterns. The adrenal glands produce cortisol, which exerts a negative feedback both on CRH production by the hypothalamus and on ACTH production by the pituitary. The secretion of GH is activated by growth hormone-releasing hormone (GHRH) and inhibited by somatostatin (SST), which originate in the hypothalamus. GH stimulates the production of insulin-like growth factor 1 (IGF-1). Both GH and IGF-1 exert negative feedback on GH secretion.

aquaporin in the collecting ducts of the kidneys. When the receptor is activated, aquaporin molecules translocate from intracellular storage vesicles to the cytoplasmic membrane, where they allow water molecules to permeate from the nascent urine back into the kidney cells and from there into the interstitial space. This movement of water is driven by the unusually high osmolality of the interstitial fluid in the kidneys. Water is retained, and a smaller volume of more concentrated urine is excreted.

Inadequate or lacking secretion of vasopressin causes excretion of large volumes of dilute urine, a condition that is called *diabetes insipidus* and treated with hormone substitution therapy. In this treatment, it is desirable to selectively stimulate the V_2 receptors in order to restore fluid retention but to avoid the vasoconstriction, which might unduly raise the blood pressure. The synthetic peptide desmopressin deviates from vasopressin in two positions (Figure 7.3). These modifications make it an V_2-selective agonist.

Oxytocin H₂N-Cys—Tyr—Ile—Gln—Asn—Cys—Pro—l-Leu—Gly-CO-NH₂

Vasopressin H₂N-Cys—Tyr—Phe—Gln—Asn—Cys—Pro—l-Arg—Gly-CO-NH₂

Desmopressin H-Cys—Tyr—Phe—Gln—Asn—Cys—Pro—d-Arg—Gly-CO-NH₂

Figure 7.3 Amino acid sequences of oxytocin, vasopressin, and the synthetic vasopressin analog desmopressin. All amino acid residues are in the L configuration, except the D-arginine in desmopressin. In oxytocin and vasopressin, the amino group at the C-terminus is derived from a glycine that becomes oxidatively truncated after proteolytic cleavage of the precursor polypeptide.

A different set of specific changes to the vasopressin molecule can convert it from an agonist into an antagonist. Both V_1- and V_2-selective peptide antagonists have been obtained. For both receptors, nonpeptide antagonists are also available. Vasopressin antagonists are used in conditions such as congestive heart failure, in which the underlying pathophysiology and the treatment with various drugs of different types may conspire to induce a state of increased vascular resistance and fluid retention.

Oxytocin induces contraction of smooth muscle cells in the uterus and other organs. Its most conspicuous action is to induce labor at the time of birth. When a mother nurses her baby, her pituitary gland releases pulses of oxytocin that induce contraction of myoepithelial cells in the milk ducts and so support the outflow of milk. In both men and women, oxytocin is secreted during orgasm and contributes to both the pleasurable experience and, in men, to the ductal motility involved in ejaculation.

Oxytocin is used as a drug to induce or reinforce labor, usually when the amnion sac has broken, but labor does not set in spontaneously; this is necessary to avert the danger of ascending uterine infection. It also is used to staunch excessive uterine bleeding following delivery.

7.2.2 Growth hormone

Growth hormone (GH) is a peptide hormone with a molecular weight of 22 kDa, and a minor form of 20 kDa that arises through alternative splicing. Its receptor is located in the cell membrane. On the intracellular side, the receptor is bound to Janus kinase, a tyrosine kinase that phosphorylates a spectrum of substrate proteins overlapping that of the insulin receptor (Section 10.3.1). GH causes some of its effects in the target tissues directly, whereas others are mediated by insulin-like growth factor 1 (IGF-1), which is produced in the liver and in other organs in response to GH. The receptor for IGF-1 resembles the insulin receptor.

Both GH and IGF-1 promote body growth. Longitudinal bone growth occurs mostly at the epiphyseal lines during adolescence, and excessive secretion of growth hormone in children accordingly causes gigantism. In adults, the epiphyseal lines are closed, and bones exposed to high levels of GH will assume plumper shapes without longitudinal growth; this condition is referred to as *acromegaly*.

Aside from its regulation by the hypothalamus, GH is also regulated by the current nutrient supply. Somewhat counterintuitively, GH secretion is stimulated by low blood glucose concentration and inhibited by high glucose levels. GH itself raises blood glucose. Failure of glucose to suppress GH secretion occurs in GH-overproducing tumors of the anterior pituitary; this is used in the diagnosis of such tumors. GH-producing tumors often remain responsive to hormonal control; they will cease to produce GH and may shrink under therapy with somatostatin or its analogs. The effect of excessive GH secretion can be countered with pegvisomant, a recombinant GH variant containing several strategic amino acid substitutions that turn it into a receptor antagonist. It is derivatized with polyethylene glycol (PEG) to extend its lifetime in the circulation.

Recombinant GH is used in children who lack sufficient secretion due to developmental defects of the pituitary. GH also increases muscle mass, which has motivated some professional athletes to inject themselves with recombinant GH. Such illicit use can be detected in serum samples by the absence of the minor 20-kDa splicing form in the recombinant hormone preparation.

GH is homologous to another pituitary hormone, prolactin, and like the latter promotes lactation. In some countries, cows are routinely injected with growth hormone to increase milk production.

7.3 THYROID GLAND HORMONES

The major hormones produced by the thyroid gland are tetraiodothyronine (thyroxine, T_4) and triiodothyronine (T_3; Figure 7.5C). In addition, specialized cells in the thyroid gland produce calcitonin, a peptide hormone unrelated to T_3 and T_4 that participates in the regulation of blood calcium levels (see Section 7.5). In this section, we will consider only T_3 and T_4.

Thyroid hormones have an essential role in fetal and postnatal development, and accordingly thyroid hormone deficiency becomes most severely manifest in infants. Thyroid hormones are also required in metabolic regulation throughout adult life. Their major mode of action is transcriptional regulation through thyroid hormone receptors, which belong to the nuclear hormone receptor family.

Various diseases cause either excessive or lacking thyroid hormone function. Most of these conditions can be treated with drugs.

7.3.1 Biosynthesis of thyroid hormones

The biosynthesis of T_3 and T_4 is activated by thyroid stimulating hormone (TSH), which is released by the anterior pituitary gland, and itself is controlled by the hypothalamic thyrotropin releasing hormone (TRH). Both TSH and TRH are peptide hormones and act through G protein-coupled receptors. Apart from increasing the activity of enzymes and transporters involved in thyroid

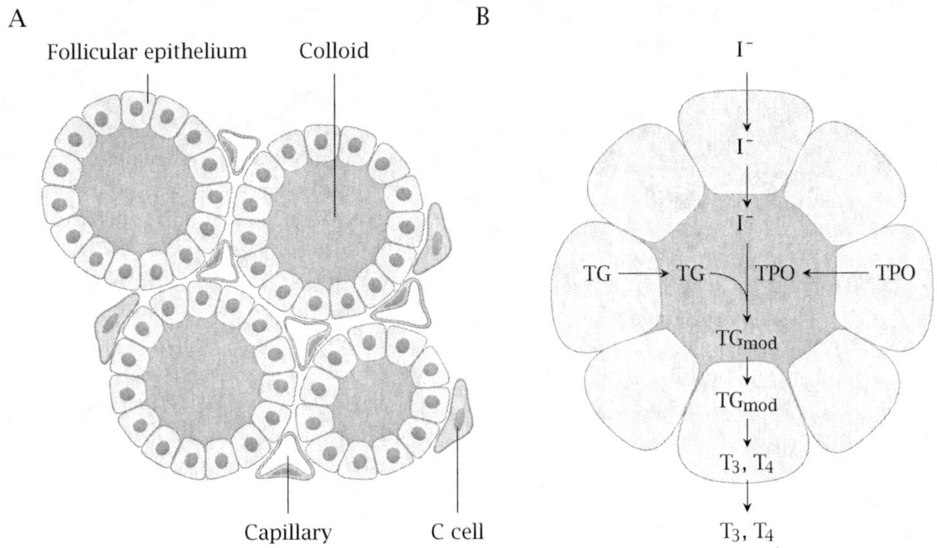

Figure 7.4 Tissue structure of the thyroid gland, and localization of hormone synthesis. **A**: Follicular epithelial cells are arranged in follicles that enclose a central lumen filled with protein-rich colloid. C cells are located outside the follicles; they produce the peptide hormone calcitonin. **B**: Epithelial cells synthesize thyroglobulin (TG) and thyroid peroxidase (TPO) and secrete them into the follicle lumen. They also accumulate iodide and transport it into the follicle lumen, where it is used by thyroid peroxidase for the iodination of tyrosine side chains (see Figure 7.5). The modified thyroglobulin (TG_{mod}) is taken up into the follicle cells again via endocytosis and proteolytically degraded, which yields the free hormones triiodothyronine (T_3) and thyroxine (T_4).

hormone biosynthesis, TSH also stimulates the proliferation of thyroid gland cells.

The epithelial cells of the thyroid gland surround small cavities or follicles. The crucial steps in the biosynthesis of T_3 and T_4 occur within the protein-rich colloid that is produced by the epithelia and secreted into the follicle lumen (Figure 7.4). In the first stage, iodine is incorporated into several tyrosine residues in the precursor protein thyroglobulin. This reaction involves H_2O_2 and is mediated by thyroid peroxidase.[1] The side chain of one mono- or di-iodinated tyrosine is then transferred to the phenolic hydroxyl group of another. The follicle cells then take up the modified thyroglobulin, break it down, and release the hormones (Figure 7.5).

The thyroid gland releases both T_3 and T_4. In the periphery, T_4 is converted to T_3 by enzymatic deiodination. T_3 binds to thyroid hormone receptors with much higher affinity than T_4; the latter therefore serves mainly as a reservoir of T_3. Degradation of T_3 and T_4 variously involves deiodination, oxidative deamination, and conjugation.

[1] The iodination reaction catalyzed by thyroid peroxidase is mechanistically similar to those that occur in commonly used in vitro protocols for labeling proteins with radioactive iodine. Such protocols use either lactoperoxidase [15] or chemical oxidation agents to convert ^{125}I or ^{131}I to hypoiodite, which then reacts spontaneously with tyrosine residues.

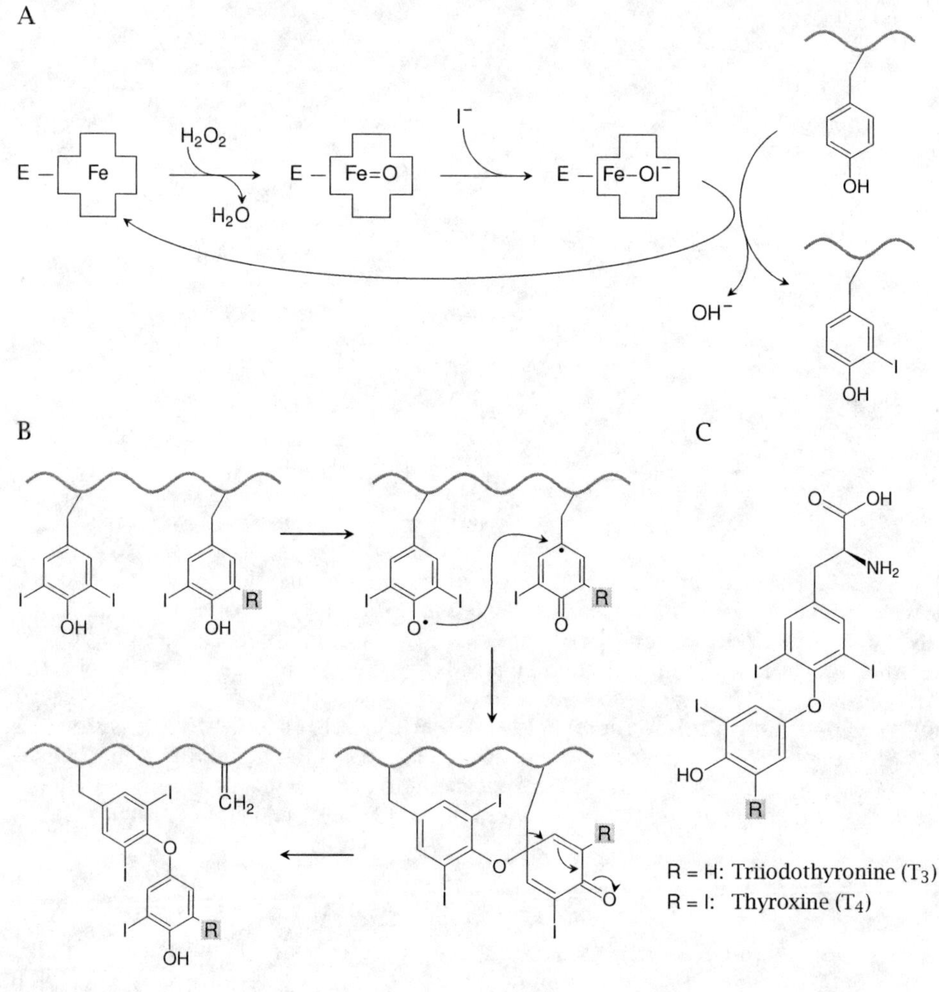

Figure 7.5 Biosynthesis of the thyroid hormones. **A:** Iodination of tyrosyl residues in thyroglobulin. Thyroid peroxidase contains a heme moiety that generates hypoiodite from iodide and H_2O_2. The hypoiodite then reacts with tyrosine to form monoiodotyrosine. Diiodotyrosine is formed in the same manner. **B:** The coupling reaction between two iodinated tyrosyl side chains is also catalyzed by thyroid peroxidase. The enzyme abstracts an electron from each of the substrate side chains, which then undergo radical recombination [23]. The reaction leaves a dehydroalanine residue in the protein backbone [11]. C: Thyroxine (T_4) and triiodothyronine (T_3) are released from thyroglobuline by proteolysis.

7.3.2 Mode of action and functional effects of thyroid hormones

The TRHA gene encodes the thyroid hormone receptor TRα, whereas the TRHB gene encodes TRβ_1 and TRβ_2. These receptors bind to thyroid hormone response elements (TREs) in the DNA. TREs occur in several variations, but an important one consists of two copies of the nucleotide sequence AGGTCA, separated by several intervening nucleotides. The thyroid hormone receptors often form heterodimers with the retinoid X receptor that also cooperates with other hormone receptors (Figure 7.1).

The receptor dimers can bind to DNA with or without thyroid hormone bound. The hormone-bound receptor typically increases transcription. However, feedback inhibition of TSH synthesis in the pituitary gland is due to transcriptional repression by T_3 and so is an example of negative regulation by ligand-bound receptor. Transcriptional repression also occurs with unliganded receptor. Its significance is illustrated by the fact that knockout mice devoid of both TRα and TRβ are less sick than mice that express the receptors but lack the hormones [12].

The number of genes that are subject to transcriptional regulation by thyroid hormones is large, and the contribution of any individual gene and its change in expression to the overall physiological effect is difficult to establish. Lack of thyroid hormones, or hypothyroidism, inhibits growth and disrupts development of the nervous system.[2] An excessive level of thyroid hormones (hyperthyroidism) increases the metabolic rate as well as the heart rate and blood pressure. One underlying mechanism is the increased expression of mitochondrial uncoupling proteins, which increase the rate of cellular respiration by short-circuiting the proton gradient across the inner mitochondrial membrane and so produce heat without ATP synthesis.

Thyroid hormones lower blood lipid levels, and at least in animal experiments they also increase glucose tolerance. Apparently, these effects can be separated from unwanted effects on heart rate and blood pressure by selective activation of the TRβ receptor. TRβ-selective agonists like KB-141 (Figure 7.6) may become useful in the treatment of atherosclerosis and diabetes mellitus [5].

In addition to the nuclear hormone receptors, cell surface receptors for T_4 have been characterized [3]. Their relative significance in physiology and pathophysiology remains to be established, and they are not targeted by drugs in current clinical use.

7.3.3 Hypothyroidism

Congenital hypothyroidism is most commonly caused by deficient development of the thyroid gland. At one case per 3000 births, it is among the most common inborn diseases, and in developed countries newborns are routinely screened for it. In adults, a leading cause of hypothyroidism is inflammation, due either to an autoimmune reaction (Hashimoto's disease) or to virus infections. Hypothyroidism is treated with thyroxine, which is converted in vivo to T_3.

7.3.4 Hyperthyroidism

Autoimmune disease is also a common cause of hyperthyroidism. In this condition, which is known as Graves' disease, autoantibodies activate the TSH

[2] The role of thyroid hormones in ontogenetic development is most striking in frogs, in which they drive the metamorphosis of the water-bound tadpole to the amphibious adult shape. Lung development is part of this metamorphosis. Thyroid hormones appear to promote lung maturation in mammalian fetuses also [6].

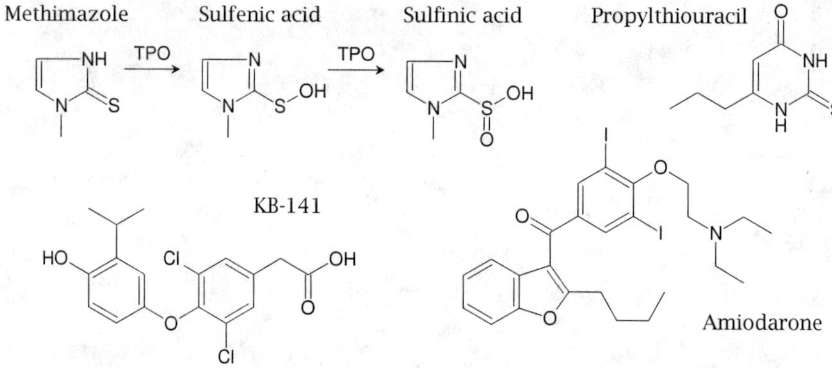

Figure 7.6 Drugs that influence thyroid hormone function. Propylthiouracil and methimazole are mechanism-based inhibitors of thyroid peroxidase. Shown next to methimazole are the putative oxidation products that are formed by thyroid peroxidase (TPO) and inactivate the enzyme [8]. KB-141 is a TRβ-selective agonist [26]. Amiodarone is an antiarrhythmic drug that, as a side effect, inhibits the enzymatic deiodination of thyroid hormones.

receptors, thereby inducing uncontrolled activation and proliferation of the thyroid gland cells. The autoimmune reaction ultimately destroys the gland, and hyperthyroidism gives way to hypothyroidism. Another prominent cause of hyperthyroidism are hormone-producing benign or malignant tumors.

Hyperthyroidism can be treated with thyrostatic drugs such as propylthiouracil and methimazole (Figure 7.6). These drugs act as suicide inhibitors of thyroid peroxidase, which converts them to sulfenic and sulfinic acids that then bind and inactivate the enzyme [8].

A more incisive form of therapy is the application of $^{131}I^-$, which is used on thyroid gland tumors (see end of section 12.2) and also in Graves' disease. Thyroid gland cells accumulate the isotope and are killed by its soft, short-range β radiation. A crucial feature of this form of radiotherapy is that it will also reach tumor metastases, as long as the tumor cells retain the ability to accumulate iodine. The effectiveness of radioiodine therapy can be increased by simultaneous application of TSH, which stimulates the cellular uptake of iodine.

Exhaustive $^{131}I^-$ therapy will destroy both tumorous and normal thyroid gland cells, and patients will subsequently require thyroxine substitution.

7.3.5 Iodine deficiency

A shortage of dietary iodine limits the output of thyroid hormones, which in turn disinhibits the pituitary production of TSH. The permanently stimulated thyroid gland becomes visibly enlarged, resulting in goiter. The enlarged gland usually produces enough T_3 and T_4 to avoid manifest hypothyroidism, so that the enlargement of the gland (goiter) as such is the leading symptom. Goiter has become rare in developed countries because of the widespread supplementation of foods with iodide or iodate, both of which can be taken up by the thyroid gland by specific transporters and utilized for hormone synthesis.[3]

7.3.6 Interference of drugs with thyroid hormone action

Drugs can interfere with thyroid hormones in several ways. Lithium inhibits the release of thyroid hormones from the gland. This can cause symptoms of hypothyroidism in psychiatric patients on lithium therapy. On the other hand, lithium can be used therapeutically in Graves' disease. In radioiodine therapy, inhibition of hormone secretion with lithium translates into a measurably increased retention of radioactivity within the thyroid gland [4,9].

The enzymatic deiodination of circulating T_4 to T_3 is inhibited by the antiarrhythmic drug amiodarone (Figure 7.6). Since T_3 is much more active than T_4, this inhibition can cause hypothyroidism.

In the bloodstream, thyroid hormones are mostly bound to a special transport protein, thyroxine binding globulin (TBG). Many drugs can compete for binding to TBG. This is of little physiological consequence, as it is only the unbound fraction of the hormone that is active; however, it can distort the results of laboratory assays that detect the total of bound and free thyroid hormones.

7.4 STEROID HORMONES

Steroid hormones are produced by the adrenal glands as well as the testes and ovaries. They are synthesized from cholesterol, mostly by mitochondrial cytochrome P450 enzymes. They control a wide range of functions, including sexual organ development and function, electrolyte balance, inflammatory activity, and metabolism. Each of these physiological functions is relevant to drug therapy with steroid hormones.

Steroid hormone receptors belong to the nuclear hormone receptor family (see Table 7.1). Each individual steroid hormone binds preferentially to one of these receptors, which determines the functional family it belongs to. This specificity of steroid hormones for their receptors is not absolute, and some level of cross-activation exists, particularly at high steroid concentrations. Receptor specificity can be higher with synthetic hormone analogs, which therefore are often preferable for drug therapy.

Before they bind to their cognate hormones, steroid hormone receptors reside in the cytosol, where they exist in a complex with molecular chaperones and so-called immunophilin proteins. Hormone binding leads to dissociation of these complexes, formation of receptor homodimers, their translocation into the nucleus, and DNA binding.

Apart from their often superior specificity, synthetic steroids also address the issue of bioavailability. Natural steroid hormones undergo inactivation through cytochrome P450-mediated and conjugations in the liver, and therefore undergo a strong first-pass effect after oral ingestion. Synthetic agonists

[3]Other ion species may inhibit iodide uptake; this is the case with perchlorate (ClO_3^-) and isothiocyanate (NCS^-). Pertechnetate (TcO_4^-), on the other hand, is a substrate for active uptake; this is used in diagnostic γ imaging of the thyroid gland.

with greater metabolic stability are available for all classes of steroid hormone receptors.

In addition to the transcriptional effects that are mediated by binding of the nuclear hormone receptors to their cognate DNA response elements, various other mechanisms of steroid hormone signaling have been described. Hormone-bound glucocorticoid receptor directly binds and inhibits the proinflammatory transcriptional regulators AP-1 and NF-κB. Several ion channels and other membrane proteins are subject to direct regulation by steroid hormones [13].[4]

7.4.1 Adrenal steroids

The adrenal gland consists of the medulla, which produces epinephrine and norepinephrine (Section 6.8.2), and the cortex, which produces steroid hormones. The latter comprise glucocorticoids, mineralocorticoids, and adrenal androgens. The biosynthetic pathways of the adrenal steroids are shown in Figure 7.7.

Synthesis of glucocorticoids and mineralocorticoids is driven by the hypophyseal adrenocorticotropic hormone (ACTH), whereas production of mineralocorticoids is driven by angiotensin (see Figure 1.5).

Glucocorticoids. The most important endogenous glucocorticoid is cortisol. The DNA consensus sequences that are recognized by the ligand-bound glucocorticoid receptor (GR) homodimers are referred to as *glucocorticoid response elements* (GREs). Binding of the receptor to GRE either induces or represses transcription of genes in the vicinity. A prominent effect of GRE-mediated transcriptional induction is the increased expression of enzymes of gluconeogenesis, which increases the availability of blood glucose. Adrenergic receptors are also induced, which amplifies the insulin-antagonistic effect of epinephrine and raises the blood pressure. Symptomatic diabetes and hypertension therefore accompany high levels of glucocorticoid activity, which may result from either hormone-producing tumors or prolonged drug therapy with glucocorticoids.

Hormone-bound glucocorticoid receptor can also signal in an GRE-independent manner. An important example is its direct inhibitory interaction with the proinflammatory transcription factors AP-1 [25] and NF-κB [22]. Through this so-called trans-repression, glucocorticoids reduce the expression of cyclooxygenase 2, inducible nitric oxide synthase (iNOS), and several peptide mediators of inflammation. Conversely, some of these mediators—interleukins 1, 2 and 6, as well as tumor necrosis factor TNF-α—induce the secretion of the hypothalamic corticotropin releasing hormone (CRH) and of ACTH. This makes cortisol an essential part of a negative feedback loop that contains the intensity of

[4] There exists a class of steroid mediators, the so-called *neurosteroids*, whose primary targets are the $GABA_A$ receptor and other ion channels [21]. As well, the primary target of the steroidal plant glycoside ouabain is Na^+/K^+-ATPase (Section 6.2). Considering that steroid hormones are derived from cholesterol, which in animals is an essential cell membrane constituent, it appears possible that they initially evolved as regulators of membrane protein function and later acquired transcriptional regulation as an additional mode of action. While estrogens signal mostly through their nuclear receptor, a GPCR with high affinity for estrogens has recently been characterized. The role of this receptor and its interaction with nuclear receptor-mediated signaling is still tentative [14].

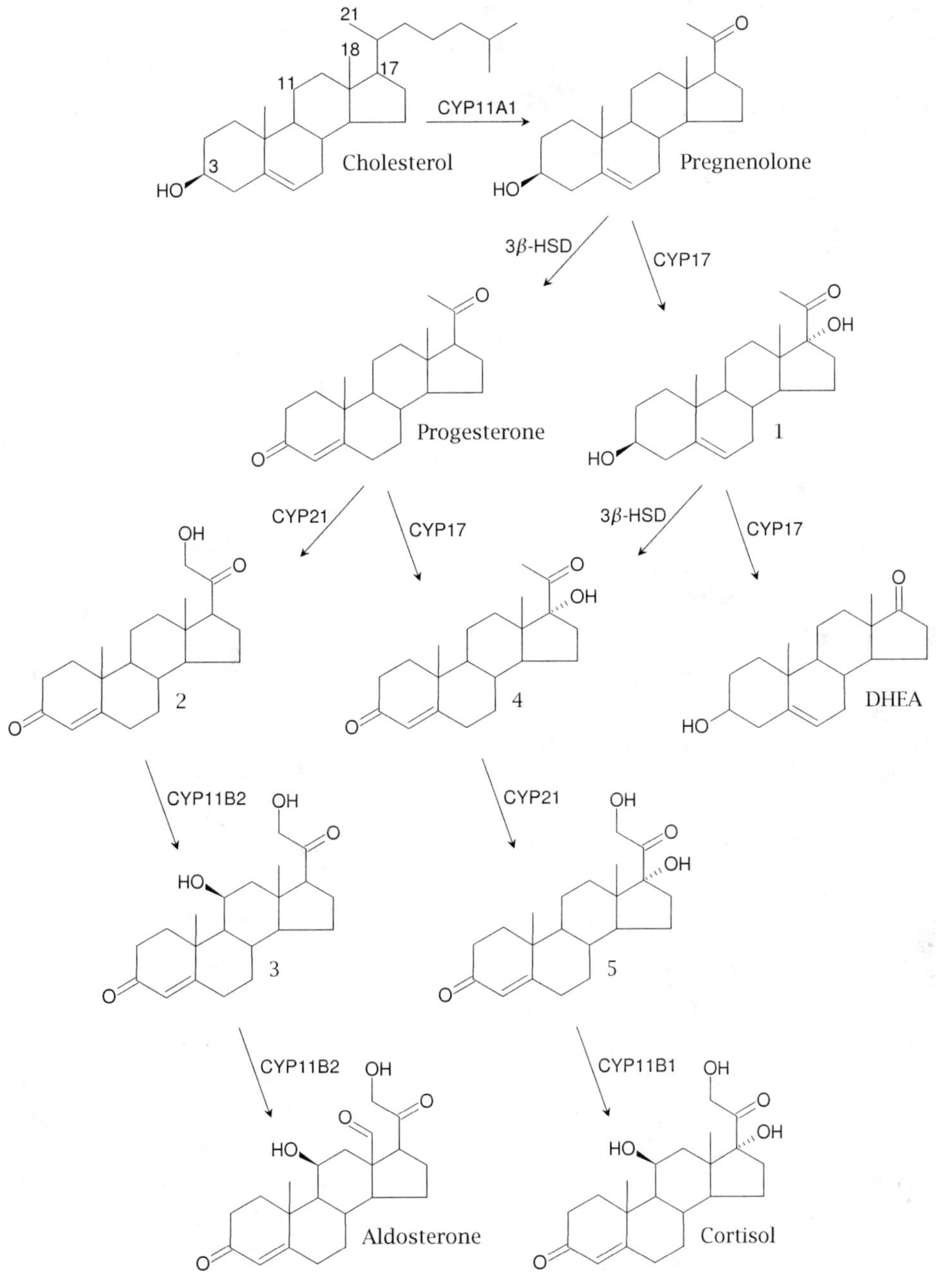

Figure 7.7 Biosynthesis of the adrenal steroid hormones cortisol, aldosterone, and dehydroepiandrosterone (DHEA) in the cortex of the adrenal glands. Enzymes: CYP11A1, cholesterol side chain cleavage enzyme; CYP11B1, steroid 11β-hydroxylase; CYP11B2, aldosterone synthase; CYP17, steroid 17α-hydroxylase; CYP21, steroid 21-hydroxylase; 3β-HSD, 3 β-hydroxysteroid dehydrogenase. All enzymes except 3β-HSD belong to the cytochrome P450 family. The numbers of the carbon atoms that are modified by the enzymes are indicated in the cholesterol molecule. Numbered intermediates: (1) 17α-hydroxypregnenolone, (2) 11-deoxycorticosterone, (3) corticosterone, (4) 17α-hydroxyprogesterone, (5) 11-deoxycortisol.

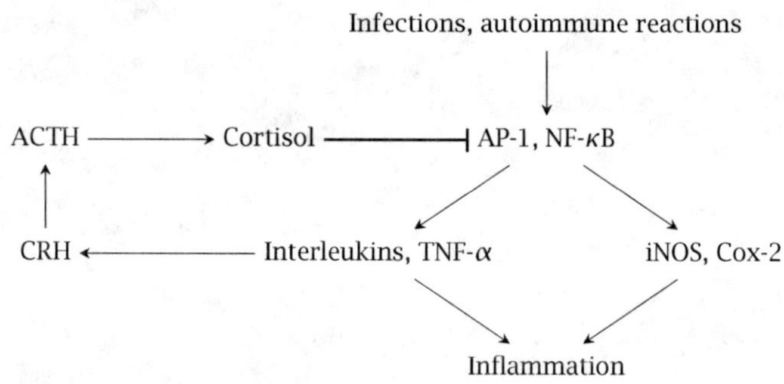

Figure 7.8 Role of cortisol in the control of inflammatory activity. Glucocorticoid receptor bound to cortisol inhibits the proinflammatory transcription factors AP-1 and NF-κB through trans-repression. Interleukins 1, 2 and 6 as well as TNF-α are induced by AP-1 and NF-κB and stimulate the secretion of CRH.

inflammation (Figure 7.8). The anti-inflammatory activity of glucocorticoids is superior to that of cyclooxygenase inhibitors (see Chapter 9) and is widely used in the treatment of arthritis, asthma and autoimmune diseases.

When studied in vitro, cortisol activates both the GR and the mineralocorticoid receptor (MR), and both with similar activity. How, then, does cortisol selectively activate the GR in vivo? Cells that express the MR also express the enzyme 11β-hydroxysteroid dehydrogenase, which converts the 11β-hydroxy group of the cortisol molecule to a keto group. The resulting molecule, cortisone, is inactive at both the GR and the MR. The major mineralocorticoid, aldosterone, occurs mostly in the hemiacetal form that exposes no free 11β-hydroxy group and therefore is exempt from the enzymatic inactivation (Figure 7.10A).

Even in the presence of 11β-hydroxysteroid dehydrogenase, some cortisol molecules escape inactivation, and therefore at high dosages cortisol will activate the MR to some degree. For sustained therapy, cortisol analogs with greater selectivity for the GR have been developed, for example, prednisolone and dexamethasone. Like cortisol, these drugs activate both GRE-dependent transcriptional regulation and the trans-repression pathway of GR-mediated signaling, as do all other glucocorticoid drugs in current clinical use. Intriguingly, however, experimental GR agonists have been developed that preferentially activate trans-repression (Figure 7.9). In animal models, some of these drugs exhibit anti-inflammatory activity similar to that of dexamethasone [24]. Since the metabolic and cardiovascular side effects of GR agonists depend mostly on transcriptional induction via GRE, such compounds may be better suited for long term anti-inflammatory treatment.

Glucocorticoids are essential for our ability to cope with stress of all kinds, and deficient endogenous production of cortisol is a serious, sometimes fatal situation. It can arise through the destruction of the adrenal glands (Addison's disease), or through the sudden discontinuation of prolonged glucocorticoid therapy that has severely suppressed endogenous production via negative feedback on CRH and ACTH secretion. Therefore, long-term corticosteroid therapy

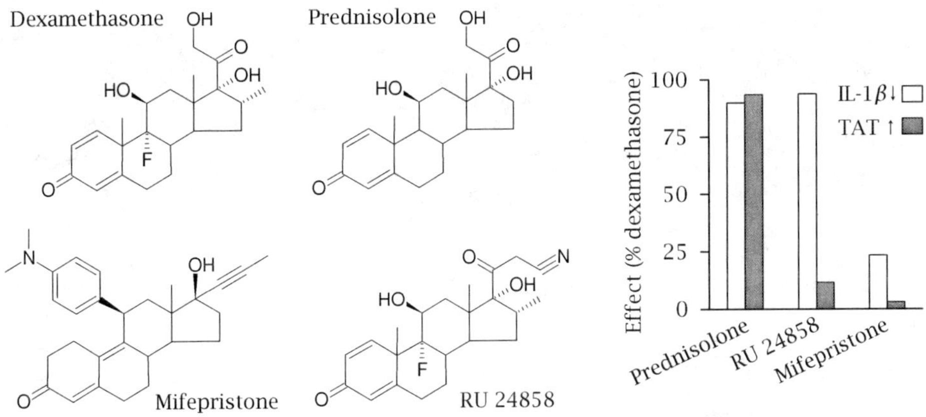

Figure 7.9 Structures of glucocorticoid receptor agonists and of the antagonist mifepristone, and their transcriptional effects. Tyrosine transaminase (TAT) is part of a degradation pathway that provides substrate carbon for gluconeogenesis and is transcriptionally induced via GRE. Interleukin 1β (IL-1β) is down-regulated via trans-repression of AP-1. Traditional GR agonists such as dexamethasone and prednisolone activate both GRE-mediated responses and trans-repression, whereas the experimental agonist RU 24858 activates AP-1 trans-repression only. Note that mifepristone is an antagonist of the progestin receptor also. (Figure redrawn from original data in Ref. [24].).

must always be terminated by way of slow successive dosage reductions that give the adrenal glands time to recover.

Mineralocorticoids. The major mineralocorticoid is aldosterone (Figure 7.7). The mineralocorticoid receptor (MR) is expressed mostly in epithelial tissues that are involved in ion transport, in particular the kidney tubules and collecting ducts. Activation of the MR causes transcriptional induction of membrane proteins that promote Na^+ reuptake from the nascent urine, such as Na^+ channels at the apical side and Na^+/K^+-ATPase at the basolateral side of the epithelial cells. Reuptake of Na^+ also promotes reuptake of water through osmosis.

Aldosterone synthesis is driven mostly by angiotensin and persists in the absence of ACTH. The aldosterone antagonist spironolactone (Figure 7.10) has long been used in a variety of diseases that involve sodium and water retention. It has several side effects that result from activation of progestin receptors (gynaecomastia or breast growth, in men) and androgen receptors (hirsutism or increased body hair growth, in women). The more recently introduced antagonist eplerenone is reportedly less hampered by these side effects [1]. In adrenal insufficiency, fludrocortisone can be used for aldosterone replacement therapy.

7.4.2 Gonadal steroid hormones

Gonadal steroids include androgens, estrogens, and progestins. Androgens are predominant in men and drive male sexual function, while estrogens and progestins predominate in women and sustain female sexual function. Nevertheless, men also require estrogens, and women require androgens.

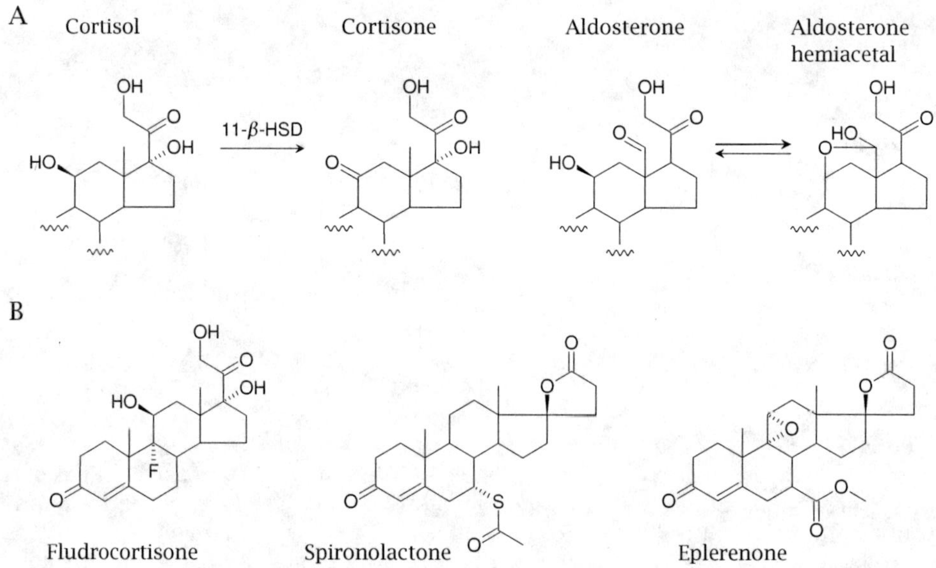

Figure 7.10 Mineralocorticoid receptor ligands. **A:** Role of 11-β-hydroxysteroid dehydrogenase (11-β-HSD) in selective mineralocorticoid receptor activation by aldosterone. Both cortisol and aldosterone activate the MR. Cells that express the MR also express 11-β-HSD, which dehydrogenates cortisol to cortisone, which is inactive. Aldosterone is protected from this reaction because it mostly exists in the hemiacetal form. (Squiggly lines indicate partially rendered structures.) **B:** Structures of the mineralocorticoid receptor agonist fludrocortisone and of the antagonists spironolactone and eplerenone.

Secretion of gonadal steroids is controlled by pituitary hormones. Follicle-stimulating hormone (FSH) promotes formation of estrogens, whereas luteinizing hormone (LH) promotes formation of androgens and progestins. The placenta produces human chorionic gonadotropin (hCG) that has LH activity and sustains the production of elevated levels of estrogens and progestins by the ovaries throughout pregnancy.

The most important progestin is progesterone, which also occurs as an intermediate in adrenal steroid synthesis (Figure 7.7). Androgens and estrogens are synthesized in the gonads from dehydroepiandrosterone (Figure 7.11), which in turn is produced in either the adrenal glands or the gonads themselves.

Androgens. The most important androgen is testosterone. Inside some target tissues, it is converted by 5α-reductase to dihydrotestosterone, which binds more avidly to the androgen receptor. The transient production of testosterone in male fetuses drives the male differentiation of the sexual organs. Its level of production remains low throughout childhood and then rises steeply in puberty, during which it drives maturation of the sexual organs and the expression of the secondary male body characteristics. Testosterone is produced throughout adulthood at slowly declining levels. It sustains spermatogenesis, libido, and male potency. It also increases muscle strength and promotes ag-

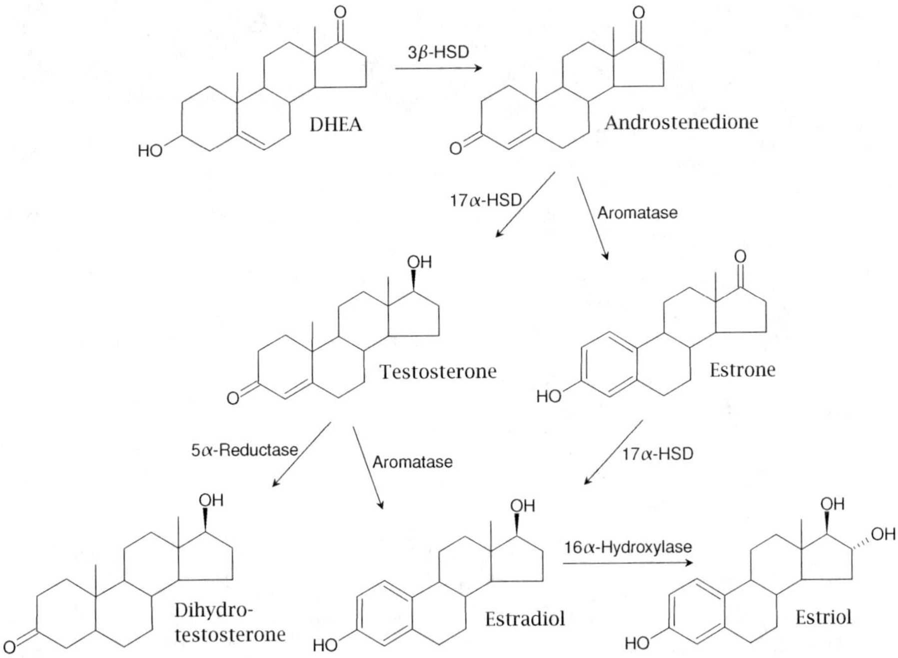

Figure 7.11 Biosynthesis of androgens and estrogens. Abbreviations: DHEA, dehydroepiandrosterone; 3β-SDH and 17α-SDH, 3β- and 17α-hydroxysteroid dehydrogenase, respectively. Aromatase and 16α-hydroxylase are cytochrome P450 enzymes.

gressive behavior. The level of testosterone inside the testicles is much higher than in the rest of the body; this is believed to be required for spermatogenesis.

Testicular androgen production depends on stimulation by LH, and androgens in turn exert feedback inhibition on gonadotropin-releasing hormone (GnRH), LH and FSH. Application of androgens will therefore suppress testicular testosterone production, and in the process also inhibit spermatogenesis. This can in principle be exploited for male contraception. However, the considerable side effects of this approach [16] have so far prevented its adoption in practice.

Infertility due to suppression of spermatogenesis is also observed in males who use androgens to increase muscle mass or improve athletic performance; this is usually reversible after discontinuation. However, concerns have been voiced about more serious side effects such as prostate cancer, particularly since such drug use will typically occur without medical guidance and may involve excessive dosages.

Natural androgens are subject to the first-pass effect and have low bioavailability. Synthetic analogs have been created that are metabolized more slowly and can be used orally (Figure 7.12). Various testosterone esters can be applied by injection. Endogenous androgen production can be stimulated with hCG. This is not relevant to regular treatment regimens but reportedly has been used by athletes in order to evade detection of androgen abuse.

Androgen activity can be reduced with receptor antagonists or with 5α-reductase inhibitors, which block the conversion of testosterone to the more potent dihydrotestosterone. Both types of drugs are used in the treatment of androgen-dependent prostate cancers. A related form of therapy is that with GnRH and analogs, which indirectly lower the secretion of endogenous androgens (see Section 14.4.1).

Estrogens and progestins. In men, sexual hormone secretion essentially maintains a steady state. In women, the menstrual cycle, pregnancy, and birth require more complicated regulatory mechanisms. Each menstrual cycle sees the maturation of an individual oocyte and its surrounding hormone-producing follicle in the ovaries. The estrogens produced by this follicle exert feedback inhibition of pituitary FSH and LH secretion by the pituitary throughout most of the cycle. However, in midcycle, the pituitary responds to estrogens with a transient spike of LH that triggers the release of the egg from the follicle (ovulation). The follicle then is transformed into the corpus luteum, which sustains the production of elevated levels of estrogens and progestins for the next two weeks, inducing proliferation of the mucous membrane in the womb.

If the egg is fertilized and implanted during the second half of the cycle, the developing placenta begins producing human chorionic gonadotropin (hCG), a peptide hormone that is analogous to LH and sustains the corpus luteum throughout pregnancy. If no implantation occurs, the corpus luteum regresses, the production of estrogens and progestins subsides, and the hormone-dependent mucous membrane of the womb is cast off in the menstrual bleeding.

As with other steroid hormones, natural estrogens and progestins have poor bioavailability, and synthetic derivatives with greater metabolic stability are used as drugs. By far the most common application of estrogens and progestins, alone or in combination, is for oral contraception, which works by way of suppressing pituitary LH and FSH secretion, thus preventing maturation of the oocyte and the follicle. Estrogens and progestins are also used to treat a variety of disturbances that arise from the lack of endogenous hormone production in postmenopausal women. In particular, estrogens are often prescribed to combat osteoporosis, a condition in which thinning of the tissue structure of bones renders them prone to fractures and compression (see Section 7.5).[5]

In addition to inducing and sustaining the growth of the uterine mucous membrane, estrogens and progestins also drive the proliferation of mammary gland tissue. Breast cancers frequently remain dependent on estrogen and progestin stimulation, and hormone receptor antagonists are commonly used in their treatment (see Section 12.2). On the other hand, oral contraceptives, and to a greater extent postmenopausal hormone substitution, are associated with increased risk of breast cancer.

Apart from their use as antitumor drugs, estrogen and progestin receptor antagonists are also used to manipulate reproductive function. The progestin and glucocorticoid receptor antagonist mifepristone is used for abortion in the

[5]Estrogens affect bone growth in both sexes. In men who lack aromatase activity, the absence of estrogens prevents closure of the epiphyseal lines, which causes excessive longitudinal bone growth.

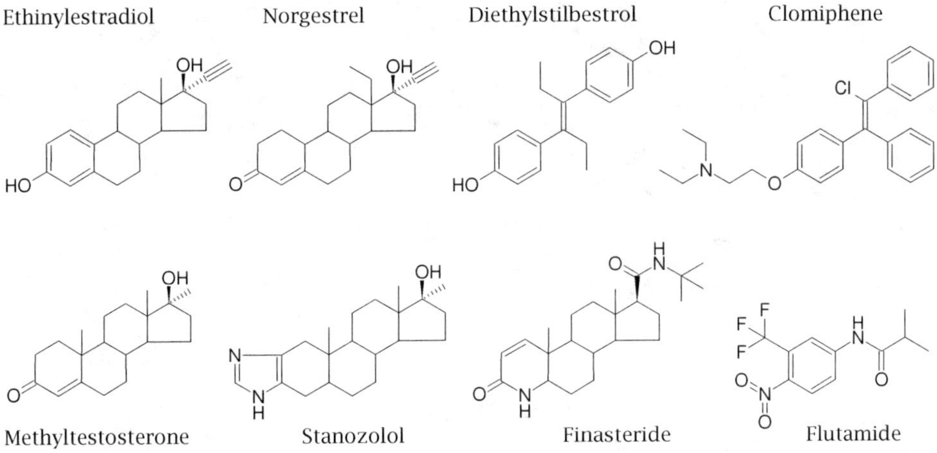

Figure 7.12 Synthetic analogs of gonadal steroids. Ethinylestradiol, norgestrel, and methyltestosterone are orally available estrogen, progestin, and androgen analogs, respectively. Diethylstilbestrol has estrogen activity but was withdrawn because it induced cancers. Clomiphene inhibits the hypophyseal feedback of gonadal steroids and is used in fertility treatment. Stanozolol is an oral androgen that is popular among body builders. Finasteride is a 5α-reductase inhibitor, and flutamide is an androgen receptor antagonist; both are used in the treatment of prostate cancer.

early stages of pregnancy. The antagonist clomiphene acts more strongly in the pituitary than in peripheral tissues. It increases pituitary secretion of LH and FSH, which can be used to treat anovulation. The increased stimulation of the ovaries may result in the simultaneous maturation of multiple oocytes and follicles, which in turn may lead to multiple pregnancies.

7.5 ENDOCRINE CONTROL OF BONE MINERALIZATION

The extracellular bone material is a composite of collagen fibers, which endow it with tensile strength, and a crystalline mineral that gives it compressive strength. This mineral consists mostly of hydroxyapatite [$Ca_5(PO_4)_3OH$] but also contains some other ions such as carbonate, fluoride and magnesium.

The bone matrix also serves as a reservoir for soluble calcium and phosphate, which fulfill essential roles in cell excitability and metabolism, and therefore must be kept within a narrow concentration range at all times. This regulation is effected by hormones that simultaneously control the intestinal uptake and renal elimination of calcium and phosphate, as well as their deposition or mobilization in the bone matrix. Accordingly, bone mineralization can be disturbed by over- or underproduction of the hormones that regulate it. Moreover, lack of soluble calcium and phosphate due to deficient intestinal uptake or excessive renal excretion will also deplete the bone mineral.

The most prevalent bone disease is osteoporosis. Its most common cause is the decline of endocrine growth stimulus in elderly people, and particularly in postmenopausal women. In osteoporosis, the entire bone substance is thinned

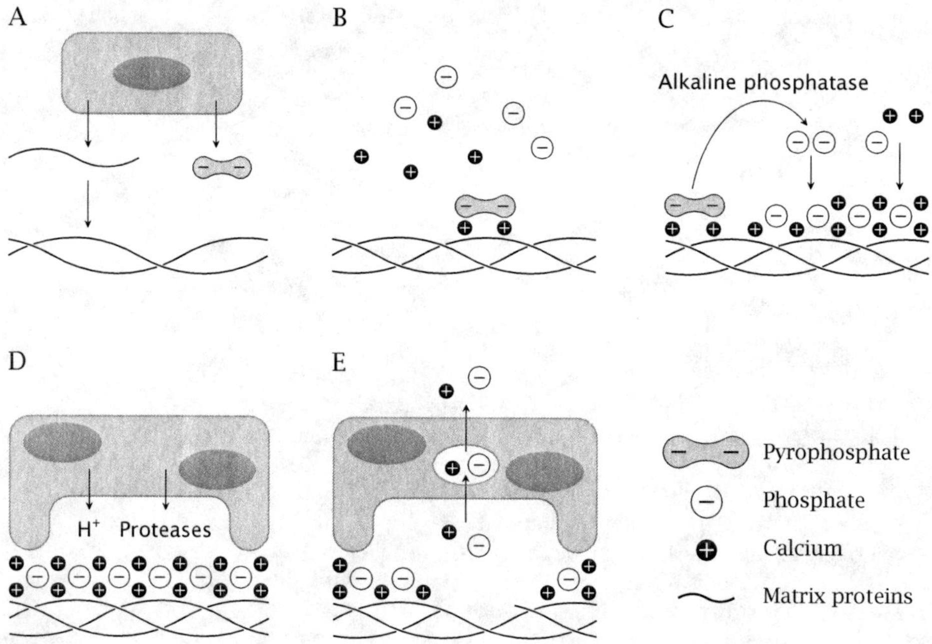

Figure 7.13 Formation (A–C) and resorption (D,E) of bone matrix. **A:** Osteoblast cells produce bone matrix proteins that self-assemble into fibrils. They also produce pyrophosphate. **B:** Pyrophosphate binds to nucleating bone mineral crystals and inhibits their premature growth. **C:** Alkaline phosphatase, which is also produced by osteoblasts, breaks up the pyrophosphate. This triggers the deposition of bone mineral (apatite). **D:** Osteoclasts bind to the surface of the bone matrix and secrete protons that dissolve the mineral and proteases that digest the matrix proteins. **E:** Calcium and phosphate ions are taken up by endocytosis and released into the circulation.

out, not just its mineral component. Nevertheless, improving bone mineralization is an important aspect of its treatment.

7.5.1 Bone tissue turnover

The ability of bone tissue to dissolve and form again is directly evident in the healing of bone fractures. These are a rare occurrence; however, like other composite materials, the bone matrix is prone to sustain microfractures under mechanical stress, and therefore replacement of bone tissue is going on all the time.

New bone matrix is deposited by one specialized cell type, the osteoblasts, while bone matrix dissolution is carried out by another cell type, the osteoclasts (Figure 7.13). Osteocytes remain permanently embedded in the bone substance. They act as sensors of bone stress and lesions, and in response initiate the process of bone "remodeling," that is, the localized and orderly resorption and replacement of bone tissue.

In the formation of new bone tissue, the production of collagen and other extracellular matrix macromolecules precedes mineralization. The deposition of apatite is a facile process, since the extracellular concentrations of calcium and phosphate are above their solubility limit. Premature precipitation is kept in check by pyrophosphate [27]. The deposition of bone mineral is triggered by alkaline phosphatase, which is secreted by osteoblasts and cleaves pyrophosphate [2].

The solubility of hydroxyapatite is increased at low pH. Osteoclasts secrete protons, which dissolve the hydroxyapatite. The effect is localized by the adhesion of osteoclasts to the bone matrix surface; the patch underneath the osteoclast cell is sealed off and selectively acidified. Dissolution of the mineral is followed by the proteolytic digestion of collagen by acidic extracellular proteases.

Osteoblasts and osteoclasts are grouped into functional units, and they regulate each other's activities. Osteoclast precursor cells express the receptor for activation of NF-κB, or RANK, which is activated by a membrane protein found in osteoblasts, RANKL (RANK ligand). This interaction induces differentiation and activation of the osteoclasts. Bone tissue and mineral turnover is also subject to regulation by several hormones and cytokines. One mechanism of hormonal regulation is the induction of RANK.

7.5.2 Parathyroid hormone and calcitonin

Parathyroid hormone (PTH) is a peptide hormone produced by the parathyroid glands, which are four small appendages of the thyroid gland. Its purpose is the short-term control of plasma calcium concentration, which must be kept stable within a narrow range for the sake of proper function of the excitable cells. PTH acts on cognate GPCRs. It affects the activities of several organs: (1) it causes the release of calcium and phosphate from the bone, (2) increases the reabsorption of calcium but reduces that of phosphate in the kidneys, and (3) activates the final step in the metabolic activation of vitamin D, which augments the effect of the latter on calcium uptake (see Section 7.5.3). All these effects of PTH increase the plasma concentration of calcium, which in turn inhibits the secretion of PTH itself. This negative feedback is mediated by the calcium-sensing receptor, an interesting and unusual GPCR that responds directly to extracellular calcium and is expressed by the parathyroid gland cells.

Excessive production of PTH occurs in adenomas (benign tumors) of the parathyroid glands. This condition is referred to as *primary hyperparathyroidism* and causes excessively high plasma levels of calcium. In contrast, in secondary hyperparathyroidism, abnormally high levels of PTH result from the parathyroid glands "just doing their job" of trying to normalize an excessively low plasma calcium level, which is due to renal loss or intestinal malabsorption of calcium. In both primary and secondary hyperthyroidism, one of the therapeutic options is the drug cinacalcet (Figure 7.14), which is an allosteric agonist of the calcium-sensing receptor. Cinacalcet therefore augments the inhibitory effect of calcium on the secretion of parathyroid hormone.

Considering that PTH mobilizes calcium from the bone, it is somewhat surprising that its intermittent use *increases* the bone density in osteoporosis, a condition that involves bone demineralization. An N-terminal fragment of PTH that retains full biological activity is quite effective in this application [17].

Calcitonin is another peptide hormone that is involved in the regulation of calcium. It is produced in the C cells of the thyroid gland (Figure 7.4). Calcitonin promotes deposition of calcium in the bone and is regarded as an antagonist to PTH. As with the latter, hormone secretion is regulated by calcium-sensing receptors. However, in this case, receptor activation increases hormone secretion, which is likely due to the coupling of the receptor to different G proteins. Calcitonin is used in the treatment of osteoporosis, too, but remarkably its effectiveness is lower than that of PTH.

7.5.3 Vitamin D

The name Vitamin D refers to cholecalciferol and ergocalciferol, which are related to cholesterol and ergosterol, respectively, as well as to the products of their metabolic activation. Ergocalciferol and cholecalciferol are functionally equivalent, and both occur in the diet.

Metabolically activated vitamin D stimulates a cognate nuclear hormone receptor that is required for the expression of intestinal transport proteins for calcium and phosphate. An overt lack of vitamin D causes a shortage of calcium and phosphate, which prevents proper bone mineralization. This renders the bone fragile and soft, which leads to fractures and bone deformities. The disease is known as rickets and is treated—or, preferably and more commonly, prevented—with vitamin D supplementation.

Strictly speaking, cholecalciferol is not really a vitamin, since it *can* be formed endogenously from 7-dehydrocholesterol, which in turn is an intermediate of cholesterol synthesis (Figure 7.14). However, this reaction is not enzymatically catalyzed. Instead, it is a photochemical process that requires one ultraviolet photon per molecule and therefore occurs only in the skin. The turnover of this reaction depends on the exposure of the skin to sunlight, and it is reduced by skin pigmentation.[6] Therefore, dietary supplementation is advisable for people living in moderate or cold climes in order to guarantee a sufficient supply.

To become biologically active, the product of the photochemical reaction has to undergo two subsequent hydroxylations, which are carried out by cytochrome P450 enzymes. The first hydroxylation occurs in the liver and is swift, and most of the vitamin D in the human body is found as the product of this reaction (25-hydroxy-cholecalciferol). The second hydroxylation occurs in the kidney and is regulated by PTH. The reaction product—1,25-dihydroxy-cholecalciferol, or calcitriol—exercises negative feedback on the secretion of PTH in the parathyroid gland.

[6]The essential role of cholecalciferol is responsible for the variation of human skin colors. While dark pigment protects the skin from damage by UV irradiation, the UV photons swallowed up by the pigment are lost to the synthesis of calciferol. When *Homo sapiens* left Africa for less sunny climates, the shortage of sunshine created a selective pressure for lighter skin, which increases the availability of photons for calciferol synthesis.

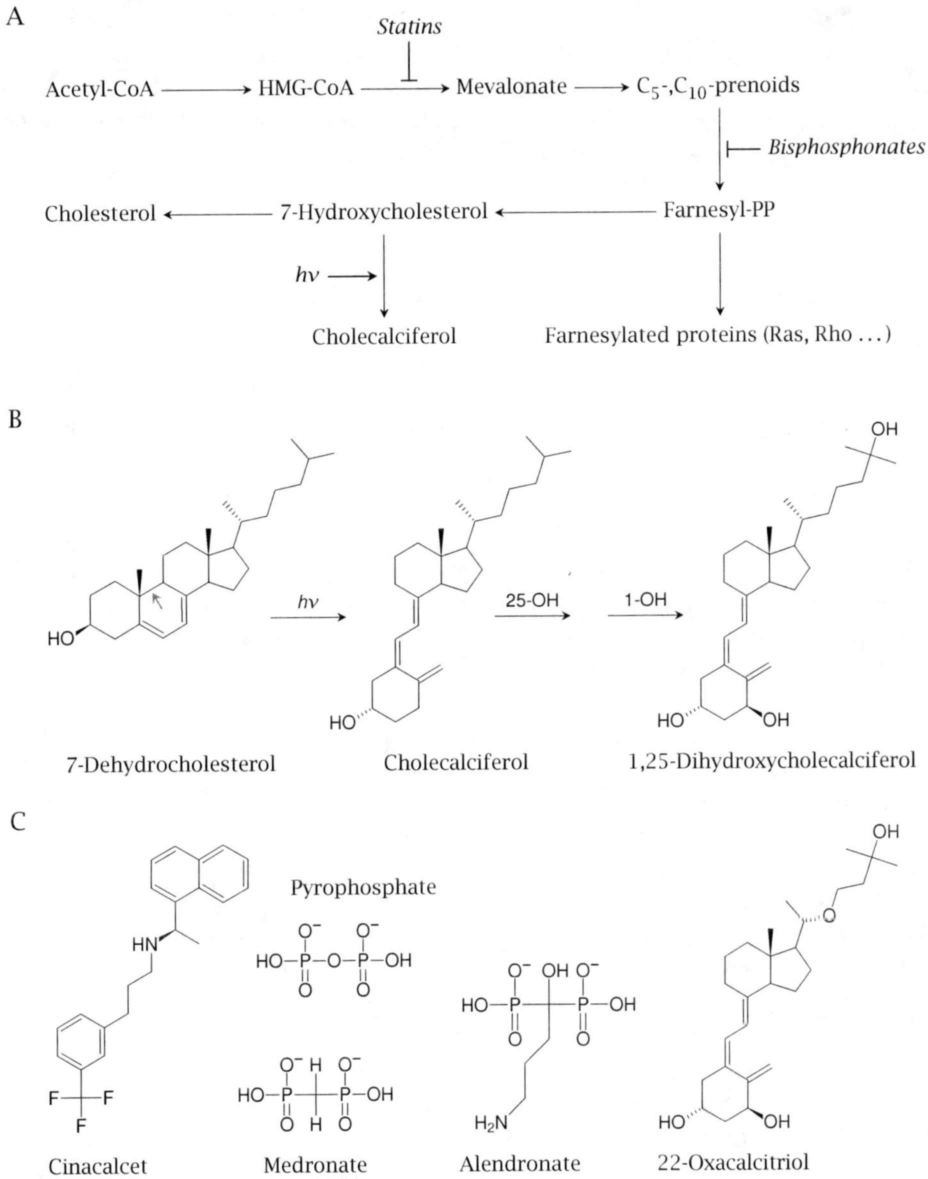

Figure 7.14 Sites of action of statins and bisphosphonates (A), endogenous biosynthesis of calcitriol (B), and drugs that influence calcium balance and bone mineralization (C). **A**: In the sterol synthesis pathway, both statins and bisphosphonates act upstream of protein farnesylation and of cholecalciferol synthesis. **B**: Photochemical breakage of a bond (arrow) in 7-dehydrocholesterol produces cholecalciferol. Two subsequent enzymatic hydroxylations yield 1,25-dihydroxycholecalciferol (calcitriol). **C**: Cinacalcet is an allosteric agonist of the calcium-sensing receptor in the parathyroid glands. It is used to suppress parathyroid hormone secretion in primary and secondary hyperparathyroidism. The bisphosphonates medronate and alendronate inhibit bone resorption. Pyrophosphate (not a drug) is shown for comparison. 22-Oxacalcitriol is an analog of calcitriol that is used to inhibit parathyroid hormone secretion.

Calcitriol and several equivalent compounds are available for oral and parenteral therapy. Some analogs, for example 22-oxacalcitriol (Figure 7.14C), inhibit the secretion of PTH but do not have pronounced effects on calcium and phosphate uptake, and are preferentially used in chronic kidney disease associated with secondary hyperparathyroidism.

7.5.4 Bisphosphonates

Bisphosphonates structurally resemble pyrophosphate but have a carbon atom in the middle, which renders them resistant to hydrolysis. Their mode of action is different from pyrophosphate, however, and rather than inhibiting bone mineralization, they promote it; they are widely used in the treatment of osteoporosis.

Bisphosphonates bind to the surface of bone mineral, and thus may directly slow down its dissolution. When they are dissolved along with the mineral, they are taken up by the osteoclast cells, to which they are thereby preferentially targeted. Inside the cell, bisphosphonates inhibit farnesylpyrophosphate synthetase. Farnesylpyrophosphate is required for the posttranslational modification and activation of some membrane-associated proteins that are essential for osteoclast function. The reduced osteoclast activity will then inhibit further bone demineralization.

Farnesylpyrophosphate is also an intermediate in the cholesterol synthesis pathway. This pathway begins with hydroxymethylglutaryl-CoA (HMG-CoA) reductase, which is inhibited by the so-called statin drugs (see Section 10.4). From this, one might expect some functional similarity between statins and bisphosphonates, and indeed some effects of statins on bone metabolism have been observed and discussed along these lines [10,18]. Furthermore, 7-dehydrocholesterol, the precursor of photochemically formed cholecalciferol, is downstream of both enzymes, from which one might expect an interference of both bisphosphonates and statins with vitamin D supply (Figure 7.14A). Surprisingly, a clinical study on atorvastatin actually found the opposite to be true [19].

Bisphosphonates are also used in metastatic bone cancer. The intent of this application is not curative; instead, the purpose is to slow the loss of bone substance that occurs within the metastatic tumor foci and is due to the overstimulation of osteoclasts by inflammatory cytokines. The dosages used in this application are much higher than those used in osteoporosis and induce apoptosis of the osteoclasts.

STUDY QUESTIONS

✎ **7.1** In Section 7.4.1, it was mentioned that some experimental glucocorticoid receptor agonists preferentially activate trans-repression, while having little transcriptional activity. How can this behavior be explained?

✎ **7.2** Hydroxyapatite, $Ca_5(PO_4)_3OH$, is the major component of bone mineral and is dissolved by weak or strong acid. In the enamel of the teeth, hydroxyapatite is partly replaced by fluoroapatite, $Ca_5(PO_4)_3F$, which gives the teeth greater resistance to acid.

(1) Why is fluoroapatite is more resistant to acid than hydroxyapatite? (2) Increasing the incorporation of fluoride into the tooth mineral is a proven strategy to increase their resistance to caries, which is caused by the acids that result from the bacterial fermentation of sugars. Between the fluoridation of tap water and the direct application of fluorides to the surface of the teeth, which one would you expect to be more effective?

REFERENCES

[1] M. Abuannadi and J. H. O'Keefe, "Review article: eplerenone: an underused medication?," *J. Cardiovasc. Pharmacol. Ther.* 15 (2010), 318–325.

[2] W. N. Addison et al., "Pyrophosphate inhibits mineralization of osteoblast cultures by binding to mineral, up-regulating osteopontin, and inhibiting alkaline phosphatase activity," *J. Biol. Chem.* 282 (2007), 15872–15883.

[3] J. J. Bergh et al., "Integrin $\alpha_V\beta_3$ contains a cell surface receptor site for thyroid hormone that is linked to activation of mitogen-activated protein kinase and induction of angiogenesis," *Endocrinology* 146 (2005), 2864–2871.

[4] F. Bogazzi et al., "Treatment with lithium prevents serum thyroid hormone increase after thionamide withdrawal and radioiodine therapy in patients with Graves' disease," *J. Clin. Endocrinol. Metab.* 87 (2002), 4490–4495.

[5] G. Bryzgalova et al., "Anti-obesity, anti-diabetic, and lipid lowering effects of the thyroid receptor β subtype selective agonist KB-141," *J. Steroid Biochem. Mol. Biol.* 111 (2008), 262–267.

[6] L. Chan et al., "Antenatal triiodothyronine improves neonatal pulmonary function in preterm lambs," *J. Soc. Gynecol. Investig.* 5 (1998), 122–126.

[7] Y. Chen and M. A. Young, "Structure of a thyroid hormone receptor DNA-binding domain homodimer bound to an inverted palindrome DNA response element," *Mol. Endocrinol.* 24 (2010), 1650–1664.

[8] D. R. Doerge, "Mechanism-based inhibition of lactoperoxidase by thiocarbamide goitrogens. Identification of turnover and inactivation pathways," *Biochemistry* 27 (1988), 3697–3700.

[9] S. Dunkelmann et al., "Lithium as an adjunct to radioiodine therapy in Graves' disease for prolonging the intrathyroidal effective half-life of radioiodine. Useful or not?," *Nuklearmedizin* 45 (2006), 213–218; quiz N51–2.

[10] I. R. Garrett and G. R. Mundy, "The role of statins as potential targets for bone formation," *Arthritis Res.* 4 (2002), 237–240.

[11] J. M. Gavaret, H. J. Cahnmann, and J. Nunez, "Thyroid hormone synthesis in thyroglobulin. The mechanism of the coupling reaction," *J. Biol. Chem.* 256 (1981), 9167–9173.

[12] S. Göthe et al., "Mice devoid of all known thyroid hormone receptors are viable but exhibit disorders of the pituitary-thyroid axis, growth, and bone maturation," *Genes Dev.* 13 (1999), 1329–1341.

[13] J. Haller, E. Mikics, and G. B. Makara, "The effects of non-genomic glucocorticoid mechanisms on bodily functions and the central neural system. A critical evaluation of findings," *Front. Neuroendocrinol.* 29 (2008), 273–291.

[14] L. Kang et al., "Involvement of estrogen receptor variant ER-alpha36, not GPR30, in nongenomic estrogen signaling," *Mol. Endocrinol.* 24 (2010), 709–721.

[15] R. P. Magnusson, A. Taurog, and M. L. Dorris, "Mechanisms of thyroid peroxidase- and lactoperoxidase-catalyzed reactions involving iodide," *J. Biol. Chem.* 259 (1984), 13783–13790.

[16] E. Mommers et al., "Male hormonal contraception: a double-blind, placebo-controlled study," *J. Clin. Endocrinol. Metab.* 93 (2008), 2572–2580.

[17] R. M. Neer et al., "Effect of parathyroid hormone (1-34) on fractures and bone mineral density in postmenopausal women with osteoporosis," *N. Engl. J. Med.* 344 (2001), 1434–1441.

[18] S. Omoigui, "The Interleukin-6 inflammation pathway from cholesterol to aging--role of statins, bisphosphonates and plant polyphenols in aging and age-related diseases," *Immun. Ageing* 4 (2007), p. 1.

[19] J. L. Pérez-Castrillón et al., "Effects of Atorvastatin on vitamin D levels in patients with acute ischemic heart disease," *Am. J. Cardiol.* 99 (2007), 903–905.

[20] F. Rastinejad et al., "Structural determinants of nuclear receptor assembly on DNA direct repeats," *Nature* 375 (1995), 203–211.

[21] D. S. Reddy, "Neurosteroids: endogenous role in the human brain and therapeutic potentials," *Prog. Brain Res.* 186 (2010), 113–137.

[22] R. I. Scheinman et al., "Characterization of mechanisms involved in transrepression of NF-kappa B by activated glucocorticoid receptors," *Mol. Cell. Biol.* 15 (1995), 943–953.

[23] A. Taurog, M. Dorris, and D. R. Doerge, "Evidence for a radical mechanism in peroxidase-catalyzed coupling. I. Steady-state experiments with various peroxidases," *Arch. Biochem. Biophys.* 315 (1994), 82–89.

[24] B. M. Vayssière et al., "Synthetic glucocorticoids that dissociate transactivation and AP-1 transrepression exhibit antiinflammatory activity in vivo," *Mol. Endocrinol.* 11 (1997), 1245–1255.

[25] H. F. Yang-Yen et al., "Transcriptional interference between c-Jun and the glucocorticoid receptor: mutual inhibition of DNA binding due to direct protein-protein interaction," *Cell* 62 (1990), 1205–1215.

[26] L. Ye et al., "Thyroid receptor ligands. 1. Agonist ligands selective for the thyroid receptor β_1," *J. Med. Chem.* 46 (2003), 1580–1588.

[27] B. Zimmermann, "Effects of pyrophosphate on desmal and endochondral mineralization and TNAP activity in organoid culture," *Ann. Anat.* 190 (2008), 167–177.

CHAPTER 8

PHARMACOLOGY OF NITRIC OXIDE

Nitric oxide (NO) is a mediator that is quite different from other hormones and transmitters. Three key properties of NO are important to its unique mode of signal transmission:

1. NO is a very small molecule and permeates cell membranes with ease; its membrane permeability is comparable to that of oxygen.

2. It binds very avidly to heme, as both O_2 and carbon monoxide (CO) do as well. Its affinity for heme is exceeds that of O_2 but is lower than that of CO. Binding of NO to heme is at the heart of its major established signaling mechanism, the activation soluble guanylate cyclase.

3. NO is a radical (·N=O) and therefore quite reactive. It can react with molecular oxygen and with various reactive oxygen species. The ensuing products in turn may react with amino acid side chains in proteins, in particular with cysteine and tyrosine residues.

While one might expect thiol modification by NO to be fairly nonselective, it appears that individual cysteine residues differ enough in their reactivity to cause preferential modification of relatively few proteins, and thus to make selective thiol modification viable as a regulatory mechanism. Regulatory effects of cysteine modification by NO have been demonstrated in several proteins that have established roles in signal transduction.

Biochemical Pharmacology. By M. Palmer, A. Chan, T. Dieckmann, J. Honek

8.1 CHARACTERIZATION OF NITRIC OXIDE AS A BIOLOGICAL SIGNALING MOLECULE

The discovery that nitric oxide is a physiological mediator was made during an investigation of the nature of the *endothelium-derived relaxing factor* (EDRF). The release of EDRF can be observed when acetylcholine is applied to the aorta of experimental animals (Figure 8.1). In the aorta, as in other blood vessels, both the smooth muscle layer and the endothelium contain parasympathetic, cholinergic nerve terminals, the acetylcholine receptors of which are of the muscarinic type.

After cutting the aorta into strips, one can remove the endothelium mechanically or through limited enzymatic proteolysis. Aortic strips both with or without endothelium will respond with contraction to noradrenaline or α-selective adrenergic agonists. However, if acetylcholine is applied subsequently, only the strip that retains its endothelium will respond with relaxation, whereas the denuded one will stay contracted [4]. Thus, the endothelium is required for the relaxation of the smooth muscle to occur. If a denuded strip is placed next to one retaining the endothelium and the entire assembly is exposed first to norepinephrine and then to acetylcholine, the denuded strip will still relax. This indicates that direct contact between endothelial and muscle cells is not required for the action of EDRF, which therefore must be a diffusible substance.

Approximately at the same time that EDRF was characterized, it was found that a variety of drugs containing nitro or nitrate groups caused vasodilation through the release of nitric oxide. The identity of EDRF and NO was initially suspected from circumstantial evidence: Both NO and EDRF bind to and are inactivated by hemoglobin. Furthermore, the effects of both are augmented by superoxide dismutase, which scavenges superoxide anions by disproportionation into O_2 and H_2O_2. Since superoxide reacts very rapidly with NO, superoxide dismutase will increase the lifetime and biological activity of NO. The identity was finally established by the direct detection of endogenously produced NO in biological samples.

While the formation of endogenous nitric oxide is now quite well understood, the mode of action of several nitrate-containing drugs is still not entirely clear. It appears that not only NO itself but also other species such as NO^- can be released, and that these may contribute to and modify the vascular reaction to these drugs. Moreover, the reactions that lead to the release of NO and related species are incompletely characterized.

8.2 NITRIC OXIDE SYNTHASE AND ITS ISOFORMS

NO is generated in vivo by nitric oxide synthase (NOS; Figure 8.2). This enzyme is located in the cytosol and uses arginine, molecular oxygen, and NADPH as substrates. The reaction occurs in two major steps, with N-hydroxyarginine (NOHA) as an intermediate product, and it consumes 1.5 moles NADPH per mole of NO released. The reaction also yields citrulline, which can be converted back to arginine through reactions that are part of the urea cycle.

NOS is a dimer. Each monomer possesses several redox coenzymes that together form a little electron transport chain of their own. The molecule con-

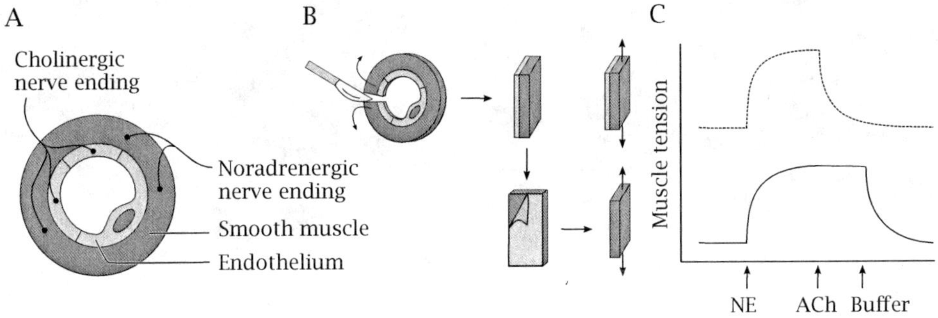

Figure 8.1 Vascular relaxation in response to acetylcholine. **A**: Both the muscular and the endothelial layers of blood vessels are innervated by cholinergic nerve endings. **B**: Preparation of aortic strips. A slice is cut from an aorta and then opened with a radial cut. The endothelium may then be peeled away in order to study its effect on the muscular layer. The resulting aortic strip is mounted between to hooks to measure its contractile force. **C**: Acetylcholine (ACh) reverses contraction induced by norepinephrine (NE) in an aortic strip only if it retains the endothelium (top) but not when the endothelium is stripped away. The responsible *endothelium-derived relaxing factor* (EDRF) has been identified as nitric oxide.

tains two major domains. The N-terminal reductase domain contains two flavin coenzymes (FAD and FMN), which accept electrons from NADPH and deliver them one at a time to the C-terminal oxygenase domain. The oxygenase domain contains a heme group, which is directly involved in substrate oxidation, and yet another redox coenzyme, tetrahydrobiopterin, whose role in the reaction is not clearly understood.[1] The function of the NOS reductase domain resembles that of cytochrome P450 reductase. The two also share significant sequence homology. In contrast, the NOS oxygenase domain is not homologous to cytochrome P450 enzymes.

At the interface between the reductase and the oxygenase domains, NOS contains a binding site for calmodulin. Binding of calmodulin is required for the two domains to align properly so that electrons will flow from the reductase to the oxygenase domain. NOS therefore is inactive in the absence of calmodulin.

There are several subtypes of NOS: endothelial NOS (eNOS), neuronal NOS (nNOS), and inducible NOS (iNOS).[2] The three forms occur in different tissues, and the NO released in these different locations serves different purposes. Endothelial NOS is responsible for the relaxation of blood vessels, which was discussed above. It is found in both arteries and veins; NO-mediated relaxation of arteries will decrease flow resistance, while venous relaxation will increase volume capacity of the vasculature. In combination, these two effects will cause a strong reduction of blood pressure.

[1]NOS depleted of tetrahydrobiopterin produces reactive oxygen species instead of NO, which is not a very healthy thing to do. Therapeutic supplementation of tetrahydrobiopterin to prevent this has been tried [14].

[2]There is also a mitochondrial form of NOS. This enzyme appears to be concerned mainly with the regulation of the respiratory chain, which is mediated by the binding of NO to heme cofactors in the electron transport chain [21].

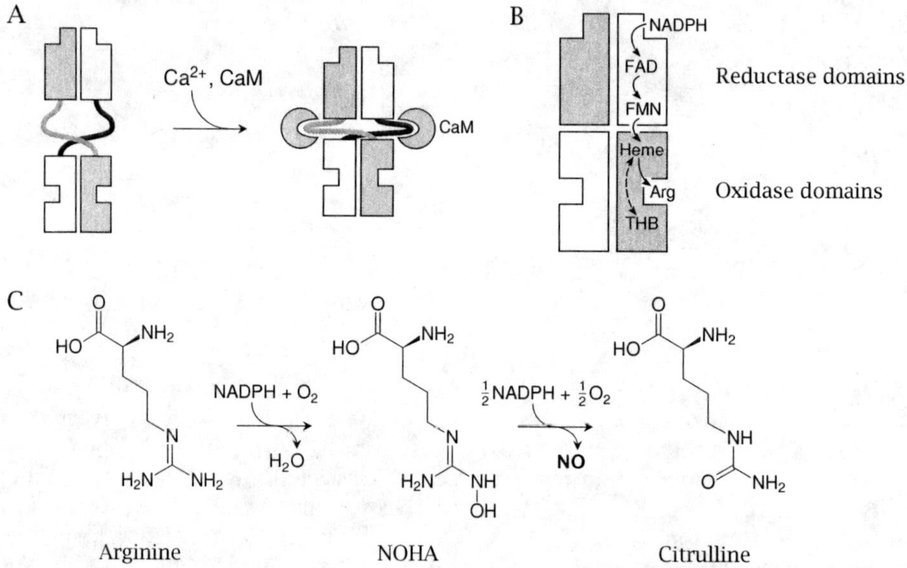

Figure 8.2 The nitric oxide synthase reaction. **A:** Binding of calmodulin is required for close, productive contact between the reductase and the oxidase domains of NOS dimer. **B:** In the activated enzyme, electrons flow from the soluble cosubstrate NADPH via FAD and FMN to heme, which engages directly in arginine oxidation. Tetrahydrobiopterin (THB) assists in the latter reaction. The electron flow occurs between the reductase of one enzyme monomer and the oxidase domain of the other. **C:** The two monooxygenase steps of the NOS reaction. Arginine is converted to N-hydroxyarginine (NOHA), which is then cleaved to release NO and citrulline.

Neuronal NOS is found in the nervous system, where NO serves as a neurotransmitter. NO-mediated signaling between neurons works in much the same way as between endothelial and smooth muscle cells; the molecule easily diffuses out of the originating cell into the target cell and binds directly to intracellular receptor molecules.

Inducible NOS is most strongly expressed in macrophages, although it can also be induced in virtually any other cell type. Nitric oxide released by iNOS serves not only as a messenger but also as an antimicrobial effector mechanism. In this latter capacity, it cooperates with other antimicrobial molecules such as reactive oxygen species.

In addition to the tissues implied by the enzyme nomenclature, NOS is expressed in many other tissues as well. While current NO-related pharmacotherapy is concerned mainly with the control of blood pressure by nitric oxide, additional physiological roles of NO continue to be discovered and may in time lead to new applications in pharmacotherapy. For example, the discovery of multiple regulatory roles of NO in apoptosis [16] has led to the currently experimental evaluation of NO-releasing drugs in tumor prevention and therapy.

All three forms of the enzyme share a high degree of structural and functional similarity. They differ, however, in their affinity for calmodulin, which results in different modes of regulation. Calmodulin will bind to eNOS and nNOS only in the presence of elevated concentrations of Ca^{2+}. Upon return of cytosolic Ca^{2+} to resting levels, eNOS and nNOS will therefore revert to the inactive state. In contrast, iNOS binds calmodulin so avidly that the two proteins remain associated even at resting levels of cytosolic Ca^{2+}. Therefore, iNOS is active regardless of Ca^{2+} fluctuations in the cell, and transcriptional induction or repression becomes its major mechanism or regulation.

The two different regulatory mechanisms seem to match the different functional roles of NO release. The modulation of eNOS and nNOS by Ca^{2+} and calmodulin allows for rapid and dynamic control, as seems appropriate in the communication between cells. In contrast, the activation of inflammatory reactions and the killing of ingested microbes by phagocytes is a more protracted activity, for which the slower and more sustained activation of iNOS seems adequate.

An additional regulatory mechanism for NOS activity is the negative feedback by its product (NO), which is exerted through *S*-nitrosylation of a strategic cysteine in the enzyme. Protein-*S*-nitrosylation also contributes to the regulation of other proteins and is discussed in more detail below.

8.3 BIOCHEMICAL MECHANISMS OF NO SIGNALING

Nitric oxide that is produced inside endothelial cells or nerve cells crosses cell membranes with ease and diffuses into the neighboring smooth muscle cells or nerve cells, respectively. The major established mechanism of signaling involves the binding of NO to heme groups, in particular that of soluble guanylate cyclase. In addition, *S*-nitrosylation of specific cysteine residues contributes to the regulation of several effector molecules.

8.3.1 NO effects mediated by soluble guanylate cyclase

Inside the target cells, NO binds and activates soluble guanylate cyclase (sGC), an enzyme that converts GTP to guanosine monophosphate (cGMP; Figure 8.3). This reaction is analogous to that of adenylate cyclase, which forms cyclic adenosine monophosphate (cAMP) from ATP. However, the molecular mechanism of sGC regulation by NO is quite unusual: sGC contains a heme group, which in contrast to other heme-containing enzymes has a purely regulatory role and does not take part in the actual catalysis. Binding of NO to the heme will break the bond between the heme iron to a histidine residue on the opposite face of the heme ring, which in turn triggers the conformational change that activates the enzyme.

Cyclic GMP is a second messenger with a variety of effector mechanisms. The foremost one is the activation of a cognate protein kinase, variously referred to as protein kinase G (PKG) or cGMP-dependent kinase (cGK). In addition, cGMP opens ligand-gated cation channels in sensory cells, neurons, and smooth muscle cells. A prominent function of cGMP-gated channels is positive feedback of synaptic transmission, which occurs when a postsynaptic cell releases NO upon its presynaptic nerve terminals. Finally, cGMP activates phosphodi-

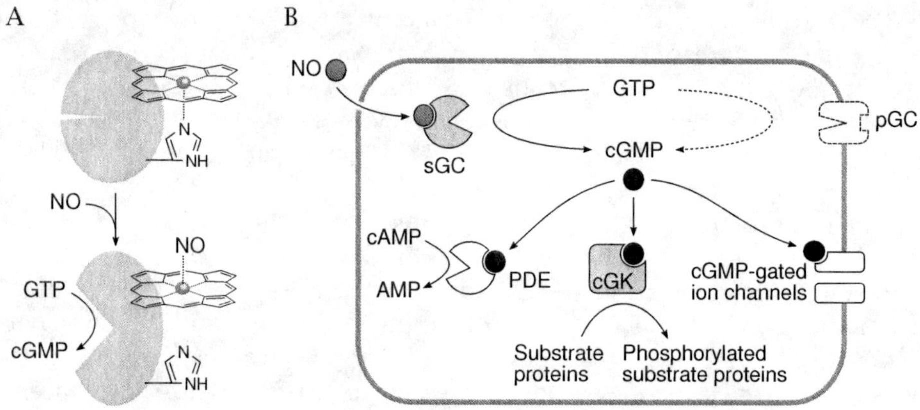

Figure 8.3 Mechanism of signaling by NO and cGMP. **A:** NO binds to the heme group of soluble guanylate cyclase (sGC), dislodging a strategic histidine residue. This causes allosteric activation of the enzyme. **B:** Regulatory effects of cGMP. Activation of cGMP-dependent protein kinases is the most important single mechanism. Phosphodiesterases (PDEs) are activated, too, reducing levels of cAMP and cGMP. Actuation of cGMP-gated potassium channels affects the membrane potential. Membrane-bound receptor or *particulate* guanylate cyclases (pGCs) provide an alternate means of cGMP production that is independent of NO.

esterase (PDE), which will inactivate cAMP by cleavage to AMP. Thus, cGMP is somewhat of an antagonist of cAMP.[3]

How does cGMP bring about relaxation of vascular smooth muscle? Several mechanisms have been proposed. Most of these involve changes in protein phosphorylation, which may be caused either by PKG directly or by secondary protein kinases. The contraction of smooth muscle is ultimately controlled by the phosphorylation of the myosin regulatory light chain. The extent of this phosphorylation depends on the regulatory states of both myosin light chain kinase (MLCK) and of myosin light chain phosphatase. PKG phosphorylates the phosphatase, thereby increasing its activity (Figure 8.4).

Another important effector molecule is the IP_3 receptor channel in the endoplasmic reticulum. Phosphorylation of this channel inhibits the release of calcium into the cytosol in response to inositol triphosphate, which in vascular smooth muscle cells is released downstream of α-adrenergic receptor activation. A lower level of cytosolic Ca^{2+} translates into a reduced level of MLCK and therefore myosin light chain activation. The regulatory subunit of the IP_3 receptor that is targeted by phosphorylation is essential for cGMP-dependent relaxation [5], which demonstrates a major role of this effector mechanism.

There is also evidence indicating that phospholipase C is inhibited by PKG, which would lead to inhibited formation of IP_3, thus intercepting another stage of α-adrenergic signalling, and that Ca^{2+}-transporting ATPases in both the ER and the cytoplasmic membrane may be activated. Since Ca^{2+}-ATPases remove

[3] There are numerous variants of phosphodiesterase with distinct or overlapping specificities for cAMP and cGMP, and cGMP also promotes its own inactivation through phosphodiesterase activation. Protein kinase G also occurs in several isoforms.

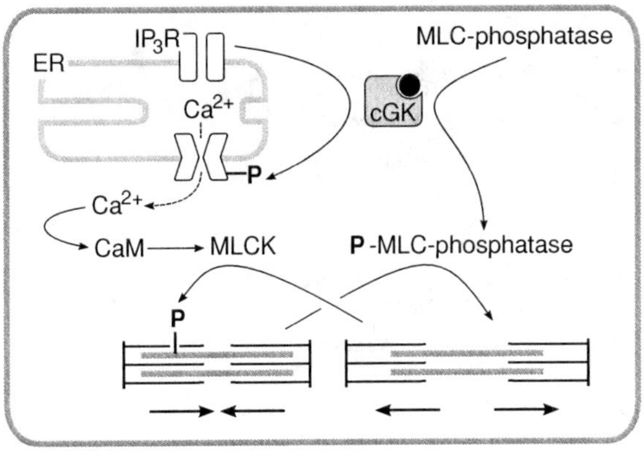

Figure 8.4 Relaxation of smooth muscle cells mediated by cGMP-dependent protein kinase (cGK). cGK activates myosin light chain phosphatase, which results in the dephosphorylation of myosin and actomyosin relaxation. cGK also phosporylates a regulatory subunit of the IP_3 receptor channel. This reduces the efflux of calcium from the ER and therefore the calmodulin-dependent activation of myosin light chain kinase (MLCK).

Ca^{2+} from the cytosol, this would further reduce the amount of Ca^{2+} available to support muscle contraction.

All these mechanisms appear plausible and would work in parallel to reduce smooth muscle contractility. It is difficult to assess their relative contributions to cGMP- and NO-mediated signaling; however, given the very strong vascular relaxation brought about by NO, there is room for more than one mechanism to contribute significantly.

Termination of guanylate cyclase activation requires that NO dissociate from the sGC heme group. Free NO may be disposed of through binding to other heme groups that don't have a regulatory function, such as those in hemoglobin. Binding to hemoglobin may be followed by reaction with oxygen, which will permanently inactivate NO.[4] Alternatively, NO may be transferred from heme to protein cysteine sulfhydryl groups, leading to the formation of nitrosyl groups. Hemoglobin contains a cysteine residue that participates in this reaction quite readily; this helps to free the heme group and to restore hemoglobin function.

8.3.2 Regulation of protein activity by cysteine *S*-nitrosylation

Protein *S*-nitrosylation not only serves to inactivate NO but also is an additional mechanism of NO-mediated signaling. The mechanism of the reaction between free NO and protein sulfhydryl groups may be as outlined in Figure 8.5A [23,6]. In this scheme, *S*-nitrosylation reduces one equivalent of molecular oxygen to superoxide, which then reacts with a second molecule of NO to generate peroxynitrite. This mechanism accounts for the observation that the reaction is dependent on the concentration of free oxygen.

[4]NO bound to nonoxygenated heme may be released again, and therefore hemoglobin has also been considered as a carrier rather than a sink for NO. Given the vast amount of hemoglobin present in the circulation, a significant transfer of NO from hemoglobin to regulatory binding sites would require the affinity of NO to be substantially lower for the former than for the latter.

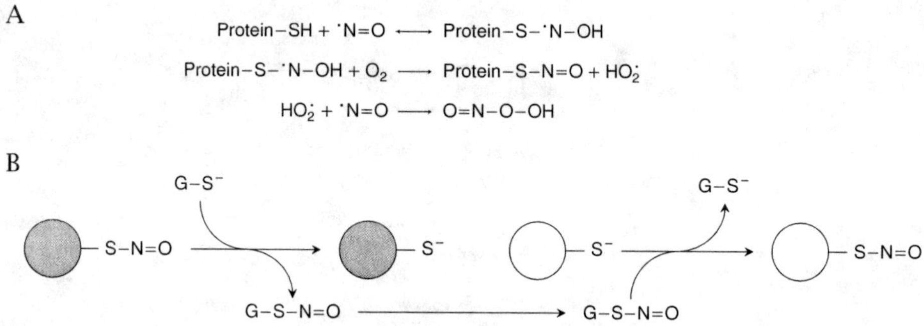

Figure 8.5 *S*-Nitrosylation of cysteine residues in proteins by NO. **A:** Reaction scheme. Two molecules of NO and one molecule of O_2 produce one nitrosylated cysteine residue and one molecule of peroxynitrite. **B:** Transfer of *S*-nitrosyl groups between proteins by glutathione (G-S⁻).

The peroxynitrite generated in the second step is very reactive and may oxidize other sites in proteins, or it may give rise to *O*-nitrosylation of protein tyrosine side chains. Nitroso groups can readily be transferred from one thiol to another. Since glutathione is the most abundant sulfhydryl compound in the cell, it will function as the major scavenger, carrier and reservoir for *S*-nitroso groups (Figure 8.5B). Since the extent of glutathione nitrosylation will reflect the global cellular NO concentration, the latter would appear to be the most significant parameter controlling regulatory protein *S*-nitrosylation.

Since S-nitrosylation is a nonenzymatic reaction, one might presume it to be a rather indiscriminate process that would randomly affect all free sulfhydryl groups in the cell to a similar extent. In this case, it could hardly function as a very meaningful signal but would rather amount to noise or pollution.[5] However, experimental studies have shown a surprisingly high degree of selectivity in vivo. As an example, Figure 8.6 summarizes one study on the selectivity of *S*-nitrosylation in brain tissue and the identity of the proteins affected by it. In the experiment shown, subunit 1 of the NMDA receptor, which is an ionotropic glutamate receptor, stands out as the most intensely nitrosylated species. This receptor has indeed been shown to be inhibited by *S*-nitrosylation; the functionally most important cysteine residue is located in the 2A chain [2].

Protein-*S* nitrosylation is important also outside the brain. Nitrosylation of caspases [10] contributes to the regulation of apoptosis. The ryanodine receptor in the ER membrane is activated by *S*-nitrosylation of a strategically located cysteine [19]; this would increase the availability of Ca^{2+} in the cytosol and therefore promote contraction of vascular smooth muscle. On the other hand, Ca^{2+}-dependent K^+ channels are activated by nitrosylation, which will hyperpolarize the cytoplasmic membrane and promote relaxation [1]. Since eNOS and nNOS are regulated by Ca^{2+} and calmodulin, the modulation of calcium channels such as the NMDA and ryanodine receptors will also cause feedback effects on NO release.

[5]High concentrations of nitric oxide indeed give rise to *nitrosative stress* and play a role in tissue destruction as part of inflammatory disease.

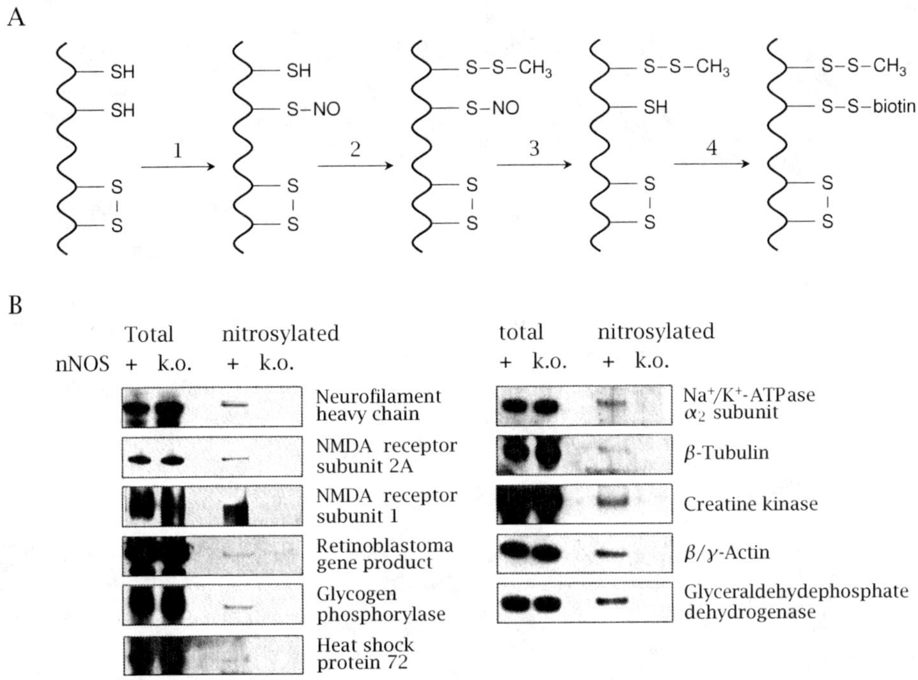

Figure 8.6 Detection of protein *S*-nitrosylation in mouse brains through nNOS activation. **A:** Experimental strategy for selective detection of *S*-nitrosylated proteins. (1) NOS activation causes nitrosylation of some SH groups; (2) remaining free SH groups are converted to disulfides; (3) S-NO groups are selectively reduced with ascorbic acid; (4) the reduced SH groups are labeled with biotin. The labeled proteins can be detected or purified by binding of biotin to streptavidin. **B:** Analysis of individual proteins by Western blot. With each protein, the extent of *S*-nitrosylation is represented by the intensity of the third sample relative to the first one. *S*-nitrosylation is caused by nNOS, as it is absent in nNOS knockout (k.o.) mice. (Panel B reproduced with permission from Ref. [8].)

The selectivity of protein *S*-nitrosylation could be thermodynamically or kinetically controlled. Reversible exchange with glutathione should allow protein cysteine residues to equilibrate between their free and nitrosylated states, and repeated transfer should allow NO to finally reach those cysteine residues that bind it most avidly. On the other hand, a role for kinetic regulation is suggested by the preferential nitrosylation of NOS itself and of proteins interacting with NOS or located in its vicinity, such as cyclooxygenase 2 [9] and NMDA receptor.

According to the reaction scheme in Figure 8.5, the reactive species in the transfer of nitroso groups between thiols is the thiolate anion. The pK_a value of a given protein thiol group may be modulated by metal ions or basic and acidic amino acid side chains in its vicinity. Thiol groups with low pK_a values indeed appear to be preferentially nitrosylated [7]. Cysteine residues in the active sites of enzymes often have low pK_a values, suggesting that they should be susceptible to nitrosylation.

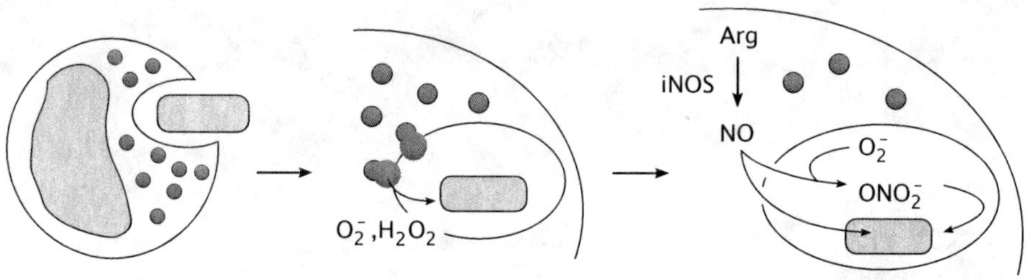

Figure 8.7 Role of inducible NOS (iNOS) in the killing of microbes by phagocytes. Bacteria are ingested by phagocytes and wind up inside intracellular vesicles called *phagosomes*. Peroxisomes fuse to the phagosomes, releasing superoxide (produced by NADPH oxidase) and H_2O_2 (produced by superoxide dismutase). At the same time, iNOS is activated. NO enters the phagosomes and either reacts with superoxide to form peroxynitrite, or diffuses across the microbial cell wall to wreak havoc inside.

8.4 THE BIOLOGICAL FUNCTION OF iNOS

While protein nitrosylation can serve as a signal at moderate concentrations of NO, at higher concentrations it will become destructive. This is exploited by macrophages, which use NO in conjunction with reactive oxygen species to kill ingested microbes (Figure 8.7). Macrophages are the immune system's only effective weapon against hardy microbes such as *Mycobacterium tuberculosis*, which are completely resistant to the other major phagocytic cell type, the neutrophil granulocytes. While reactive oxygen species have a substantial bactericidal effect[6] in the absence of NO, the latter enhances the ability of macrophages to kill bacteria. This may be partially due again to the ability of NO to easily cross membranes, which enables it to penetrate the interior of the microbial cell; most other effector molecules, including the very toxic superoxide anion, cannot do this. NO also reacts with molecular oxygen or reactive oxygen species to yield peroxynitrite or N_2O_3, which broadens the spectrum of antibacterial compounds generated within and without the macrophage or the microbial cell.

Why does NO release due to inflammation matter here? In states of severe infection, patients may develop *septic shock*, a state of pathologically low blood pressure in which minimal perfusion of some vital organs is no longer guaranteed and which often is the ultimate cause of death. One of the underlying mechanisms of septic shock is the excessive release of NO from macrophages into the circulation, causing vasodilation. Accordingly, one therapeutic strategy under evaluation for the treatment of septic shock is pharmacological inhibition of iNOS.

8.5 NO-RELEASING DRUGS

We have seen that NO acts on intracellular targets to intercept the signals of mediators such as norepinephrine that act on receptors at the cell surface.

[6] H_2O_2 reacts, for example, with chloride to form HOCl (hypochloric acid), which is known as bleach and used for disinfection. Formation of HOCl in granulocytes is catalyzed by myeloperoxidase. People lacking this enzyme suffer from a quite severe form of immune deficiency, in particular with respect to bacterial infections.

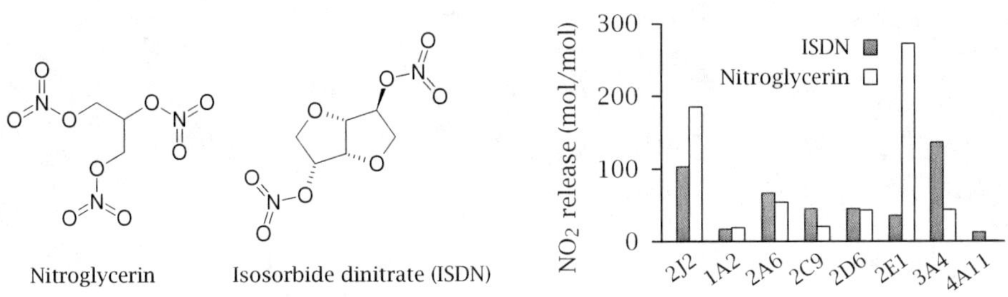

Figure 8.8 Structures of the organic nitrate drugs nitroglycerin and isosorbide dinitrate, and release of NO from them by human cytochrome P450 isoforms. The enzymes were recombinantly expressed in yeast. The released NO was chemically converted to nitrite, which in turn was quantified by HPLC. (Figure redrawn from original data in Ref. [13].)

Accordingly, NO-releasing drugs are very powerful vasodilators. They also have a fairly prompt onset of action and thus are useful when the blood pressure must be lowered, or vasospasms must be relieved immediately.

Nitric oxide itself is a gas and in principle can be applied by inhalation. Because of its reactivity, NO would then mainly be expected to act within the lungs, and inhalation therapy has accordingly been used in diseases that involve pathological restrictions of pulmonary blood flow. However, in animal experiments, quite substantial effects of NO inhalation have been observed outside the lung as well, such as the reduction of *reperfusion injury* in myocardial infarction.[7] More commonly, however, NO is applied in the form of precursor drugs, which contain oxidized nitrogen in various forms.

The most widely used NO-releasing drugs are organic nitrates. The first one of these to be used clinically, glyceryl trinitrate or nitroglycerine, is more widely known in another application, which helped Mr. Alfred Nobel make his fortune—and still maintain his good conscience, since he believed that this would be so horrible a weapon that humankind would henceforth abstain from warfare (a hope now widely held for nuclear weapons). The medical effect of nitroglycerine was initially noted in the form of the *monday headache.* Nobel's factory workers, returning to work on a Monday, experienced strong headaches that faded away with continuous exposure to nitroglycerine fumes during the workweek, only to reappear the next Monday after withdrawal during the weekend.[8] Nitroglycerin—applied sublingually as a spray for rapid uptake and avoidance of liver first-pass effect—is still the standard treatment of acute episodes of *angina pectoris.* This condition results from vasoconstriction in the coronary arteries, which leads to hypoxia in the heart muscle and acute pain,

[7]The concept of reperfusion injury implies that destruction of tissue downstream of occluded arteries occurs to a substantial extent not while perfusion is interrupted but only after it is restored, for example, by dissolution of the occluding blood clot with tissue plasminogen activator. Reperfusion injury involves circulating leukocytes and platelets, and it is possible that inhaled NO acts on these cells while they are passing through the lungs.

[8]A nitrate headache can also be nicely observed when using organic nitrate-containing paint diluents indoors. Headache is often linked to disturbed vasomotor function, and some of the drugs used in migraine, such as ergotamine, are in fact vasoconstrictors.

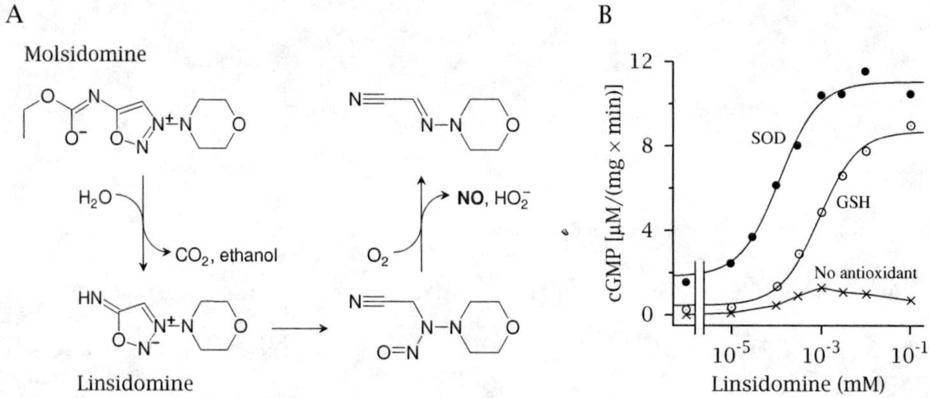

Figure 8.9 NO release by molsidomine. **A:** Mechanism. Enzymatic hydrolysis of molsidomine yields linsidomine, which decomposes spontaneously. The final step depends on O_2 and also yields superoxide, which can combine with NO into peroxynitrite. **B:** In order to render NO released by linsidomine available for the activation of guanylate cyclase, superoxide must be scavenged, for example by superoxide dismutase (SOD) or glutathione(GSH). (Plot redrawn from original data in Ref. [18].)

and is associated with coronary atherosclerosis. Among the first patients to benefit from nitroglycerin treatment of angina pectoris was Nobel himself.

Organic nitrates such as nitroglycerin and isosorbide dinitrate, when orally applied, undergo a pronounced first pass effect in the liver, which is consistent with the observation that metabolism involves cytochrome P450 enzymes (see Figure 8.8) and glutathione-*S*-transferases. In addition, a variety of other enzymes such as cytochrome P450 reductases, glyceraldehyde-3-phosphate dehydrogenase, glutathione-*S*-transferase and even nonenzyme hemoproteins such as hemoglobin have been implicated. Because of the limited stability of NO, precursor drugs should be most effective when undergoing metabolism close to the site of action. Interestingly, nitroglycerin is metabolized and releases NO in tissue slices obtained from blood vessels. In particular, cytochrome P450 type 2J2 is quite strongly expressed in vascular tissues and is fairly active toward both nitroglycerin and isosorbide dinitrate (Figure 8.8).

Cytochrome P450 contains heme, and since NO binds to heme groups, one might expect cytochrome P450 over time to be inactivated by NO. This may be a factor in the frequently observed development of *nitrate tolerance*, which necessitates the temporary suspension of treatment with organic nitrates. In keeping with this assumption, nitrosothiol compounds,[9] which do not require activation by cytochrome P450, remain effective after development of tolerance to nitroglycerin in animal experiments [12]. Nitrosothiols may release NO through enzymatic and nonenzymatic reactions, and accordingly they could in principle be used as NO-releasing drugs. The nitroso group can easily be attached to drug molecules containing sulfhydryl groups, and several experimental bifunctional

[9]It has been proposed that even EDRF itself, as obtained from cultured cells, is not free NO but instead an *S*-nitroso compound such as *S*-nitrosocysteine [15].

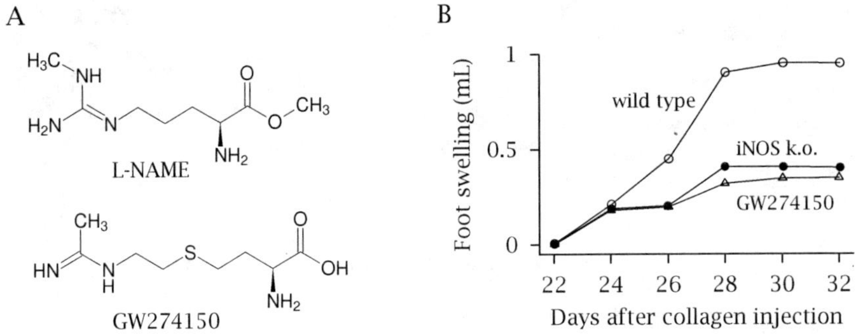

Figure 8.10 NOS inhibitors. **A:** L-*N*-methylarginine methyl ester (L-NAME) inhibits all NOS isoforms, whereas GW274150 is an iNOS-selective inhibitor. **B:** Partial suppression of collagen-induced arthritis in mice by GW274150. Bovine collagen type II was injected on day 0. The swelling of paws was evaluated in comparison to untreated wild type mice and iNOS-knockout mice. The immune reaction that gives rise to the arthritis and the symptomatic swelling take 3 weeks to develop. (Figure redrawn from original data in Ref. [3].)

drugs that combine NO release with another, complementary mode of action have been prepared. An example is *S*-nitroso-captopril [17], which combines the effect of enalaprilate (see Figure 1.5) with that of NO release. Clinical use of nitrosothiols is, however, hampered by limited stability.[10]

Molsidomine and sodium nitroprusside contain NO in alternative precursor forms. Molsidomine is activated toward NO release by a cytochrome P450-independent pathway that involves enzymatic hydrolysis in the intestine or liver, followed by oxygen-dependent non-enzymatic decay (Figure 8.9). The NO released apparently forms mostly *S*-nitrosoglutathione in vivo [18]. The fact that NO release from molsidomine does not involve a hemoprotein may account for the fact that it is less prone to induce tolerance than the organic nitrates [11].

Sodium nitroprusside [$Na_2Fe(CN)_5NO$] is a salt that must be applied by intravenous infusion. It readily reacts with free thiols to release NO, without requiring enzymatic activation. As such, sodium nitroprusside is one of the strongest vasodilating agents known. Since the NO released is also fairly quickly inactivated (or, at least, dissipated in large reservoirs such as hemoglobin), its clinical effect is also short-lived and therefore will closely follow the rate of infusion. This makes nitroprusside very suitable for the continuous titration of fluctuating blood pressure in critical care. It is also used to reverse the diffuse, severe counterregulatory vasoconstriction that occurs in advanced states of circulatory shock.

Metabolism of NO-releasing drugs will not produce NO only but may also release other oxidation stages of nitrogen such as HNO, HNO_2, and, as discussed above, $HONO_2$. These products will also arise endogenously after NOS activa-

[10]Bifunctional NO donors that contain nitro groups have also been prepared; for example, to overcome the adverse effects of cyclooxygenase inhibitors on mucosal perfusion, nitrated versions of several such inhibitors have been developed [20].

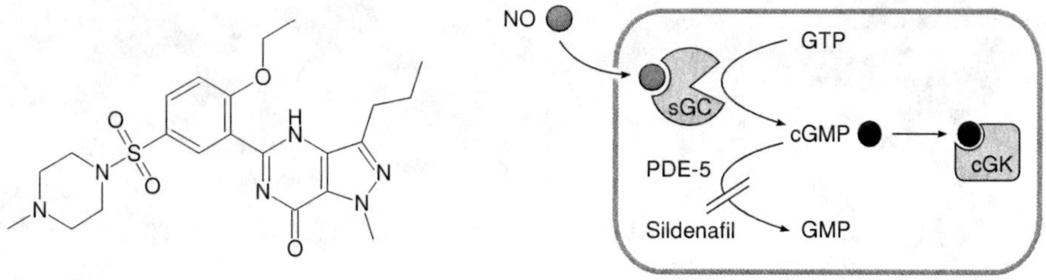

Figure 8.11 Structure and mode of action of sildenafil (Viagra™). Inhibition of phosphodiesterase (PDE) 5 reduces the rate of cGMP breakdown and raises the levels of cytosolic cGMP. This mode of action is complementary to that of NO, which increases cGMP synthesis.

tion; however, the proportions produced by drugs will likely differ from those of endogenous production. The physiological implications of such variation are not fully understood.

8.6 NOS INHIBITORS

As mentioned above, excessive activation of iNOS in severe infections may cause system-wide blood vessel relaxation, leading to septic shock. Treatment with NOS inhibitors has been investigated for a considerable time but is still in the experimental stage. A clinical trial with the inhibitor *N*-methylarginine indicated stabilization of blood pressure; however, overall mortality increased, causing the trial to be aborted [22].

The logical strategy to minimize the side effects of inhibitor treatment is to make inhibition selective for iNOS. Such inhibitors are also of potential utility in the treatment of chronic inflammatory diseases such as rheumatoid arthritis. One iNOS-selective inhibitor, GW274150 has been tested and found beneficial in animal models of arthritis (Figure 8.10).

8.7 PHOSPHODIESTERASE INHIBITORS

Finally, a discussion of nitric oxide pharmacology would not be complete without mentioning one of the most successful drugs of recent times—namely sildenafil, more widely known under its trade name, Viagra™. Sildenafil inhibits phosphodiesterase subtype 5, which is selective for cGMP (Figure 8.11). Removal of cGMP by this enzyme inactivates the major signaling mechanism of NO.

Sildenafil was not developed with its currently popular use in mind; instead, the goal was a new kind of vasodilator. Its effect on penile erection was unexpected and gave rise to the discovery that NO actually is one of the transmitters that trigger this process.

In theory, it should be possible to further augment the effect of sildenafil with nitrate drugs, shouldn't it? So it is. However, this cannot be recommended, as it has resulted in heart attacks, a circumstance that in some jurisdictions has prompted the training of prostitutes in the use of defibrillators. Sildenafil may also cause disturbances of color vision; this is due to the role of cGMP-gated channels downstream of rhodopsin activation in the sensory cells of the retina.

STUDY QUESTIONS

8.1 While EDRF was originally identified as NO, a subsequent report has suggested that EDRF is in fact an endogenous nitrosothiol compound [15]. What implications would this have for the distribution of EDRF across cell barriers?

8.2 In addition to NO, other endogenous gases have more recently been implicated in cardiovascular regulation. What is the evidence for H_2S as a physiological cardiovascular regulator? Do you see possibilities for the direct interactions of NO and H_2S?

REFERENCES

[1] V. M. Bolotina et al., "Nitric oxide directly activates calcium-dependent potassium channels in vascular smooth muscle," *Nature* 368 (1994), 850–853.

[2] Y. B. Choi et al., "Molecular basis of NMDA receptor-coupled ion channel modulation by S-nitrosylation," *Nat. Neurosci.* 3 (2000), 15–21.

[3] S. Cuzzocrea et al., "Beneficial effects of GW274150, a novel, potent and selective inhibitor of iNOS activity, in a rodent model of collagen-induced arthritis," *Eur. J. Pharmacol.* 453 (2002), 119–129.

[4] R. F. Furchgott and J. V. Zawadzki, "The obligatory role of endothelial cells in the relaxation of arterial smooth muscle by acetylcholine," *Nature* 288 (1980), 373–376.

[5] A. Geiselhöringer et al., "IRAG is essential for relaxation of receptor-triggered smooth muscle contraction by cGMP kinase," *EMBO J.* 23 (2004), 4222–4231.

[6] A. J. Gow, D. G. Buerk, and H. Ischiropoulos, "A novel reaction mechanism for the formation of S-nitrosothiol in vivo," *J. Biol. Chem.* 272 (1997), 2841–2845.

[7] D. T. Hess et al., "Protein S-nitrosylation: purview and parameters," *Nat. Rev. Mol. Cell Biol.* 6 (2005), 150–166.

[8] S. R. Jaffrey et al., "Protein S-nitrosylation: a physiological signal for neuronal nitric oxide," *Nat. Cell Biol.* 3 (2001), 193–197.

[9] S. F. Kim, D. A. Huri, and S. H. Snyder, "Inducible nitric oxide synthase binds, S-nitrosylates, and activates cyclooxygenase-2," *Science* 310 (2005), 1966–1970.

[10] J. Li et al., "Nitric oxide reversibly inhibits seven members of the caspase family via S-nitrosylation," *Biochem. Biophys. Res. Commun.* 240 (1997), 419–424.

[11] R. Messin et al., "Efficacy and safety of molsidomine once-a-day in patients with stable angina pectoris," *Int. J. Cardiol.* 98 (2005), 79–89.

[12] M. R. Miller et al., "Novel S-nitrosothiols do not engender vascular tolerance and remain effective in glyceryltrinitrate-tolerant rat femoral arteries," *Eur. J. Pharmacol.* 408 (2000), 335–343.

[13] Y. Minamiyama et al., "Cytochrome P450 is responsible for nitric oxide generation from NO-aspirin and other organic nitrates," *Drug Metab. Pharmacokinet.* 22 (2007), 15–19.

[14] A. L. Moens and D. A. Kass, "Tetrahydrobiopterin and cardiovascular disease," *Arterioscler. Thromb. Vasc. Biol.* 26 (2006), 2439–2444.

[15] P. R. Myers et al., "Vasorelaxant properties of the endothelium-derived relaxing factor more closely resemble S-nitrosocysteine than nitric oxide," *Nature* 345 (1990), 161–163.

[16] S. Y. Olson and H. J. Garbán, "Regulation of apoptosis-related genes by nitric oxide in cancer," *Nitric Oxide* 19 (2008), 170–176.

[17] J. W. Park, "Dual role of S-nitrosocaptopril as an inhibitor of angiotensin-converting enzyme and a nitroso group carrier," *Biochem. Biophys. Res. Commun.* 189 (1992), 206-210.

[18] A. Schrammel et al., "Activation of soluble guanylyl cyclase by the nitrovasodilator 3-morpholinosydnonimine involves formation of S-nitrosoglutathione," *Mol. Pharmacol.* 54 (1998), 207–212.

[19] J. Sun et al., "Cysteine-3635 is responsible for skeletal muscle ryanodine receptor modulation by NO," *Proc. Natl. Acad. Sci. U. S. A.* 98 (2001), 11158–11162.

[20] K. Takeuchi et al., "Effect of nitric oxide-releasing aspirin derivative on gastric functional and ulcerogenic responses in rats: comparison with plain aspirin," *J. Pharmacol. Exp. Ther.* 286 (1998), 115–121.

[21] L. B. Valdez, T. Zaobornyj, and A. Boveris, "Mitochondrial metabolic states and membrane potential modulate mtNOS activity," *Biochim. Biophys. Acta* 1757 (2006), 166–172.

[22] D. Watson et al., "Cardiovascular effects of the nitric oxide synthase inhibitor NG-methyl-L-arginine hydrochloride (546C88) in patients with septic shock: results of a randomized, double-blind, placebo-controlled multicenter study (study no. 144-002)," *Crit. Care Med.* 32 (2004), 13–20.

[23] D. A. Wink et al., "Reaction kinetics for nitrosation of cysteine and glutathione in aerobic nitric oxide solutions at neutral pH. Insights into the fate and physiological effects of intermediates generated in the NO/O2 reaction," *Chem. Res. Toxicol.* 7 (1994), 519–525.

CHAPTER 9

EICOSANOID MEDIATORS AND RELATED DRUGS

Eicosanoid mediators are derived from arachidonic acid (eicosatetraenoic acid) and related polyunsaturated fatty acids. These fatty acids occur as constituents of phospholipids in cellular membranes. The major classes of eicosanoids are prostaglandins, thromboxanes, leukotrienes, and endocannabinoids. In contrast to most other hormones or transmitters, eicosanoid mediators are not stored prior to release; rather, the lipid precursors are mobilized and the mediators are synthesized on demand. An outline of synthetic pathways is given in Figure 9.1, and some structures are shown in Figure 9.2.

Eicosanoids are very widespread in the mammalian organism; almost all cells synthesize them. Their actions are mediated through cognate G protein-coupled receptors, and in some cases through nuclear hormone receptors [10]. They are usually of short duration and occur within a short distance of the site of release, and the eicosanoids have accordingly been called "local hormones". They are inactivated through enzymatic oxidation and hydrolysis either locally or in the lung.[1]

Among the numerous physiological functions that are affected by eicosanoids, pharmacological interest focuses mostly on pain, inflammation, and

[1]The lung is strategically placed for rapid clearance of circulating mediators, since the entire blood volume passes through the lung on each round trip through the circulation.

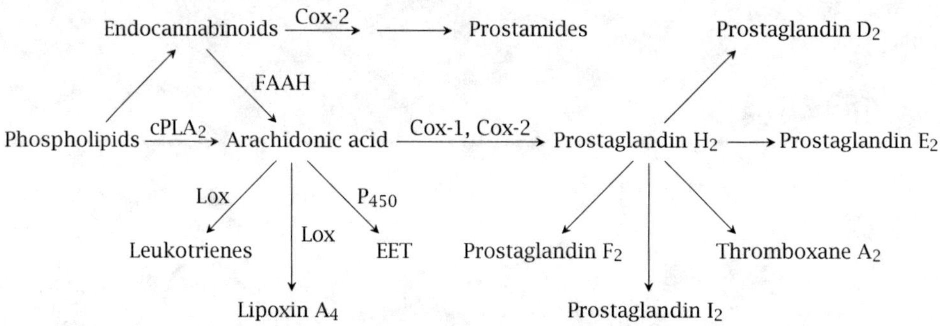

Figure 9.1 Metabolic pathways and key enzymes in eicosanoid mediator synthesis. Release of arachidonic acid from membrane phospholipids by cytoplasmic phospholipase A_2 ($cPLA_2$) feeds cyclooxygenases 1 and 2 (Cox-1, Cox-2). The prostaglandin H_2 formed by cyclooxygenases is converted to other prostaglandins and to thromboxanes by specific synthases. Arachidonic acid is also a substrate for lipoxygenases (Lox) and for cytochrome P450 enzymes (P450), which produce leukotrienes, lipoxins, and epoxytrienoic acids (EET), respectively. Endocannabinoids, which are formed through separate pathways, are cleaved to arachidonic acid by fatty acid amide hydrolase (FAAH). Alternatively, they may be metabolized first by Cox-2 and subsequently by one of the prostaglandin synthases, giving rise to prostamides.

blood coagulation. Others, such as the regulation of blood flow in the kidneys and the gastric mucous membrane, are relevant because they give rise to potentially serious side effects. As you can see from table 9.1, many eicosanoid mediators have multiple physiological roles, which already leads us to expect that drugs that interfere with eicosanoid receptors and metabolism will be prone to side effects. This is certainly true for inhibitors of the enzyme cyclooxygenase, which controls the synthesis of all prostaglandins and thromboxanes (Figure 9.1). Nevertheless, among all drugs that affect eicosanoid signaling, cyclooxygenase inhibitors are currently the most commonly used ones.

9.1 BIOSYNTHESIS OF EICOSANOIDS

For all eicosanoids except the endocannabinoids, the first and rate-limiting step in biosynthesis is the release of arachidonic acid from phospholipids in intracellular membranes, in particular the ER and nuclear membranes. This reaction is catalyzed by phospholipase A_2 (PLA_2). This enzyme occurs in multiple forms that differ in size, catalytic mechanism, and regulation. Some of these isoforms reside in the cytosol, whereas others are extracellularly secreted; both kinds are involved and transcriptionally upregulated in inflammation. The most important single form is a calcium-dependent, cytosolic enzyme ($cPLA_2$). Calcium mediates the binding of $cPLA_2$ to negatively charged phospholipids in the nuclear and ER membranes (Figure 9.3). Once bound, the enzyme releases arachidonic acid and related polyunsaturated fatty acids from phosphatidylinositol and other membrane phospholipids. In the remainder of this chapter, we will mostly consider eicosanoid mediators derived from arachidonic acid; however, most pathways and signaling functions also apply to the derivatives

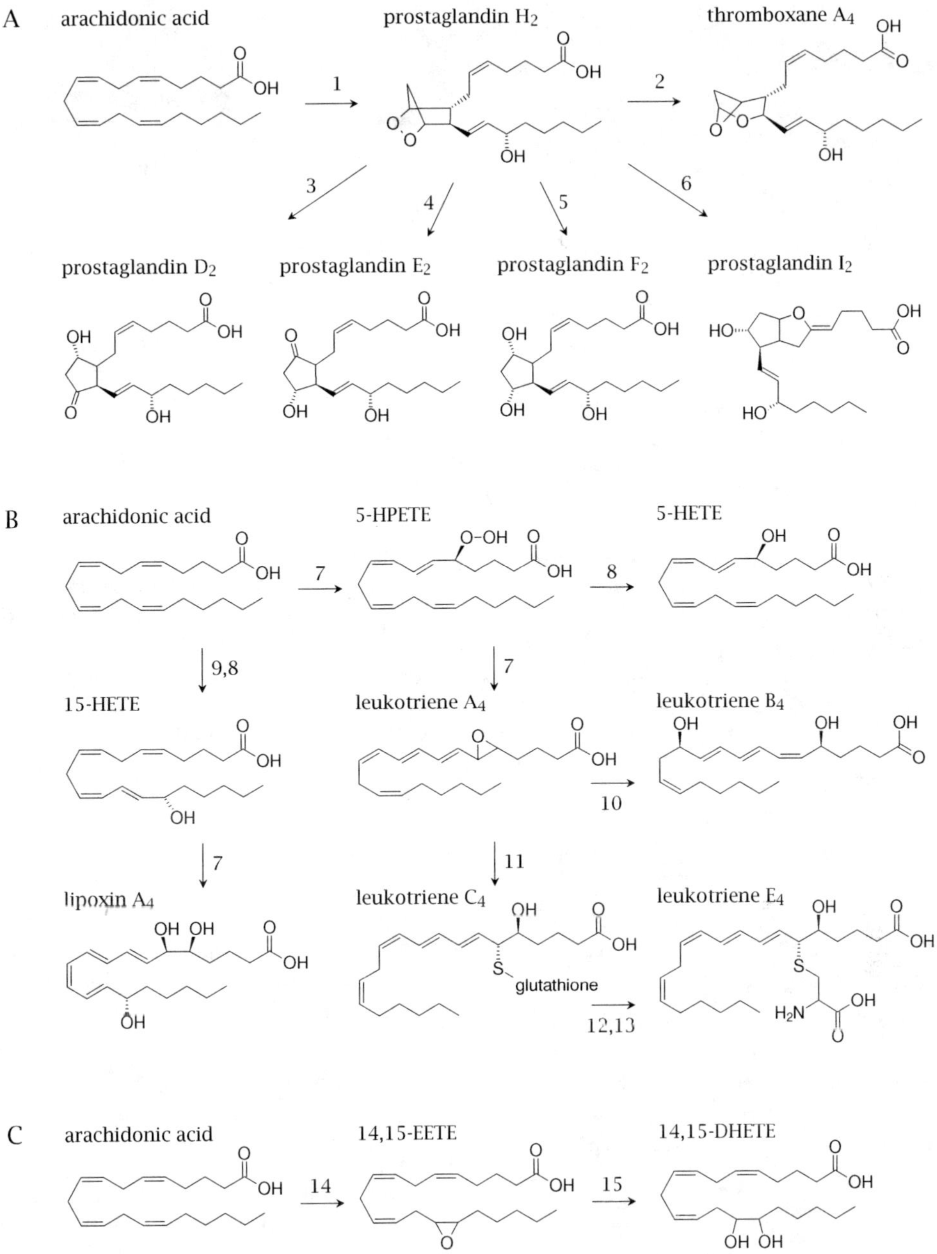

Figure 9.2 Biosynthesis and structures of major eicosanoid mediators derived from arachidonic acid. **A**: Mediators derived via cyclooxygenase/prostaglandin H_2. **B**: Mediators derived via lipoxygenase reactions. **C**: Mediators derived via cytochrome P450. Enzymes: Cyclooxygenase (1); thromboxane A synthase (2); prostaglandin D,E,F and I synthases (3–6); 5-lipoxygenase (7); HPETE peroxidase (8); 15-lipoxygenase (9); leukotriene A_4 hydrolase (10); glutathione-*S*-transferase (11); peptidases (12,13); cytochrome P450 (14); soluble epoxide hydrolase (15).

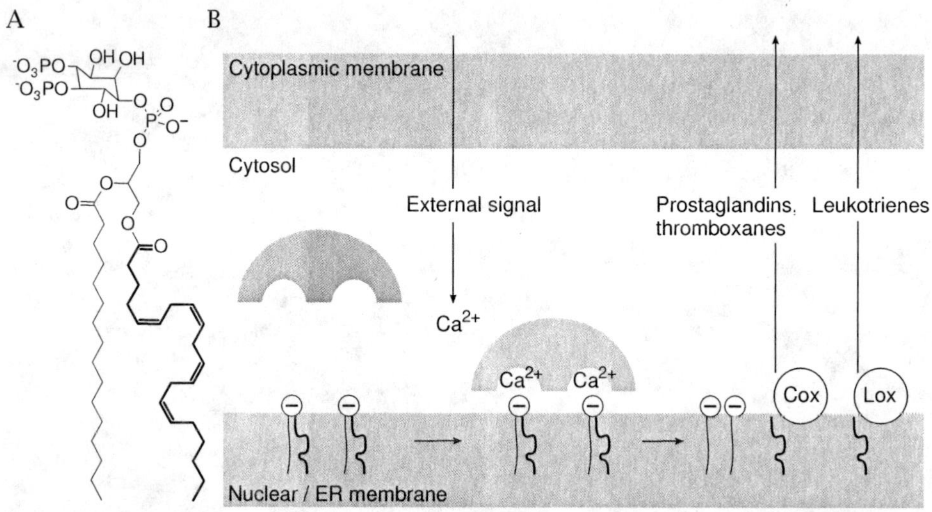

Figure 9.3 Initiation of prostaglandin and leukotriene synthesis by $cPLA_2$. **A:** Structure of phosphatidylinositol, one of the major lipid reservoirs of arachidonic acid. The arachodonyl residue is highlighted. **B:** Increased levels of cytosolic calcium, induced for example by the phospholipase C pathway, mediate binding of $cPLA_2$ to negatively charged lipids on the nuclear and ER membranes. Arachidonic acid is subsequently released from phosphatidylinositol and other phospholipids. It is converted to eicosanoid precursors by membrane-associated cyclooxygenases (Cox) and lipoxygenases (Lox). Lipoxygenases show a similar mode of calcium-dependent binding as $cPLA_2$ (not shown).

of the related substrates eicosatrienoic acid, eicosapentaenoic acid and docosahexaenoic acid.

Most of the arachidonic acid released by phospholipase A_2 released is metabolized by two types of membrane-associated enzymes, cyclooxygenases and lipoxygenases. Cyclooxygenase, also called *prostaglandin H synthase*, converts arachidonic acid first to prostaglandin G_2 (PGG_2) and then to PGH_2. The latter is the common precursor of all other prostaglandins and of the thromboxanes. For this reason, cyclooxygenase is the single most important enzyme in eicosanoid metabolism, and it is also currently the most important drug target. The reaction mechanism of cyclooxygenase and some of its inhibitors are considered in some detail below.

Lipoxygenases introduce hydroperoxy groups or epoxy groups into the arachidonic acid molecule. There are various isoforms that are named according to the position of the double bond that gets modified in the substrate molecule. They provide the precursors of leukotrienes and of lipoxins, respectively, which are important regulators of inflammation. Arachidonic acid may also undergo oxidation by cytochrome P450. This will yield epoxyeicosatrienoic acids, of which there are four different isomers, corresponding to the four double bonds in arachidonic acid. Hydrolysis of these products by epoxide hydrolase yields dihydroxytrienoic acids.

9.2 THE CYCLOOXYGENASE REACTION

Because the central role of its product, prostaglandin H_2, as a precursor of multiple eicosanoid mediators,[2] cyclooxygenase is the major traditional drug target in prostaglandin metabolism. The synthesis of PGH_2 occurs in two separate successive reactions, for which there are two separate active sites on the enzyme molecule. The first reaction—also referred to as the *cyclooxygenase* reaction—introduces two peroxy groups, one endoperoxide that forms a ring with carbons 9–11 and a hydroperoxide that is attached to position 15 (Figure 9.4). The intermediate product, prostaglandin G_2, dissociates from the first active site and then binds to the second active site of the same or another enzyme molecule. There, the hydroperoxide is reduced to a simple hydroxy group at the expense of two molecules of reduced glutathione. This second step is catalyzed by heme and is called the *peroxidase* reaction.

While there is no direct path for the substrate molecule from the peroxidase active site to the cyclooxygenase active site, the two sites are located close to each other, and this proximity is important in the *priming* of the cyclooxygenase site. The first step in the cyclooxygenase reaction is initiated by the radical form of a tyrosine residue in the active site; in Cox-1, this tyrosine occupies position 385. This tyrosyl radical will not exist in a newly translated enzyme molecule, and once it is formed, it may be lost again through capture of an electron from somewhere else than the substrate. Therefore, a mechanism is needed for its formation and regeneration. This mechanism is provided by an electron conduit between the two active sites. The heme radical cation, which forms as an intermediate during the peroxidase reaction, can indirectly abstract an electron from the tyrosine hydroxyl group, which thus may act as a reductant in place of one of the two glutathione molecules normally functioning as cosubstrates.

The peroxidase reaction can proceed in the absence of cyclooxygenase activity if its substrate, prostaglandin G_2, is provided by another enzyme molecule, and this reaction can then prime or regenerate the tyrosyl radical. In accord with this model, a sample of cyclooxygenase that was expressed recombinantly and in the absence of substrate will initially be inactive toward arachidonic acid. However, once the first enzyme molecules have been activated, presumably by reactive oxygen species, the enzyme activity will quickly increase according to "burst" kinetics, due to the cascading activation of more and more enzyme molecules by PGG_2 [16].

9.3 CYCLOOXYGENASE ISOFORMS AND INHIBITORS

Most cyclooxygenase inhibitors in current clinical use, including all of those shown in Figure 9.5, bind noncovalently in the cyclooxygenase site of the enzyme. The figure also illustrates the orientation of the natural substrate arachi-

[2]The numerical subscripts in the names of prostaglandins and leukotrienes denote the number of carbon-carbon double bonds, which varies depending on the particular precursor fatty acid. The mediators mentioned in this chapter (such as PGH_2 and LTD_4) derive from arachidonic acid, unless otherwise stated.

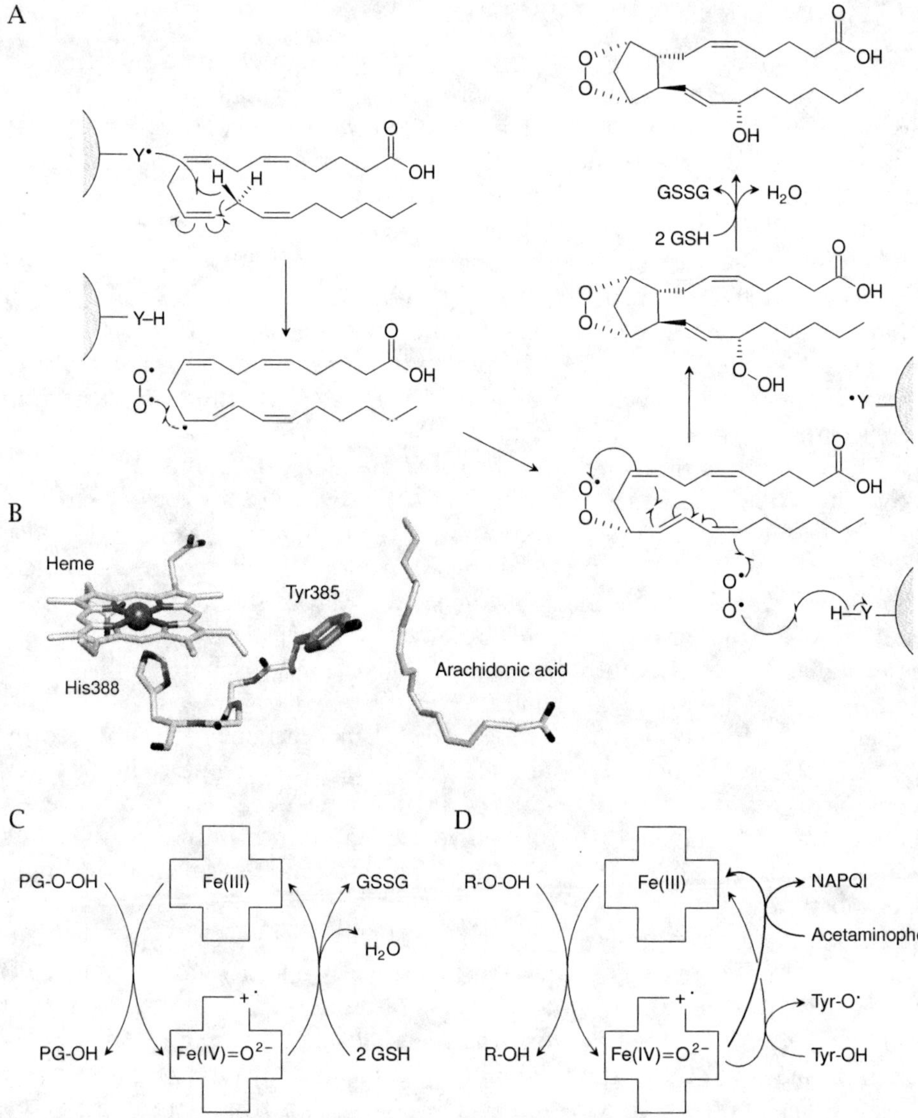

Figure 9.4 The cyclooxygenase reaction. **A:** Reaction intermediates. The reaction begins with the abstraction of a single electron from arachidonic acid by a tyrosyl radical (Tyr385 in Cox-1). Two peroxy groups are then introduced, which gives prostaglandin G_2. The latter intermediate leaves the cyclooxygenase site and binds to the peroxidase site, where it is reduced to prostaglandin H_2 by glutathione. **B:** Tyr385 is part of the cyclooxygenase site but is connected to the heme in the peroxidase site via a stretch of just three amino acid residues, which act as an electron conduit between tyrosine 385 and heme. (Rendered from 1diy.pdb [31].) **C:** Glutathione-dependent reduction of PGG_2 (PG–O–OH) to PGH_2 (PG–OH) in the peroxidase site. **D:** Priming of the tyrosine 385 radical, and mode of action of acetaminophen. The abstraction of an electron from tyrosine 385 begins with the oxidation of heme by PGG_2 or another peroxide. Heme then in turn obtains an electron from tyrosine 385 via the conduit shown in B. Acetaminophen intercepts this process by reducing heme itself [39,5].

A

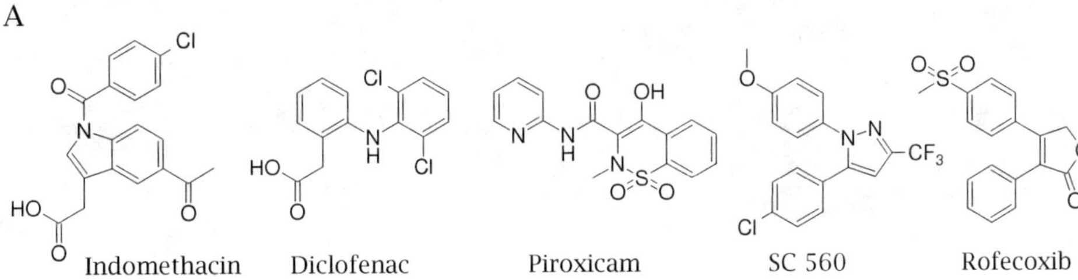

B

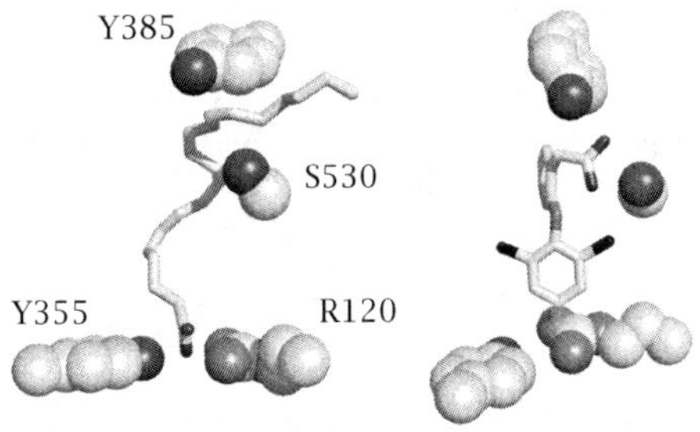

C

Cox mutant	R120A	Y355F	S530A
	$IC_{50, mutant} / IC_{50, wild type}$		
Indomethacin	>240	>240	1.1
Diclofenac	3.3	1.8	>650
Piroxicam	>109	>109	>109

Figure 9.5 Cyclooxygenase inhibitors. **A:** Structures of several nonselective inhibitors, the Cox-1 selective inhibitor SC 560, and the Cox-2 inhibitor rofecoxib. **B:** Interaction of arachidonic acid and of diclofenac with key residues in the cyclooxygenase active site. The carboxyl group of arachidonic acid (left) binds to tyrosine 355 and arginine 120. The carboxyl group of indomethacin binds in the same location (not shown). In contrast, the carboxyl group of diclofenac (right) is hydrogen-bonded to serine 530 and to the catalytic tyrosine 385. **C:** Effects of mutagenesis on the susceptibility of Cox-1 to inhibition by diclofenac, indomethacin and piroxicam. An increase in the inhibitor concentration required for 50% inhibition (IC_{50}) relative to the wild-type enzyme indicates reduced inhibition. (Data from Ref. [42]. Structures rendered from 1pxx.pdb [42] and 1diy.pdb [31].)

donic acid in the active site. In Cox-1, substrate binding is stabilized by hydrogen and ionic bonds to residues arginine 120 and tyrosine 355. Serine 530 is located across from the catalytic tyrosine 385 and restricts the space available to the substrate. Like arachidonic acid, indomethacin has a single carboxyl group, and it also interacts with arginine 120 and tyrosine 355; substitution of these residues with apolar residues greatly diminishes the inhibitory potency of indomethacin. In contrast, the carboxyl group of diclofenac binds to serine 530, and replacement of this residue but not of arginine 120 or tyrosine 355 renders the enzyme insensitive to this inhibitor. All three residues are important with a third inhibitor, piroxicam [42].

Cyclooxygenase (Cox) occurs in two isoforms, Cox-1 and Cox-2. Both are homologous and catalyze the very same reaction, yet their active sites are sufficiently different to permit some inhibitors to discriminate between them. Traditional cyclooxygenase inhibitors inhibit both Cox-1 and Cox-2 and include molecules such as acetylsalicylic acid, indomethacin, and diclofenac. An inhibitor that is selective for Cox-2 is rofecoxib (Figure 9.5). -

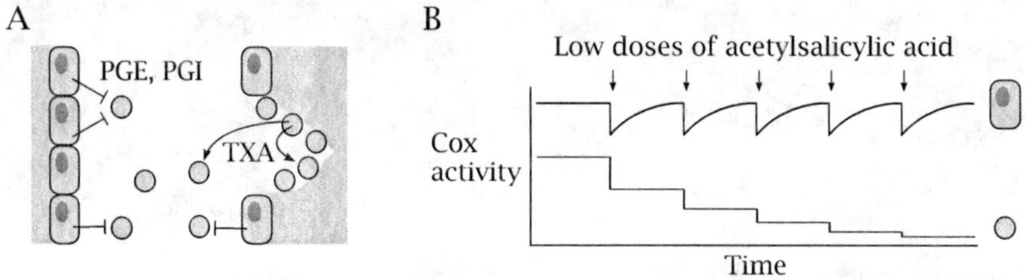

Figure 9.6 Role of eicosanoids in thrombocyte aggregation, and rationale for low-dose acetylsalicylic acid treatment. **A:** Thrombocyte activity is suppressed by the intact vascular endothelium through sustained secretion of PGE and PGI. At sites of lesions, however, exposure of collagen sets off thrombocyte adhesion and aggregation, which is amplified through the secretion of thromboxanes by the platelets themselves. **B:** Effects of low-dose acetylsalicylic acid treatment. In endothelial cells, covalently inactivated cyclooxygenase molecules will be replaced by newly synthesized ones after each dosage. In contrast, thrombocytes lack protein synthesis, so that the effect of repeated doses will be cumulative.

Cox-1 is constitutively expressed and has been credited with most of the housekeeping functions of eicosanoids such as the control of calcium turnover in the bone and maintenance of stomach mucous membrane integrity. It is also involved in the synthesis of thromboxanes in thrombocytes and of prostacyclin (PGI) in endothelial cells, which have antagonistic roles in thrombocyte aggregation and activation. Cox-2 is inducible and strongly expressed in inflammatory cells. Because of this, it has been considered the main culprit in the release of prostaglandins at sites of inflammation, which in turn has motivated the development of selective inhibitors of Cox-2. Selective Cox-2 inhibitors such as rofecoxib indeed show good activity in arthritis and a lower incidence of "traditional" side effects such as gastric ulcers. However, they turned out to cause even more harmful side effects such as myocardial infarction, apparently by upsetting the balance of prostaglandin I and thromboxane A, which controls thrombocyte activity, and they have therefore been withdrawn from the market. Indeed, as early as 1995, experiments with Cox knockout mice indicated that inactivation of Cox-2 was not without serious side effects [34]. Selective inactivation of Cox-1 caused inhibition of thrombocyte function [25], which might have suggested the possibility of excess thrombocyte activity with selective inhibition of Cox-2.[3]

Acetylsalicylic acid also binds in the cyclooxygenase site, where it causes acetylation of serine 530. The plasma half-life of acetylsalicylic acid itself is only about 15 minutes, since the drug is quite rapidly hydrolyzed to acetic acid and salicylic acid. However, the covalent modification of Cox achieved early on will persist after elimination of acetylsalicylic acid, so that the clinical effect of this drug will outlast its elimination.

[3]Inactivation of Cox-1 resulted in a very strong reduction of prostaglandin synthesis in the stomach. Surprisingly, this did not give rise to spontaneous ulcers, and the rate of ulcers induced by indomethacin was even reduced in Cox-1 knockout mice.

The covalent, irreversible mode of action of acetylsalicylic acid is important in its use for inhibiting thrombocyte aggregation. This has proved effective in lowering the incidence of myocardial infarction and stroke in patients with atherosclerosis (see Section 10.4.4).

Thrombocyte aggregation is promoted by thromboxanes, which are synthesized and released by the thrombocytes themselves. Aggregation is inhibited by prostaglandins I and E, which are released by endothelial cells (Figure 9.6). As all of these are derived via Cox-1, it is necessary to inhibit Cox-1 in thrombocytes only but not in endothelial cells. How can such selectivity possibly be achieved?

The solution to this problem lies in the different lifetimes of Cox-1 in the two cell types. In endothelial cells, the enzyme is turned over within hours; covalently inactivated enzyme molecules will therefore have been replaced by newly translated ones by the time of the next drug application. Thrombocytes, however, don't have a nucleus and thus lack protein synthesis; irreversibly inactivated enzyme molecules will therefore never be replaced. In low-dose aspirin therapy, the dosage of acetylsalicylic acid is chosen so as to maintain the enzyme activity in the endothelium yet efficiently inhibit it in the thrombocytes. Increasing the dosage will *reduce* the beneficial effect, since it will achieve no further effect on thromboxane synthesis but will reduce the endothelial synthesis of PGI and PGE.

The cyclooxygenase active site of Cox-2 is more spacious than that of the Cox-1 isoform [28]. As a consequence, Cox-2 accepts not only free arachidonic acid as a substrate but also amides or esters of it, which are known as *endocannabinoids*; this is discussed further in Section 9.8. It also has a surprising consequence for the effect of acetylsalicylic acid. While Cox-1 is completely inactivated by acetylation of serine 530, Cox-2 can still bind and metabolize arachidonic acid. However, instead of converting it into prostaglandin G_2, acetylated Cox-2 may produce 15-epi-lipoxin A_4, which is similar to the physiological mediator lipoxin A_4 (see Figure 9.2B) and stimulates the same receptor. This is believed to contribute to the antiinflammatory effect of acetylsalicylic acid [6,41].

The inhibitor acetaminophen does not act through binding of the cyclooxygenase site. Instead, it reduces the oxidized state of heme [39], thereby preventing the latter from priming or regenerating the radical form of tyrosine 385 (see Figure 9.4C).[4] The inhibitory effect of acetaminophen can be countered by a higher concentration of PGG_2 or other peroxides, which revert heme to the oxidized state. Different intracellular concentrations of peroxides may account for the varying effectiveness of acetaminophen in different tissues. Acetaminophen works rather well in the central nervous system to suppress pain

[4] The physiological reductant of the peroxidase site, glutathione, evidently does not prevent regeneration of the tyrosyl 385 radical. This suggests that some conformational switch links the two active sites and restricts access of glutathione to the peroxidase site when tyrosine 385 has become reduced. Acetaminophen then would bypass this conformational restriction when reducing heme and so forestall reoxidation of tyrosine 385.

Table 9.1 Physiological roles of eicosanoid mediators and related drug effects.

Function	Relevant mediators	Drug effects
Regulation of blood flow in kidneys and other organs	Diverse prostaglandins, thromboxanes, and epoxyeicosatrienoic acids	Severe organ damage after chronic use or abuse of cyclooxygenase inhibitors
Deposition and mobilization of calcium in bone tissue	Prostaglandins E, F	Osteoporosis after prolonged use of glucocorticoids
Perfusion of gastric mucosa, protective mucus secretion	Prostaglandins E, F, I	Erosive gastritis, gastric ulcers after prolonged use of cyclooxygenase inhibitors or steroids; protection from ulcers by prostaglandin analog enprostil [44]
Activity of smooth muscle in uterus and bronchi	Contraction—prostaglandins D, F, leukotriene E; relaxation—prostaglandin E	Induction of labor by intravenous application of prostaglandins and synthetic analogs; relief of menstrual discomfort by cyclooxygenase inhibitors; therapeutic use of leukotriene receptor blockers in asthma
Thrombocyte activation and aggregation	Promoted by thromboxanes; inhibited by prostaglandin I	Therapeutic inhibition with low-dose aspirin and with thromboxane synthase inhibitors or receptor antagonists
Fever, pain	Prostaglandins E, D	Suppression with cyclooxygenase inhibitors
Proliferation of colon cancer cells	Promoted by thromboxane A, prostaglandins E, I	Preventive use of cyclooxygenase inhibitors
Chemotaxis of neutrophile granulocytes (PMN)	Promoted by leukotrienes and thromboxanes, inhibited by lipoxin A_4	Therapeutic use of leukotriene receptor or synthesis blockers in chronic inflammation; anti-inflammatory activity of aspirin-induced lipoxins
Chemotaxis of basophile and eosinophile granulocytes	Prostaglandin D	Experimental use of prostaglandin D receptor antagonists in asthma

and fever. In contrast to many other cyclooxygenase inhibitors, however, it does not suppress the inflammatory activity of leukocytes, which contain high levels of peroxides. Acetaminophen also fails to inhibit aggregation of thrombocytes, which contain a relatively high amount of the lipoxygenase product 12-hydroperoxyeicosatetraenoic acid (12-HPETE [5]).

It is noteworthy that the oxidation product of acetaminophen that is produced by cyclooxygenase, N-acetylimidoquinone (NAPQI), is the same one that also results from oxidation by cytochrome P450 and is responsible for the liver toxicity caused by excessive dosages of acetaminophen (see Figure 4.9).

9.4 PHOSPHOLIPASE A_2 INHIBITORS

Inhibition of cyclooxygenase reduces the utilization of arachidonic acid released by $cPLA_2$. The excess may end up being converted to leukotrienes [11], which may then aggravate inflammatory symptoms, as has been observed in allergic asthma. The lysophospholipid remnant that results from arachidonate release by $cPLA_2$ can also develop potent signaling activity. For example, in thrombocytes, the glyceryl-2-OH group of 1-*O*-alkyl-glycerophosphocholine released by $cPLA_2$ can subsequently be acetylated, giving rise to a potent thrombocyte-aggregating and pro-inflammatory lipid mediator called *platelet-activating factor* (PAF).

Both of these proinflammatory effects should be avoided by inhibition of $cPLA_2$ itself instead of cyclooxygenase. Genetic $cPLA_2$ knockout mice are viable and show reduced inflammatory reactions in response to experimental antigen challenges [47], suggesting that pharmacological inhibitors of $cPLA_2$ should be useful. Currently, however, specific inhibitors for $cPLA_2$ do not exist. For example, the mechanism-based inhibitor methyl arachidonylfluorophosphonate [27] also inhibits cannabinoid receptors [13] and anandamide amidase [8].

Inhibition of $cPLA_2$ contributes to the antiinflammatory activity of the glucocorticoids. It arises through the increased expression of annexin 1. Like $cPLA_2$, annexin 1 binds to negatively charged phospholipids in a calcium-dependent manner, and competition for membrane binding has been suggested as the mechanism of $cPLA_2$ inhibition [2]. On the other hand, a more specific and direct mode of inhibition is suggested by co-immunoprecipitation of $cPLA_2$ and annexin 1 [23]. Glucocorticoids also induce annexins other than annexin 1, and these recruit different binding partners that cause further effects, including apoptosis and phagocytosis. It is therefore unclear to what extent the potent antiinflammatory action of glucocorticoids arises from $cPLA_2$ inhibition.

The pancreas produces a large amount of soluble phospholipase A_2 ($sPLA_2$) for secretion into the small intestine. This enzyme plays a role in severely destructive inflammation of the pancreas itself, which occurs in acute or chronic pancreatitis. In an animal model of pancreatitis, the $sPLA_2$ inhibitor varespladib (S-5920) showed a significant therapeutic benefit [46].

9.5 DERIVATIVES OF PROSTAGLANDIN H_2 AND RELATED DRUGS

An obvious approach to improve on the low selectivity of the cyclooxygenase inhibitors is to target the specific synthases that convert prostaglandin H_2 to the various other prostaglandins and to thromboxane A_4, respectively. Receptor antagonists for some of these mediators have been developed as well. Most of these drugs are still experimental, but some have shown significant potential in clinical or preclinical trials.

There are two varieties of prostaglandin D synthase: lipocalin-type (L-PGDS) and hematopoietic type (H-PGDS). The latter is found in mast cells and releases prostaglandin D as part of allergic reactions. Experimental inhibitors for this enzyme have been developed [1,17]. Prostaglandin D acts on two major receptors, and again one of these, the CRTH2 receptor,[5] has been implicated in its allergic and inflammatory effects. Selective inhibitors of this receptor exist. Prostaglandin E synthase inhibitors and receptor blockers exist, too; several of these have shown potential in animal models of pain, inflammation, and tumor progression [49,24,37].

The drug targets in this group that have received the most attention are the synthase and receptors for thromboxane A. Thromboxanes promote both thrombocyte aggregation and leukocyte activity (that is, inflammation). Inhibitors should be useful in preventive treatment of atherosclerosis and thrombosis in a way similar to low-dose aspirin, but one would expect them to be more selective, be less dose-dependent, and possibly possess higher efficacy than the latter. For the dual receptor and synthase antagonist picotamide, significant benefits have indeed been described in a clinical study [36].

The thromboxane receptor is also the target of competitive regulation by an intrinsic mediator, 14,15-epoxy-eicosatrienoic acid (see Figure 9.2C). Considering the great number of eicosanoids mediators, this recently described mechanism [3] may be of significantly wider scope than currently appreciated.

9.6 LIPOXYGENASES, LEUKOTRIENES, AND RELATED DRUGS

Lipoxygenases introduce hydroperoxy or epoxy groups into the arachidonic acid molecule. The various lipoxygenase isoforms are distinguished by numerical prefixes, which correspond to the position of the double bond that is modified. Both 5- and 15-lipoxygenase are involved in the synthesis of lipoxin A_4 (Figure 9.2), which contributes to the termination of inflammation. As discussed above, this receptor is also activated by aspirin-induced epi-lipoxins.

The dominant role of 5-lipoxygenase (5-Lox), however, is in the synthesis of leukotrienes, which are potent proinflammatory mediators and play a major role in allergic asthma and in autoimmune diseases such as rheumatoid arthritis. In leukotriene biosynthesis, 5-Lox converts arachidonic acid first to 5-hydroperoxyeicosatetraenoic acid and then to its epoxy derivative leukotriene A_4 (Figure 9.2B). As in the cyclooxygenase reaction, the initial step consists in the abstraction of a hydrogen atom from the substrate. The electrophile, however, is a ferric iron ion, which is not embedded in a heme group but instead is chelated by several histidine side chains of the enzyme. Hydrogen abstraction

[5]The acronym CRTH stands for chemoattractant receptor-homologous (receptor). This receptor shares similarity with those for the *chemokines*, a family of peptide inflammatory mediators, rather than with other prostaglandin receptors.

from the substrate is mediated by a water molecule that is also coordinated by the catalytic iron.

5-Lox can be inhibited by the iron-chelating drug zileuton. A dual inhibitor of 5-Lox and Cox-2 is tepoxalin. Curcumin, a natural polyphenol that occurs in the Indian spice plant turmeric, inhibits 5-Lox as well as $cPLA_2$ [18]. Both effects should inhibit inflammation, and indeed turmeric has been used in traditional medicine for this purpose. Similarly, the reduction in thrombocyte activity and arterial thrombosis associated with regular consumption of red wine and green tea has been attributed to their content of polyphenols.

Like $cPLA_2$, 5-Lox binds to the nuclear membrane when cytosolic calcium is increased. On the membrane, it associates with 5-lipoxygenase-activating protein (FLAP), an integral membrane protein that channels the substrate fatty acids released by $cPLA_2$ to 5-Lox. FLAP itself is targeted by several experimental inhibitors [12].

Leukotriene A_4 is converted to leukotriene B_4 by leukotriene A_4 hydrolase, and there are experimental inhibitors for the enzyme as well as for both types of leukotriene B receptors. Formation of the cysteine-containing leukotrienes C_4, D_4, and E_4 is initiated by glutathione-*S*-transferase. Receptor antagonists for cysteinyl-leukotrienes are used in the treatment of allergic asthma; an example is the drug montelukast.

9.7 EICOSANOIDS SYNTHESIZED BY CYTOCHROME P450

Some cytochrome P450 isoforms such as 2C9 and 2J2 are expressed in vascular endothelia and convert arachidonic acid to 14,15-epoxy-eicosatrienoic acid (14,15-EETE; Figure 9.2C). The latter causes vasorelaxation, which correlates with the activation of calcium-gated potassium channels [26]. When released in the brain, 14,15-EETE also suppresses pain perception, which is mediated by endogenous opioids [45]. The mediator is degraded by soluble epoxide hydrolase. Experimental inhibitors of this enzyme have shown promise in animal models of atherosclerosis.

9.8 ENDOCANNABINOIDS AND RELATED DRUGS

Endocannabinoids are esters or amides of arachidonic acid. Their name derives from the fact that they act on the same receptors as the active constituents of the hemp or cannabis plant, such as tetrahydrocannabinol (THC). The endocannabinoid receptors, CB1 and CB2, are GPCRs. The CB1 receptor occurs predominantly in the central nervous system, where it regulates, through dampening synaptic transmission, diverse functions such as mood, motor control, and pain. The CB2 receptor is found in the periphery on cells other than neurons. Nevertheless, it does contribute to the control of pain perception, apparently through triggering the release of endorphin [22]. CB2 receptor knockout mice show a loss of bone mass [38], suggesting that CB2 receptor agonists may be useful in the treatment of osteoporosis.

In addition to the activation of CB1 and CB2 receptors, endocannabinoids and agonists such as THC can also interact with ion channels. The allosteric activation of glycine receptors [15] has been shown to cause pain suppression independently of CB1 or CB2 receptor activation [48]. Some endocannabinoids also act on the vanilloid receptors, which are ligand-gated cation channels that also open in response to heat and are involved in the perception of heat and pain. Additional receptor types have been postulated [30].

While cannabis is most widely known for the psychotropic effects that are the objective of its abuse, endocannabinoid receptor agonists thus have a much wider range of therapeutic applications, including the treatment of pain, epilepsy, and nausea.

9.8.1 Biosynthesis and degradation of endocannabinoids

Metabolic pathways for synthesis and degradation have been established for two endocannabinoids, 2-arachidonylglycerol and N-arachidonylethanolamine, which is also known as anandamide (Figure 9.7). Other endocannabinoids include N-arachidonylglycine and -dopamine (NADA). The latter is a particularly potent agonist at the vanilloid VR1 receptor and is found in both the peripheral and the central nervous systems; it may be the major endogenous ligand for that receptor [20].

The synthesis of anandamide[6] is initiated by a rise of intracellular Ca^{2+}, which in nerve cells may result from activation of ionotropic channels. Ca^{2+} activates an acyltransferase that removes arachidonic acid from one phospholipid molecule and attaches it to the free amino group of phosphatidylethanolamine (PE). Cleavage of *N*-arachidonyl-PE by a phospholipase D then releases anandamide.

2-Arachidonylglycerol is formed from diacylglycerol, which arises from the phospholipase C pathway that is activated by many different G protein-coupled receptors. The phospholipase C pathway also raises intracellular Ca^{2+}. Therefore, both ionotropic receptors and G protein-coupled receptors may activate the synthesis of endocannabinoids in nerve cells.

Inactivation of endocannabinoids occurs through cellular uptake, followed by enzymatic degradation. Cellular uptake is susceptible to specific inhibition, which suggests the existence of a specific transporter protein [33], although none has been molecularly characterized so far. Intracellular breakdown of anandamide is mediated by fatty acyl amide hydrolase (FAAH); in FAAH knockout mice, anandamide accumulates to very high levels [7]. FAAH and monoacylglycerol lipase share in the degradation of 2-arachidonyl-glycerol. While both mediators are also substrates for cyclooxygenase 2, inhibition of the latter enzyme does not significantly increase their tissue levels [35].

9.8.2 Role of endocannabinoids in synaptic feedback inhibition

Endocannabinoids serve a unique regulatory function in synaptic transmission (Figure 9.8). They are produced by the postsynaptic cell in response to transmitter stimulation. After release into the synaptic cleft, they activate CB1 receptors

[6]In case you thought, "enough with the Greek names already": This name derives from the Sanskrit word *ananda*, which means "bliss".

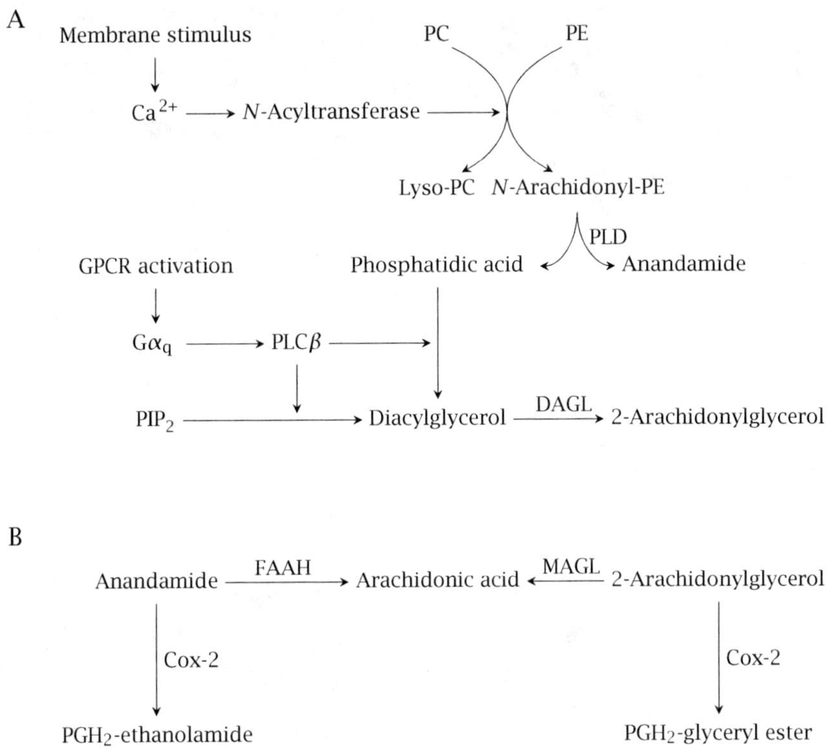

Figure 9.7 Metabolism of the endocannabinoids anandamide (*N*-arachidonylethanolamine) and 2-arachidonylglycerol (2-AG). **A**: Biosynthesis. Increased Ca^{2+} levels activate an *N*-acyltransferase that transfers arachidonic acid from another phospholipid molecule to the amino group of phosphatidylethanolamine (PE); anandamide is then released by phospholipase D (PLD). 2-AG is produced by diacylglycerol lipase (DAGL) downstream of the phospholipase C pathway. **B**: Degradation. Anandamide is hydrolyzed by fatty acid amide hydrolase (FAAH) and 2-AG by monoacylglyerol lipase (MAGL). Both mediators can also be metabolized by cyclooxygenase 2.

on the presynaptic cell. The $G\beta\gamma$ subunit of the $G_{q/11}$ protein inhibits calcium channels and activates potassium channels. This results in a decrease of membrane excitability and transmitter release. Release of endocannabinoids by the postsynaptic cell therefore creates a negative-feedback loop that tempers synaptic transmission. This mode of action ties in with the observation that CB1 agonists suppress pain and inhibit the neuronal overexcitability characteristic of epilepsy.

9.8.3 Significance of cyclooxygenase 2 activity on endocannabinoids

Cyclooxygenase 2 converts anandamide to the ethanolamide analog of prostaglandin H_2, which can then be further converted to other prostaglandin analogs by the various prostaglandin synthases. These derivatives of anandamide are referred to as *prostamides*; similar products are also formed from 2-arachidonylglycerol. Prostamide receptors exist but are incompletely characterized; antagonists are useful in the treatment of ocular hypertension (glaucoma).

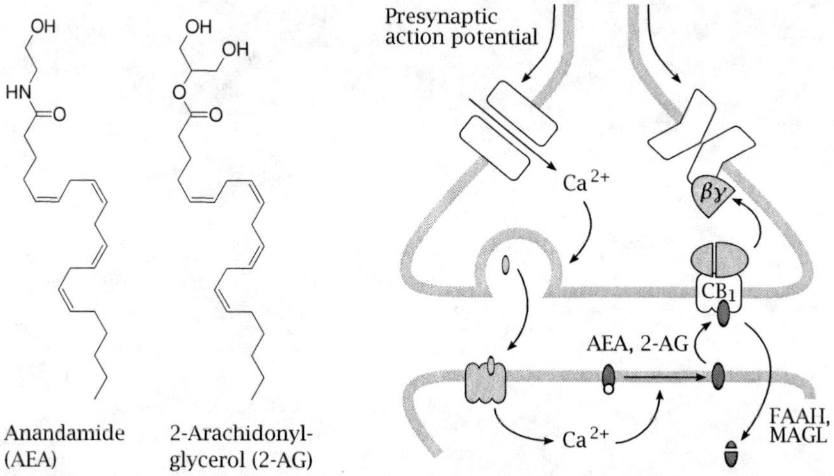

Figure 9.8 Feedback inhibition of synaptic transmission by endocannabinoids. Stimulation of the postsynaptic cell by neurotransmitters triggers synthesis of anandamide or 2-arachidonyl-glycerol, which are released into the synaptic cleft and activate CB1 receptors on the presynaptic cells. This causes release of G_q $\beta\gamma$-dimers, which inhibit Ca^{2+} channels and activate K^+ channels, thereby reducing membrane excitability and transmitter release. The action of endocannabinoids is terminated by cellular reuptake and degradation by FAAH and MAGL (see Figure 9.7).

In direct contradistinction to 2-arachidonylglycerol itself, the prostaglandin E_2 analog derived from it by Cox-2 and prostaglandin E synthase *triggers* pain [19]. This suggests that inhibited metabolism of endocannabinoids may contribute to the pain-suppressing effects of cyclooxygenase inhibitors. Tipping the balance in favor of anandamide may also play a role in the addictive potential of cyclooxygenase inhibitors. Mutations that decrease the activity of fatty acyl amide hydrolase are associated with an increased rate of abuse of alcohol and several other drugs in humans [43], and CB1 receptor knockout mice show reduced alcohol consumption [21]. This effect is likely related to the role of endocannabinoid receptors in the reward circuit of the CNS (see Section 6.8.2).

9.8.4 Drugs interacting with the endocannabinoid system

The most widely known receptor agonist is Δ^9-tetrahydrocannabinol (THC), which is found in marijuana and activates both CB1 and CB2 receptors. Ajulemic acid resembles THC but contains a carboxylate group and shows reduced penetration of the blood–brain barrier. It therefore lacks the psychotropic effects of THC yet still has analgesic activity.

Some synthetic receptor agonists and antagonists are selective for CB1 or CB2 subtypes, respectively. Receptor stimulation can be indirectly increased by inhibitors of fatty acyl amide hydrolase and of cellular endocannabinoid uptake. Figure 9.9 shows some example structures.

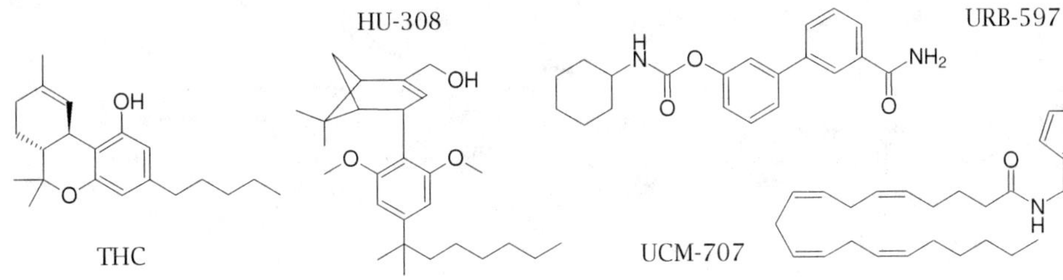

Figure 9.9 Drugs interacting with the endocannabinoid system. Δ^9-Tetrahydrocannabinol (THC) occurs in cannabis and stimulates both CB1 and CB2 receptors. HU-308 is a CB2-selective agonist. URB-597 inhibits fatty acyl amide hydrolase, whereas UCM-707 inhibits cellular uptake of endocannabinoids.

9.9 THE ROLE OF POLYUNSATURATED FATTY ACIDS IN EICOSANOID SIGNALING

Eicosapentaenoic and docosahexaenoic acid are more highly unsaturated analogues of arachidonic acid. They differ from it by the possession of a double bond between the third and fourth carbon atoms counted from the far end and thus are referred to as "ω-3" fatty acids. This particular double bond cannot be formed in human metabolism, and therefore ω-3 fatty acids have to be acquired from the diet. Algae and fish living in cold waters require them in high levels in order to maintain membrane fluidity in spite of low temperatures. Fish oil therefore is a rich source of ω-3 fatty acids, and for this reason is used as a dietary supplement.

Numerous benefits have been claimed to result from a diet enriched in ω-3 fatty acids, including lower levels of serum cholesterol and triglycerides, inhibition of blood coagulation and inflammation, and reduction of sudden cardiac deaths. Decreased mortality due to polyunsaturated fatty acid medication has been substantiated in a sizable collective of patients who had suffered a myocardial infarction and were at increased risk of repeat infarctions and cardiac arrhythmia [32].

Some of the beneficial effects of polyunsaturated fatty acids are likely due to activation of peroxisome proliferator-activated receptors. Another conceivable mode of action of eicosapentaenoic acid is that it substitutes for arachidonic acid as a precursor of eicosanoid mediators, for example, replacing thromboxane A_2 with thromboxane A_3, which may have a lower proinflammatory and procoagulatory activity. However, at least with thromboxane A, this substitution does not appear to be very effective [14]. On the other hand, cultured smooth muscle cells preloaded with eicosapentaenoic or docosahexaenoic acid show a stunted proliferative response to thromboxane A_2 [40], suggesting an effect downstream rather than upstream of the thromboxane receptor.

The suppression of cardiac arrhythmias by eicosapentaenoic or docosahexaenoic acid is apparently related to a direct effect on voltage-gated sodium channels. A membrane environment enriched in these fatty acids, presumably containing them within phospholipid molecules, affects these channels in a manner similar to lidocaine (see Section 6.6.1; [4]).

STUDY QUESTIONS

9.1 It has been observed in cell culture that lipoxygenase products inhibit fatty acyl amide hydrolase (FAAH [29]). Could this observation be relevant to the effect of cyclooxygenase inhibitors?

9.2 Cyclooxygenases 1 and 2 (as well as 12/15-lipoxygenase) have been reported to catalytically degrade nitric oxide, which is used by the enzymes as a reducing cosubstrate (the substrate is a peroxide, for example, prostaglandin G in the case of cyclooxygenase). What could this mean for the effectiveness of cyclooxygenase inhibitors of those enzymes with respect to inflammation?

REFERENCES

[1] K. Aritake et al., "Structural and functional characterization of HQL-79, an orally selective inhibitor of human hematopoietic prostaglandin D synthase," *J. Biol. Chem.* 281 (2006), 15277–15286.

[2] B. C. Bastian et al., "Inhibition of human skin phospholipase A2 by 'lipocortins' is an indirect effect of substrate/lipocortin interaction," *J. Invest. Dermatol.* 101 (1993), 359–363.

[3] D. J. Behm et al., "Epoxyeicosatrienoic acids function as selective, endogenous antagonists of native thromboxane receptors: identification of a novel mechanism of vasodilation," *J. Pharmacol. Exp. Ther.* 328 (2009), 231–239.

[4] G. E. Billman, J. X. Kang, and A. Leaf, "Prevention of sudden cardiac death by dietary pure omega-3 polyunsaturated fatty acids in dogs," *Circulation* 99 (1999), 2452–2457.

[5] O. Boutaud et al., "Determinants of the cellular specificity of acetaminophen as an inhibitor of prostaglandin H(2) synthases," *Proc. Natl. Acad. Sci. U. S. A.* 99 (2002), 7130–7135.

[6] N. Chiang et al., "Activation of lipoxin A(4) receptors by aspirin-triggered lipoxins and select peptides evokes ligand-specific responses in inflammation," *J. Exp. Med.* 191 (2000), 1197–1208.

[7] B. F. Cravatt et al., "Supersensitivity to anandamide and enhanced endogenous cannabinoid signaling in mice lacking fatty acid amide hydrolase," *Proc. Natl. Acad. Sci. U. S. A.* 98 (2001), 9371–9376.

[8] D. G. Deutsch et al., "Methyl arachidonyl fluorophosphonate: a potent irreversible inhibitor of anandamide amidase," *Biochem. Pharmacol.* 53 (1997), 255–260.

[9] W. A. van der Donk, A. Tsai, and R. J. Kulmacz, "The cyclooxygenase reaction mechanism," *Biochemistry* 41 (2002), 15451–15458.

[10] I. Dussault and B. M. Forman, "Prostaglandins and fatty acids regulate transcriptional signaling via the peroxisome proliferator activated receptor nuclear receptors," *Prostaglandins Other Lipid. Mediat.* 62 (2000), 1–13.

[11] P. X. Elsas et al., "Cysteinyl leukotrienes mediate the enhancing effects of indomethacin and aspirin on eosinophil production in murine bone marrow cultures," *Br. J. Pharmacol.* 153 (2008), 528–535.

[12] J. F. Evans et al., "What's all the FLAP about?: 5-lipoxygenase-activating protein inhibitors for inflammatory diseases," *Trends Pharmacol. Sci.* 29 (2008), 72–78.

[13] S. R. Fernando and R. G. Pertwee, "Evidence that methyl arachidonyl fluorophosphonate is an irreversible cannabinoid receptor antagonist," *Br. J. Pharmacol.* 121 (1997), 1716–1720.

[14] T. Hamazaki et al., "The infusion of trieicosapentaenoyl-glycerol into humans and the in vivo formation of prostaglandin I3 and thromboxane A3," *Biochem. Biophys. Res. Commun.* 151 (1988), 1386–1394.

[15] N. Hejazi et al., "Delta9-tetrahydrocannabinol and endogenous cannabinoid anandamide directly potentiate the function of glycine receptors," *Mol. Pharmacol.* 69 (2006), 991–997.

[16] M. E. Hemler, G. Graff, and W. E. Lands, "Accelerative autoactivation of prostaglandin biosynthesis by PGG2," *Biochem. Biophys. Res. Commun.* 85 (1978), 1325–1331.

[17] M. Hohwy et al., "Novel prostaglandin D synthase inhibitors generated by fragment-based drug design," *J. Med. Chem.* 51 (2008), 2178–2186.

[18] J. Hong et al., "Modulation of arachidonic acid metabolism by curcumin and related β-diketone derivatives: effects on cytosolic phospholipase A_2, cyclooxygenases and 5-lipoxygenase," *Carcinogenesis* 25 (2004), 1671–1679.

[19] S. S. Hu et al., "Prostaglandin E_2 glycerol ester, an endogenous COX-2 metabolite of 2-arachidonoylglycerol, induces hyperalgesia and modulates NFκB activity," *Br. J. Pharmacol.* 153 (2008), 1538–1549.

[20] S. M. Huang et al., "An endogenous capsaicin-like substance with high potency at recombinant and native vanilloid VR1 receptors," *Proc. Natl. Acad. Sci. U. S. A.* 99 (2002), 8400–8405.

[21] B. L. Hungund et al., "Cannabinoid CB1 receptor knockout mice exhibit markedly reduced voluntary alcohol consumption and lack alcohol-induced dopamine release in the nucleus accumbens," *J. Neurochem.* 84 (2003), 698–704.

[22] M. M. Ibrahim et al., "CB2 cannabinoid receptor activation produces antinociception by stimulating peripheral release of endogenous opioids," *Proc. Natl. Acad. Sci. U. S. A.* 102 (2005), 3093–3098.

[23] K. M. Kim et al., "Annexin-I inhibits phospholipase A2 by specific interaction, not by substrate depletion," *FEBS Lett.* 343 (1994), 251–255.

[24] T. Kitamura et al., "Combined effects of prostaglandin E receptor subtype EP1 and subtype EP4 antagonists on intestinal tumorigenesis in adenomatous polyposis coli gene knockout mice," *Cancer Sci.* 94 (2003), 618–621.

[25] R. Langenbach et al., "Prostaglandin synthase 1 gene disruption in mice reduces arachidonic acid-induced inflammation and indomethacin-induced gastric ulceration," *Cell* 83 (1995), 483–492.

[26] B. T. Larsen et al., "Epoxyeicosatrienoic and dihydroxyeicosatrienoic acids dilate human coronary arterioles via BK(Ca) channels: implications for soluble epoxide hydrolase inhibition," *Am. J. Physiol. Heart Circ. Physiol.* 290 (2006), H491–H499.

[27] Y. C. Lio et al., "Irreversible inhibition of Ca^{2+} -independent phospholipase A2 by methyl arachidonyl fluorophosphonate," *Biochim. Biophys. Acta* 1302 (1996), 55–60.

[28] C. Luong et al., "Flexibility of the NSAID binding site in the structure of human cyclooxygenase-2," *Nat. Struct. Biol.* 3 (1996), 927–933.

[29] M. Maccarrone et al., "Anandamide and 2-arachidonoylglycerol inhibit fatty acid amide hydrolase by activating the lipoxygenase pathway of the arachidonate cascade," *Biochem. Biophys. Res. Commun.* 278 (2000), 576–583.

[30] K. Mackie and N. Stella, "Cannabinoid receptors and endocannabinoids: evidence for new players," *AAPS J.* 8 (2006), E298–E306.

[31] M. G. Malkowski et al., "The productive conformation of arachidonic acid bound to prostaglandin synthase," *Science* 289 (2000), 1933–1937.

[32] R. Marchioli et al., "Efficacy of n-3 polyunsaturated fatty acids after myocardial infarction: results of GISSI-Prevenzione trial. Gruppo Italiano per lo Studio della Sopravvivenza nell'Infarto Miocardico," *Lipids* 36 Suppl (2001), S119–S126.

[33] S. A. Moore et al., "Identification of a high-affinity binding site involved in the transport of endocannabinoids," *Proc. Natl. Acad. Sci. U. S. A.* 102 (2005), 17852–17857.

[34] S. G. Morham et al., "Prostaglandin synthase 2 gene disruption causes severe renal pathology in the mouse," *Cell* 83 (1995), 473–482.

[35] G. G. Muccioli et al., "Identification of a novel endocannabinoid-hydrolyzing enzyme expressed by microglial cells," *J. Neurosci.* 27 (2007), 2883–2889.

[36] G. G. Neri Serneri et al., "Picotamide, a combined inhibitor of thromboxane A2 synthase and receptor, reduces 2-year mortality in diabetics with peripheral arterial disease: the DAVID study," *Eur. Heart J.* 25 (2004), 1845–1852.

[37] N. Niho et al., "Suppression of azoxymethane-induced colon cancer development in rats by a prostaglandin E receptor EP1-selective antagonist," *Cancer Sci.* 96 (2005), 260–264.

[38] O. Ofek et al., "Peripheral cannabinoid receptor, CB2, regulates bone mass," *Proc. Natl. Acad. Sci. U. S. A.* 103 (2006), 696–701.

[39] M. Ouellet and M. D. Percival, "Mechanism of acetaminophen inhibition of cyclooxygenase isoforms," *Arch. Biochem. Biophys.* 387 (2001), 273–280.

[40] R. Pakala, R. Pakala, and C. R. Benedict, "Thromboxane A2 fails to induce proliferation of smooth muscle cells enriched with eicosapentaenoic acid and docosahexaenoic acid," *Prostaglandins Leukot. Essent. Fatty Acids* 60 (1999), 275–281.

[41] M. J. Paul-Clark et al., "15-epi-lipoxin A4-mediated induction of nitric oxide explains how aspirin inhibits acute inflammation," *J. Exp. Med.* 200 (2004), 69–78.

[42] S. W. Rowlinson et al., "A novel mechanism of cyclooxygenase-2 inhibition involving interactions with Ser-530 and Tyr-385," *J. Biol. Chem.* 278 (2003), 45763–45769.

[43] J. C. Sipe et al., "A missense mutation in human fatty acid amide hydrolase associated with problem drug use," *Proc. Natl. Acad. Sci. U. S. A.* 99 (2002), 8394–8399.

[44] S. J. Sontag et al., "Healing of NSAID-induced gastric ulcers with a synthetic prostaglandin analog (enprostil)," *Am. J. Gastroenterol.* 89 (1994), 1014–1020.

[45] M. Terashvili et al., "Antinociception produced by 14,15-epoxyeicosatrienoic acid is mediated by the activation of β-endorphin and met-enkephalin in the rat ventrolateral periaqueductal gray," *J. Pharmacol. Exp. Ther.* 326 (2008), 614–622.

[46] Y. Tomita et al., "Effect of a selective inhibitor of secretory phospholipase A_2, S-5920/LY315920Na, on experimental acute pancreatitis in rats," *J. Pharmacol. Sci.* 96 (2004), 144–154.

[47] N. Uozumi et al., "Role of cytosolic phospholipase A2 in allergic response and parturition," *Nature* 390 (1997), 618–622.

[48] W. Xiong et al., "Cannabinoid potentiation of glycine receptors contributes to cannabis-induced analgesia," *Nat. Chem. Biol.* 7 (2011), 296–303.

[49] D. Xu et al., "MF63 [2-(6-chloro-1H-phenanthro[9,10-d]imidazol-2-yl)-isophthalonitrile], a selective microsomal prostaglandin E synthase-1 inhibitor, relieves pyresis and pain in preclinical models of inflammation," *J. Pharmacol. Exp. Ther.* 326 (2008), 754–763.

CHAPTER 10

INTERMEDIATE METABOLISM, DIABETES AND ATHEROSCLEROSIS

In this chapter, we will look at the effects of drugs on intermediate metabolism and their use in hereditary enzyme defects, diabetes mellitus, and atherosclerosis. The opposite case, namely the effect of metabolism on drugs, is covered in Chapter 4.

10.1 HEREDITARY ENZYME DEFECTS

Enzyme defects in intermediate metabolism are fairly rare diseases. The most common one is phenylketonuria, which affects only about 1 in 10,000 people. Most enzyme defects are recessive; that is, they will be clinically manifest only if both copies of the gene encoding the enzyme in question are deficient.[1] Very often, enzyme defects become manifest after birth. In the uterus, the fetus is connected to the maternal circulation via the placenta; any soluble metabolite that the fetus does not process itself will cross the placenta and be processed by the mother's metabolism. Once the child is born and disconnected from this carefree state of permanent hemodialysis, it has to cope with all metabolites

[1]Some enzymes are encoded on the X chromosome, of which there are two copies in female body cells but only one in male ones. Such enzyme defects will be more readily manifest in males than females. An example is glucose-6-phosphate dehydrogenase deficiency.

Biochemical Pharmacology. By M. Palmer, A. Chan, T. Dieckmann, J. Honek

on its own, and any enzyme defects will cause the corresponding substrates to accumulate.[2] Often, it is the toxicity of the accumulated substrate, rather than the lack of product of the deficient reaction, that causes the pathogenic effect.

Drug therapy is available with only few hereditary enzyme defects; some examples are discussed below. In many cases, dietary restrictions can limit the accumulation of toxic metabolites. In an increasing number of diseases, enzyme replacement therapy is used. Gene therapy—the introduction of intact copies of the deficient gene into the genome of the patient—offers a longer-lasting, or potentially permanent therapeutic effect, but is currently still largely experimental. Diseases that affect mostly one organ are often treated with organ transplantation; examples are bone marrow transplants in adenosine deaminase deficiency and liver transplants in Crigler–Najjar syndrome. The transplanted cells or organs will continue to express the intact genes they came with; one could therefore say that transplantation is a form of gene therapy.

10.1.1 Phenylketonuria

Phenylketonuria is due to a homozygous defect for phenylalanine hydroxylase, which converts the amino acid phenylalanine to tyrosine (Figure 10.1). Phenylalanine accumulates, and some of it is transaminated to phenylpyruvate, a ketone compound whose observation in the urine gave the disease its name. Accumulation of phenylalanine in the brain causes neurological deficits and mental retardation, possibly due to interference with serotonin synthesis [11]. Toxic levels of phenylalanine accumulate only after delivery, but the level observed at birth is sufficiently high to allow immediate laboratory diagnosis. Because of its relatively high incidence, all newborns are screened for the phenylketonuria.

Treatment consists mostly in a phenylalanine-controlled diet. Since phenylalanine is an essential amino acid, the diet must maintain a level of the amino acid high enough to support protein synthesis but low enough to avoid toxicity.[3] Compliance with the diet is most crucial during childhood and adolescence. In addition, female patients must obey the diet very strictly during pregnancy. Otherwise, the placental substrate exchange will *not* work to protect the fetus; on the contrary, even fetuses that express functional phenylalanine hydroxylase will be overwhelmed by the excess phenylalanine transferred from the mother. This can cause severe neurological deficits that will already be manifest at birth.

[2]A transient version of this effect can be observed in most newborns in the form of neonatal jaundice. This is due to an initially low level of expression of UDP-glucuronosyltransferase 1A1, which is required for the conjugation of bilirubin, the degradation product of heme. Unconjugated bilirubin cannot be effectively excreted, which leads to jaundice. Excessive accumulation of bilirubin can lead to toxicity within the central nervous system (*kernicterus*). This complication can be prevented with *phototherapy*: Absorption of blue light by bilirubin in the skin induces a cis–trans transition to an isomer that is more readily excreted in its unconjugated form. Expression of the enzyme reaches sufficient levels after a few weeks, and jaundice subsides.

[3]Since patients cannot convert phenylalanine to tyrosine, the latter amino acid becomes essential, which, however, does not seem to pose problems in practice.

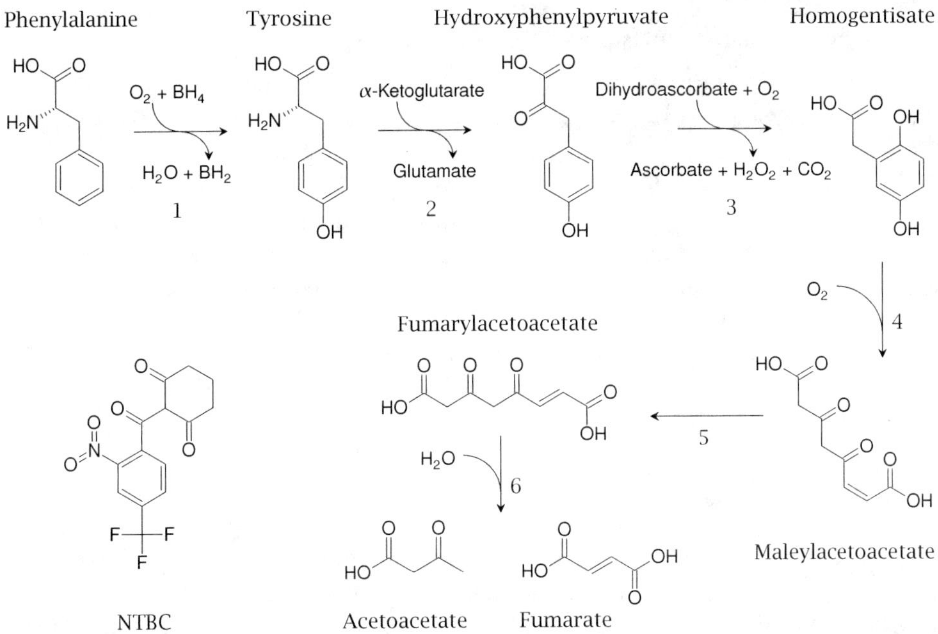

Figure 10.1 Pathway for degradation of phenylalanine and of tyrosine, and structure of NTBC {2-[2-nitro-4-(trifluoromethyl)benzoyl]cyclohexane-1,3-dione}, an inhibitor of *p*-phenylpyruvate dioxygenase (3) that is used in tyrosinemia type I. BH_2 and BH_4: di- and tetrahydrobiopterin, respectively. Enzymes: (1) phenylalanine hydroxylase; (2) tyrosine transaminase; (3) *p*-phenylpyruvate dioxygenase; (4) homogentisate dioxygenase; (5) maleylacetoacetate isomerase; (6) fumarylacetoacetate hydrolase.

Phenylalanine hydroxylase requires tetrahydrobiopterin. Some cases of phenylketonuria are due to a reduced affinity of the enzyme for this cosubstrate, rather than to an actual loss of catalytic activity. These patients can be treated with high dosages of tetrahydrobiopterin.

10.1.2 Tyrosinemia type I

Tyrosine can be completely degraded in human metabolism. The gene defect in tyrosinemia concerns the last enzyme of this pathway, fumarylacetoacetate hydrolase. This causes all preceding metabolites to accumulate, up to tyrosine itself. The toxic one is maleylacetoacetate, which can react covalently with nucleophiles, including proteins and nucleic acids, leading to cell damage. The activity of the tyrosine degradation pathway is highest in the liver, which accordingly sustains the most severe damage, often giving rise to liver cancer.

Pharmacological treatment uses NTBC, an inhibitor of *p*-hydroxyphenylpyruvate dioxygenase (Figure 10.1). This enzyme is positioned upstream of maleylacetoacetate, and its inhibition therefore reduces formation of the toxic metabolite. NTBC therapy is supplemented with dietary restriction of phenylalanine and tyrosine. Even so, the harmful effects of the enzyme defect are not fully suppressed, and the best long-term treatment therefore is liver transplantation.

10.1.3 Urea cycle enzyme defects

The urea cycle produces urea, which is the major vehicle for the elimination of excess nitrogen from the body. Excess nitrogen accrues from to the degradation of amino acids. It is transported from peripheral organs to the liver in the form of glutamine. Glutaminase cleaves glutamine into glutamate and ammonia, which is incorporated into carbamylphosphate and as such enters the urea cycle. Glutamate can provide the second nitrogen for urea synthesis via transamination of oxaloacetate. Figure 10.2 gives an overview of the urea cycle and the related pathways.

Hereditary enzyme defects are known for all enzymes in the cycle. The problems caused by these defects are twofold: (1) elimination of excess nitrogen is impaired, and (2) the level of free ammonia is increased, which causes neurotoxicity.

Since the urea cycle receives is substrate nitrogen from the degradation of amino acids, one obvious and important therapeutic principle is the restriction of dietary protein. Pharmacological treatment follows the principle of alternative pathway therapy [4]. The idea is to provide an alternate route for the elimination of excess nitrogen. This can be accomplished by supplying organic acids as substrates for conjugation with glycine or glutamine (see Section 4.5.6); the conjugates are then eliminated with the urine. Because glutamine functions as the major carrier of excess nitrogen in normal metabolism, it is a logical target for this therapeutic approach. An organic acid suitable for oral application is phenylbutyrate. It undergoes β oxidation to phenylacetyl-CoA, which is then conjugated with glutamine. Phenylbutyrate is combined with benzoate, which undergoes conjugation with glycine and thus further drains the liver's nitrogen pool [14].

In addition to organic acids, defects of urea cycle enzymes other than arginase are also treated with citrulline or arginine, which both reduce free ammonia. Citrulline is used with enzyme defects that are located upstream of itself within the cycle, and so do not interfere with its conversion to arginine. Arginine is used in enzyme defects that disrupt this conversion, except in arginase deficiency [4].

Citrulline contains only one of the two nitrogen atoms that will eventually be found in urea. Utilization of exogenous citrulline by the urea cycle will therefore result in a net elimination of nitrogen. In contrast, arginine already contains both nitrogens destined for urea synthesis; exogenous arginine thus cannot directly improve the nitrogen balance. It follows that the reduction of ammonia by arginine must be due to some regulatory effect. The exact nature of this regulatory mechanism is an open question.

A possible answer involves the anaplerotic conversion of glutamate to ornithine (reactions 10 and 11 in Figure 10.2). The throughput of this pathway must in some way be controlled by the current level of urea cycle intermediates. This may occur through inhibition either by arginine directly or by ornithine, which is produced from arginine by arginase. In either case, a low level of arginine would disinhibit synthesis of ornithine from glutamate. Since glutamate

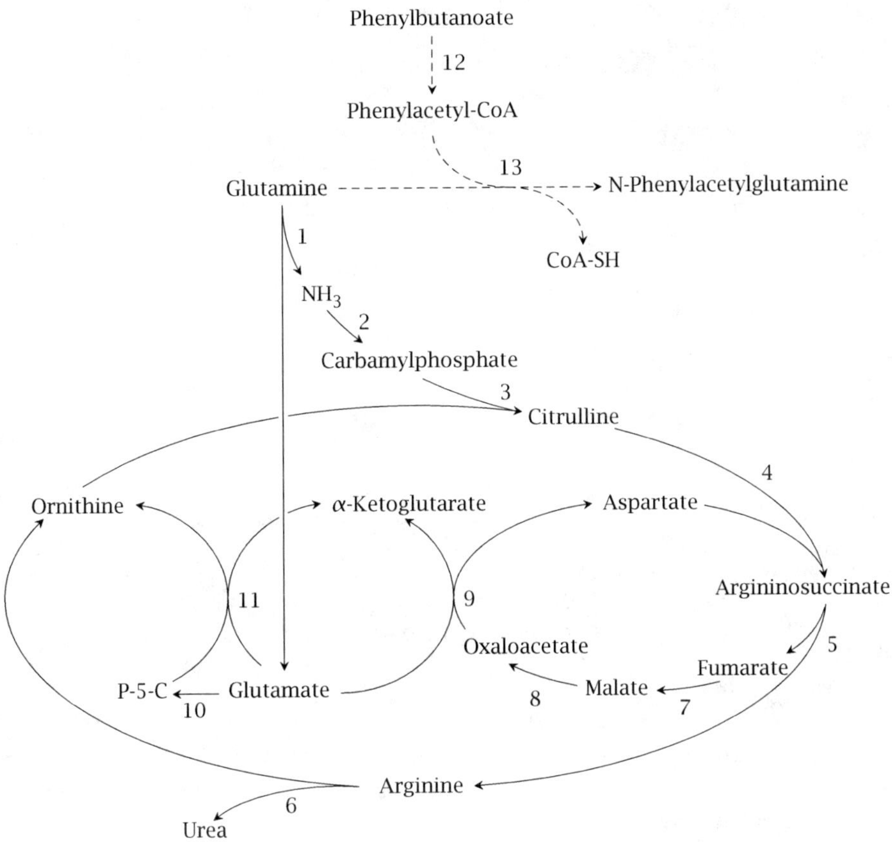

Figure 10.2 Overview of the urea cycle and related pathways. Reactions 2–6 form the urea cycle. Reactions 1 and 7–9 supply the urea cycle with nitrogen from glutamine, which serves as the major nitrogen carrier. Reaction 10 yields Δ^1-pyrroline-5-carboxylate (P-5-C), and together with reaction 11 replenishes ornithine. Urea cycle enzyme defects can be treated with phenylbutanoate, which, via transformations 12 and 13, yields N-phenylacetylglutamine. This conjugate is eliminated with the urine and thus provides an alternative vehicle for nitrogen excretion. Enzymes: (1) glutaminase; (2) carbamylphosphate synthetase; (3) ornithine transcarbamylase; (4) argininosuccinate synthase; (5) argininosuccinase; (6) arginase; (7) fumarase; (8) malate dehydrogenase; (9) aspartate aminotransferase; (10) Δ^1-pyrroline-5-carboxylate synthase; (11), ornithine aminotransferase; (12) enzymes of β-oxidation; (13) glutamine *N*-phenylacetyltransferase.

inhibits glutaminase [8], consumption of glutamate would then stimulate its own replenishment by glutaminase, and with it the release of ammonia.

10.1.4 Adenosine deaminase deficiency

When purine nucleotides and nucleosides are present in excess, they can either undergo complete degradation to uric acid or be partially degraded and later be restored by salvage enzymes. Adenosine deaminase (ADA) functions in the degradation of adenosine and deoxyadenosine. When this enzyme is ge-

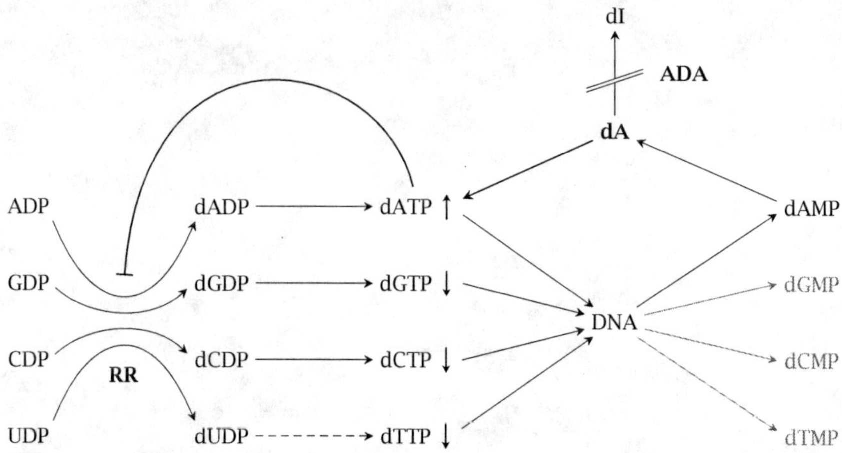

Figure 10.3 The pathogenetic mechanism of adenosine deaminase (ADA) deficiency. Deoxyadenosine (dA) that forms in DNA degradation is converted by ADA to deoxyinosine (dI). If ADA is deficient, dA accumulates and is phosphorylated to dATP, which is an allosteric inhibitor of ribonucleotide reductase (RR). The latter enzyme is required in the production of all deoxyribonucleotides. The resulting imbalance in the deoxyribonucleotide supply interferes with DNA synthesis, which promotes apoptosis in T lymphocytes.

netically deficient, the block of deoxyadenosine (dA) degradation results in its conversion to dATP by salvage nucleoside and nucleotide kinases. The accumulating dATP inhibits ribonucleotide reductase (Figure 10.3), which suppresses the synthesis of deoxyribonucleotides other than dATP. This interferes with DNA synthesis and promotes apoptosis.

The salvage nucleoside kinases that phosphorylate deoxyadenosine to dATP are particularly strongly expressed in lymphocytes. Since lymphocytes also enter apoptosis very readily, they are killed at a fast clip in ADA deficiency. The depletion of lymphocytes leads to severe combined immunodeficiency syndrome, that is to defects of both cellular and humoral immunity. Other tissues are not significantly affected.

Since lymphocytes originate in the bone marrow, the problem can be addressed by allogenic bone marrow transplantation, which, however, is a severe intervention and requires a compatible donor. As an alternative, gene therapy to introduce functional ADA genes into the patients' own bone marrow stem cells is currently in clinical development [1].

Enzyme replacement therapy is carried out with ADA enzyme obtained from cattle. Nucleotides can enter and leave cells by facilitated diffusion, so that they can encounter the enzyme in the extracellular space. The enzyme therefore does not have to enter the cells to be effective, which makes this form of therapy relatively straightforward.

Since the enzyme used for therapy is not of human origin, repeated application is prone to induce antibodies, which will inactivate the enzyme and may cause allergic reactions. In order to reduce the immunogenicity of the enzyme and to extend its lifetime in the circulation, it is covalently modified with polyethylene glycol (PEG). PEGylation can also be used for the protection of liposomal drug carriers (see Section 14.3.1).

The phosphorylation of deoxyadenosine is catalyzed both by deoxyadenosine kinase and, although that enzyme's name does not suggest it, also by deoxycytidine kinase. In cell culture experiments, the lymphotoxic effect of ADA deficiency can be suppressed by pharmacological inhibition of both kinases [19]. The applicability of this approach in vivo remains to be elucidated.

10.1.5 Lysosomal storage diseases

Breakdown of many macromolecules of both intra- and extracellular origins occurs in phagolysosomes, which form through the fusion of lysosomes with other vesicles that in turn arise by phagocytosis or cellular autophagy. When lysosomal enzymes from such degradative pathways are deficient, the corresponding substrates accumulate, which typically results in organ enlargement and in loss of organ function. Depending on the particular substrate and enzyme, the defect can be manifest in a specific organ, or it may affect a range of organs or tissues.

Gene therapy is still mostly experimental, and drug therapy is applicable in only a few cases. The currently most useful approach is enzyme replacement therapy. Lysosomal enzymes are active only in the acidic milieu of the phagolysosomes, which is also the place where the substrates are transported to. Therefore, therapeutically applied enzymes must enter the phagolysosomes as well in order to be useful. In several cases, this is facilitated by the glycosyl moieties of the enzymes, which are recognized by cellular carbohydrate receptors that mediate endocytosis.

Pompe disease. The defective enzyme in this disease is acid maltase, which cleaves α-1,4-glucosyl bonds and functions in the lysosomal degradation of glycogen particles. The defect affects mostly striated muscle, that is, skeletal muscle and the heart. Depending on the specific mutation present, the enzyme defect can vary in severity, and muscle weakness can accordingly become clinically manifest in early childhood or only during adolescence.

The disease is treated with enzyme replacement therapy using the recombinantly expressed human enzyme. Cellular uptake and lysosomal targeting are mediated by mannose-6-phosphate receptors on the target cells. For expression of the enzyme, cell cultures have been genetically engineered so as to maximize the number of mannose-6-phosphate residues incorporated into the enzyme molecule during posttranslational modification [23].

Gaucher disease. The enzyme defect in Gaucher disease concerns glucocerebrosidase, which cleaves the glucose headgroup from glucocerebroside, a sphingolipid that occurs in cell membranes (Figure 10.4A). The deficiency affects primarily phagocytes in the spleen, liver, and bone marrow. Liver and spleen are enlarged, often enormously so, and show impaired function. Only a small fraction of the surplus organ mass consists of accumulated lipid; the patho-

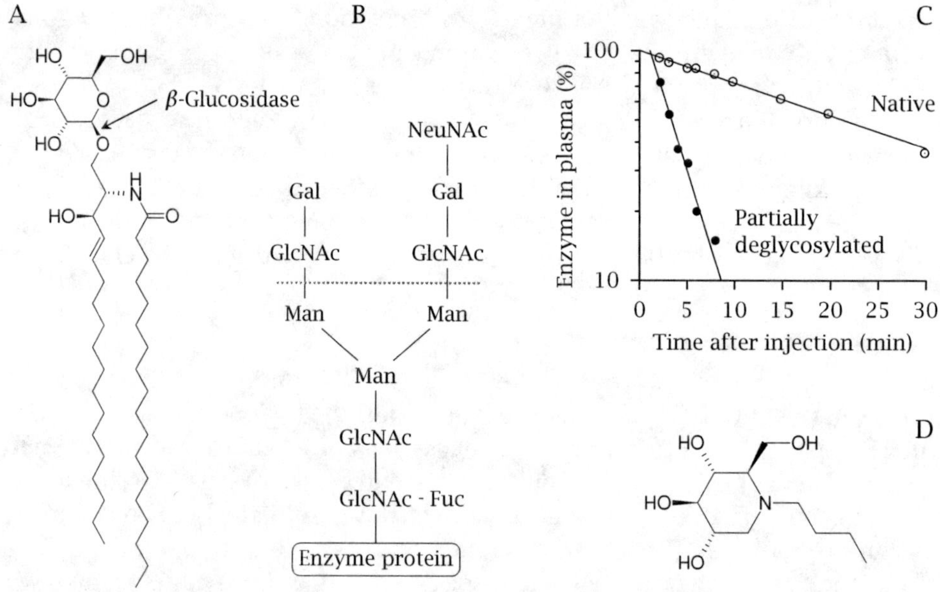

Figure 10.4 Biochemistry of Gaucher disease. **A:** Lysosomal *β*-glucosidase or glucocerebrosidase cleaves the glucose moiety from glucocerebroside (glucosylceramide), a membrane lipid. Deficiency of this enzyme causes accumulation of glucocerebroside in macrophages. **B:** Structure of the glycosyl moiety on *β*-glucosidase. Sugars: NeuNAc, *N*-acetylneuraminic acid; Gal, galactose; GlcNAc, *N*-acetylglucosamine; Man, mannose; Fuc, L-Fucose. The dotted line indicates the extent of partial deglycosylation applied in (C). **C:** Plasma clearance of native and partially deglycosylated enzyme. The rate of disappearance from the plasma reflects uptake by macrophages via mannose receptors. (Figure redrawn from original data in Ref. [3].) **D:** Structure of miglustat, an inhibitor of glucocerebroside synthesis that is used in the treatment of Gaucher disease.

genetic mechanism that leads to the additional organ enlargement is not exactly understood.

Since macrophages originate in the bone marrow, transplantation of bone marrow is an effective and established therapeutic option, and gene therapy of bone marrow stem cells an experimental one. Enzyme therapy again requires uptake of the enzyme molecules into the phagolysosomes. Macrophages have multiple classes of cell surface receptors that trigger endocytosis. One class are *lectins*, which bind to the cell surface oligosaccharides of microbes and initiate their ingestion. A mannose-specific lectin mediates endocytosis of glucocerebrosidase. To facilitate this interaction, the enzyme is partially deglycosylated so as to expose the mannose residues contained within its native glycosyl moiety (see Figure 10.4B and C).

In addition to enzyme therapy, Gaucher patients are being treated with inhibitors of glucocerebroside synthesis such as miglustat (Figure 10.4D) and eliglustat. Complete inhibition of glucocerebroside synthesis would be toxic, so this strategy does not obviate the need for enzyme therapy. Serendipitously, miglustat was also found to inhibit spermatogenesis in mice, raising hopes for a new approach for male contraception. This effect, however, could not be substantiated in men [2].

10.2 GOUT

Gout results from the accumulation of uric acid, the terminal product of purine nucleotide degradation in humans (Figure 10.5). Monosodium urate is poorly soluble, and at elevated levels it forms crystals in connective tissues, particularly within joints. Urate crystals trigger inflammation by a recently elucidated mechanism [22], which leads to the characteristic gouty arthritis. While inflammation occurs acutely and intermittently, the underlying deposition of urate crystals is a long-term process.

Uric acid is eliminated primarily via the kidneys. It is subject to glomerular filtration and to both reuptake and active secretion by tubular epithelial cells. The most important variable is the rate of reuptake, which is mediated by the URAT1 exchange transporter. Several drugs and metabolites interact with this transporter [13]. Typically, when present in the lumen of the tubule, they will compete with urate reuptake; this is the mode of action of *uricosuric* drugs such as benzbromarone or probenecid. On the other hand, when present inside the tubular epithelial cells, they may become substrates for exchange and therefore increase the rate of urate reuptake, as is the case with the metabolites of pyrazinamide (see Figure 10.6).

10.2.1 Diet and gout

Diet is the most important factor in the causation of gout. Nucleic acids contained in food are degraded in the small intestine by pancreatic DNAse, RNAse and phosphatase. The resulting ribo- and deoxyribonucleosides are taken up by active transport. The sugar moieties are utilized, while the pyrimidine and purine bases are mostly degraded and excreted. Gout is promoted by food and beverages that are rich in nucleic acids or nucleosides, such as, for example, meat and beer. Accordingly, dietary restrictions placed on such foods are a cornerstone of therapy. However, some less obvious dietary factors can also play a role.

Fructose has been linked to increased uric acid production both statistically [9] and experimentally [16,25]. The mechanism is as follows. Fructokinase, the first enzyme in fructose degradation, produces fructose-1-phosphate more rapidly than it is utilized. Accumulation of fructose-1-phosphate sequesters phosphate, which inhibits the regeneration of ATP. AMP builds up and enters degradation by AMP deaminase or 5′-nucleotidase (see Figure 10.5).

The transiently sequestered phosphate will eventually be brought back into circulation by the degradation of fructose-1-phosphate, which starts with aldolase B. A homozygous lack of aldolase B causes *hereditary fructose intolerance*. The severe liver cell damage characteristic of this disease is again due to phosphate sequestration, which, however, is much more pronounced in this case. While heterozygous carriers of the enzyme defect don't suffer acute liver

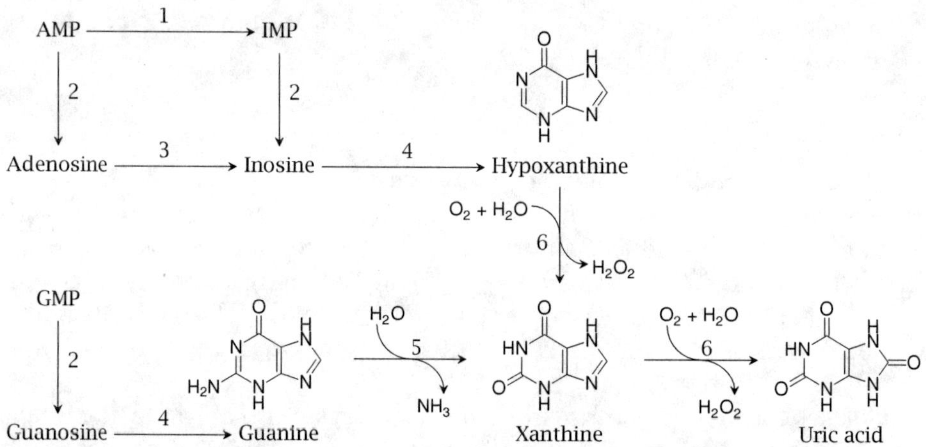

Figure 10.5 Purine nucleotide degradation. Enzymes: (1) AMP deaminase; (2) alkaline phosphatase or 5′-nucleotidase; (3) adenosine deaminase; (4) purine nucleoside phosphorylase; (5) guanase; (6) xanthine oxidase. Note that xanthine oxidase performs two successive reactions.

damage, they still have an increased likelihood to develop gout in response to a diet rich in fructose or sucrose.

Since gout is typically associated with an overly rich diet, one may be surprised to learn that it also afflicts patients with *anorexia nervosa* [20], who compulsively starve themselves to an often dangerous degree. Gout in anorectics is not due to increased synthesis of uric acid but rather to its increased retention. In starvation, triglycerides are broken down and converted to acetoacetate and β-hydroxybutyrate, the so-called *ketone bodies*, which serve as surrogate substrates for glucose. Ketone bodies function as exchange substrates for URAT1 and so increase uric acid reuptake. In addition, the acidosis (accumulation of acids) that goes with ketone body formation lowers the pH value of the urine, which promotes urate reuptake by way of nonionic diffusion.

Alcoholic beverages are often assumed to promote gout, but this is not uniformly the case; beer does it, while wine does not [10], suggesting that ingredients other than alcohol itself are important. At high dosages, alcohol promotes acidosis by stimulating both lactate and ketone body formation, which should again promote gout through increased renal retention of uric acid.

10.2.2 Pharmacotherapy of gout

Drug treatment of gout follows three principles: (1) enhancement of renal uric acid elimination, (2) inhibition of uric acid synthesis, and (3) anti-inflammatory treatment of acute episodes of gout. The molecular structures of several drugs are shown in Figure 10.6.

As stated before, uricosuric drugs inhibit the tubular reuptake of uric acid by the URAT1 transporter, which increases the efficiency of renal elimination. Synthesis of uric acid by xanthine oxidase is inhibited by allopurinol. Degrada-

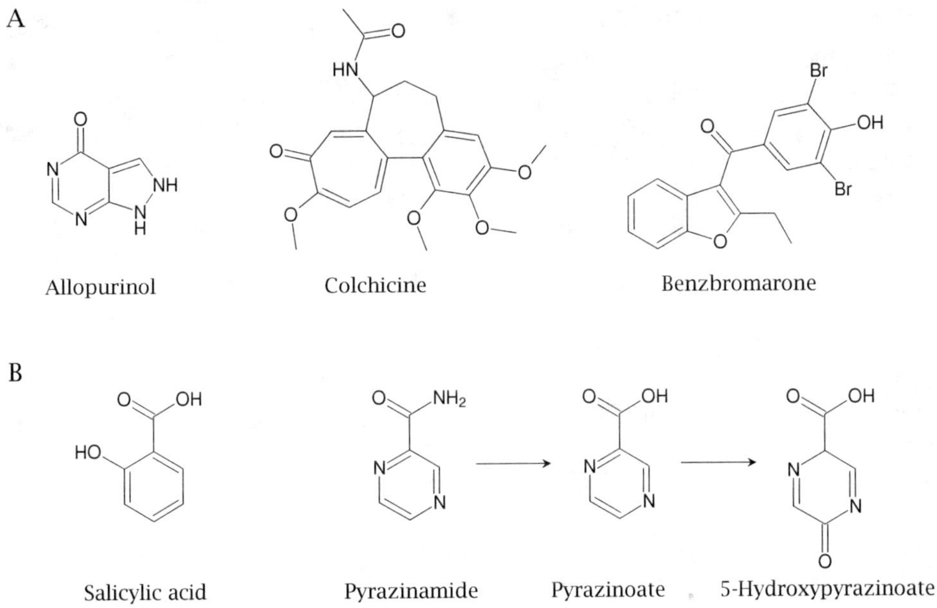

Figure 10.6 Drugs used in the treatment of gout (**A**) and drugs that promote gout (**B**). Allopurinol inhibits xanthine oxidase and thereby the formation of uric acid. Colchicine is an alkaloid that inhibits actin polymerization and exerts anti-inflammatory effects. Benzbromarone inhibits the tubular reuptake of uric acid and so increase its renal elimination. Salicylic acid as well as the metabolites of the tuberculostatic drug pyrazinamide, pyrazinoic acid and 5-hydroxypyrazinoic acid, act as exchange substrates for uric acid in tubular reuptake and therefore reduce its renal elimination.

tion then stops at the level of hypoxanthine or xanthine, which are more readily excreted, presumably because they are exempt from tubular reuptake.

Treatment of acute inflammatory episodes uses colchicine, cyclooxygenase inhibitors, and corticosteroids. Colchicine is an alkaloid from autumn crocus (*Colchicum autumnale*). It inhibits actin polymerization and with it cellular motility, including the migration and secretory activity of leukocytes. The molecular mode of action resembles the cancer drug vinblastine (see Section 12.4.4). As suggested by this similarity, colchicine treatment can cause considerable toxicity.

10.2.3 Acute urate nephropathy

A condition that is related to gout but much more acute is known as *tumor lysis syndrome.* It is particularly common in acute lymphomas and leukemias, in which no surgery is performed that would remove the bulk of the malignant cell mass before application of cytotoxic chemotherapy (see Chapter 12). The rapid destruction of a large number of tumor cells by chemotherapy, followed by the degradation of the nucleic acids contained within, produces a flood of uric acid that precipitates within the tubules of the kidney. This acute urate nephropathy is dangerous and often lethal. To prevent it, chemotherapy of leukemia is routinely combined with allopurinol. However, under this regimen, xanthine may reach excessive levels and also precipitate within the kidneys, again causing an acute and potentially lethal nephropathy [27].

A therapeutic option that avoids accumulation of xanthine is enzyme therapy with urate oxidase, which catalyzes the conversion of uric acid to allantoin, a more water-soluble substance that is also excreted with the urine. Urate oxidase occurs in many animal species other than primates.[4] A microbial urate oxidase preparation is commercially available under the name *rasburicase.*

10.3 DIABETES MELLITUS

Literally translated, the term *diabetes mellitus* means "honey-sweet passing through," and it denotes both the profuse flow of urine and its sweet taste, which was discovered by Thomas Willis in the 17th century and until not too long ago figured prominently in the diagnosis of the disease.[5] The characteristic signs of diabetes are due to elevated blood glucose levels. Since the capacity of the kidneys for tubular reuptake of glucose is limited, excess glucose in the primary filtrate passes down the nephron, taking water along with it by way of osmosis.

The two major forms of diabetes are both related to insulin, the major hormone controlling glucose transport and metabolism. In type 1 diabetes, the insulin-producing β cells of the pancreatic islets are destroyed by an immunological cross-reaction, which is triggered by infection with certain serotypes of Coxsackievirus. According to its typical age of onset, type 1 diabetes is also named *juvenile diabetes.* Type 2 diabetes is more common and usually of later onset. Here, the β cells are not destroyed, and in the early stage insulin secretion may be even higher than normal. The disease is caused by a reduced cellular sensitivity to the hormone. The underlying molecular mechanism of insulin insensitivity is not yet fully understood, but one aspect is the reduced degree of activation of the insulin receptor in response to hormone binding.

10.3.1 Pathophysiology of insulin deficiency

The insulin receptor is a receptor tyrosine kinase, which, in response to the binding of insulin, first undergoes autophosphorylation and then begins to phosphorylate a series of proteins that as a group are simply named *insulin receptor substrates.* Phosphorylated insulin receptor substrates then bind to several adapter proteins. This triggers a number of phosphorylation cascades, at the end of which the activities of glucose-utilizing enzymes and glucose transporters are upregulated, whereas pathways that provide glucose are inhibited. Regulation by insulin occurs both through the phosphorylation of existing enzyme molecules and at the transcriptional level.

[4] The human genome contains a pseudogene for urate oxidase, indicating that at some point in evolution we lost the ability for allantoin synthesis. The complete pathway comprises two enzymes in addition to urate oxidase [29], but the reactions catalyzed by these occur also without enzymatic catalysis.

[5] In the likewise tastefully named *diabetes insipidus,* the excessive urine flow is due to the lack of antidiuretic hormone, which is produced in the posterior hypophyseal gland (Section 7.2.1). Glucose is not involved in the pathogenesis of this disease.

Glucose transport. Uptake of glucose in the small intestine and the kidney tubules is mediated by sodium-coupled glucose transporters (SGLTs), whereas all other transport of glucose occurs through facilitated diffusion by simple glucose transporters (GLUTs). Insulin affects a subset of the GLUT transporters. Nerve cells, red blood cells, and some other cell types strictly require glucose and import it with or without insulin. In other tissues, and in particular in skeletal muscle and fat tissue, GLUT transporters are active only in the presence of insulin. The regulation of transporters occurs through their reversible endocytosis. If insulin activity is lacking, these major tissues cease to consume glucose, which causes it to accumulate in the blood.

Metabolism. Dysregulation of metabolic pathways further amplifies the accumulation of glucose and also causes additional metabolic turmoil. In the liver, glucose production is ramped up by the activation of glycogen breakdown and gluconeogenesis. Muscle cells break down protein and release amino acids, which are utilized for gluconeogenesis in the liver. Fat cells break down triglycerides and release fatty acids, which the liver converts to ketone bodies.

The acidosis that accompanies ketone body formation,[6] as well as the increased osmotic activity of the blood plasma due to the high glucose levels can impair the function of the brain, resulting in diabetic coma. This can occur in fresh, undiagnosed and untreated cases of diabetes, as well as in patients who are adequately treated but whose metabolism becomes derailed by infectious diseases or other causes.

Long-term complications. Long-term complications of diabetes are ascribed mostly to the effects of chronically elevated blood glucose. The lens of the eye suffers cataract; this is due to insulin-independent uptake of glucose and its subsequent conversion to sorbitol by aldose reductase. A similar pathogenic mechanism has also been assumed for other kinds of diabetic organ damage such as *polyneuropathy*, that is, degeneration of peripheral nerves. This hypothesis has motivated the development of aldose reductase inhibitors, but these have not demonstrated a major clinical benefit.

Excess blood glucose can be converted to triglycerides and cholesterol in the liver, which will give rise to elevated blood lipoproteins. Elevated LDL, possibly rendered more pathogenic through chemical modification by free glucose, accelerates the manifestation of atherosclerosis.

10.3.2 Insulin therapy

Insulin therapy is always necessary in type 1 diabetes. Type 2 diabetics frequently require it, too, although in the early stages dietary carbohydrate restriction and oral antidiabetic drugs may be sufficient. The various forms and modes of application of insulin are considered in Section 14.4.2.

[6]The two major ketone bodies are β-hydroxybutyrate and acetoacetate. The latter can decarboxylate to acetone. At plasma concentrations below 10 mM, acetone has antiepileptic activity. Such concentrations can be reached with ketogenic diet, which is useful in epilepsy that is refractory to conventional drug treatment [21]. At somewhat higher concentrations, acetone is narcotic [7]. The possibility that acetone narcosis is a causative factor in diabetic ketoacidotic coma has been considered and dismissed [31].

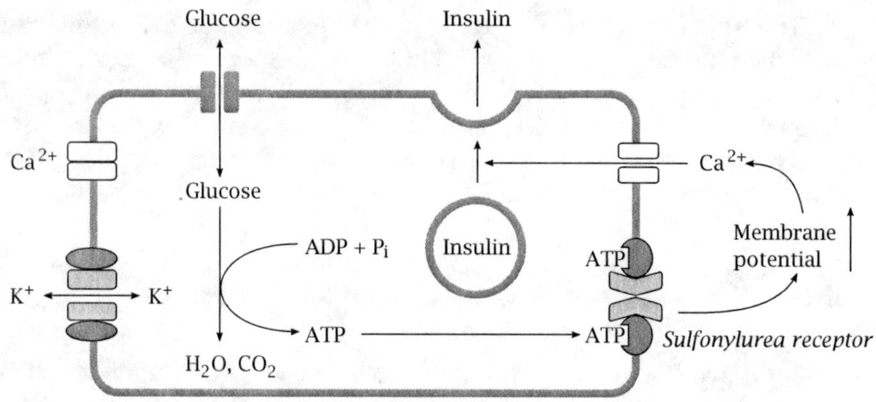

Figure 10.7 Role of the sulfonylurea receptor in the secretion of insulin from pancreatic β cells. The β cells themselves take up glucose in an insulin-independent fashion. When glucose is high, its degradation increases the level of ATP, which activates the sulfonylurea receptor. This closes the associated K_{ir} channel, raising the membrane potential. Activation of Ca_V channels then triggers insulin release.

10.3.3 Oral antidiabetic drugs

These are used exclusively in type 2 diabetes. Sulfonylurea derivatives promote insulin secretion from pancreatic β cells; they activate the sulfonylurea receptor, which links the increase of intracellular glucose turnover to insulin secretion (Figure 10.7; see also Section 6.6.3). Their mode of action is allosteric; therefore, they amplify rather than substitute the glucose signal, so that the physiological regulatory loop remains intact.

Thiazolidinediones, such as rosiglitazone (Figure 10.8), activate peroxisome proliferator-activated receptor γ (PPARγ), a nuclear hormone receptor with complex transcriptional effects that result in an improved insulin sensitivity, but also in numerous side effects that have caused the withdrawal of several drugs from the market.

The biguanide metformin causes activation AMP-activated protein kinase, a regulatory enzyme that promotes insulin-independent cellular uptake of glucose and its degradation by glycolysis. The mode of this activation is not entirely clear. Metformin partially inhibits complex I in the respiratory chain (NADH dehydrogenase [26]). This should increase the [AMP]/[ATP] ratio and so indirectly activate the enzyme (Figure 10.9). Inhibition of NADH dehydrogenase will raise the NADH level, which would promote the reduction of pyruvate to lactate. Lactic acidosis is indeed a known side effect of metformin, which suggests that inhibition of NADH dehydrogenase is significant in vivo. However, the connection between the inhibition of NADH dehydrogenase and the activation of AMP-activated protein kinase depicted in Figure 10.9 has been disputed, on the grounds that a measurable change in the [AMP]/[ATP] ratio after metformin application could not be observed [15].

Acarbose (Figure 10.8) is an inhibitor of α-glucosidase or maltase, which is located at the intestinal epithelium and cleaves maltose and other glucose

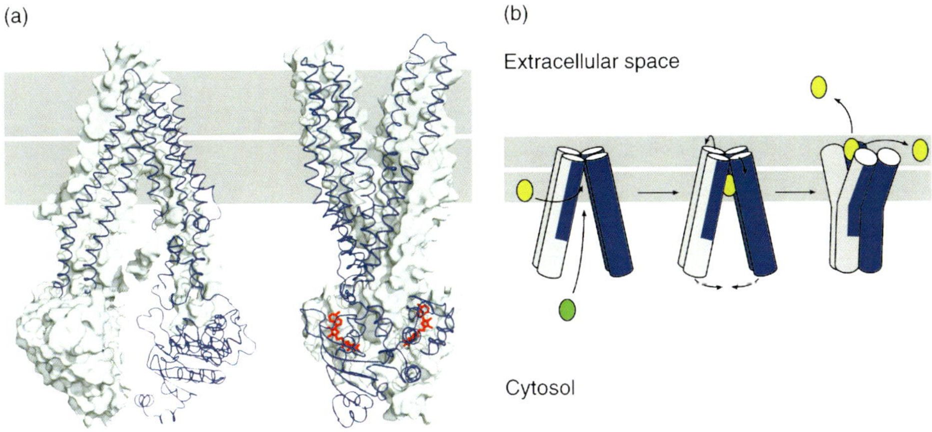

Structure and function of ABC transporters. In both **A** and **B**, one of the two subunits is rendered in blue, the other in gray. Two molecules of a non-hydrolyzable ATP analog bound to the outward-facing conformation are rendered in red. See legend to Figure 3.5.

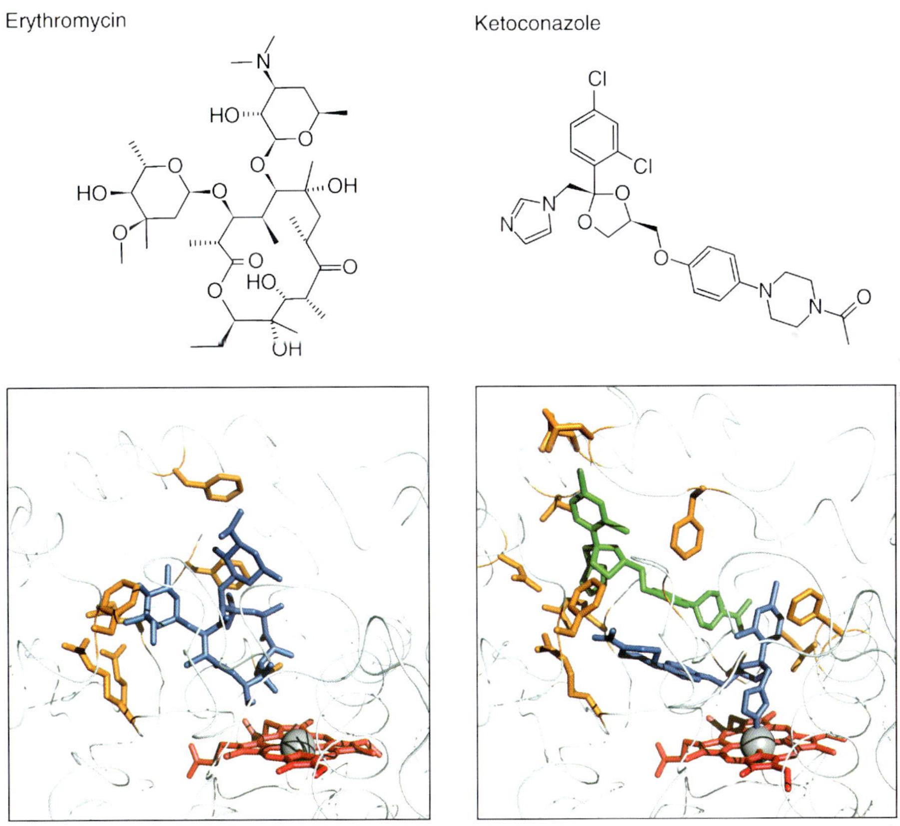

Binding of erythromycin and ketoconazole to cytochrome P450 3A4. The Heme is rendered in magenta. Erythromycin and one ketoconazole molecule are rendered in blue, and a second ketoconazole molecule is shown in green. Amino acid residues within 3.5 nm of the bound drugs are in yellow. See legend to Figure 4.5.

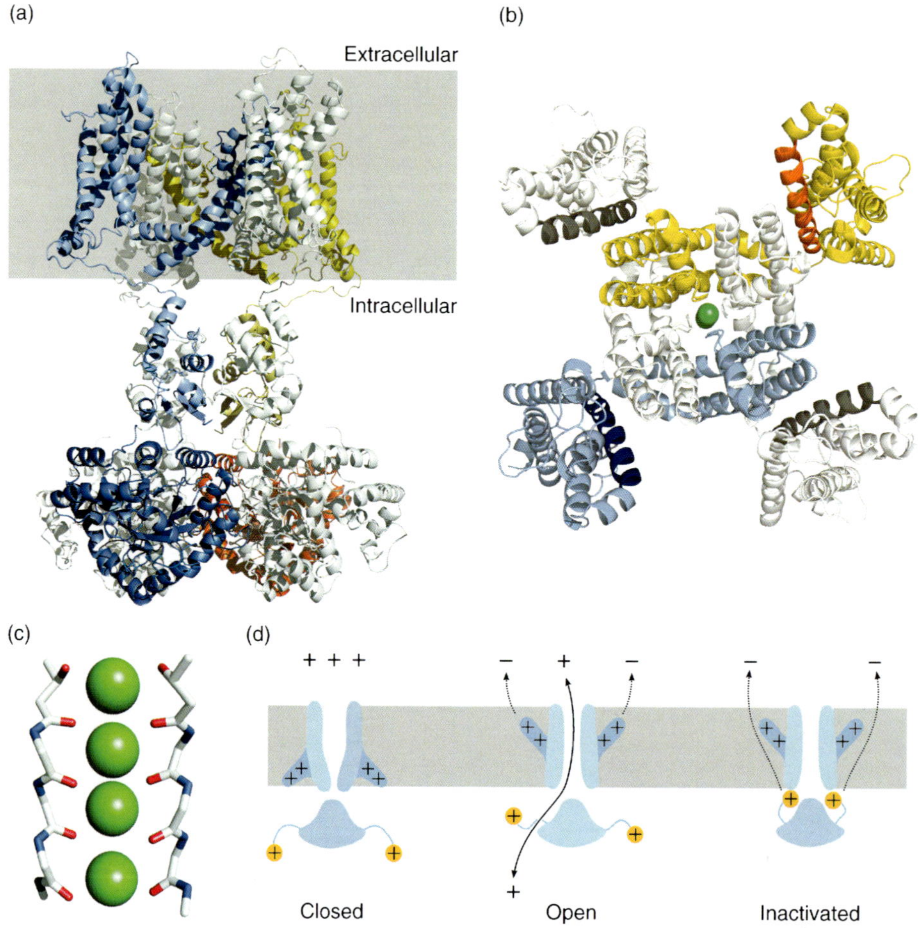

Structure and function of a voltage-gated potassium channel. In **A**, the gray rectangle delineates the membrane-embedded portion. In both **A** and **B**, one of four identical subunits is rendered in blue, and a second one in yellow. **A**: Overall structure. The gray rectangle delineates the membrane-embedded portion. The N-terminal inactivation domain that is located in the cytosolic portion (see D) is missing from this structure. **B**: View from the extracellular side. The membrane-embedded portion consists of an inner layer, which contains the selectivity filter (see C) and an outer layer that mediates gating (see D). An arginine-rich helix that occurs in each subunit and forms the voltage sensor of the opening gate is rendered in darker tones. **C**: Structure of the K^+ selectivity filter. Backbone oxygen atoms tightly enclose the dehydrated K^+ ions. Each of the four subunits contributes one backbone segment, but for clarity only two segments are shown. **D**: The voltage-dependent gating mechanism. When the membrane potential is reversed, the membrane-embedded voltage sensor moves outward, which loosens up the inner layer and opens the channel. A separate inactivation gate, which is flexibly attached to the cytosolic portion of the channel and also carries a positive net charge, then responds and moves to plug the channel. See legend to Figure 6.4.

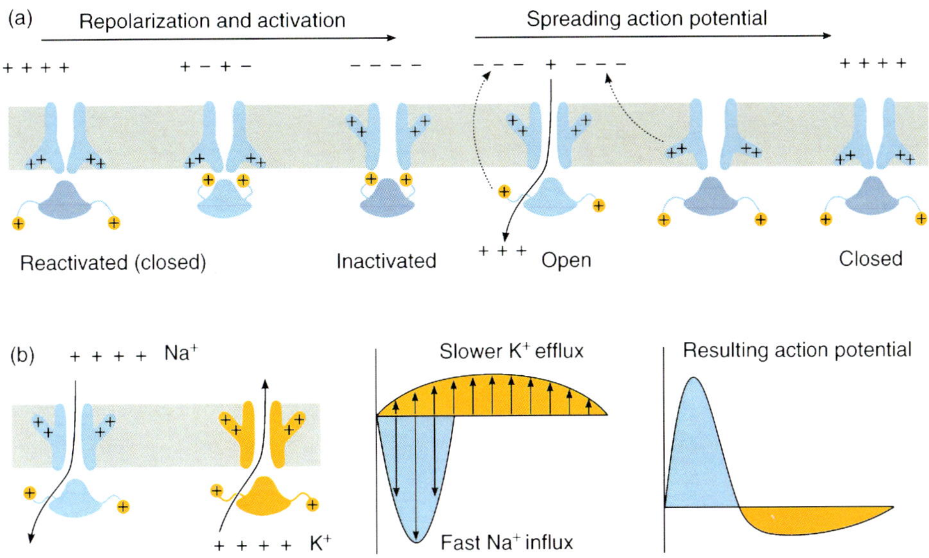

Propagation of action potentials by voltage-gated channels. See legend to Figure 6.5.

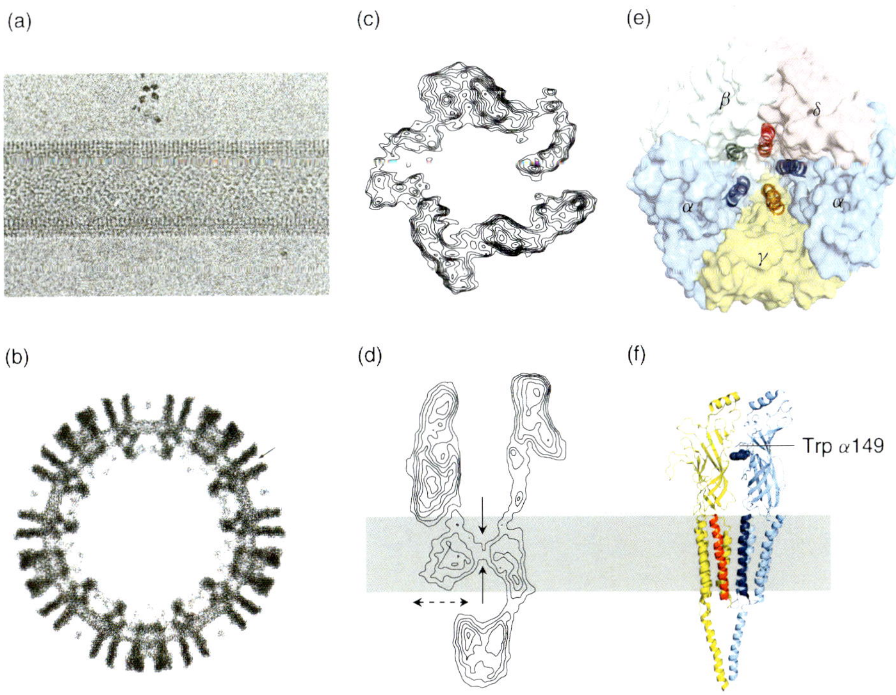

Structure of the nicotinic acetylcholine receptor, as elucidated by electron crystallography. See legend to Figure 6.13.

For the next 4 figures, please note: Where individual atoms are highlighted in color, oxygen is red, nitrogen is blue, phosphorus is yellow, hydrogen is white and halogens are green. Carbon may be white or gray.

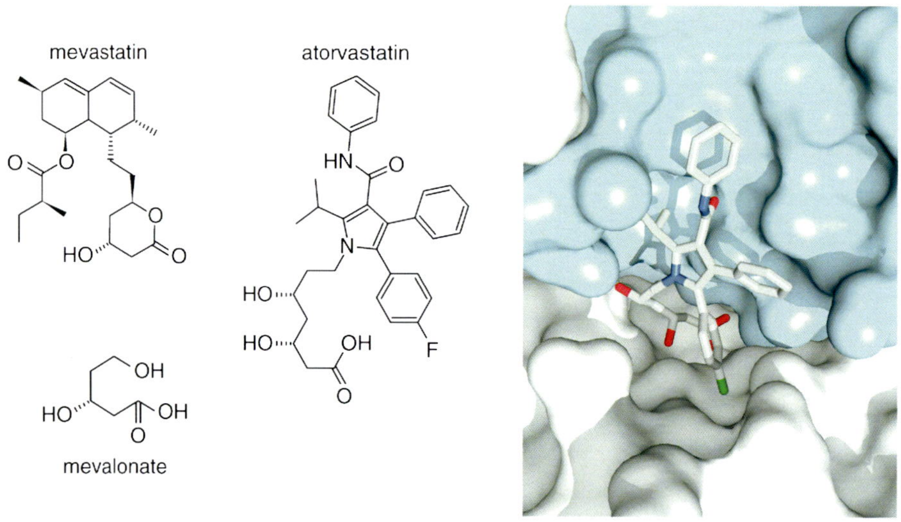

Inhibition of HMG-CoA reductase with statins. The two separate enzyme subunits that participate in forming the active site are shown in white and blue, respectively. The orientation of the bound atorvastatin molecule is similar to that of the covalent structure schematic. See legend to Figure 10.13.

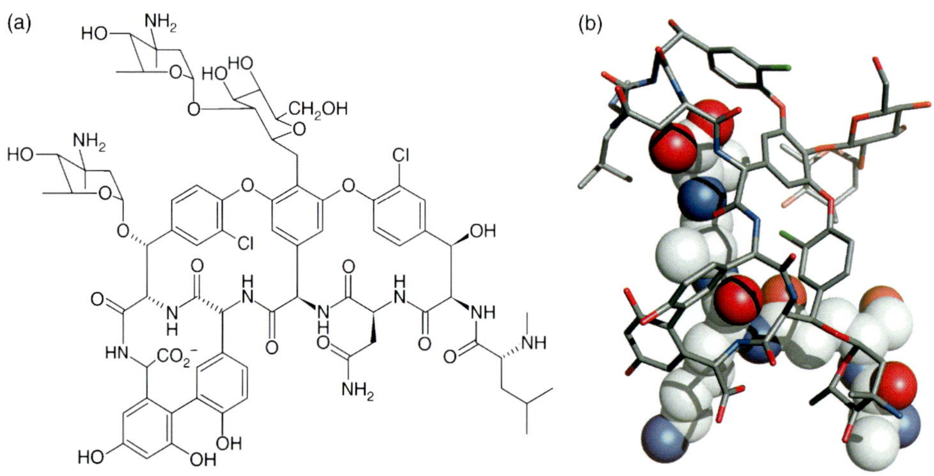

Structure and function of the glycopeptide antibiotic A82846B. **A**: Structural formula. **B**: A82846B (sticks) bound to the peptide moiety of lipid II (spheres). The two linked D-alanine residues of lipid II are pointing upwards. See legend to Figure 11.6.

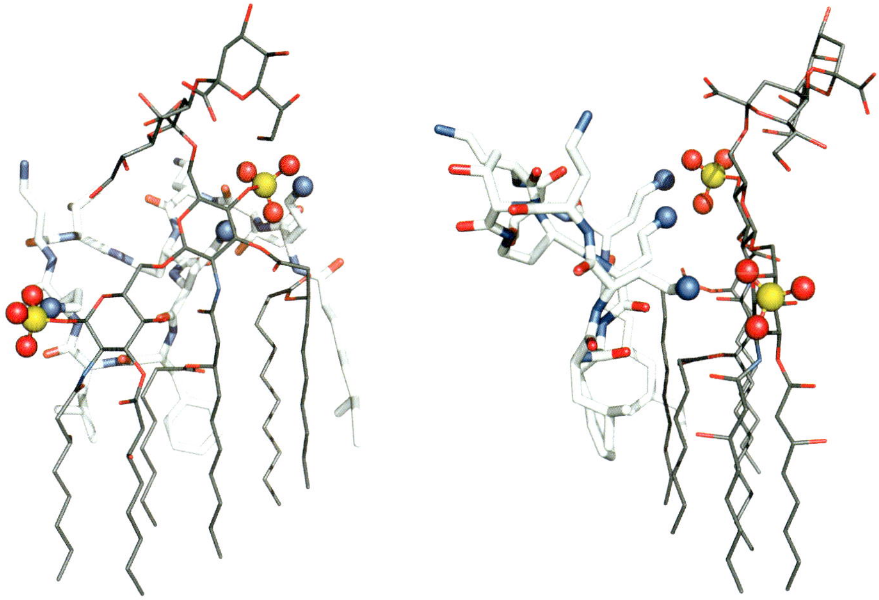

Structure of polymyxin B (thick lines), bound to its target lipid A (thinner lines). The same structure is shown from two different angles. Three amino groups of the drug that pair with the two phosphates of lipid A are highlighted, as are the phosphates themselves. See legend to Figure 11.13.

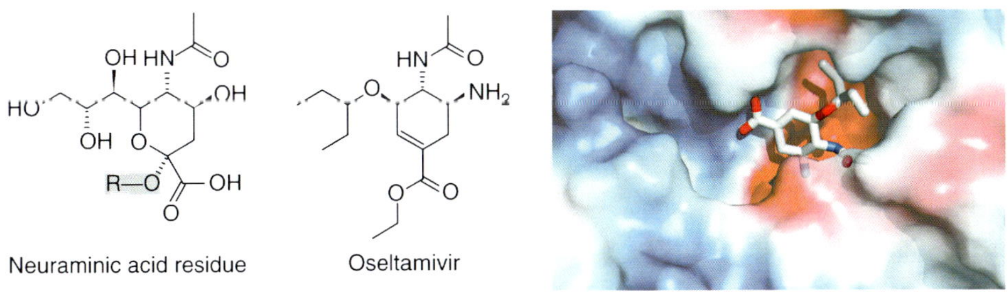

Binding of the inhibitor oseltamivir to influenzavirus neuraminidase. The coloring of the protein surface reflects positive (blue) or negative (red) electrostatic potential. See legend to Figure 11.23.

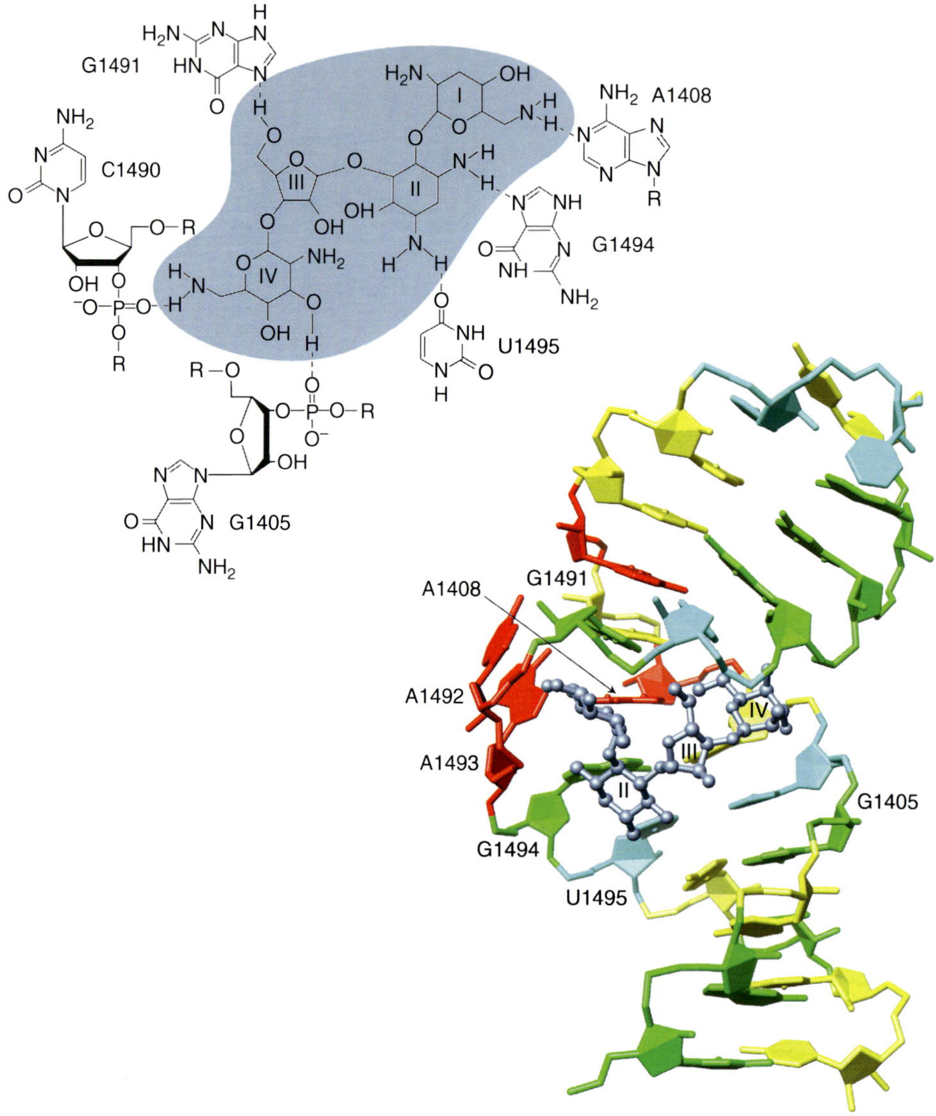

Paromomycin and its interactions with the ribosomal A-site RNA. Paromomycin is highlighted in blue. The bases in the RNA are color-coded: A, red; C, yellow; G, green; U, cyan. See legend to Figure 13.3.

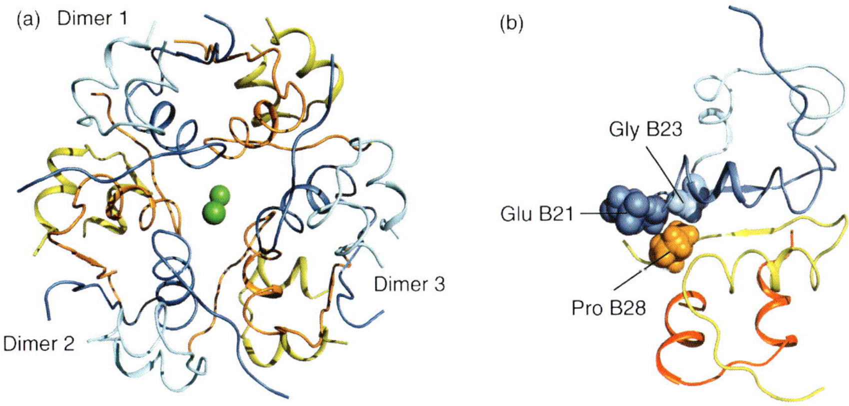

Structure of the insulin hexamer (**A**) and of one of three dimers in the hexamer (**B**). In **B**, several residues that stabilize the dimer interaction are highlighted. See legend to Figure 14.14.

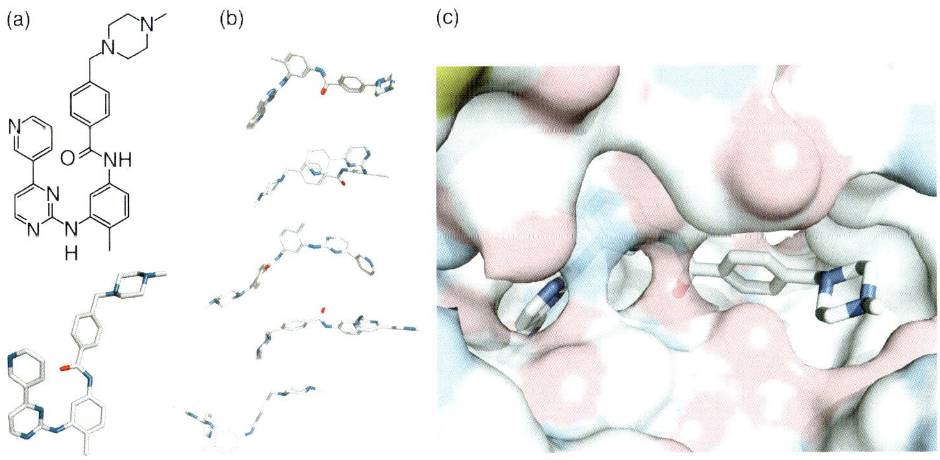

An example in silico docking experiment with the tyrosine kinase inhibitor imatinib. Red and blue hues on the protein surface indicate locations of oxygen and nitrogen atoms. See legend to Figure 15.7.

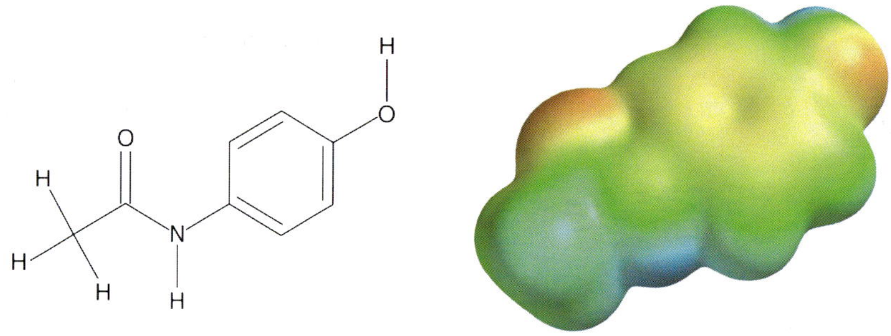

Electrostatic potential mapped onto the electronic density for acetaminophen. Blue color indicates a δ^{+}, red a δ^{-} partial charge. Other color values are intermediate. See legend to Figure 15.8.

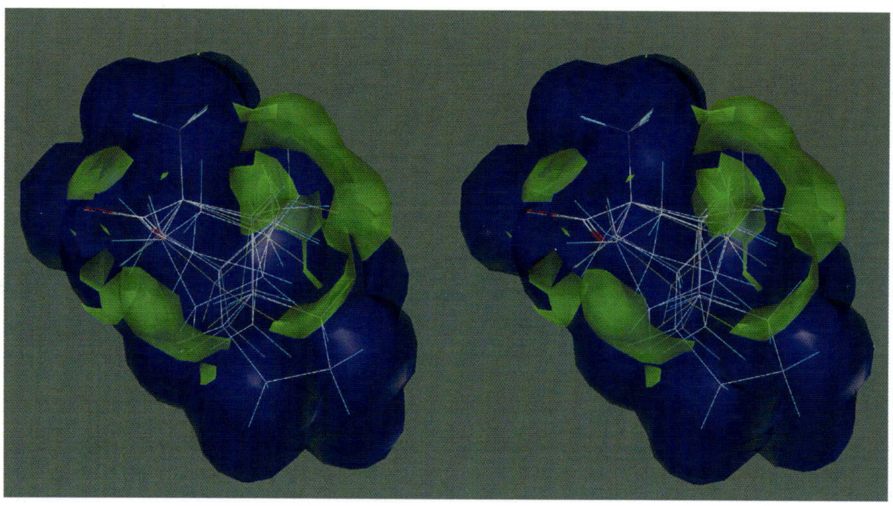

Hypothetical pharmacophore for inhibitors of the target enzyme ATP:L-Methionine S-Adenosyltransferase. The volume that envelopes all active compounds is outlined in blue; the volumes of substituents that prevent binding extend out and prevent binding are indicated in green. See legend to Figure 15.9.

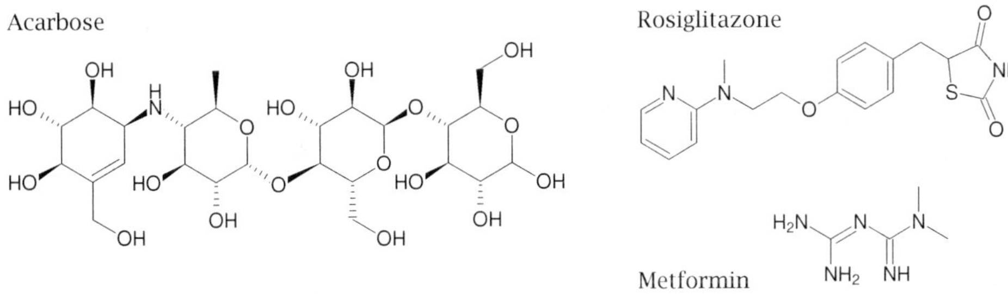

Figure 10.8 Oral antidiabetic drugs. Acarbose is an inhibitor of intestinal maltase. Rosiglitazone is a peroxisome proliferator-activated receptor γ agonist. Metformin activates AMP-activated protein kinase.

oligosaccharides to glucose. The undigested oligosaccharides cannot be taken up and thus become fodder for colon bacteria; this can lead to symptoms similar to those of lactose intolerance. Analogous inhibitors have also been developed for sucrase. Patients who manage to adhere to starch- and sugar-restricted diets can do without drugs of this type.

10.3.4 Role of C-peptide for long-term diabetic complications

Insulin contains consists of two peptide chains, the A and the B chains, which are held together by disulfide bonds. These two peptides are formed by proteolytic cleavage of proinsulin inside the secretory vesicles within the β cells. This cleavage also produces C-peptide, which is secreted along with insulin. Traditionally, this peptide has been considered dead freight. Since C-peptide is absent from therapeutic insulin preparations, its plasma level can be used as a diagnostic parameter to gauge the rate of remaining endogenous insulin secretion in type 2 diabetics who also inject insulin. However, some experimental data suggest that C-peptide may have a signaling function of its own, and small-scale clinical studies suggest that injection of C-peptide helps to avoid diabetic long-term complications that affect the peripheral nerves and the kidneys [24]. The receptor and exact mode of signaling remain to be established.

10.3.5 Symptomatic diabetes

Diabetes can also occur when hormones that are antagonists of insulin are present in excess. The most common case is the application of high dosages of glucocorticoids, which are widely used for immunosuppression. Other important cases are benign or malignant tumor that produce epinephrine, growth hormone, thyroid hormones, glucagon, or again glucocorticoids. Treatment will address the underlying cause where possible.

10.4 ATHEROSCLEROSIS

Atherosclerosis is a degenerative disease of the arterial blood vessels. In its pathogenesis, two causative factors converge: (1) hypercholesterolemia, that is, elevated levels of blood plasma cholesterol, and (2) elevated blood pressure. Both of these factors are quite common in Western societies, and so

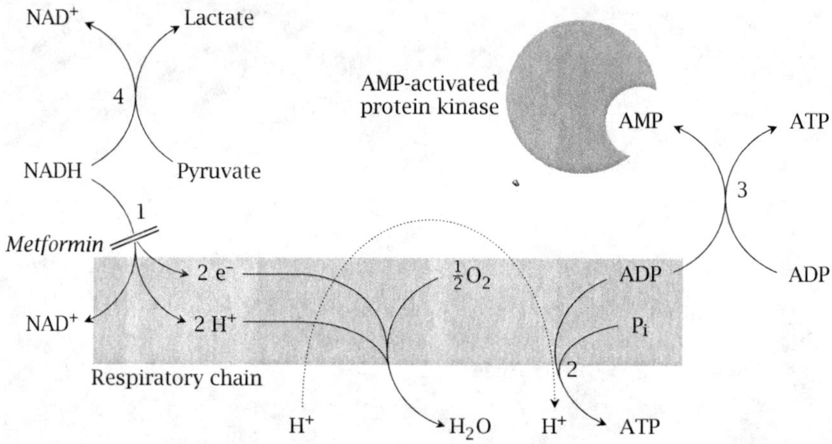

Figure 10.9 Hypothetical mode of action of metformin. Partial inhibition of NADH dehydrogenase (1, [26]) slows the respiratory chain and thereby the regeneration of ADP to ATP by ATP synthase (2). Adenylate kinase (3) converts ADP to AMP, which binds to AMP-activated protein kinase. The activated kinase stimulates glucose transport and catabolic pathways. Accumulated NADH promotes reduction of pyruvate by lactate dehydrogenase (4). Depletion of pyruvate inhibits gluconeogenesis, and accumulation of lactate promotes lactate acidosis, a known side effect of metformin treatment.

is atherosclerosis, which in its various manifestations has become the most common cause of death, ahead of all cancers and leukemias combined. In virtually every autopsy of an elderly person, some degree of atherosclerosis will be evident, even if it did not cause any clinical symptoms.

10.4.1 Pathogenesis

Atherosclerotic lesions are understood to develop in several stages, which are outlined in Figure 10.10:

1. Small lesions in the endothelium of the arteries allow seepage of blood plasma. This causes the deposition of lipoproteins in the tissue layer underneath. High blood pressure promotes the formation of these initial endothelial lesions.
2. Macrophages invade the lesion. Reactive oxygen species and enzymes released by macrophages modify low-density lipoprotein (LDL) particles that were carried with the blood plasma.
3. Chemical and enzymatic modifications of LDL promote its uptake by the macrophages via *scavenger receptors.* The macrophages become overloaded with lipids and transform into foam cells.
4. Foam cells disintegrate and release the cholesterol accumulated within. Cholesterol precipitates and crystallizes; the crystals promote the release of proinflammatory cytokines [28] and so perpetuate the degenerative process. Other inflammatory cells and thrombocytes participate in progression of the lesion.

Modifications of LDL that cause it to transform macrophages into foam cells include oxidation [18], carbamylation [33], and enzymatic modification

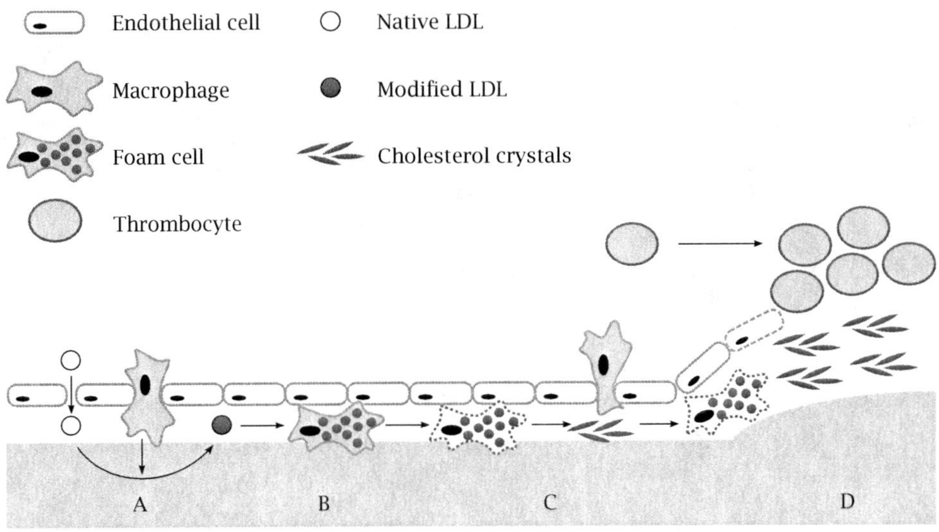

Figure 10.10 Development of an atherosclerotic lesion. **A:** Small defects in the endothelium cause seepage of LDL into the subendothelial tissue and transmigration of macrophages. Reactive oxygen species and enzymes released by macrophages modify the LDL. **B:** Modified LDL is taken up by macrophages via scavenger receptors. Lipid overload causes degeneration of macrophages to foam cells. **C:** Foam cells die and release accumulated cholesterol, which forms crystalline deposits. These stimulate macrophages to release inflammatory cytokines and intensify inflammation. **D:** In an advanced lesion, cells in the muscular layer proliferate, progressively constricting the artery. When the endothelium covering the lesion erodes, thrombocytes and plasmatic factors initiate blood clotting, causing acute obstruction.

[5]. All of these modifications have also been shown to occur in vivo and thus may contribute to the development of atherosclerosis. Their prevention would appear to be a promising strategy for pharmacotherapy; for example, LDL oxidation could be targeted with the lipophilic antioxidant vitamin E. However, no such strategies have demonstrated clinical benefits in humans so far.

Advanced atherosclerotic lesions become clinically manifest through occlusion of the affected arteries. Occlusion can develop gradually, through the thickening of the arterial wall as extracellular detritus and proliferating cells build up within, or acutely, through the formation of a blood clot atop a degenerative lesion that has destroyed its covering endothelial layer. Acute thrombotic occlusion becomes manifest as myocardial infarction, as a stroke, or as an analogous event in another organ. Gradual occlusion causes symptoms such as intermittent claudication or angina pectoris. In both mechanisms of occlusion, the symptoms are due to the disruption of oxygen supply to the tissues downstream. If occlusion is complete, as it is in an infarction, there will typically be irreversible tissue damage.

Degenerated arteries may also rupture. Complete rupture, with hemorrhage and compression of the surrounding brain tissue, is responsible for a significant proportion of all cases of stroke. In *aortic dissection*, a partial rupture occurs within the wall of the aorta. Blood then forces its way into the breach and tears off the inner layer, pushing it against the opposite wall and causing the main lumen to collapse. The consequence again is obstruction of the arterial blood flow.

Pharmacological prevention and therapy of atherosclerosis focuses mostly on three aspects of its pathogenesis: (1) controlling LDL-cholesterol, (2) controlling blood pressure, and (3) controlling blood coagulation and thrombocyte aggregation. The inflammatory and proliferative aspects of pathogenesis offer potential drug targets, too, but these are not yet part of mainstream pharmacotherapy.

10.4.2 Transport and metabolism of cholesterol

Cholesterol is an essential component of animal cell membranes, and it is also the biosynthetic precursor of steroid hormones and of bile acids. It is both acquired from the diet and synthesized endogenously in the liver (Figure 10.11). Cholesterol is transported via various types of lipoproteins, which are essentially droplets of lipids stabilized in aqueous suspension by a coat of phospholipids and proteins. Chylomicrons carry cholesterol from the intestine to the liver, while LDL carries it from there to the cells in the periphery. The endocytosis of LDL by peripheral cells is mediated by the cognate LDL receptor and is subject to feedback inhibition by intracellular cholesterol. Cells can offload excess cholesterol to high-density lipoprotein (HDL), which carries it back to the liver.

Receptor-mediated LDL endocytosis and export of cholesterol to HDL also occur in macrophages. What sets these cells apart is the possession of scavenger receptors, which mediate the uptake of chemically or enzymatically modified LDL. This scavenging function is part of the physiological role of macrophages; however, since it is not subject to feedback inhibition by intracellular cholesterol, it can result in cholesterol overload. Macrophages stuffed with lipids acquire the aspect of foam cells.

The transformation of macrophages into foam cells by modified LDL and their subsequent degeneration is a key aspect of development of the atherosclerotic lesion. Reduction of plasma LDL levels therefore is an important goal in the treatment of atherosclerosis. On the other hand, a high level of HDL reflects on the capacity for removal of excess cholesterol and is considered beneficial. Several types of peroxisome proliferator-activated receptor agonists influence the balance between LDL and HDL synthesis by transcriptional regulation [32]. However, most drugs in clinical use don't directly affect this balance and instead address the uptake or synthesis of cholesterol.

Intestinal uptake of cholesterol. The uptake of cholesterol by the epithelial cells in the small intestine is illustrated in Figure 10.12A. Uptake at the luminal side occurs through endocytosis, which is controlled by a cholesterol-sensing membrane protein, NPC1L1. This process is not very specific for cholesterol and will also cause uptake of other sterols, such as those contained in plants or mushrooms. Discrimination between cholesterol and plant sterols occurs at the

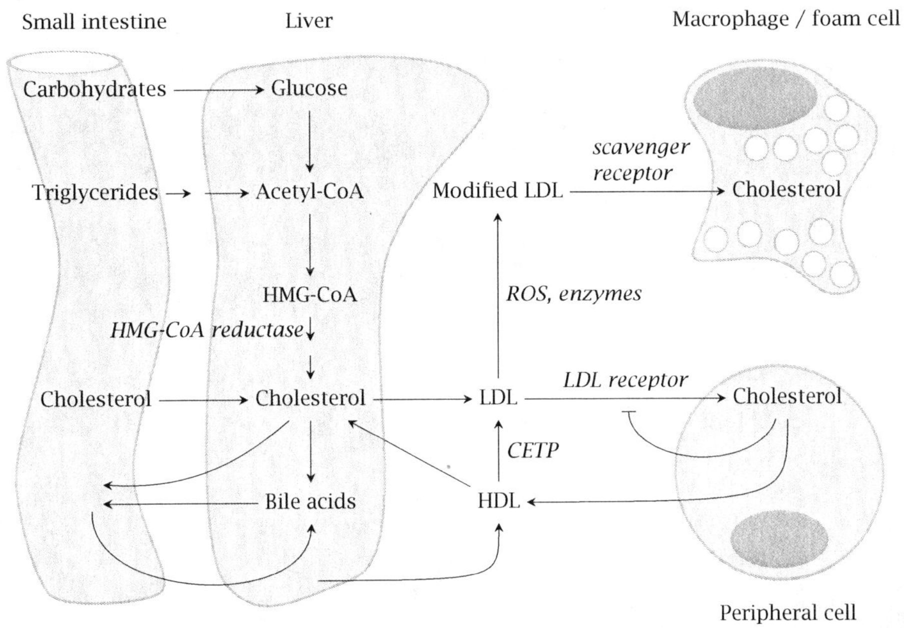

Figure 10.11 Transport and metabolism of cholesterol. The pool of cholesterol in the liver is filled by uptake of dietary cholesterol and by synthesis from other foodstuffs via HMG-CoA. Cholesterol is supplied via LDL to peripheral tissues, which take it up by endocytosis via LDL receptors (LDL-R). Uptake is subject to feedback inhibition by cholesterol. Excess cholesterol is returned to the liver via HDL, but en route some of it is transferred back from HDL to LDL by cholesterol-ester transfer protein (CETP). The liver also uses cholesterol to synthesize bile acids, which are secreted into the bile along with some excess cholesterol. Bile acids undergo reuptake at the end of the small intestine and return to the liver. Within atherosclerotic lesions, LDL undergoes modification by reactive oxygen species (ROS) or enzymes. Modified LDL enters macrophages via scavenger receptors, overloads them with cholesterol, and converts them into foam cells.

next stage. Cholesterol is transferred to the endoplasmic reticulum, where it is esterified by acyl-CoA:cholesterol acyltransferase (ACAT) and then incorporated into nascent chylomicrons. The transfer to the ER is mediated by the somewhat misleadingly named microsomal triglyceride transport protein (MTTP). The same protein also transfers plant sterols back to the apical membrane, where they are extruded by the ABC transporter ABC5/8.

Some drugs that inhibit this uptake mechanism are shown in Figure 10.12B. Since endocytosis is not specific for cholesterol, plant sterols such as sitosterol can be used to competitively inhibit cholesterol uptake.[7] The NPC1L1 protein is directly inhibited by ezetimibe, which is in clinical use but has yet to demonstrate a major therapeutic benefit. Inhibitors of MTTP and ACAT are at the experimental stage. Considering that cholesterol occurs only in meat and other

[7]Genetic defects of ABC5/8 cause *sitosterolemia*. Sitosterol and other plant sterols are absorbed and deposited in connective tissue, and atherosclerosis is promoted. Quite obviously, patients with this condition should *not* be treated with sitosterol.

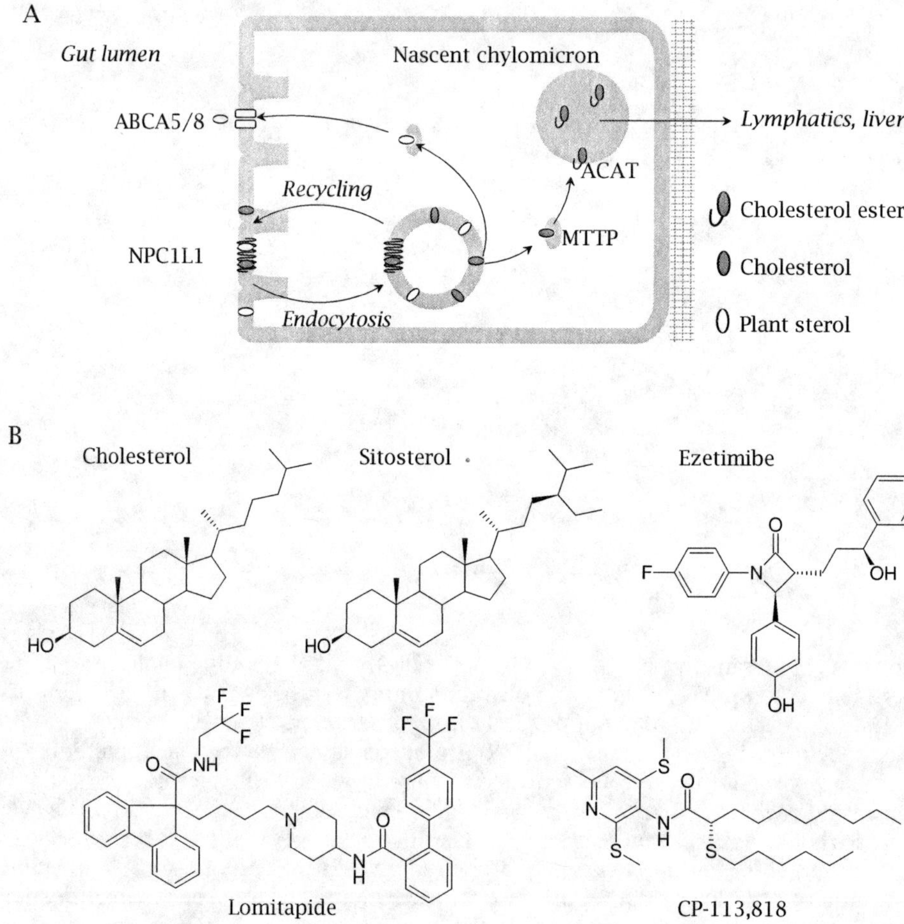

Figure 10.12 Intestinal cholesterol uptake. **A**: Mechanism of cholesterol uptake. The membrane protein NPC1L1 controls uptake of cholesterol and of plant sterols by endocytosis. The microsomal triglyceride transport protein (MTTP) shuttles cholesterol to the ER, where it is esterified by acyl-CoA:cholesterol acyltransferase (ACAT) and incorporated into nascent chylomicrons. Plant sterols such as sitosterol are returned to the intestine through an ABC transporter. Chylomicrons are released on the basolateral side and make their way to the liver via the lymphatics and the blood circulation. **B**: Structure of cholesterol, and of drugs that inhibit its intestinal uptake. Sitosterol is a natural plant sterol that competes with cholesterol for uptake by endocytosis. Ezetimibe is an inhibitor of NPC1L1. Lomitapide and CP-113,818 are experimental inhibitors of MTTP and of ACAT, respectively.

animal products, it is clear that intestinal uptake can readily be limited by a sensible choice of diet, and the effort expended on drugs of this type reflects more on the widespread consumption of overly rich food than on medical cogency.

Synthesis and turnover of cholesterol. While dietary restriction of cholesterol is feasible and advisable, it is usually not sufficient to control LDL-cholesterol, since cholesterol is also synthesized endogenously from acetyl-CoA. The rate-limiting step is catalyzed by hydroxymethylglutaryl-CoA (HMG-CoA)

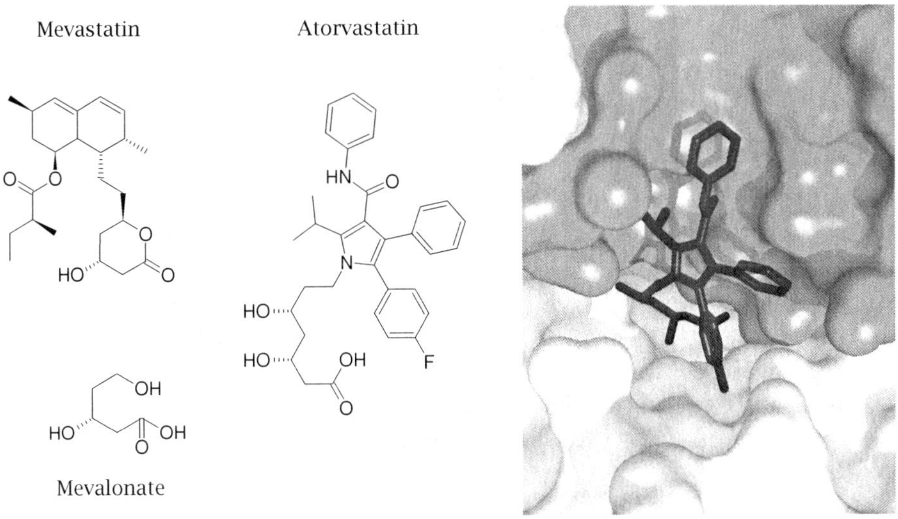

Figure 10.13 Inhibition of HMG-CoA reductase with statins. Mevastatin, a natural compound, was the first statin to be isolated. Atorvastatin is a synthetic inhibitor; it is shown on the right within the active site of HMG-CoA reductase (rendered from 1hwk.pdb [17]). Both inhibitors contain a moiety that resembles mevalonate, the product of HMG-CoA reductase. A colored version of this figure is shown in the insert in the middle of this book.

reductase, which is also the major drug target in the biosynthetic pathway. The first HMG-CoA reductase inhibitor or *statin* was the antibiotic mevastatin [12], which structurally resembles mevalonate, the product of the enzyme reaction. Several synthetic drugs were derived from this template. The availability of the crystal structure of HMG-CoA reductase has allowed the development of novel inhibitors such as atorvastatin (Figure 10.13).

Cholesterol is the precursor of bile acids, which are synthesized in the liver and secreted into the bile; they are necessary for the solubilization of fat during digestion. Bile acids are almost quantitatively recovered in the lowermost section of the small intestine. Capture of bile acids with nonabsorbable cationic polymeric particles prevents their reuptake and carries them toward the natural exit; the deficit will have to be replaced by synthesis from cholesterol. This is an indirect strategy for cholesterol depletion that is often used in combination with statins.

10.4.3 Control of blood pressure

Blood pressure is subject to multiple regulatory mechanisms, and pharmacological intervention addresses several of these. Several types of antihypertensive drugs have been discussed in other chapters, in the context of the underlying biochemistry and physiology. NO-releasing drugs are covered in Section 8.5. Angiotensin peptidase and receptor antagonists are discussed in Section 1.3.2. Several types of drugs that modulate the autonomic nervous system are covered in Chapter 6.

A class of drugs that are not covered in this book but are nevertheless important in the treatment of blood pressure are diuretics, which inhibit various renal ion transporters. Their effects are mediated by a decreased volume and the sodium concentration in the extracellular fluid.

10.4.4 Control of blood coagulation

The physiological role of blood coagulation is the emergency repair of wounds. When blood vessels are cut open, blood comes into contact with tissues other than the endothelium of the blood vessels. Such contact activates both the thrombocytes (platelets) and a cascade of blood plasma proteases and related proteins, termed *coagulation factors.* Blood coagulation can also be triggered when an atherosclerotic lesion has eroded the endothelial layer that used to cover it, and so exposes the tissue underneath. Accordingly, in patients with advanced atherosclerosis, inhibition of blood coagulation is an important strategy to lower the risk of stroke and myocardial infarction.

Blood coagulation and fibrinolysis. Figure 10.14 summarizes the process of blood coagulation and identifies some of the drug targets that are used to inhibit it. The activation of thrombocytes starts with binding to exposed collagen fibers in the lesion. Activated thrombocytes release ADP, which activates more thrombocytes through cognate P2Y receptors. They also synthesize and release thromboxane A_2, which propagates thrombocyte activation through thromboxane (TP) receptors. Both P2Y and TP receptors are GPCRs. Thrombocyte activation is also promoted through the binding to fibrinogen and several other plasma proteins. The interaction with fibrin is mediated by GPIIA/IIIB receptors.

Activated thrombocytes "scramble" the phospholipids of their cytoplasmic membranes; that is, lipids that normally reside in the inner leaflet are translocated to the outer leaflet. Phosphatidylserine molecules that become exposed in this way create binding sites for coagulation factors on the cell surface. This binding interaction is mediated by calcium, and to engage in it, the coagulation factors require calcium-chelating groups in the form of γ-glutamyl residues. The posttranslational modification that forms these γ-glutamyl residues is the target of warfarin therapy.

The activation of the plasma coagulation cascade also starts at the site of the lesion. Tissue factor, a cell surface protein, binds and activates factor VII. Subsequent activation of further coagulation factors takes place on the surface of activated thrombocytes. The activated factors form the catalytic *prothrombinase* complex, which proteolytically activates prothrombin to thrombin. Thrombin activates fibrinogen to fibrin, which then aggregates into a cobweb structure that together with the enmeshed blood cells constitutes the blood clot. Thrombin also reinforces the coagulation cascade through the cleavage of several upstream factors and of protease-activated receptors on thrombocytes.

Fibrinolysis is the reversal of blood clot formation. Tissue plasminogen activator (tPA) cleaves plasminogen, which releases the protease plasmin. Cleavage of fibrin by plasmin dissolves the blood clot. Physiologically, fibrinolysis occurs

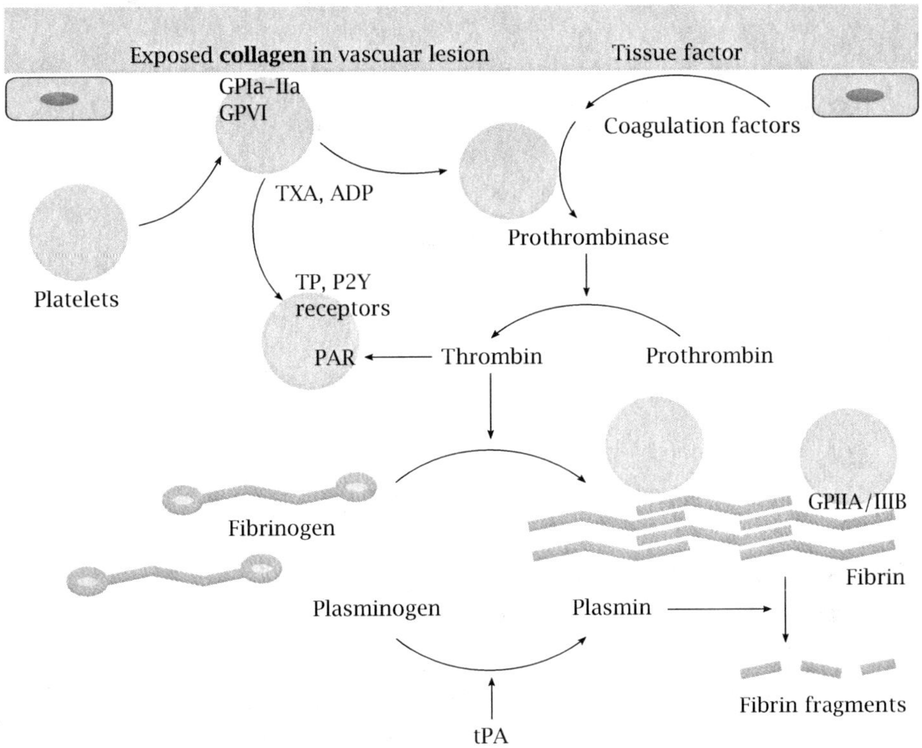

Figure 10.14 Overview of blood coagulation. Lesions of the vascular endothelium expose collagen fibers. Platelets (thrombocytes) bind to collagen via their cognate receptors (GP1a-IIa, GPVI) and become activated, resulting in the secretion of ADP and thromboxane A (TXA). These mediators activate further platelets through P2Y and TP receptors, respectively. Plasmatic coagulation factors activate one another by proteolysis at the site of the lesion and on activated platelets, culminating in the activation of thrombin (factor II). Thrombin cleaves fibrinogen to fibrin, which forms the clot. Cleavage of protease-activated receptors (PARs) on platelets further amplifies platelet activation. Platelets adhere to fibrinogen or fibrin via GPIIA/IIIA receptors. The fibrin clot is ultimately dissolved by plasmin, which is proteolytically released from plasminogen by tissue plasminogen activator (tPA) and similar enzymes.

quite slowly, but therapeutic application of recombinant tPA can be used to activate fibrinolysis and dissolve blood clots rapidly.

Drug targets in blood coagulation. Both the platelets and the plasma cascade offer drug targets that can be used for the partial inhibition of blood coagulation.

Thrombocyte activation. Inhibitors of thromboxane synthesis and of thromboxane receptors are considered in Section 9.3. P2Y receptors, which are activated by ADP, can be inhibited by thienopyridines such as ticlopidine and clopidogrel. Several thromboxane and ADP antagonists are in common clinical use. Despite their unusual mode of activation, protease-activated receptors

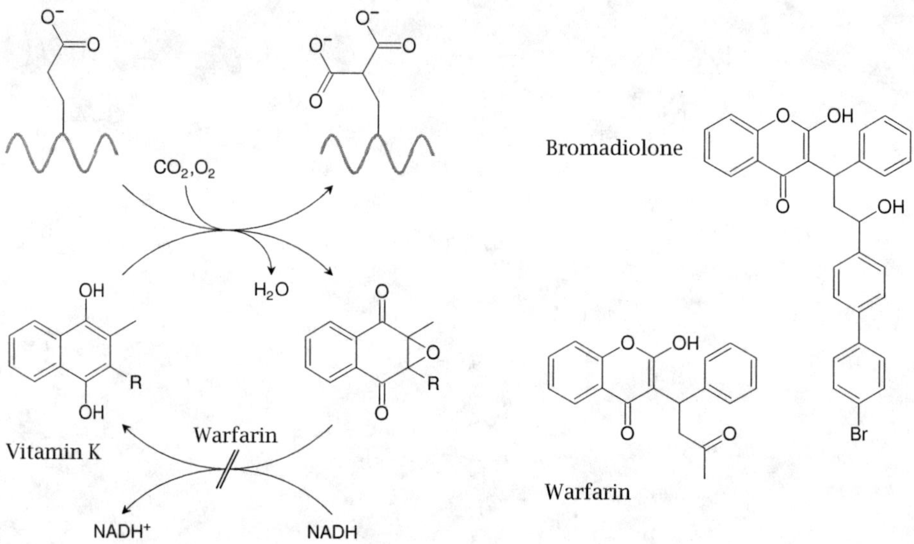

Figure 10.15 Mode of action of warfarin. Several coagulation factors (II, VII, IX, and X) require posttranslational modification of glutamate residues to γ-glutamate. The enzyme, γ-glutamylcarboxylase, requires reduced vitamin K and converts it to an epoxide. The latter is reduced by vitamin K epoxide reductase, which is inhibited by the drug warfarin, and also by rodenticides such as bromadiolone. (In the structures of vitamin K and its epoxide, R represents a side chain containing several isoprenoid residues.)

are in principle amenable to inhibition with small molecules, as discussed in Section 5.4.3.

The currently available GPIIA/IIIB receptor blockers are either peptides or monoclonal antibodies. While not suitable for long-term therapy, they are used intravenously during invasive procedures to reduce the risk of thromboembolic complications. The GPIIA/IIIB receptors, as well as the GPVI and the GPIa-IIa receptors, are structurally similar to immunoglobulin receptors. Like the latter, they are activated by crosslinkage to one another as a consequence of binding to the same macromolecular target, and they signal through downstream protein kinases. The Syk kinase (*SyK* for spleen tyrosine kinase), which is involved in the signaling pathways of immune cells, is also activated by GPVI on thrombocytes [30]. Small molecule inhibitors of Syk are available [6] and may be a therapeutic option for short-term treatment, although long-term usage would likely be hampered by undesirable interference with immune cell function.

Activation of the coagulation cascade. Several coagulation factors contain γ-glutamyl residues that enable them to chelate calcium and bind to one another and to the surfaces of activated thrombocytes. These γ-glutamyl residues are formed through posttranslational modification, in a reaction that requires vitamin K. In this reaction, vitamin K is turned over to an epoxide, which is subsequently reduced back to the native form. Warfarin inhibits this reduction, and therefore prevents the reuse of vitamin K (Figure 10.15). It does not affect the function of existing γ-glutamyl-containing protein molecules, and therefore its effect sets in only as those protein molecules are turned over. Complete inhibition of glutamyl-γ-carboxylation would be deleterious, and accordingly warfarin has a low therapeutic index and requires frequent drug monitoring. Related compounds such as bromadiolone are used as rodenticides.

Another important drug that interferes with the activation of the coagulation cascade is heparin, a sulfated, acidic glycosaminoglycan. It occurs naturally in human tissues, but free levels in the blood are normally low. It binds and activates antithrombin III, which is a suicide substrate for several activated coagulation factors, including thrombin. Heparin does not have a prominent place in the management of coagulation in atherosclerosis, but it is widely used in the prevention and treatment of venous thrombosis and pulmonary embolism in surgical patients.

Fibrinolysis. Human tissue plasminogen activator (tPA), expressed recombinantly in cell culture, is the most important therapeutic agent in the acute treatment of stroke and myocardial infarction. If timely administered, it can prevent or very significantly reduce the extent of tissue damage that would otherwise result from the vascular occlusion. The activation of plasminogen to plasmin by tPA occurs through proteolytic cleavage. An older agent with the same mode of action and also of human origin is urokinase. The bacterial protein streptokinase binds and activates plasminogen without cleaving it. While streptokinase is also clinically effective, it is immunogenic and therefore prone to allergic reactions, and therefore is no longer in common use.

STUDY QUESTIONS

10.1 In Section 10.3.1, we considered that one long-term complication of diabetes mellitus is the formation of cataract due the activity of the sorbitol pathway (see reaction scheme below).

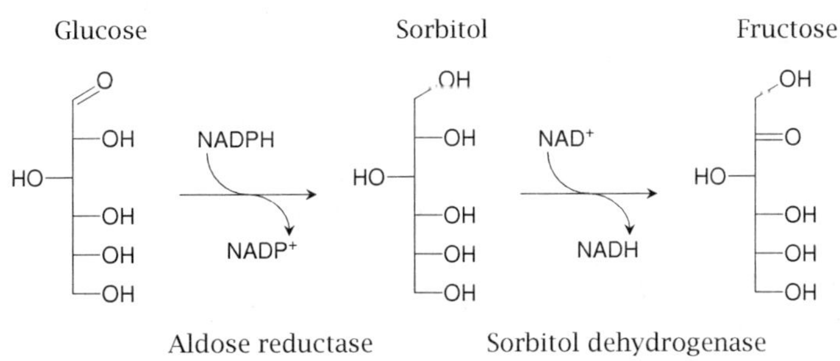

One theory to explain the pathogenic effect of increased flow through the sorbitol pathway is that it produces excess NADH, which in turn induces pseudo-hypoxic conditions in the cell. Questions:

1. How does this theory fit with the presumed mode of action of metformin? How would metformin interact with the increased flow through the sorbitol pathway?

2. While sorbitol dehydrogenase reduces one molecule of NAD^+ to NADH, the preceding aldose reductase reaction oxidizes one molecule of NADPH to $NADP^+$. Are there metabolic pathways that could use NADH to reduce $NADP^+$, thus canceling out the net effect of the sorbitol pathway on the redox state of the cell?

✎ **10.2** In Section 10.1.3, we considered a possible mechanism for the therapeutic effect of arginine in urea cycle enzyme defects. Assuming that this effect indeed consists in a negative feedback on the replenishing reactions that convert glutamine to ornithine, how could we determine whether it is arginine itself or its conversion product ornithine that exercises this inhibition? Could we learn something from clinical observations?

✎ **10.3** Some animals, such as reptiles, employ uric acid instead of urea as the main carrier for nitrogen elimination. Does this observation suggest a possible strategy for the drug therapy of urea cycle enzyme defects?

REFERENCES

[1] A. Aiuti et al., "Gene therapy for immunodeficiency due to adenosine deaminase deficiency," *N. Engl. J. Med.* 360 (2009), 447–458.

[2] J. K. Amory et al., "Miglustat has no apparent effect on spermatogenesis in normal men," *Hum. Reprod.* 22 (2007), 702–707.

[3] J. A. Barranger and E. O'Rourke, "Lessons learned from the development of enzyme therapy for Gaucher disease," *J. Inherit. Metab. Dis.* 24 Suppl 2 (2001), 89–96; discussion 87–8.

[4] M. L. Batshaw, R. B. MacArthur, and M. Tuchman, "Alternative pathway therapy for urea cycle disorders: twenty years later," *J. Pediatr.* 138 (2001), S46–54; discussion S54–S55.

[5] S. Bhakdi et al., "Beyond cholesterol: the enigma of atherosclerosis revisited," *Thromb. Haemost.* 91 (2004), 639–645.

[6] S. Braselmann et al., "R406, an orally available spleen tyrosine kinase inhibitor blocks fc receptor signaling and reduces immune complex-mediated inflammation," *J. Pharmacol. Exp. Ther.* 319 (2006), 998–1008.

[7] E. Browning, *Toxicity of industrial organic solvents,* Chemical Publishing, 1953.

[8] J. A. Campos-Sandoval et al., "Expression of functional human glutaminase in baculovirus system: affinity purification, kinetic and molecular characterization," *Int. J. Biochem. Cell Biol.* 39 (2007), 765–773.

[9] H. K. Choi and G. Curhan, "Soft drinks, fructose consumption, and the risk of gout in men: prospective cohort study," *Br. Med. J.* 336 (2008), 309–312.

[10] H. K. Choi et al., "Alcohol intake and risk of incident gout in men: a prospective study," *Lancet* 363 (2004), 1277–1281.

[11] H. C. Curtius et al., "Serotonin and dopamine synthesis in phenylketonuria," *Adv. Exp. Med. Biol.* 133 (1981), 277–291.

[12] A. Endo, "A gift from nature: the birth of the statins," *Nat. Med.* 14 (2008), 1050–1052.

[13] A. Enomoto et al., "Molecular identification of a renal urate anion exchanger that regulates blood urate levels," *Nature* 417 (2002), 447–452.

[14] F. Feillet and J. V. Leonard, "Alternative pathway therapy for urea cycle disorders," *J. Inherit. Metab. Dis.* 21 Suppl 1 (1998), 101–111.

[15] S. A. Hawley et al., "The antidiabetic drug metformin activates the AMP-activated protein kinase cascade via an adenine nucleotide-independent mechanism," *Diabetes* 51 (2002), 2420–2425.

[16] P. U. Heuckenkamp and N. Zöllner, "Fructose-induced hyperuricaemia," *Lancet* 1 (1971), 808–809.

[17] E. S. Istvan and J. Deisenhofer, "Structural mechanism for statin inhibition of HMG-CoA reductase," *Science* 292 (2001), 1160–1164.

[18] H. Itabe, "Oxidative modification of LDL: its pathological role in atherosclerosis," *Clin. Rev. Allergy Immunol.* 37 (2009), 4–11.

[19] M. L. Joachims et al., "Inhibition of deoxynucleoside kinases in human thymocytes prevents dATP accumulation and induction of apoptosis," *Nucleos. Nucleot. Nucleic Acids* 27 (2008), 816–820.

[20] M. Kishibe, H. Sakai, and H. Iizuka, "Chronic tophaceous gout secondary to self-induced vomiting in anorexia nervosa," *J. Dermatol.* 37 (2010), 578–580.

[21] S. S. Likhodii et al., "Anticonvulsant properties of acetone, a brain ketone elevated by the ketogenic diet," *Ann. Neurol.* 54 (2003), 219–226.

[22] F. Martinon et al., "Gout-associated uric acid crystals activate the NALP3 inflammasome," *Nature* 440 (2006), 237–241.

[23] A. J. McVie-Wylie et al., "Biochemical and pharmacological characterization of different recombinant acid α-glucosidase preparations evaluated for the treatment of Pompe disease," *Mol. Genet. Metab.* 94 (2008), 448–455.

[24] L. Nordquist and M Johansson, "Proinsulin C-peptide: friend or foe in the development of diabetes-associated complications?," *Vasc. Health Risk Manag.* 4 (2008), 1283-1288.

[25] R. D. Oberhaensli et al., "Study of hereditary fructose intolerance by use of 31P magnetic resonance spectroscopy," *Lancet* 2 (1987), 931–934.

[26] M. R. Owen, E Doran, and A. P. Halestrap, "Evidence that metformin exerts its anti-diabetic effects through inhibition of complex 1 of the mitochondrial respiratory chain," *Biochem. J.* 348 Pt 3 (2000), 607–14.

[27] J. L. Potter and A. A. Silvidi, "Xanthine lithiasis, nephrocalcinosis, and renal failure in a leukemia patient treated with allopurinol," *Clin. Chem.* 33 (1987), 2314–2316.

[28] K. Rajamäki et al., "Cholesterol crystals activate the NLRP3 inflammasome in human macrophages: a novel link between cholesterol metabolism and inflammation," *PLoS One* 5 (2010), e11765.

[29] I. Ramazzina et al., "Completing the uric acid degradation pathway through phylogenetic comparison of whole genomes," *Nat. Chem. Biol.* 2 (2006), 144–148.

[30] J. C. Spalton et al., "The novel Syk inhibitor R406 reveals mechanistic differences in the initiation of GPVI and CLEC-2 signaling in platelets," *J. Thromb. Haemost.* 7 (2009), 1192–1199.

[31] M. J. Sulway and J. M. Malins, "Acetone in diabetic ketoacidosis," *Lancet* 2 (1970), 736–740.

[32] J. M. Wallace et al., "Effects of peroxisome proliferator-activated receptor alpha/delta agonists on HDL-cholesterol in vervet monkeys," *J. Lipid Res.* 46 (2005), 1009-1016.

[33] Z. Wang et al., "Protein carbamylation links inflammation, smoking, uremia and atherogenesis," *Nat. Med.* 13 (2007), 1176–1184.

CHAPTER 11

CHEMOTHERAPY OF INFECTIOUS DISEASES

Microbial infections are the most widespread form of disease. To this day, tuberculosis remains the single most frequent cause of death in the world, and other scourges such as malaria and HIV likewise claim thousands of victims every day. One should realize that the magnitude of the problem is primarily due to socioeconomic reasons. By far the most victims of tuberculosis live in third-world countries. In developed countries, hygiene and surveillance, immunization, and proper nutrition have done far more to quell infectious diseases than the availability of specific drug treatment. These achievements are fully reversible, as illustrated by the greatly increased incidence of tuberculosis in Russia after the collapse of the Soviet Union and its systems of health care and surveillance [37]. Even in the United States, the lack of robust general health insurance prevents too many patients from fully complying with proper therapy regimens, which favors the emergence and spreading of resistant strains of *Mycobacterium tuberculosis* [36] or HIV. Similarly, the spreading of resistant germs in hospitals is facilitated by the less than perfect observance of hygiene, which itself is promoted by shortages of staff and training brought on by the current overemphasis on cost-cutting in many Western countries.

It can therefore be said that infectious diseases flourish not so much due to lack of knowledge but rather due to lack of its application. Nevertheless, there remain significant unsolved fundamental problems, and even with the strictest adherence to best practices of drug use, it would be impossible to

Biochemical Pharmacology. By M. Palmer, A. Chan, T. Dieckmann, J. Honek

entirely prevent the emergence and proliferation of microbial resistance. The development of antimicrobial drugs thus is caught in an everlasting arms race with microbial evolution.

In a sense, control and treatment of infectious diseases is quite simple: The microbes must be killed, or at least be denied the ability to multiply.[1] Some important means of infection control are not applied directly to humans and therefore can be very aggressive. For example, surgical instruments can be sterilized by autoclaving, ionizing radiation, or treatment with ethylene oxide. Skin or wound disinfection still permits the use of non-selective agents such as alcohols, organic mercury compounds, or iodine. However, when it comes to systemic drug therapy, we are limited by the requirement for *selective toxicity*—we must kill only the microbes but not harm the host.

11.1 PATHOGENIC MICROBES: DIVERSITY AND SELECTIVE TOXICITY

The only thing common to all pathogenic microbes is their pathogenicity; otherwise, they come from all walks of life. Traditionally, they are grouped into the following classes: (1) bacteria, which alone are prokaryotic, (2) fungi and (3) parasites, both of which are eukaryotic, and (4) viruses, which do not have a cellular structure whatsoever.

Mammalian cells are eukaryotic; the inherent relatedness to eukaryotic pathogens (Figure 11.1) means that selective toxicity is more difficult to achieve with the latter than with bacteria. Compared to either pro- or eukaryotic pathogens, most viruses have a very simple structure, with sometimes only a handful of genes and gene products, which limits the number of potential drug targets.

The principle of selective toxicity is epitomized by penicillin, which is nontoxic for humans even at levels that exceed the bactericidal concentrations by many orders of magnitude. Penicillin is not toxic to humans simply because its target in the bacterial cell has no counterpart in human cells. The drug inhibits the synthesis of the murein layer of the bacterial cell wall. However, a drug that interacts with an exclusively prokaryotic target may still be toxic for humans. This is illustrated by polymyxin, which disrupts the outer membrane of Gram-negative bacteria by disorganizing the lipopolysaccharide that it consists of. While lipopolysaccharide does not occur in human cell membranes, the drug still is known to cause nephro- and neurotoxicity in humans.[2]

On the other hand, many drugs that are fairly safe and effective against bacterial or other pathogens act on targets that *do* have counterparts in mammalian cells. For example, the enzyme dihydrofolate reductase is essential in bacteria as well as humans and eukaryotic pathogens. The bacterial enzyme can be inhibited with trimethoprim, whereas the enzymes occurring in the eukary-

[1]Most microbes have multifaceted mechanisms of pathogenicity, and targeting individual toxins or mechanisms of invasion will not suffice to control the infection. Exceptions are some particularly potent toxins such as cholera or anthrax toxin, the neutralization of which will effectively disarm the corresponding bacterial pathogens.

[2]It has been reported that polymyxin toxicity was less severe in a more recent group of patients than in older studies [11]. As in the case of vancomycin, this might be related to improved methods for purification of the drug.

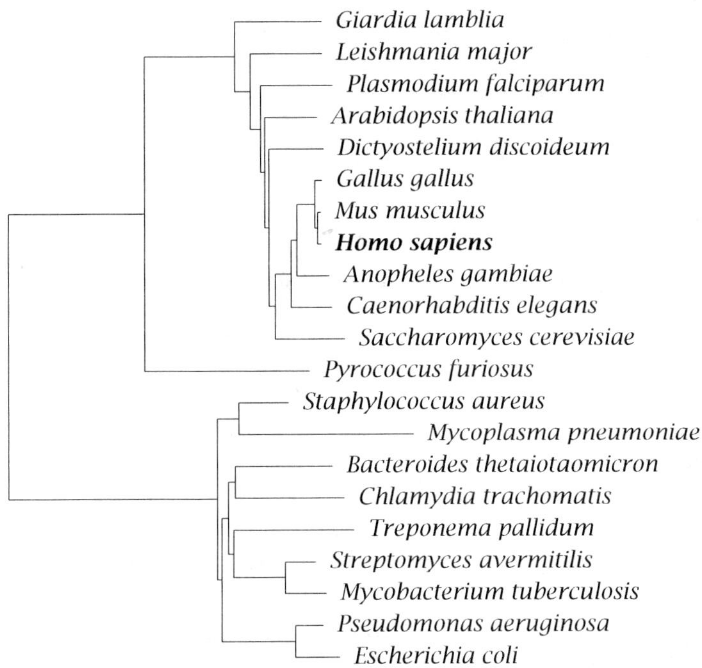

Figure 11.1 A phylogenetic tree showing *Homo sapiens* in the company of some pathogens and reference organisms. *Arabidopsis* is a plant, and *Saccharomyces* is a yeast; both are more closely related genetically to humans than are *Plasmodium*, *Leishmania* and other single-celled eukaryotic parasites. The nematode *Caenorhabditis* is related to human worm pathogens such as *Wuchereria* or *Ascaris*. *Pyrococcus* belongs to the Archaebacteria; these organisms are not human pathogens but are more closely related to us than are the Eubacteria, which include all pathogenic bacteria. (Figure created from data obtained via the web service described in Ref. [22].)

otic parasites *Toxoplasma gondii* and *Plasmodium falciparum* are susceptible to inhibition by pyrimethamine (Figure 11.2). The human enzyme is not inhibited by these but is susceptible to methotrexate, which is used in cancer and immunosuppressive therapy (Figure 12.10).

In recent years, many bacterial genome sequences have been determined, and numerous genes have been experimentally characterized as *essential*, that is, indispensable for bacterial life. If such an essential gene does not have a counterpart in the human genome, it deserves evaluation as a target for antibacterial therapy. While this is in principle a straightforward approach to target identification, it may be unnecessarily narrow, given that many of the antimicrobial drugs in current use act on targets that are not exclusive to microbes.

11.2 PHARMACOKINETIC CONSIDERATIONS

Paracelsus' principle that the dosage makes the poison applies to bacteria and other pathogens as well—only a proper concentration of poison at the site of infection will actually kill the bug, and therefore both the site of infection

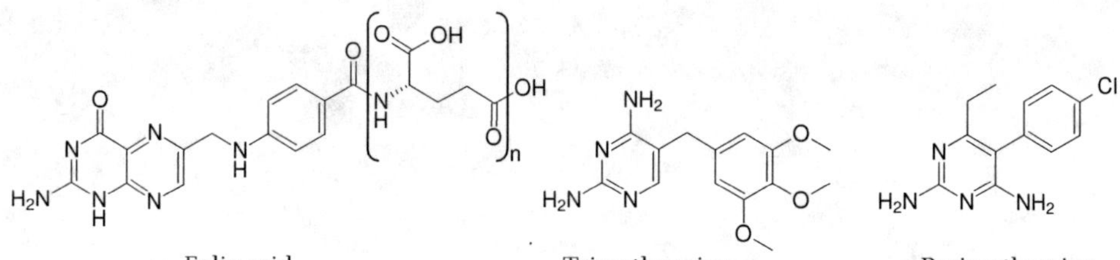

Figure 11.2 Structures of folic acid and two inhibitors of dihydrofolate reductase. Trimethoprim inhibits the bacterial enzyme, whereas pyrimethamine inhibits those of *Plasmodium* and *Toxoplasma gondii* parasites. The human enzyme is not significantly inhibited by either.

and the pharmacokinetic properties of drugs are important determinants in the choice of therapy. One distinction that is important in this context is that between intracellular and extracellular pathogens. While many species of pathogenic bacteria live exclusively in the extracellular space, others have evolved mechanisms for invading and persisting inside host cells, where they are to some degree protected both from the host's immune system and from drugs. Therefore, the selection of antibiotics for the treatment of intracellular pathogens such as *Chlamydia* or *Listeria* must take into account their ability to reach the intracellular space.

While the blood-brain barrier protects the brain tissue from many drugs in the general circulation, this protection will also be extended to pathogens that have established themselves inside the central nervous system.[3] This is of particular concern with drugs that have a low therapeutic index such as the antifungal drug amphotericin B. Examples are *Aspergillus fumigatus* mold infections in patients recovering from bone marrow transplants or in similar states of extreme immunosuppression. The infection is usually acquired by inhalation of the spores and first afflicts the lungs. This condition is serious enough, but it can be treated with amphotericin B. However, once the mold establishes itself within the brain, the prognosis becomes practically hopeless. Similarly, abscesses, that is, sites of infections containing a mass of dead tissue that is no longer perfused, require surgical cleanup in addition to chemotherapy.

The targets of antimicrobial drugs may be located at the cell surface or inside the microbial cell. In the latter case, bacterial cell walls may also present very significant obstacles to the distribution of chemotherapeutic agents.

11.3 RESISTANCE TO ANTIMICROBIALS

There is no better illustration of evolution at work than the observation of pathogenic microbes becoming resistant to even the most sophisticated and powerful chemotherapeutic agents. This problem applies to all classes of pathogens: Bacteria become resistant to penicillins and sulfonamides, malaria

[3]The blood-brain barrier will become somewhat leaky, like other capillary walls, in meningitis, which is a diffuse infection of the cerebrospinal fluid space and adjacent soft tissues. However, in case of more circumscribed lesions such as brain abscesses, this does not occur to the same extent.

parasites become resistant to chloroquine and pyrimethamine, and HIV strains become resistant to protease inhibitors. Resistance may arise by various mechanisms: (1) target site alteration, (2) compensatory overexpression of the target protein, (3) acquisition or increased expression of enzymes that inactivate the drug, and (4) acquisition, mutation or increased expression of transport proteins, leading to reduced uptake or increased active extrusion of drugs by the microbial cell.

Target site alteration often requires merely a point mutation that endows an enzyme with increased discrimination between inhibitor and substrate. Good examples are mutants of bacterial DNA topoisomerase II that confer resistance to inhibitors such as ciprofloxacin, and point mutations in HIV protease. Target sites that are not proteins can also be modified, as exemplified by resistance to the antibiotics vancomycin and polymyxin.

Malaria parasites may acquire resistance to the drug pyrimethamine by compensatory overexpression of dihydrofolate reductase, its target enzyme. β-Lactamases, which inactivate penicillins and cephalosporins, and various classes of enzymes that inactivate aminoglycosides may be plasmid-borne and acquired by genetic transfer. The expression of extrusion pumps is often increased by a factor of >100 in multiresistant bacterial strains.

Countermeasures can delay but not entirely avoid the emergence and spreading of resistance. They include (1) avoidance of unnecessary drug application, (2) use of combination therapy, and (3) rigorous hygiene.

Avoiding the unnecessary use of antibiotics is easier in theory than in practice, since clinical urgency often forbids waiting for conclusive and specific evidence of an infection. In combination therapy, the idea is to deny the microbe the benefit of acquiring resistance against a single antibiotic by simultaneously applying a second one that remains effective.

The significance of hygiene is highlighted by the fact that most infections with highly resistant bacteria are acquired in hospitals. Most of these cases are contracted in regular wards and intensive care units rather than operation theaters. The mostly successful avoidance of infection in operation theaters illustrates both the power of hygiene and the discipline and drudgery required to make it work.

11.4 ANTIBACTERIAL CHEMOTHERAPY

While there are some classes of completely synthetic antibacterial agents, most drugs in use today are *antibiotics*, that is, natural compounds of microbial origin or derivatives thereof. Antibiotics are isolated from both prokaryotic and eukaryotic soil microorganisms, which use them as chemical weapons in their struggle against one another for the same ecological niche. There are literally thousands of known antibiotics. Most antibiotics are toxic for both bacterial and eukaryotic cells, including human ones. Some of these, such as doxorubicin or streptozotocin, are actually used as cytotoxic agents in cancer chemotherapy (see Chapter 12). However, a still sizable number of antibiotics are indeed selectively toxic for prokaryotes and accordingly useful in the therapy of bacterial infections.[4] While penicillins and cephalosporins are produced by fungi, which are eukaryotic, many other antibiotics that are toxic for bacteria are obtained from soil-dwelling bacteria, often from *Streptomyces* species. Obviously, these bacteria must be resistant to their own poisons. These resistance mechanisms are genetically encoded and, like any other gene, potentially subject to genetic transfer. Indeed, the emergence of resistance to antibiotics in pathogenic bacteria is due in part to the spreading of resistance mechanisms that originated in the producer species or in competing soil-dwelling bacteria through gene transfer.

Antibacterial drugs are among the most frequently prescribed medicines; they probably also are the most frequently misused ones. This is due in part to the biological diversity of pathogenic bacteria. While all bacteria are prokaryotes, they differ with respect to cell structure, life cycle, and genetic equipment for resistance to antimicrobial agents. It is therefore important to isolate and characterize the bacterium responsible for a given case of infection and to use such information to guide the selection of proper chemotherapy.[5]

11.4.1 Bacterial biology

The biological differences between bacterial and mammalian cells afford a multitude of targets for selective toxicity. On the other hand, the very short generation time and the existence of numerous mechanisms of genetic transfer are responsible for the rapid emergence and spreading of bacterial resistance to antibiotics.

The bacterial cell wall. The cell wall has a dual significance for chemotherapy. On the one hand, it contains multiple drug targets, since it has no counterpart in mammalian cells; on the other hand, the cell wall restricts access of drugs to intracellular targets.

The innermost layer of the bacterial cell wall is the cytoplasmic membrane, which in many ways resembles that of mammalian cells. With most bacteria, the cytoplasmic membrane is covered by one or more additional layered structures. The composition of these structures is a major distinguishing feature between different classes of bacteria, and it also strongly influences their susceptibility to chemotherapy. The major variations in cell wall structure are schematically depicted in Figure 11.3.

In Gram-positive bacteria, the cytoplasmic membrane is covered by a thick layer of murein, which is the intensely crosslinked, sturdy peptidoglycan unique to bacteria. Other types of polysaccharides and proteins are attached to its outer surface, but there is no protective membrane around the murein layer itself. Gram-positives are therefore often quite susceptible to antibiotics that

[4]Some antibiotics are selective for eukaryotic cells. An example is the antifungal antibiotic amphotericin B.

[5]In practice, it is often necessary to begin treatment without delay. In those cases, one should secure samples before therapy is begun, base selection of initial therapy on *educated* guesses, and adapt therapy when laboratory results become available.

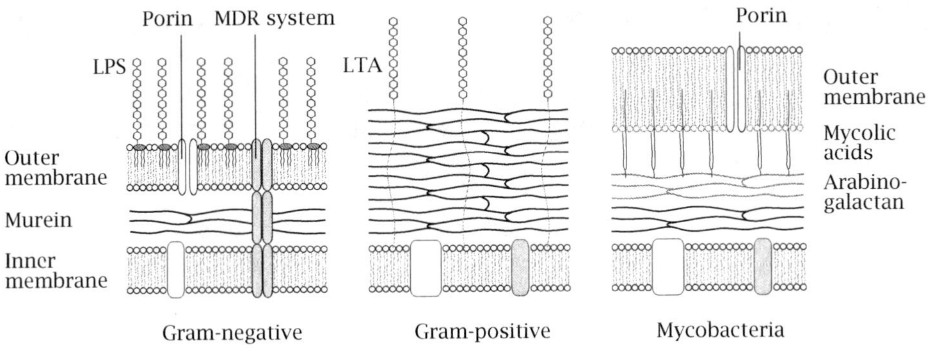

Figure 11.3 Bacterial cell wall structure. Gram-negative bacteria have a comparatively thin murein layer that is surrounded by an outer membrane. The outer leaflet of this membrane contains lipopolysaccharide (LPS), also known as *endotoxin*. Porins in the outer membrane facilitate diffusion of small polar solutes; multidrug resistance (MDR) extrude antibiotics across both membranes. Gram-positive bacteria lack an outer membrane but have a much thicker murein layer, which is decorated with lipoteichoic acids (LTA). In mycobacteria, the murein layer is surrounded by arabinogalactan polysaccharide, to which branched, long-chain fatty acids (mycolic acids) are attached; these, in turn, connect to a relatively thick outer membrane.

inhibit murein synthesis. Examples of Gram-positive pathogens are *Staphylococcus aureus* and *Listeria monocytogenes.*

Gram-negative bacteria have a thinner murein layer than Gram-positive ones. However, the entire cell is wrapped by an outer membrane, which in many cases restricts the access of drugs to the interior layers and the cytosol.[6] The space between the inner and the outer membranes is known as the *periplasm.* Important Gram-negative pathogens are gonococci and meningococci, *Salmonella* species and *Escherichia coli*, and *Pseudomonas aeruginosa.*

Mycobacteria such as *Mycobacterium tuberculosis* and *M. leprae* have a cell wall that is even more complex and less penetrable than that of Gram-negative bacteria. A thick outer membrane [17] is supported by a layer containing mycolic acids; together, these two form a thick, wax-like outer layer that efficiently excludes a wide range of antimicrobial drugs. β-Lactam antibiotics, for example, are mostly unable to penetrate this layer and are therefore inactive.[7]

Mycoplasmas don't have any additional layers outside of their cytoplasmic membranes and therefore are inherently resistant to β-lactam antibiotics. This also applies to the intracellular forms of rickettsias and chlamydias.

Kinetic aspects of transport-mediated drug resistance. Cell membranes pose barriers to drug penetration into the cytosol and, in case of Gram-negative bacteria, into the periplasmic space. However, while the rate of penetration of a drug may be low, it will be finite, so if no other mechanism of drug detoxifi-

[6]The outer membrane also restricts access of the dye crystal violet, used in the Gram staining procedure, that would otherwise stain the murein layer. The Gram staining method therefore tests for the presence of an outer membrane.

[7]Imipemem, a carbapenem antibiotic, is active on some mycobacterial species, as is cycloserine, which is not a β-lactam but inhibits murein precursor synthesis.

cation exists, any concentration gradient across the membrane will eventually vanish. The rate of equilibration will be proportional to the ratio of surface to volume of the enclosed compartment, which is inversely related to cell size. Therefore, in order to sustain a low intracellular or periplasmic drug concentration, a membrane barrier must be complemented by an additional detoxification mechanism; this can be either metabolic inactivation of the drug or active efflux.[8] Various types of efflux transporters occur in bacterial cytoplasmic membranes that operate on various intracellularly acting antibiotics such as tetracyclines, aminoglycosides, and fluoroquinolones. Some of these are structurally related to the P-glycoprotein found in mammalian cells (see Figure 3.5), while others belong to different structural families and use different modes of active transport such as proton antiport.[9] The rate of efflux across the cytoplasmic membrane will often conform to Michaelis-Menten kinetics; the concentration gradient at equilibrium will be such that the drug concentration in the cytosol will support a rate of efflux equal to that of passive influx.

In Gram-negative bacteria, the requirement for complementary drug efflux or metabolic inactivation also applies to the outer membrane and the compartment enclosed by it, that is, the periplasmic space. The requirement is accentuated by the occurrence of porins in the outer membrane. Porins facilitate entrance of many small, polar molecules. Many drugs will therefore more readily penetrate the outer than the inner membrane. Mutations in porins can increase drug resistance.

β-Lactam antibiotics, which in Gram-negative bacteria act in the periplasm, are subject to both efflux and enzymatic inactivation. Active efflux across the outer membrane is quite a remarkable feat, as there are no actively maintained ion gradients across this membrane, nor is there any ATP available in the periplasm. It is accomplished by multicomponent efflux systems that span the entire cell wall (see Figure 11.3). The cytoplasmic membrane component harnesses intracellular energy to capture, in cooperation with the periplasmic component, drug molecules from the periplasm, which are then channeled out of the cell across the outer membrane component. Periplasmic substrate capture by the multicomponent system protects against antibiotics acting in the periplasm. In addition, it amplifies the resistance afforded by efflux pumps that extrude drugs from the cytosol into the periplasm in a multiplicative manner [21,29].

The cell wall in mycobacteria has two membranes, and therefore the same considerations just outlined for the Gram-negative cell wall should apply here as well. Drug efflux across the mycobacterial outer membrane has not been experimentally characterized, however [5].

[8]In rapidly proliferating bacterial cultures, we must also consider dilution, since the increase in cellular volume due to cell division counteracts the penetration of polar drug molecules.

[9]In bacteria with aerobic metabolism, the cytoplasmic membrane fulfills the function of the inner mitochondria membrane in eukaryotic cells. The proton-motive force across the bacterial membrane drives not only ATP synthesis but also secondary active transport.

Bacterial metabolism and chemotherapy. Apart from the cell wall, there are other metabolic distinctions between bacterial and mammalian cells that can provide targets for chemotherapy.

Many bacteria are more biosynthetically self-sufficient than mammalian cells. For example, *Escherichia coli* can thrive on a minimal growth medium that contains no organic compound other than glucose, indicating that the bacterium can synthesize all those molecules that figure as vitamins or essential amino acids in mammalian metabolism. Biosynthetic enzymes exclusive to bacteria are potential drug targets, provided that the bacterial cell lacks the ability to acquire the product from the environment instead. An example is dihydropteroate synthase, which functions in the bacterial synthesis of folic acid; this enzyme is inhibited by sulfonamides (see Section 1.4).

Another interesting and practically important class of drugs that exploit metabolic diversity are nitroimidazole derivatives such as metronidazole (see Figure 11.16). These are selectively toxic for bacteria and also some eukaryotic parasites that are *anaerobic*, meaning that they thrive only in the absence of oxygen. In the strongly reducing environment that prevails inside anaerobic cells, nitroimidazoles are reduced by ferredoxin to nitro radical anions:

$$\mathrm{R{-}NO_2 + ferredoxin{-}Fe_2^{+} \longrightarrow R{-}NO_2^{-\cdot} + ferredoxin{-}Fe_3^{+}}$$

These radicals can react with DNA and cause strand breaks. Aerobic human or bacterial cells are not affected because they don't contain sufficiently strong reducing agents. In addition, any radicals that may be formed will be reoxidized by oxygen:

$$\mathrm{R{-}NO_2^{-\cdot} + O_2 \longrightarrow R{-}NO_2 + O_2^{-\cdot}}$$

The superoxide formed in the second reaction will be scavenged by peroxidase. The reaction between the nitro radicals and DNA that leads to DNA strand breakage is not understood in full detail, but it is targeted at thymine bases, and it is interesting to note that bacteria with a higher content of adenine and thymine in their DNA are more susceptible to nitroimidazole drugs [9].

Isonicotinic acid hydrazide (isoniazid, INH) is used against *Mycobacterium tuberculosis*. Inside the mycobacterial cell, it is first activated to a reactive radical by a bifunctional catalase/peroxidase enzyme, KatG. The radical then forms a covalent adduct with NAD^+ (Figure 11.4), which in turn noncovalently inhibits the reductase enzyme InhA. The latter participates in the synthesis of mycolic acids, the long chain fatty acids that occur in the mycobacterial cell wall (Figure 11.3). The successive interaction with first KatG and then InhA makes INH highly selective for mycobacterial cells. Its toxicity for humans is tolerable and caused by metabolism in the liver (see Section 4.5.4).

Bacterial genetic flexibility. One key aspect of bacterial biology that greatly accelerates the spreading of resistance to antibiotics is the exchange of DNA

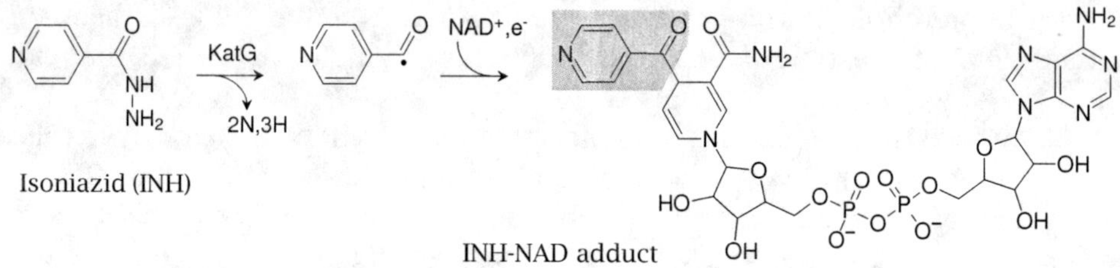

Figure 11.4 Mode of action of isoniazid (INH) on *Mycobacterium tuberculosis.* The molecule is activated by the oxidative enzyme KatG to a radicalic form, which forms an adduct with NAD^+ that inhibits InhA, an enoyl-CoA reductase involved in mycolic acid synthesis.

between bacterial strains and species. This exchange is mostly mediated by plasmids, which are DNA molecules that are propagated independently from the bacterial chromosome. Many plasmids encode an elaborate apparatus of multiple proteins that enables their own transfer from one bacterial cell to another, often crossing species boundaries in the process.

The existence of vehicles that mediate transfer of genes between cells is complemented by transposons, which are genetic elements that mediate transfer of genes between different DNA molecules within a bacterial cell. Through successive transposition events, resistance genes from multiple sources can be collected on a single plasmid molecule. Such plasmids, sometimes referred to as *R factors,* will greatly aid the survival of bacterial hosts exposed to multiple antibiotics and therefore be effectively selected for in a hospital environment. This became manifest only a few years after the introduction of antibiotic therapy, when bacterial species that had originally been susceptible began to acquire resistance to multiple antibiotics, often including ones that had not even been used on the individual patients from whom the resistant bacteria were isolated.[10]

Even without genetic transfer, the short generation time of most bacteria and the potentially large number of bacterial cells carried by an infected host promote the occurrence of resistant mutants. This problem is particularly severe in tuberculosis. While *Mycobacterium tuberculosis* actually propagates rather slowly, it is killed by antibiotics even more slowly, meaning that multiple successive generations of bacteria are exposed to them and have a chance to develop resistance. When a manifest case of tuberculosis is treated with a single antibiotic only, the emergence of resistant mutants is almost certain. This can be prevented by combining multiple drugs. Assuming that each of these drugs by itself is lethal, a bacterial cell would have to develop resistance to all of them at the same time to actually survive. With mutation rates on the order of 10^{-6} per drug and bacterial generation, the simultaneous application of

[10]Plasmids may also encode enzymes that mediate resistance to heavy metal ions; mercury, for example, is disposed of by reduction to its volatile elemental form. Apart from conferring resistance, plasmids may provide other benefits to the bacterial host, such as increased pathogenicity. Anthrax toxin, for example, is encoded by a plasmid. When "cured" of this plasmid, the dreaded *Bacillus anthracis* turns into a rather agreeable fellow.

three or preferably four drugs renders the occurrence of resistance sufficiently unlikely.[11]

Bactericidal and bacteriostatic action. Some antibiotics, such as β-lactams and aminoglycosides, kill the bacteria outright and are therefore referred to as *bactericidal*. In contrast, chloramphenicol and sulfonamides merely cause biosynthesis and proliferation to stall, but the bacteria will survive and resume growth after dilution or removal of the antibiotic. This mode of action is called *bacteriostatic*.

Very broadly speaking, a bacterial infection will overwhelm the host only if the bacteria multiply faster than they can be killed by the host's immune system. In most cases, therefore, application of bacteriostatic antibiotics will suffice to decisively tip the balance in favor of the host. However, this may not be the case in patients with severely compromised immune function. For example, patients recovering from bone marrow transplants go through a precarious period of several weeks during which they have essentially no immune system, and therefore should be treated with bactericidal antibiotics if at all possible.

The bactericidal action of some antibiotics, particularly β-lactams, applies only to actively growing bacteria. Since growth stalls under the influence of bacteriostatic agents, these will negate the effect of β-lactams. Such drug combinations should therefore be avoided.

11.4.2 Antibiotics that inhibit murein synthesis

Murein is found in the cell walls of most bacteria. Its biosynthesis starts in the cytosol, where monomeric peptidoglycan precursors are assembled on a membrane-associated lipid carrier. The final product of the cytosolic phase, known as *lipid II*, is then flipped to the outer face of the cytoplasmic membrane, and the peptidoglycan precursor moiety is incorporated into the murein layer. The process is outlined in Figure 11.5.

Fosfomycin and cycloserine (Figure 11.7) inhibit early steps in cytoplasmic precursor synthesis. Fosfomycin is an antimetabolite of phosphoenolpyruvate in the enzymatic synthesis of muramic acid (reaction 1 in Figure 11.5.). It contains an epoxy group, which covalently reacts with a cysteine residue in the enzyme's active site. Fosfomycin is taken up into the cell by active transport, piggybacking on the glycerolphosphate transporter, and cells can become resistant by inactivation of this nonessential transport protein. Cycloserine is an antimetabolite of D-alanine and inhibits both alanine racemase and D-alanine ligase (steps 2 and 3 in Figure 11.5). The lipopeptide antibiotic amphomycin inhibits transfer of the precursor to the undecaprenol phosphate carrier lipid (step 4). It inhibits similar transfer reactions in mammalian metabolism and therefore is not suitable for therapy; however, the bacterial enzyme is a target of ongoing drug development efforts. The peptide antibiotic bacitracin inhibits

[11] In a typical combination of tuberculostatic drugs, some are cheap, whereas others are expensive. When uninsured patients skip the expensive ones, this subverts the principle of combination therapy and gives *Mycobacterium tuberculosis* the opportunity to acquire resistance in a piecemeal fashion. The consequences for society are certainly more costly than subsidized therapy could ever have been.

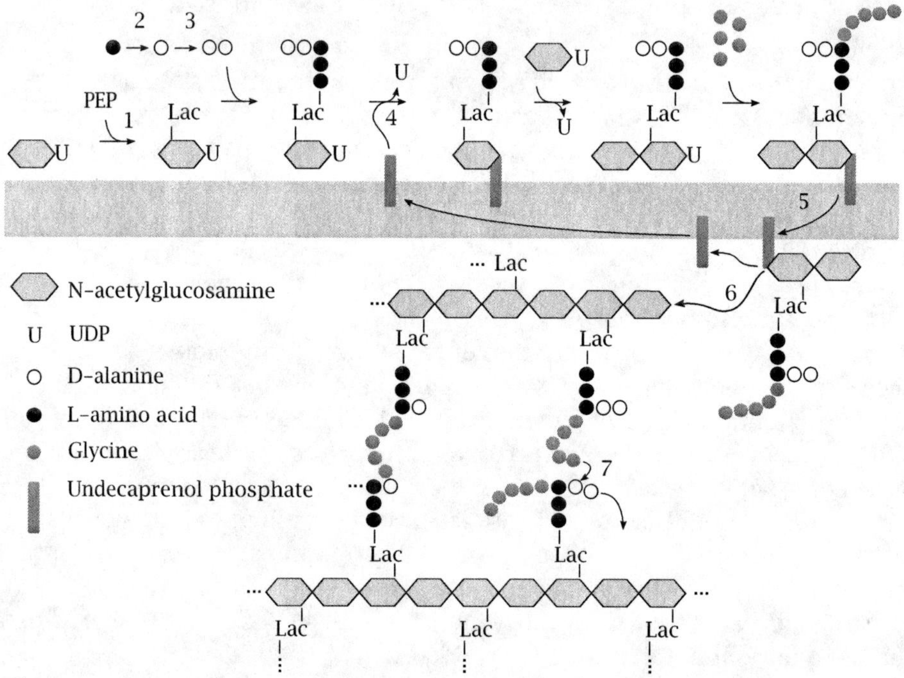

Figure 11.5 Outline of bacterial murein synthesis. Phosphoenolpyruvate (PEP) provides a lactate residue (Lac) to *N*-acetylglucosamine, which yields *N*-acetylmuramic acid (1). Onto the latter, a pentapeptide is built in successive ATP-activated reactions. The free end of this peptide contains two D-alanine residues that are supplied by alanine racemase (2) and D-alanine ligase (3). This nascent building block is transferred to the lipid carrier undecaprenol phosphate (4) and subsequently extended by another molecule of N-acetylglucosamine and five glycine residues. The completed precursor, lipid II, is flipped to the outer face of the cytoplasmic membrane (5). The glycopeptide moiety is transferred to a growing murein strand in the transglycosylase reaction (6). The final transpeptidase (7) reaction crosslinks the new subunit to an adjacent murein strand, releasing the terminal D-alanine residue.

the flipping of lipid II across the membrane (step 5). Bacitracin is suitable for local application but too toxic for systemic therapy.

The final polymerization stage of murein synthesis is particularly interesting for chemotherapy, because it occurs outside the cytoplasmic membrane. It involves two separate reactions, both of which are catalyzed by the bifunctional enzymes known as *penicillin-binding proteins*. The transglycosylase reaction (6) is inhibited by moenomycin. This lipopeptide antibiotic has a complex structure that mimics the lipid II substrate; it is not used in therapy because of its unfavorable pharmacokinetic properties.

The transpeptidation reaction (7) is inhibited by β-lactams and by glycopeptides such as vancomycin. β-Lactams react with and inactivate the enzyme. In contrast, glycopeptides such as vancomycin or A82846B (Figure 11.6) don't bind to the enzyme but instead to the substrate, that is, to the peptide moiety of the murein subunit.

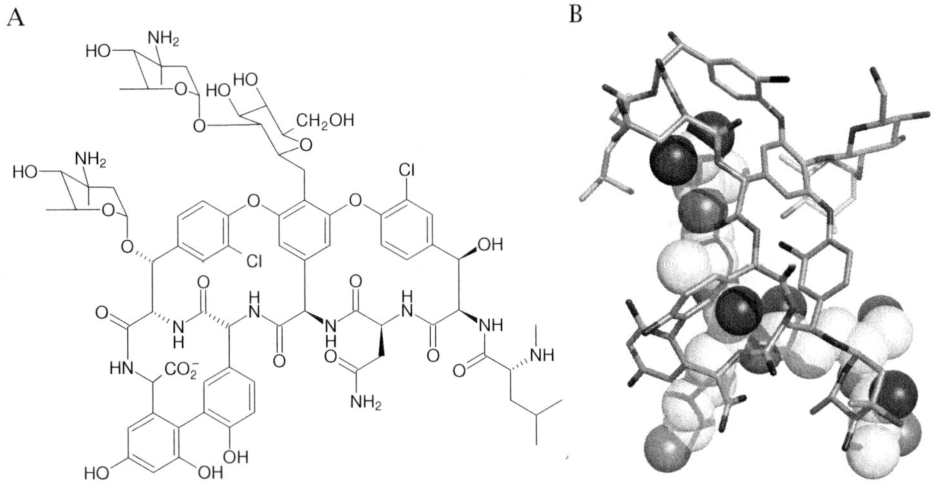

Figure 11.6 Structure and function of the glycopeptide antibiotic A82846B. **A:** Structural formula. **B:** A82846B (sticks) bound to the peptide moiety of lipid II (spheres). The two linked D-alanine residues of lipid II are pointing upwards. Rendered from 1gac.pdb [33]. A colored version of this figure is shown in the insert in the middle of this book.

Early preparations of vancomycin got a bad rap for toxicity; this appears to have been due to their high content of impurities, which earned them the nickname "Mississippi mud". Once these problems had been sorted out, vancomycin made a comeback and became a staple in the therapy of hospital-acquired infections, in particular those due to methicillin-resistant *Staphylococcus aureus* (MRSA) and *Clostridium difficile.* Considering its widespread and frequent use, development of resistance has been remarkably slow. This is probably related to its mode of action. Lipid II, the target of vancomycin, is the product of many successive enzymatic reactions, and it is much more difficult for a microbial species to replace or modify an entire pathway than to tweak the substrate selectivity of a single enzyme. However, resistant bacterial strains have begun to emerge, and other antibiotics have begun to take its place.[12]

β-Lactam antibiotics. The most widely used class of antibiotics are the β-lactam antibiotics, which comprises the penicillins, the cephalosporins, and several related compounds. These are obtained mostly through semisynthesis (see Section 15.5.1). β-lactams all act as covalent inhibitors of the muramyl-transpeptidase activity of penicillin-binding proteins. The name of the class derives from the four-membered β-lactam ring that occurs in each of these molecules. The amide bond in this ring resembles the peptide bond between the two D-alanine residues in the substrate, and it undergoes a similar reaction with the enzyme.

The active site of muramyl-transpeptidase contains a serine residue, which transiently forms an amide bond with the penultimate D-alanine residue of

[12]One common mechanism of vancomycin resistance involves the replacement of the terminal D-alanine residue by lactate, catalyzed by a dedicated enzyme.

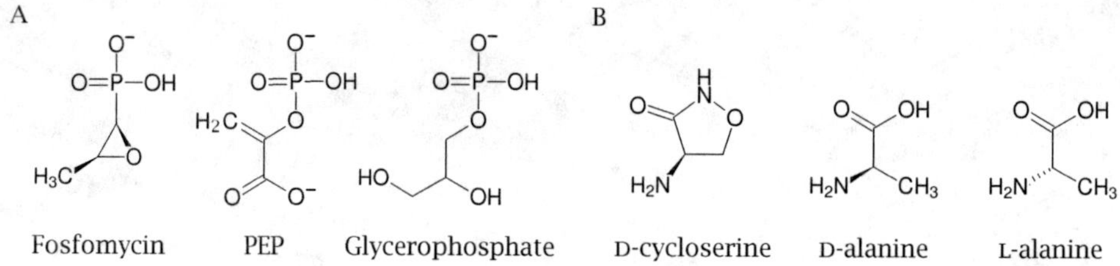

Figure 11.7 Structures of the antibiotics fosfomycin and D-cycloserine and of related metabolites. **A:** Fosfomycin is an antimetabolite of phosphoenolpyruvate (PEP) and taken up into bacterial cells by the glycerolphosphate transporter. **B:** D-cycloserine is an antimetabolite of D-alanine in the alanine racemase and D-alanine ligase reactions.

the substrate, thereby releasing the terminal D-alanine. In the second step, the amino group of the second substrate's terminal glycine substitutes and releases the catalytic serine.

The β-lactam bond in the antibiotic resembles the peptide bond between the two D-alanine residues (Figure 11.8). The four-membered ring is spring-loaded because of the strained bond angles within, which facilitates the covalent reaction of the drug with the catalytic serine. The covalent adduct typically undergoes hydrolysis with a half-life of several hours, but usually the bacterium will be dead long before this occurs.

The ring tension of the β-lactam ring that promotes its reaction with penicillin-binding proteins also makes it prone to side reactions. One side reaction is hydrolysis; aqueous solutions of β-lactams therefore are not very stable. Reaction of the β-lactam ring with free amino groups on proteins yields covalent adducts that are immunogenic, which is the reason for the relatively frequent occurrence of allergic reactions to penicillins and cephalosporins.

The first β-lactam antibiotic to be introduced into clinical use was penicillin G. It was initially reliably active against a wide variety of Gram-positive bacteria, as well as gonococci (see Figure 1.8) and *Treponema pallidum*, the causative agent of syphilis. However, microbes have come a long way since then, and only a few pathogenic bacterial species remain uniformly susceptible. On the other hand, for those strains of any species that have been confirmed as susceptible through laboratory tests, it remains unsurpassed in activity.

Most species of Gram-negative bacteria were resistant against penicillin G right from the start, because their outer membranes are not sufficiently permeable to this drug. Ampicillin shows better penetration and has extended the spectrum of activity to Several common Gram-negative pathogens such as *E. coli* and *Haemophilus influenzae.* Subsequently introduced drugs such as ticarcillin and piperacillin showed good activity even against hardy germs such as *Pseudomonas aeruginosa*, at the cost of reduced activity against Gram-positive bacteria. Their extended spectrum is due to improved permeation through porins, which actually favor polar small molecules. However, all of these drugs are now subject to widespread resistance.

Cephalosporins were introduced somewhat later than penicillins. The cephalosporin nucleus has two variable substituent positions, and a great many

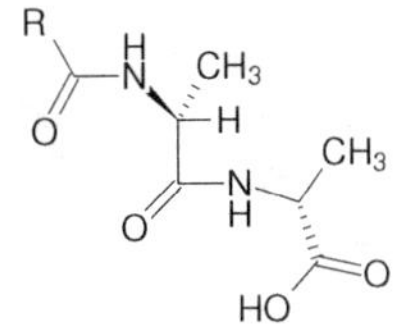

D-alanyl-D-alanine moiety

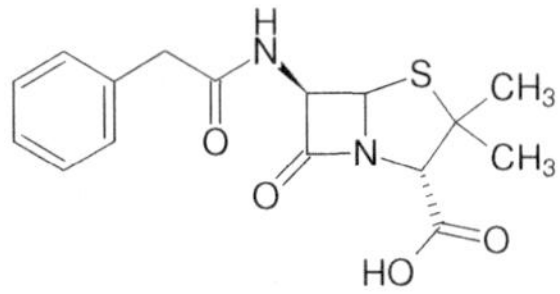

Penicillin G

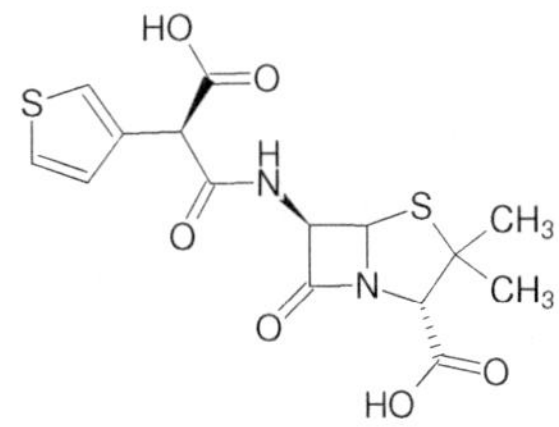

Ticarcillin

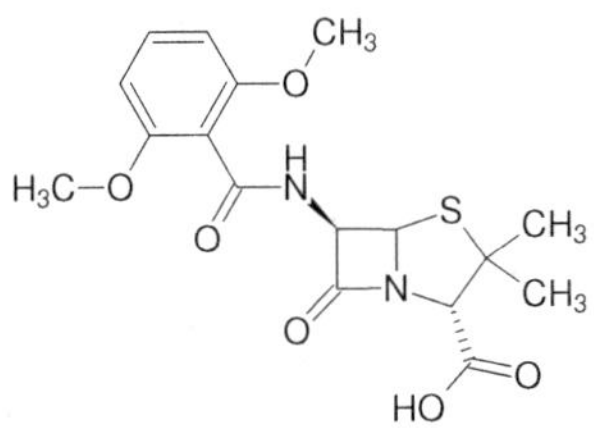

Methicillin

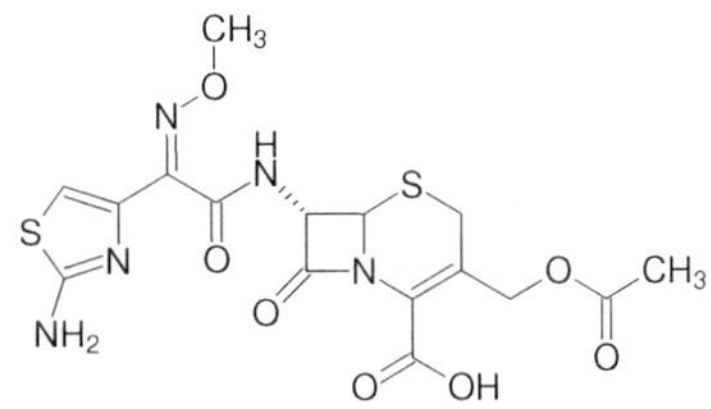

Cefotaxime

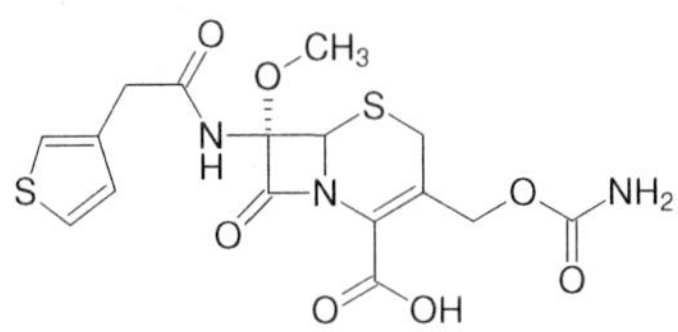

Cefoxitin

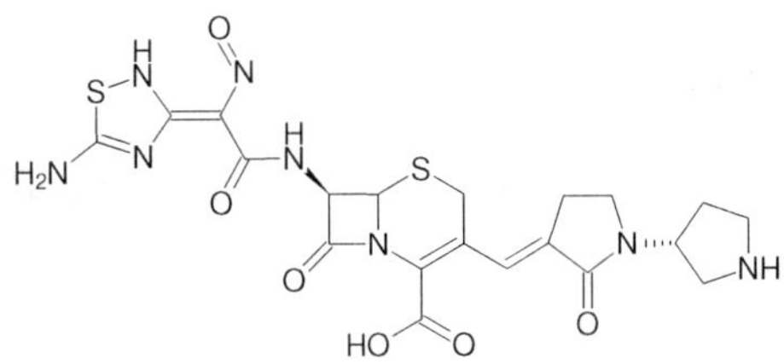

Ceftobiprole

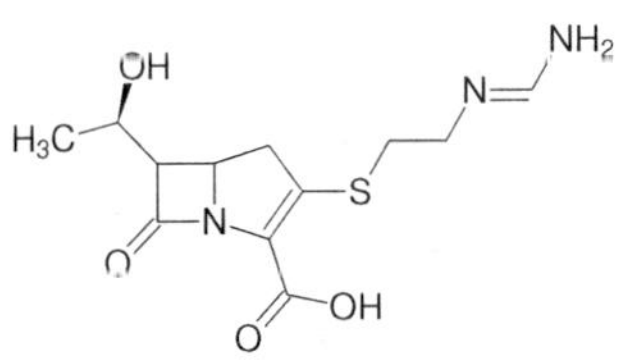

Imipenem

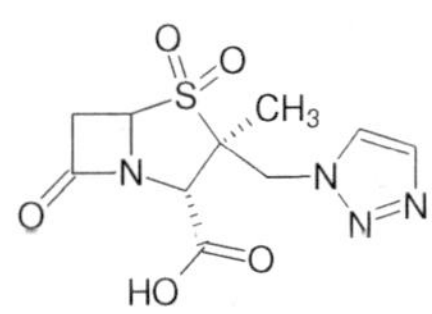

Tazobactam

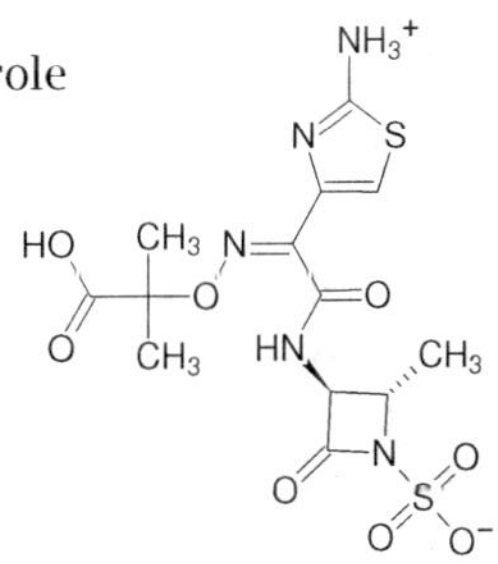

Aztreonam

Figure 11.8 Structures of the D-alanine dipeptide moiety that is the substrate of penicillin-binding proteins, as well as of various β-lactam antibiotics. Penicillin G is active mostly on Gram-positive bacteria, whereas ticarcillin is active against many Gram-negative bacteria. Methicillin is the prototype of β-lactamase-stable penicillins; the principle of β-lactamase inhibition is steric hindrance. Cefotaxim, cefoxitin, and ceftobiprole are broad-spectrum cefalosporins. Imipenem is a carbapenem, and aztreonam is a monobactam. Tazobactam is a synthetic β-lactamase inhibitor that is combined with β-lactam antibiotics in order to protect them from degradation.

different substituents have been introduced at both of these in order to improve penetration of bacterial cell walls and stability against β-lactamases. They have surpassed penicillins in usefulness against many bacterial species. Other variations of the β-lactam structure are carbapenems such as imipenem, which resemble penicillins but with a methylene group instead of the sulfur, and monobactams, in which the second ring β-lactam is missing altogether.

β-Lactamases. While active extrusion and reduced uptake through mutant porins are significant mechanisms of resistance to penicillins and cephalosporins, the single most important mechanism is inactivation of the drug molecules by β-lactamases. These enzymes are molecularly and mechanistically heterogeneous. A major distinction is that between serine amidases and metalloenzymes. The serine amidases are more common and are also more tractable, since they often are susceptible to β-lactamase inhibitors; they are found in both Gram-positive and Gram-negative organisms. While drug development efforts are underway [19], no clinically useful inhibitors exist for metalloenzymes, which are found in some Gram-negative bacteria and can confer resistance to a very broad spectrum of β-lactams.

β-Lactamases emerged in *Staphylococcus aureus* only a few years after the introduction of penicillin G. As a countermeasure, the structure of the antibiotic was modified to make it resistant to the enzyme. The bulkier substituents found in oxacillin, methicillin, and similar derivatives sterically hinder the enzyme, so that β-lactamase-producing strains were rendered susceptible again. However, those same bulky side chains may also reduce the affinity of the antibiotic to its muramyl-transpeptidase target. Indeed, with strains of *S. aureus* that do not produce β-lactamase, none of the "staphylococcal" penicillins matches the specific activity of penicillin G, which, however, is acceptable because of the generally low toxicity of the β-lactams.

The widespread use of oxacillin and similar analogs prompted the emergence of resistant bacterial strains, in particular methicillin-resistant *Staphylococcus aureus* (MRSA). By genetic transfer, these bacteria have acquired a different penicillin-binding protein that resembles one found natively in a related species (*Staphylococcus sciuri*). This protein is not susceptible to inhibition by most β-lactams, but the more recently developed cephalosporin ceftobiprole (Figure 11.8) inhibits it and offers an alternative to vancomycin in the treatment of MRSA infections [25].

Another strategy for overcoming β-lactamases that has been fairly successful is the combination of penicillins with β-lactamase inhibitors. Here, no modifications need to be made to the penicillin molecules themselves; inhibitors can be combined with any penicillin derivative. A naturally occurring β-lactamase inhibitor is clavulanic acid;[13] synthetic inhibitors with somewhat extended spectra are sulbactam and tazobactam. They have a common mode of action on β-lactamases: The initial acylation of the catalytic serine causes formation of a reactive intermediate that irreversibly traps a second serine in the active site (Figure 11.9).

The initial acylation of the catalytic serine does not suffice to inactivate the enzyme; if covalent reaction with the secondary serine does not occur in time, the opened ring will simply be removed through hydrolysis, just as with

[13]Clavulanic acid also weakly inhibits muramyl-transpeptidase and thus actually is an antibiotic; however, this activity is too low to be clinically useful.

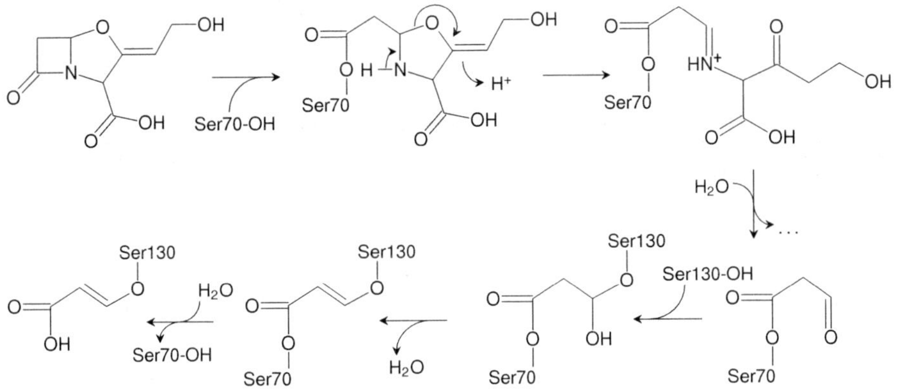

Figure 11.9 Inactivation of SHV-1 β-lactamase by clavulanic acid. The initial acylation of the catalytic serine 70 residue resembles the reaction with β-lactam antibiotics. In clavulanic acid, however, it is followed by a sequence of reactions that traps a second residue in the active site, serine 130. The active site therefore remains blocked even after serine 70 is freed by hydrolysis. Mutation of serine 130 to glycine does not compromise catalytic activity but renders the enzyme resistant to clavulanic acid. (Simplified after Ref. [43].)

substrate β-lactams. In keeping with this mechanism, mutants in which the secondary serine is changed to glycine become refractory to inhibition.

Mutations in β-lactamases can not only confer resistance to inhibitors but also extend the substrate spectrum. Evolution of high-level resistance against initially active antibiotics can be readily modeled in vitro. For example, a >100-fold increase in activity of a β-lactamase against the drug ceftazidime could be readily observed in three successive rounds of random mutagenesis and antibiotic selection [6].

11.4.3 Inhibitors of ribosomal protein synthesis

Ribosomal protein synthesis is inhibited by several classes of antibiotics, some of which are selective for prokaryotic ribosomes and therefore usable for antibacterial chemotherapy. There is, however, one caveat: Mitochondrial ribosomes resemble the bacterial ribosomes more closely than those in the cytoplasm do, and their inhibition by drugs such as chloramphenicol and gentamicin appears to mediate some of the toxicity observed with these antibiotics.

The various types of ribosomal inhibitors interact with different sites on the ribosome and accordingly interfere with protein synthesis at various stages (see Table 11.1); some of these antibiotics have been instrumental in working out the mechanistic details of ribosomal protein synthesis. An interesting example is the drug puromycin, which disrupts the protein synthesis at both prokaryotic and eukaryotic ribosomes. Puromycin mimics the business end of an aminoacylated tRNA (Figure 11.10). It is coupled to the carboxy terminus of a growing peptide chain. The ribosome then fails to attach the next amino acid, and the peptide chain is prematurely terminated and falls off the ribosome.

Table 11.1 Examples of antibiotics that inhibit ribosomal protein synthesis. All except puromycin are used in pharmacotherapy. Puromycin acts on ribosomes of both prokaryotic and eukaryotic cells and is used in biochemical research.

Name	Source	Structural class	Mode of action
Chloramphenicol	*Streptomyces venezuelae*	—	Blocks peptidyltransferase site through direct binding to ribosomal RNA
Erythromycin	*Saccharopolyspora erythraea*	Macrolide	Same as chloramphenicol
Streptomycin	*Streptomyces griseus*	Aminoglycoside	Binds to protein S12, induces mistranslation and premature termination
Amikacin	Semisynthetic derivative of kanamycin	Aminoglycoside	Blocks aminoacyl-tRNA acceptor site through direct binding to ribosomal RNA
Puromycin	*Streptomyces alboniger*	Aminonucleoside	Acts as a false substrate to induce chain termination
Linezolid	Synthetic	—	Prevents formation of initiation complex; binding site on RNA overlaps with that of erythromycin
Fusidic acid	*Fusidium coccineum*	Steroid	Binds elongation factor G

An interesting secondary effect of premature termination of translation by puromycin is the permeabilization of the cytoplasmic membrane. Membrane proteins and some proteins destined for secretion from the cell are translated directly at the cytoplasmic membrane. The ribosome attaches to the translocon, a membrane channel that guides the nascent polypeptide chain to its destination within or outside the membrane. When puromycin prematurely terminates translation and releases the peptide, the translocon remains stuck in an open conformation, which causes the membrane defect [38,39].

The antibiotic streptomycin causes mistranslation rather than premature termination (see Section 13.1.1). Nevertheless, membrane leakiness occurs with streptomycin, too; the leaks have been ascribed to misread and misfolded membrane proteins [3]. One consequence of membrane permeabilization is an increased rate of uptake of streptomycin itself and, if present, of other antibiotics. Membrane permeabilization therefore appears to be an important aspect of the bactericidal effect of streptomycin and of its synergism with other antibiotics.

While streptomycin interacts with the ribosomal protein S12, many other antibiotics (such as gentamicin, erythromycin, clindamycin and chloramphenicol) interact exclusively with the ribosomal RNA. This is discussed in more detail in Chapter 13.

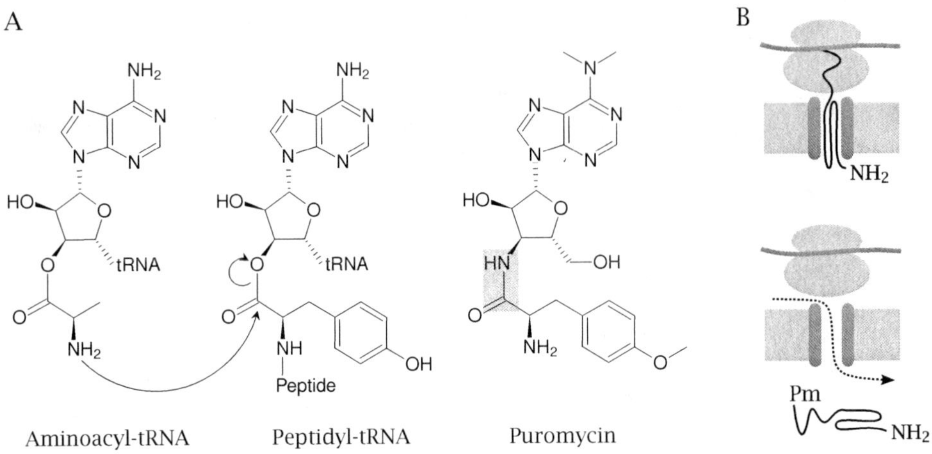

Figure 11.10 Structure and action mechanism of puromycin. **A:** The antibiotic resembles aminoacyl-tRNA and therefore is incorporated at the carboxy terminus of a growing peptide chain. However, the amide bond (highlighted) does not yield to the nucleophilic attack by the next amino acid. Peptide synthesis is terminated, and the peptide is prematurely released. **B:** Premature peptide release at the translocon pore leaves the latter in a leaky state and renders the membrane permeable toward small solutes.

Resistance to aminoglycosides can arise in a number of ways. Point mutations affecting the ribosomal RNA or ribosomal proteins may prevent binding to the target site or, with streptomycin, compensate for the functional effect of the antibiotic on the ribosome, sometimes even leading to *dependence* of the bacterium on the antibiotic. Resistance genes carried on plasmids may encode methyl-, acetyl-, and sulfotransferases that inactivate the antibiotic through covalent modification (see Figure 13.5). Alternatively, they may encode active efflux systems or enzymes that modify specific bases in the target site on the ribosome instead; such enzymes are commonly found in the soil bacteria that produce these antibiotics.

Amikacin is a semisynthetic derivative of the aminoglycoside kanamycin that has been chemically modified to forestall enzymatic modification by resistance enzymes [27]. Considering its long-standing clinical use, amikacin has stood up quite well to the development of microbial resistance.

11.4.4 Inhibitors of DNA and RNA synthesis

There are numerous antibiotics that cause direct damage to DNA. Such damage is not selective for prokaryotes in any way. While some drugs such as bleomycin or doxorubicin are used in cancer therapy (see Chapter 12), only drugs that selectively target bacterial enzymes can be used to treat infections.

Bacterial DNA polymerase III, which synthesizes DNA during replication of the genome, should in principle be a straightforward drug target. Selective inhibitors are indeed available that are both active at nanomolar concentrations

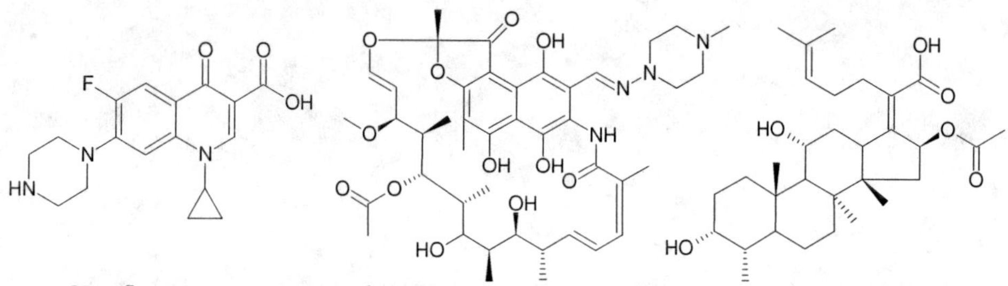

Ciprofloxacin Rifampicin Fusidic acid

Figure 11.11 Inhibitors of bacterial macromolecular synthesis. Ciprofloxacin is a synthetic inhibitor of bacterial DNA topoisomerase. Rifampicin inhibits mRNA synthesis by RNA polymerase 2. Fusidic acid inhibits protein synthesis by binding to elongation factor 2.

and effective in animal models [44]; however, they are not yet available for clinical use.

The practically most important family of inhibitors of bacterial DNA synthesis act on DNA *topoisomerase*. The role of topoisomerase in DNA replication is an indirect one. The bacterial chromosome, when fully extended, would be far longer then the entire bacterial cell, and it therefore is coiled, curled, and twisted into a formidable Gordian knot inside the cell. In order to penetrate this maze, DNA polymerase needs the help of DNA topoisomerases, which play the part of Alexander's sword, although unlike Alexander the enzymes are so considerate as to rejoin the severed ends after untangling them.

Topoisomerase I acts on a single DNA molecule, breaking one of the two strands and swiveling it around the other before rejoining (see Figure 12.12). Topoisomerase II, in contrast, acts on two molecules of DNA, breaking both strands of one of these and moving the free ends past the other before rejoining. Both natural and synthetic topoisomerase inhibitors exist; the most effective ones are fluoroquinolones such as ciprofloxacin, which inhibit topoisomerase II (Figure 11.11).[14] The antibacterial spectrum of topoisomerase inhibitors is remarkably wide, including Gram-positive, Gram-negative and atypical bacteria as well as mycobacteria. Resistance is most commonly acquired through target site alteration, that is, through mutations in the bacterial DNA topoisomerase genes.

Rifampicin binds to bacterial RNA polymerase II and inhibits the synthesis of messenger RNA. It is used mostly against *Mycobacterium tuberculosis* as well as *Mycobacterium leprae*, the causative agent of leprosy, but occasionally against Gram-positive bacteria as well. It is unsuitable for monotherapy due to rapid emergence of resistance.

11.4.5 Antibiotics that act on bacterial cell membranes

Bacterial cell membranes have in principle the appeal of being more readily accessible to drugs than intracellular targets; however, few antibiotics that act

[14]Ciprofloxacin is structurally similar to chloroquine, and it emerged from a drug development effort that initially was aimed at malaria parasites rather than bacteria.

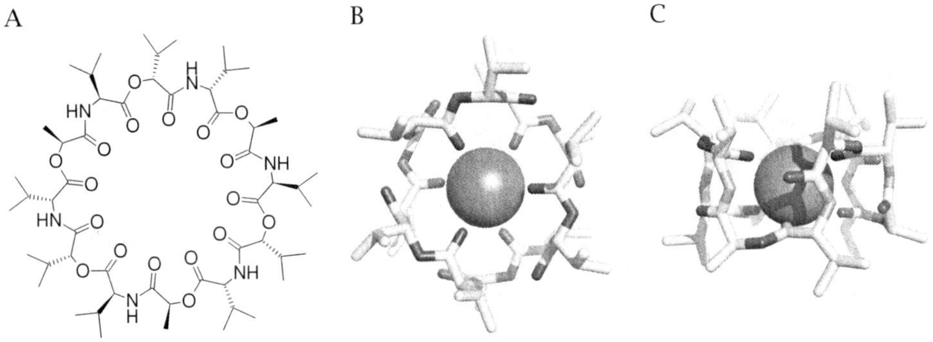

Figure 11.12 Structure and function of the potassium ionophore valinomycin. **A:** The circular molecule contains alternating ester and peptide bonds and exclusively hydrophobic side chains. **B, C:** Valinomycin wraps tightly around a potassium ion, coordinating it with the double-bonded oxygen atoms of the six ester bonds. The hydrophobic exterior of the complex enables it to diffuse across cell membranes easily. (Drawn from coordinates reported in Ref. [28].)

directly on cell membranes are sufficiently selective for bacterial ones to be of use in pharmacotherapy.

Cell membranes act as barriers, and accordingly membrane-damaging antibiotics break down this barrier function. Permeabilization can be nonselective or specific for certain solutes; it can involve structural disruption, or it can be entirely transient. An example of transient and selective permeabilization are ionophores such as valinomycin. This circular *depsipeptide*—the term denotes a peptide in which some peptide bonds are replaced by ester bonds—wraps around a potassium ion and coordinates it with its backbone oxygen atoms (Figure 11.12). Stripped of its polar hydration shell and clad instead into the hydrophobic exterior of the valinomycin molecule, the ion can efficiently partition into lipid membranes and cross them. Valinomycin acts on both bacterial and mammalian cell membranes; it therefore is not suitable for antibacterial therapy but, like other ionophores, is commonly used in experimental cell biology.

Polymyxin B and polymyxin E are cyclical peptides that, at the tip of a branch protruding from the cycle, also contain a short fatty acyl chain. They bind to the lipid A core of lipopolysaccharide (LPS) that is the predominant component of the outer membrane of Gram-negative bacteria. Adjacent LPS molecules are bound to one another both through hydrophobic forces and through calcium ions that attach to their phosphate groups. Polymyxins interfere with both these interactions (Figure 11.13), which causes the disruption of the outer membrane and subsequently of the inner membrane as well.[15]

Polymyxins are considered relatively toxic but have more recently made a comeback for use in infections with bacterial strains that are otherwise in-

[15]The inner membrane also contains some lipopolysaccharide, as it is synthesized there. The role of LPS in the damage to the inner membrane inflicted by polymyxin has apparently not been determined.

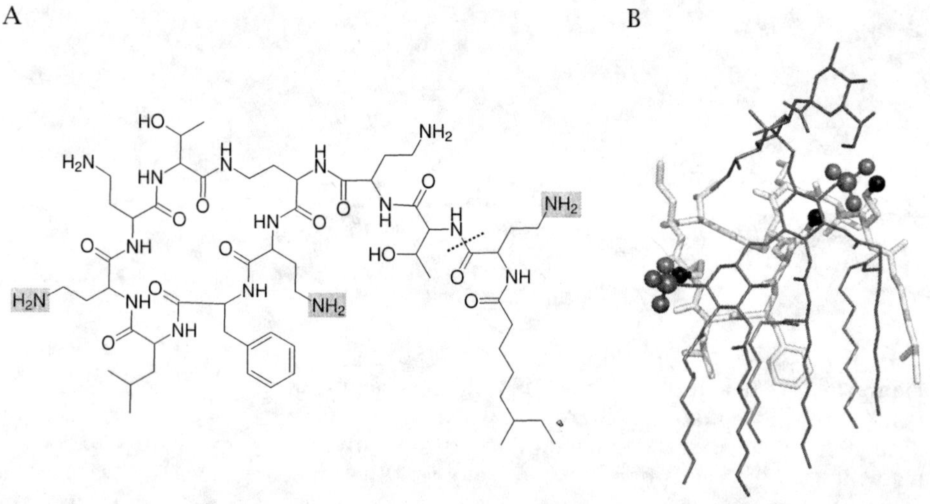

Figure 11.13 Structure and action mode of polymyxin B. The antibiotic binds to lipid A, which is the core of lipopolysaccharide (LPS), the main component of the outer membrane of Gram-negative bacteria. **A:** Structure of polymyxin B. The fragment to the right of the dotted line is missing in polymyxin nonapeptide (see text). **B:** Model of polymyxin B bound to lipid A, showing polymyxin B with thicker, lighter lines and lipid A with thinner, darker lines. Several hydrophobic side chains of polymyxin B insert between the acyl chains of adjacent lipid A molecules and disrupt the membrane. Three amino groups that pair with the two phosphates of lipid A are highlighted, as are the phosphates themselves. (Rendered from coordinates kindly provided by P. Pristovsek [32]. Several carbons have been trimmed from each of the six alkyl chains in lipid A. A colored version of panel B is shown in the insert in the middle of this book.)

tractable [11,12]. Polymyxin toxicity correlates with the presence of the fatty acyl chain. A deacylated derivative, polymyxin nonapeptide, has been prepared by limited proteolysis. It is devoid of toxicity and also no longer bactericidal; nevertheless, it still binds to LPS and increases the permeability of the outer membrane, which renders the bacterium less resistant to other antibiotics. The nonapeptide also increases bacterial susceptibility to the membrane-damaging activity of the complement system [48,20]. While these findings suggest that the nonapeptide should be useful in combination with other antibiotic agents, they have not translated into clinical application so far.

Resistance to polymyxins is mediated by covalent modifications of lipid A that mask some of its charges. Various modifying enzymes occur in Gram-negative bacteria, and resistance may arise by mutations that cause their constitutive overexpression.

The so-called *lantibiotics* such as nisin, which contain the nonstandard amino acid lanthionine, permeabilize the cytoplasmic membrane after binding to lipid II, the extracellular precursor of murein synthesis [1]. They are not used in pharmacotherapy but are useful in food processing and preservation.

11.4.6 New targets for antibacterial chemotherapy

Almost all antibiotics newly introduced into clinical practice in the last few decades have been variations of previously known molecules; they therefore

Table 11.2 Antifungal drugs for systemic use. Amphotericin and fluconazole and other triazoles have the widest spectrum of activity and are most widely used in practice. Griseofulvin is systemically applied but active only against infection of skin and nails with certain molds.

Name	Source	Class	Mode of action
Amphotericin B	*Streptomyces nodosus*	Polyene antibiotic	Ergosterol-dependent membrane permeabilization
Fluconazole	Synthetic	Triazole	Inhibition of ergosterol biosynthesis
Flucytosine	Synthetic	Nucleobase analogue	Inhibitor of thymidylate synthase (prodrug)
Caspofungin	*Glarea lozoyensis*	Echinocandin antibiotic	Inhibition of cell wall glucan synthesis
Griseofulvin	*Penicillium griseofulvum*	–	Inhibitor of tubulin polymerization
Terbinafine	Synthetic	Allylamine	Inhibitor of squalene-2,3-epoxidase

don't address any new drug targets. More recently, new prospective targets have emerged from the sequencing of whole bacterial genomes, together with genetic methods to identify *essential genes*, that is genes whose products are essential for cell viability. A good target would be one that is either unique to bacteria or sufficiently distinct from any human counterpart; it should also be conserved across a wide range of bacterial species to ensure that inhibitors will have a usefully wide spectrum of antibacterial activity.

An example is the enzyme formylmethionine deformylase, also referred to as peptide deformylase [26]. In many bacteria, protein translation starts off with *N*-formylmethionine. After translation is completed, the formyl group is cleaved from the protein's N-terminus by formylmethionine deformylase (see Figure 15.3). Many proteins will not be functional unless deformylated, and therefore inhibition of methionine deformylase is lethal. Eukaryotes, but also some bacteria, start translation with methionine directly, and therefore do not require this enzyme.

11.5 CHEMOTHERAPY OF FUNGAL INFECTIONS

The list of pathogenic fungi is quite long, but fungal infections are far less common than bacterial ones. While some fungi such as *Blastomyces dermatitidis* or *Histoplasma capsulatum* can take hold in otherwise healthy individuals, most fungal infections occur in immunocompromised patients. In these cases, the combination of immunosuppression, drug toxicity, and antibiotic resistance often poses challenging problems.

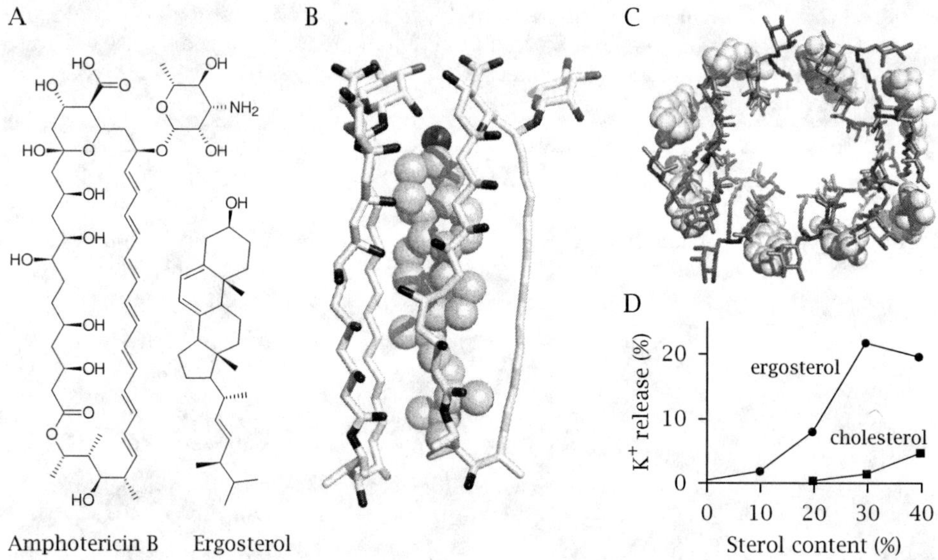

Figure 11.14 Mode of action of amphotericin B. **A**: Structures of ampotericin B and of ergosterol. **B**: Amphotericin B associates with ergosterol in the cytoplasmic membranes of fungal cells. Here, one molecule of ergosterol (spheres) is shown between two molecules of amphotericin B. **C**: Structure of a complete amphotericin B membrane pore (top view). Eight molecules each of amphotericin B and ergosterol assemble into a circular complex, the central lumen of which is lined by the amphotericin B hydroxyl groups. **D**: Selectivity of pimaricin, a polyene antibiotic related to amphotericin B, for ergosterol over cholesterol, measured as the release of K^+ ions from liposomes. (B and C rendered from molecular simulation data kindly provided by Dr. Maciej Baginski; see [2]. D replotted from data in Ref. [45].)

Because of their eukaryotic nature, fungi offer fewer suitable targets for chemotherapy than bacteria. The most commonly used antifungal drugs are polyene antibiotics—in particular, amphotericin B—and synthetic triazoles such as fluconazole (see Table 11.2). The selective toxicity of both polyenes and triazoles is based on the occurrence of ergosterol in the fungal cell membrane, where it takes the place filled by cholesterol in mammalian cells. Polyene antibiotics bind directly to ergosterol to induce membrane damage (see Figure 11.14). Specificity for ergosterol over cholesterol is not absolute, however, and significant toxicity, particularly affecting the kidneys, commonly accompanies amphotericin B treatment.[16]

Amphotericin B is not water-soluble and therefore has to be combined with some solubilizing agent. This is discussed in more detail in Section 14.5.1. Interestingly, ergosterol also occurs in some protozoal parasites such as *Leishmania*, which are therefore also susceptible to and commonly treated with amphotericin B.

[16]Another polyene antibiotic, filipin, binds cholesterol more avidly than ergosterol and therefore is useless for therapy. However, its intrinsic fluorescence can be used to visualize the distribution of cholesterol in mammalian cell membranes.

Inhibition of ergosterol biosynthesis is the therapeutic principle of triazoles such as fluconazole and ketoconazole. The target enzyme is 14-α sterol demethylase, which is a cytochrome P450 enzyme. As one might expect, drugs in this group commonly cause either induction or inhibition of hepatic CYP450 enzymes, which may lead to interaction with the elimination kinetics of other drugs and to liver toxicity (see Figure 4.5).

Combining triazoles with amphotericin B is not a good idea, since triazoles would reduce the concentration of ergosterol in the membrane and therefore the susceptibility to amphotericin B. A drug that can be used in combination with either is 5-fluorocytosine (flucytosine), which is enzymatically deaminated by many fungal species but not by human cells. The product of deamination, 5-fluorouracil, is also used in cancer chemotherapy; its mode of action—inhibition of thymidylate synthase—is discussed in Section 12.4.1.

Another class of antifungal drugs that interferes with sterol biosynthesis are the allylamines, represented by terbinafine. The enzyme inhibited by these drugs, squalene-2,3-epoxidase, is required in the biosynthesis of both ergosterol and cholesterol, and therefore occurs in both fungal and human cells. Evidently, the drug distinguishes between the human and the fungal enzyme and preferentially inhibits the latter.

11.6 CHEMOTHERAPY OF PARASITE INFECTIONS

Eukaryotic parasites are a very mixed group of pathogens with respect to both their phylogenetic relatedness and their biological complexity. The spectrum of available drug targets is accordingly heterogeneous, and there is no such thing as a broad-spectrum antiparasitic drug; most of the drugs discussed below can be used with only a few species of parasites.

A fundamental distinction among parasites is that between single-celled organisms (protozoans) and multi-celled ones (metazoans). The name *protozoa*, which is Greek for “first animals”, suggests a relatedness to higher animals that does not actually exist. As can be seen in Figure 11.1, the evolutionary distance separating them from each other and from humans is larger than that between humans and *Arabidopsis* or *Saccharomyces*.[17] Accordingly, protozoa have unique biochemical and cell-biological traits, some of which afford drug targets. A case in point is the susceptibility of *Plasmodium falciparum*, the causative agent of malaria tropica, to tetracyclines and other antibacterial agents that target the *apicosome*, an organelle of prokaryotic origin that has no counterpart in mammalian cells.

In contrast to protozoans, *metazoan* parasites are true animals, and their metabolism is more similar to that of humans. On the other hand, their more complex organization includes elements such as a nervous system that may be targeted by drugs directed at signaling molecules rather than at fundamental metabolic activities.

[17]Feel free to substitute these examples with recreational crop plants or mushrooms of your choice.

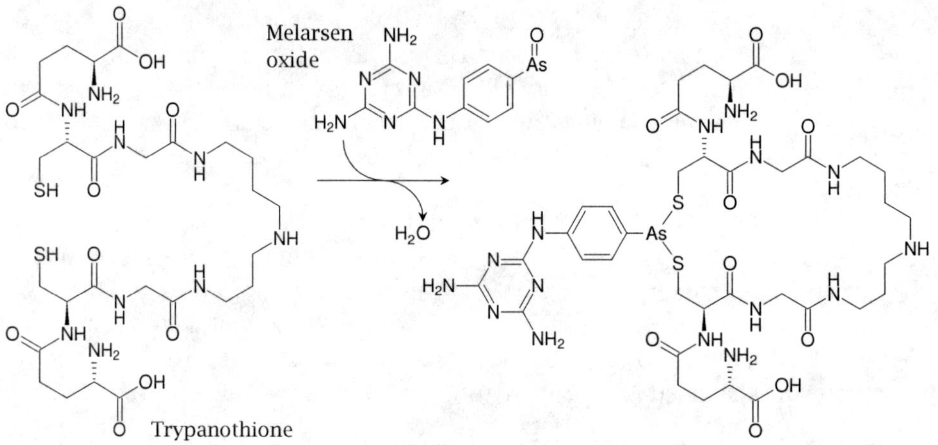

Figure 11.15 Mode of action of melarsen oxide. The oxide reacts with trypanothione, the predominant cellular thiol in *Trypanosoma* species. The adduct is a strong inhibitor of trypanothione reductase. Resistance is due to active extrusion of the complex.

11.6.1 Chemotherapy of protozoal infections

We will consider a few drugs in some detail; additional examples are summarized in Table 11.3.

The genus *Trypanosoma* includes the causative agents of sleeping sickness and of Chagas disease, which are transmitted by arthropod vectors. Several of the drugs used in the treatment of these diseases are organic derivatives of arsenic or antimony; an example is melarsen oxide. This drug combines with trypanothione, a sulfhydryl compound that in *Trypanosoma* takes the place of glutathione (Figure 11.15). Because of the divalent nature of trypanothione, this complex is fairly stable and in effect sequesters trypanothione. In addition, the complex also inhibits the enzyme trypanothione reductase, which in turn will sap the parasite's ability to protect itself from damage by radicals, such as those released by immune cells.

Malaria is caused by various *Plasmodium* species (*P. malariae*, *P. vivax*, *P. ovale*, and *P. falciparum*). After transmission of the disease by *Anopheles* mosquitoes, the parasites initially proliferate in the liver before breaking into the bloodstream and continuing to proliferate inside erythrocytes. Release of progeny parasites from consumed red cells and infection of new ones may be synchronized, resulting in characteristic periodic bouts of fever. This is not the case with *P. falciparum*, which therefore gives rise to a less regular and characteristic, yet more severe, form of the disease, known as *malaria tropica*. *P. ovale* and *P. vivax* persist in the liver for extended periods of time; treatment of these parasites must therefore be effective not only in erythrocytes but also in liver cells.

Chloroquine used to be a mainstay of malaria chemotherapy and prophylaxis. Its widespread use has, unfortunately, resulted in equally widespread resistance, which is mediated by active extrusion. Chloroquine acts only on

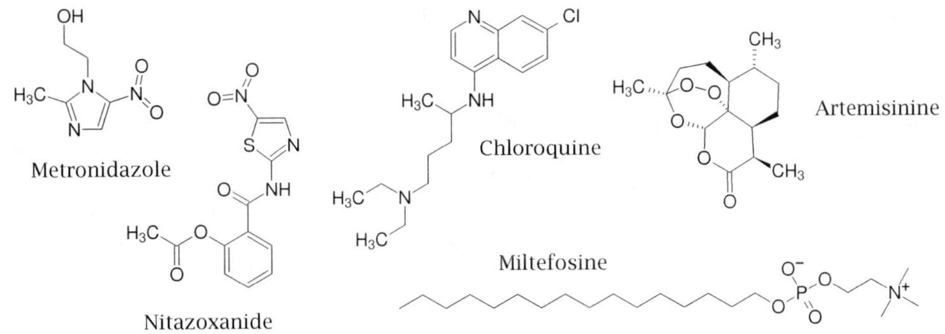

Figure 11.16 Antiprotozoal drugs. Metronidazole and nitazoxanide are reduced to toxic nitro radicals by anaerobic metabolism. Chloroquine prevents sequestration of heme and thereby indirectly causes formation of reactive oxygen species, and artemisinin inhibits sarcoendoplasmic Ca^{2+}-ATPase (SERCA); both are used against malaria. Miltefosine is a lipid analog that inhibits cytochrome *c* oxidase in *Leishmania.*

those stages of the parasite that reside in erythrocytes. There, the parasites feed on hemoglobin, the proteolytic digestion of which releases heme.[18] Free heme catalyzes the formation of reactive oxygen species from molecular O_2, which will wreak all kinds of havoc on the parasites. The latter accordingly have developed mechanisms to inactivate heme, which include heme-binding proteins as well as heme polymerization.[19] Chloroquine binds to heme and thereby prevents its sequestration; the damage to the parasite thus is again ultimately due to free radicals.

Primaquine is structurally similar to chloroquine but, in contrast to the latter, is also active on parasite stages residing in the liver. Primaquine has been reported to inhibit vesicular transport in liver cells [16]; whether this is related to the therapeutic effect on *Plasmodium* liver stages is not known with certainty.

The natural compound artemisinin has more recently gained prominence in the treatment of malaria. Reaction of the molecule's endoperoxide bridge (Figure 11.16) apparently yields reactive C-radicals, which then undergo covalent reaction with a variety of proteins and other molecules. A crucial target is the endoplasmic Ca^{2+}-ATPase (SERCA), which is selectively inhibited in the parasite but not in mammalian cells [8]. The selective susceptibility of the parasite's Ca^{2+} pump has been pinpointed to individual amino acid residues, and mutations have been characterized that confer resistance to the drug [47], adding credence to the assertion that the Ca^{2+} pump is indeed the predominant target of artemisinin. While the drug is currently isolated from the *Artemisia* plant, a more recent study has succeeded in obtaining large amounts of a precursor through metabolic engineering in yeast, which should lead to more cost-effective production [34].

[18]These heme-digesting proteases are being evaluated as possible therapeutic targets; see for example Ref. [40].

[19]The polymeric form of heme, hemozoin, can be observed microscopically in blood smears as the so-called *malaria pigment*, which aids in the diagnosis of the disease.

Table 11.3 Examples of antiprotozoal drugs, their mechanisms of action, and examples of susceptible species.

Name	Source/class	Mode of action	Comments
Artemisinin	*Artemisia annua*	Covalent inhibition of Ca^{2+}-ATPase	*Artemisia* weed was used for malaria in traditional Chinese medicine
Atovaquone	Naphthoquinone (synthetic)	Analog of ubiquinone; blocks cytochrome *b* in the respiratory chain	Active against *Plasmodium*, *Toxoplasma*, and *Pneumocystis*
Chloroquine	Quinoline (synthetic)	Inhibits polymerization of heme by *Plasmodium*, inducing oxidative damage	Resistance common; caused by increased active extrusion
Melarsoprol	Organic arsenic compound	Sequesters trypanothione	Resistance by extrusion; mechanism shared by antimony compounds
Miltefosine	Alkyl-phosphocholine	Inhibits cytochrome *c* oxidase	Originally developed as cancer drug; active against *Leishmania*
Nitazoxanide	Synthetic	Inhibitor of pyruvate oxidoreductase	Unusually broad spectrum includes both protozoans and worms
Organic bisphosphonates	Synthetic	Inhibition of hexokinase or farnesylpyrophosphate synthase	Experimental; target enzyme varies with organism and chemical structure
Pyrimethamine	Diaminopyridine (synthetic)	Inhibitor of dihydrofolate reductase	Combined with a sulfonamide or sulfone; analogous combinations used in antibacterial chemotherapy
Tetracyclines	Antibiotic	Inhibition of ribosomal protein synthesis	Selectively targets the apicoplast (organelle of prokaryotic origin) in *Plasmodium falciparum*

The drug atovaquone resembles the respiratory chain cofactor ubiquinone. It binds and inhibits the cytochrome *b* of malaria parasites and several other protozoal species with an affinity exceeding that for mammalian cytochrome *b* by 3 orders of magnitude. Toxicity for the parasite is caused by the interruption of ATP regeneration. Atovaquone is commonly applied in combination with the drug proguanil for increased effectiveness and reduced likelihood of resistance. The two drugs appear to have significant synergism [46], but the

Table 11.4 Examples of drugs used in the treatment of infections with parasitic worms.

Name	Source/class	Mode of action	Comments
Albendazole	Benznidazole (synthetic)	Inhibits tubulin polymerization	—
Ivermectin	Antibiotic	Inhibits glutamate-gated chloride channel	Targeted channel has no counterpart in humans
Diethyl-carbamazine	Synthetic	Activity depends on iNOS and cyclooxygenase activity	Exact mode of action unknown
Doxycycline	Antibiotic	Inhibits bacterial protein synthesis	Inhibits endosymbiontic bacteria of filariae, does not target the worms directly
Metrifonate	Organic phosphorus compound	Inhibitor of acetylcholinesterase	Also used to inhibit the human enzyme in Alzheimer's disease
Praziquantel	Synthetic	Induces muscle spasms in worms	Has been linked to Ca^{2+} channels, actin, and adenosine transport; primary site of action not known with certainty
Pyrantel	Synthetic	Depolarizing neuromuscular blockade	—

underlying mechanism is not clear. Proguanil undergoes metabolic activation to cycloguanil, which then inhibits dihydrofolate reductase. However, proguanil itself remains effective in plasmodia that possess a cycloguanil-resistant DHFR mutant, suggesting that proguanil has an additional mode of action. While proguanil has been reported to augment the depolarization of the mitochondrial membrane brought about by atovaquone [41], the biochemical mechanism of this interaction has not been studied in detail.

Organic bisphosphonates inhibit the enzymes hexokinase and farnesylpyrophosphate synthase in the protozoal parasites *Trypanosoma* [18] and *Toxoplasma* [23]. The relative affinity for these two enzymes, which are not closely related, varies with the structure of the organic moiety of the bisphosphonate in question. While this discovery has not yet resulted in clinical applications, the accepted use of bisphosphonates in osteoporosis—see Section 7.5.4—suggests that there would be no insurmountable problems with toxicity.

11.6.2 Chemotherapy of metazoal parasites

While this group is again somewhat heterogeneous taxonomically, metazoal parasites—mostly worms, or "helminths"—are our closest relatives among all pathogens, and therefore require the most discriminating approach to chemotherapy. Depending on their location within the human body and their own size, they may also be particularly difficult to reach by drugs. This is the case for example with *Echinococcus granulosus*, the fox tapeworm, the larval stages of which reside inside cysts within the liver and can be removed only by surgical therapy. Chemotherapy is still needed in this situation to prevent metastatic seeding of larvae released into the bloodstream during surgery. On the other hand, worms that in humans reside inside the lumen of the intestine,[20] such as the roundworm *Ascaris lumbricoides* or the beef tapeworm *Taenia saginata*, can be selectively targeted with drugs very easily and usually present no problem to chemotherapy.

Table 11.4 summarizes some drugs that are used in the treatment of worm infections.

11.7 ANTIVIRAL CHEMOTHERAPY

In contrast to all other human pathogens, viruses don't have a cellular structure. Instead, they consist of an RNA or DNA genome,[21] packaged into a shell of proteins, which is called the *capsid*. This assembly—the nucleocapsid—may or may not be surrounded by a lipid membrane envelope containing another set of viral proteins.[22] Virus particles lack any kind of metabolic or biosynthetic activity; virus multiplication therefore can proceed only inside host cells.[23]

In the life cycle of a virus, one can discern the following major stages: (1) adhesion of the virus to a specific receptor on the target cell; (2) uptake of the virus particle by the cell, often into a vesicle, from which it later escapes into the cytosol; (3) uncoating of the nucleic acids, that is shedding of the envelope, if present, and the capsid proteins; (4) replication of the genome and protein synthesis; (5) assembly of progeny virus particles from newly synthesized nucleic acids and proteins; and (6) budding, that is, the release of the newly formed virus particles from the host cell. Figure 11.17 illustrates the life

[20]The life cycle of tapeworms involves two hosts, the terminal host and the intermediate host. The parasite resides inside the intestine of the terminal host but inside muscle or other tissues in the intermediate host. Humans are terminal host to *T. saginata* but intermediate host to *E. granulosus*.

[21]Mutation rates in RNA viruses are higher in DNA viruses. This will also affect the rate of emergence of resistance to antiviral agents.

[22]Viruses without an envelope, such as poliovirus and hepatitis A virus, are typically quite hardy and persist in the environment, from which they are thus usually contracted. Enveloped viruses, in contrast, are usually more sensitive and depend on direct transmission between hosts or vectors; examples are hepatitis B and C, HIV, and influenza virus. Enveloped viruses are also susceptible to detergents or alcoholic disinfectants. These dissolve the envelope's lipid membrane and thereby strip the virus of its surface proteins that it needs to gain entry into host cells.

[23]In plants, there is another class of acellular pathogens, the *viroids*. These consist of naked nucleic acids only and do not even have a capsid, nor do they encode any proteins whatsoever. In a sense, therefore, they are the ultimate parasites.

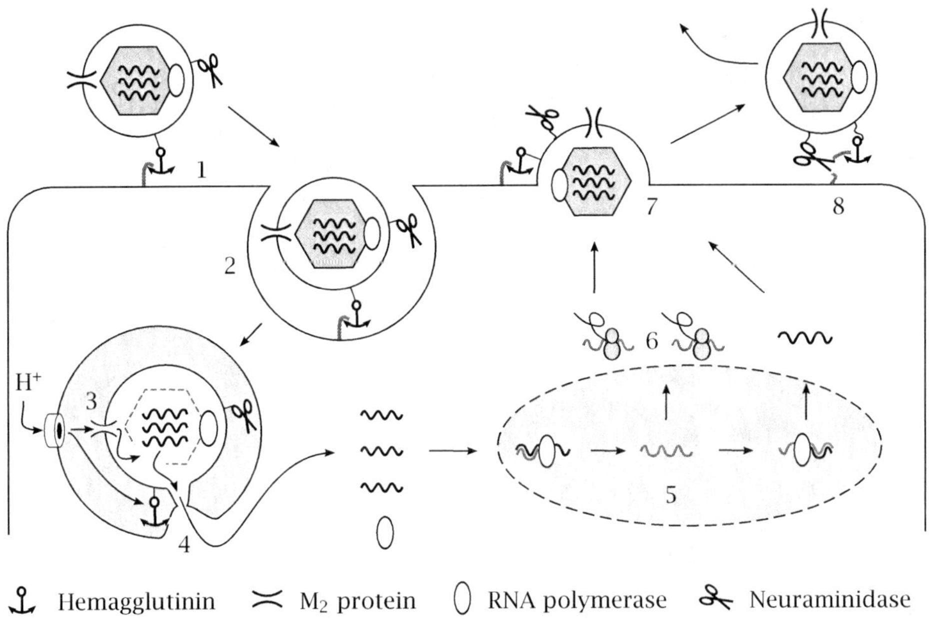

Figure 11.17 Life cycle of influenza virus. The virion attaches to the target cell via its hemagglutinin surface protein (1). It is taken up by endocytosis (2), and the cell pumps protons into the endocytotic vesicle. Protons entering the virus particle through the M_2 channel (3) protein strip the viral RNAs of the capsid proteins. Acidification of the vesicle also triggers the membrane fusion activity of hemagglutinin (4), which releases the viral RNAs and RNA polymerase from the endosome. Both enter the nucleus (5), where the viral RNA is transcribed and replicated. Plus strand RNA transcripts are translated in the cytosol (6). Progeny virions assemble and bud at the cytoplasmic membrane (7), from which they are released by neuraminidase (8).

cycle of influenzavirus. With other viruses, the individual steps may take place within different cellular compartments.

Most viruses proceed through all these stages without delay or interruption; the entire cycle then is completed within no more than a few hours. Some viruses, however, can persist inside host cells; the viral nucleic acid will then either exist independently inside the nucleus, as is the case with herpes and papilloma viruses, or be integrated into the host cell's genome, as happens with retroviruses such as human immunodeficiency virus (HIV). Such dormant stages exhibit little or no protein expression and therefore are well protected both from the immune system and from chemotherapy. Herpes simplex virus or cytomegalovirus, for example, are amenable to chemotherapy only during an acute infection or a relapse, but not in the intervals between relapses.

11.7.1 Inhibitors of virus entry into host cells

Entry of a virus into a host cell is initiated by its binding to the host cell membrane. While the viral protein involved in this interaction exists just for this purpose, the cellular receptor is a molecule that primarily has an entirely different biological role and is merely abused by the virus. Blocking adherence

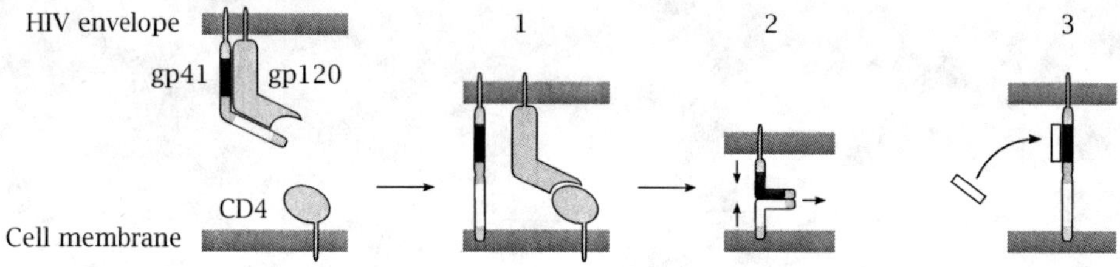

Figure 11.18 Mode of action of enfuvirtide, a peptide inhibitor of HIV fusion with target cells. Two HIV surface glycoproteins, gp41 and gp120, initially are associated with each other. Binding of gp120 to its cellular receptor CD4 and a coreceptor (not shown) releases the fusion protein gp41, the tip of which inserts into the target cell membrane (1). Two complementary heptad repeat motifs in gp41 then zip up against each other, causing the two membranes to converge (2) and eventually fuse. The soluble peptide enfuvirtide resembles one of these repeats and associates with the other to disrupt gp41 activity (3).

of a virus to its cellular receptor can, in principle, be disrupted by blocking the binding sites either on the viral surface protein or on the cellular receptor. This approach is appealing, since both proteins are readily accessible. Small molecule inhibitors of attachment are mostly experimental at present. However, antiviral antibodies act by the same principle. They may be employed just like drugs, for example, to prevent the outbreak of rabies after an animal bite.[24]

When an enveloped virus enters the host cell, the viral envelope fuses with the host cell membrane. This is initiated by the virus through a fusion protein, an integral membrane protein that is embedded in the viral envelope and binds and inserts into the host cell membrane. The process of fusion has been characterized in detail for HIV, and this has resulted in the development of enfuvirtide, a peptide inhibitor of fusion that is clinically effective. The process of HIV fusion is outlined in Figure 11.18, which is reasonably accurate except for the fact that the fusion protein gp41 is a trimer. Accordingly, oligomeric peptide inhibitors have shown superior affinity and effectiveness in preclinical experiments [7].

With herpes simplex virus, membrane fusion is suppressed by the fatty alcohol docosanol ($C_{22}H_{45}OH$). This inhibition apparently involves metabolic incorporation of the alcohol into the phospholipids of the host cell membranes [30,31]. Docosanol cannot be used systemically but is useful for local application of herpes skin lesions (cold sores).

With influenza virus, membrane fusion and uncoating are triggered by the acidification of the endocytotic vesicle by the host cell. The viral envelope contains a proton-conducting channel, the M_2 protein, which permits the protons to proceed from the endocytotic vesicle into the virion and initiate uncoating of the viral RNA molecules. Entry of protons into the virion and uncoating is

[24]Rabies immunization is typically combined passive and active, that is, both antibodies and inactivated virus are applied. Active immunization is effective even after the animal bite has occurred, because rabies virus travels to the brain by retrograde axonal transport and therefore requires an unusually long incubation time, which usually suffices for the development of antibodies.

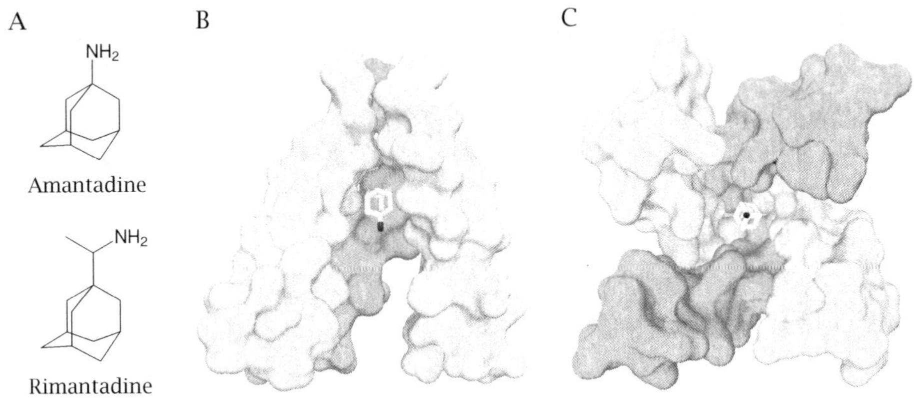

Figure 11.19 Mode of action of rimantadine and amantadine (α-adamantylamine). The influenzavirus membrane protein M_2 is a proton-activated proton channel. Inside the acidified endocytotic vesicle, the channel opens, and protons penetrating the virus particle trigger uncoating of the viral RNA. Amantadine and rimantadine block the channel. **A:** Structures of amantadine and rimantadine. **B:** Side view of amantadine bound in the cavity of the channel. One of the four subunits has been removed. **C:** View from inside the virus particle. (B and C rendered from 3c9j.pdb [42].)

inhibited by the drugs amantadine and rimantadine, which plug up the lumen of the M_2 proton channel (Figure 11.19). Acidification also triggers the membrane fusion activity of hemagglutinin, which permits the viral RNAs and RNA polymerase to enter the cytosol (Figure 11.17).

11.7.2 Inhibitors of virus genome replication

While viruses rely on cellular enzymes for most of their biosynthesis, each virus possesses a nucleic acid polymerase, which it uses for the replication of its own genomic RNA or DNA. DNA viruses possess DNA polymerases, whereas most RNA viruses contain RNA-dependent RNA polymerases. Retroviruses such as HIV are an exception; they contain a *reverse transcriptase* that transcribes the viral RNA into DNA, which then integrates into the host cell's genome.[25]

An obvious commonality of the diverse polymerases is their use of nucleotide substrates, and nucleotide analogs accordingly are the most widely used viral polymerase inhibitors. Nucleotide analogs are typically incorporated into the growing nucleic acid molecule and then act as chain terminators, for example, due to the lack of a 3′-hydroxy group on the deoxyribose-mimicking moiety,[26] as is the case with the antiretroviral drug dideoxyadenosine (Figure 11.20).

Fully phosphorylated nucleotides are subject to degradation by phosphatases and also poorly penetrate cells. Therefore, most nucleotide analogs are

[25]Integration is mediated by another viral protein, the integrase. HIV integrase inhibitors have become available [15].

[26]The very same effect underlies the use of dideoxy-nucleotides in the Sanger chain termination method of DNA sequencing.

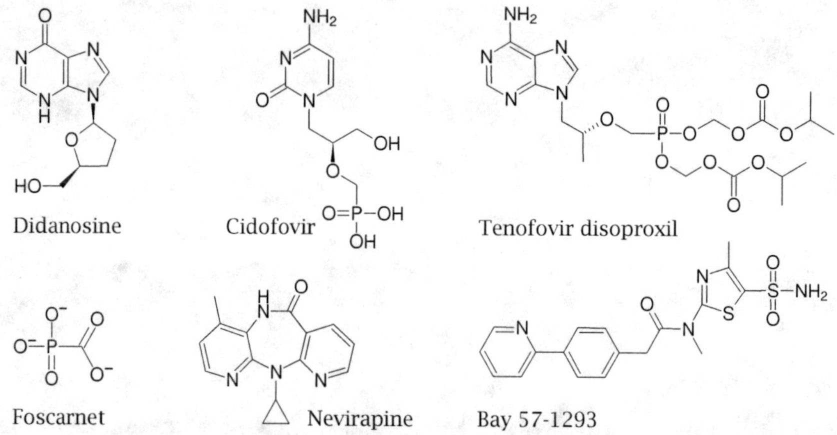

Figure 11.20 Inhibitors of virus genome replication. Didanosine and tenofovir disoproxil are chain terminators, and nevirapine is an allosteric inhibitor of HIV reverse transcriptase. Didanosine is phosphorylated by cellular kinases. Tenofovir disoproxil is a resorption ester of a phosphonate analog of adenosine. Cidofovir, which is active against many viruses, likewise carries a phosphonate group and therefore bypasses the first phosphorylation. Foscarnet also inhibits several viral polymerases; it is a hydrolysis-resistant analog of pyrophosphate. Bay 57-1293 is an experimental inhibitor of herpes virus helicase/primase. See text for further details.

actually applied as unphosphorylated nucleoside prodrugs.[27] Once inside the cell, nucleoside analogs must undergo three successive phosphorylations for conversion into their active triphosphate forms. This can be accomplished by cellular kinases or, if present, by viral ones. Viruses of the herpes group[28] contain nucleoside phosphotransferases that provide the polymerases with an increased pool of nucleotides. The inhibitor acyclovir is phosphorylated by herpes simplex virus nucleoside kinase but not by the cellular enzyme. Cellular kinases then convert the monophosphate to the triphosphate, which inhibits the viral DNA polymerase (Figure 11.22).

The sequential interaction of acyclovir with two viral enzymes, the nucleoside kinase and the DNA polymerase, results in a high degree of selective toxicity, and acyclovir has fewer side effects than many other nucleoside analogs. Unfortunately, herpes virus strains have acquired resistance due to loss of the nucleoside kinase, which is not essential for the virus, and to point mutations in the DNA polymerase that result in decreased affinity for acyclovir triphosphate.

Like acyclovir, the drug cidofovir contains an acyclic pseudo ribose moiety; however, this moiety also carries a phosphonate group, which is resistant to hydrolysis and bypasses the need for the initial phosphorylation by the viral

[27]Some nucleoside analogs will also interfere with cellular polymerases, which will be most significant in rapidly proliferating cell types such as blood cell precursors in the bone marrow. Reduced blood cell counts therefore are a prominent side effect of such drugs. Some nucleoside analogs with strong effects on cellular DNA polymerases such as cytosine arabinoside are used in tumor therapy (see Section 12.4.1).

[28]This group includes herpes simplex virus 1 and 2, varicella zoster virus, cytomegalovirus, Epstein-Barr virus, the causative agent of mononucleosis, and human herpes virus type 6.

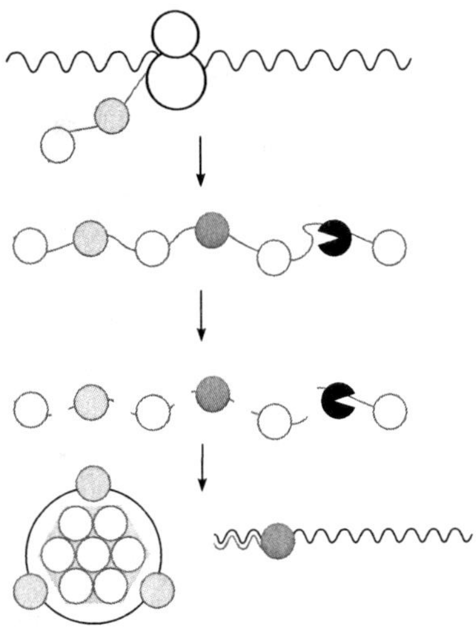

Figure 11.21 Function of virus proteases. Viruses such as hepatitis C virus and HIV translate their genome into polyproteins. One polyprotein component is a protease that cleaves itself and the other components, which then part ways to serve their respective roles in virus replication and assembly.

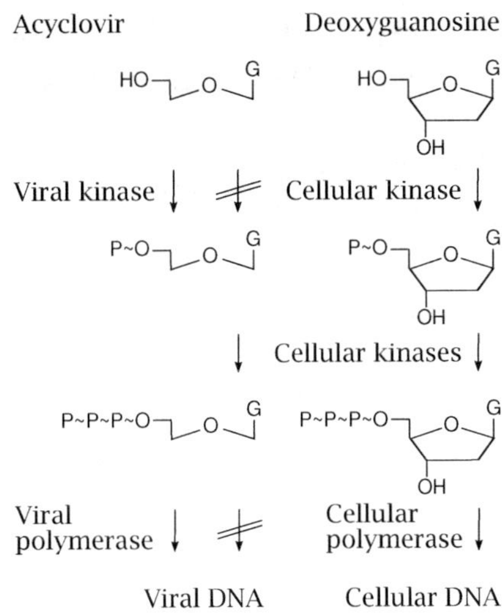

Figure 11.22 Activation of acyclovir. The free drug is monophosphorylated by herpes virus thymidine kinase but rejected by cellular nucleoside kinases. Cellular nucleotide kinases convert the monophosphate to the triphosphate, which is a substrate for the viral but not the cellular DNA polymerase.

enzyme (the attachment of two phosphate groups by cellular kinases is still required). This makes cidofovir useful not only against herpes viruses that have acquired resistance to acyclovir but also against a remarkably broad range of unrelated DNA viruses including pox viruses and hepatitis B virus. Some cidofovir congeners also are effective against HIV [4].

An extraordinarily simple molecule that acts as an inhibitor of herpes and retroviral DNA polymerases is foscarnet, or phosphonoformic acid. It resembles pyrophosphate, which is produced by all nucleic acid polymerases during each nucleotide addition cycle. To ready the enzyme for the next cycle, pyrophosphate needs to be hydrolyzed. Foscarnet occupies the binding site for pyrophosphate on the polymerase. Since it is resistant to hydrolysis, it stalls the polymerase [24].

Foscarnet and nucleoside analog phosphonates such as cidofovir are polar molecules. Organic esters of these molecules have been developed to improve their bioavailability and cellular uptake.

While single-stranded viral nucleic acids function as substrates for polymerases without any preliminary steps, the replication of double-stranded viral genomes begins with the formation of a helicase–primase complex. The drug Bay 57-1293 is an inhibitor of herpes simplex helicase.

11.7.3 Inhibitors of virus maturation

In addition to the replication of the viral genome, the formation of progeny virions involves transcription and translation of the virus genome. The viral proteins may be encoded singly and by separate genes, or they may be encoded and expressed jointly as *polyproteins*, which are then processed into the mature separate proteins by a viral protease (Figure 11.21). These proteases are again encoded by the virus and are sufficiently different from host enzymes to be suitable as drug targets for chemotherapy. The fairly rapid development of protease inhibitors for HIV (Figure 1.7), whose clinical introduction significantly improved the prognosis of HIV infection, is an impressive example of modern methods of drug discovery. The drug telaprevir inhibits the protease of hepatitis C virus.

Cytomegalovirus (CMV), which is a member of the herpes group, contains UL97, an unusual kinase that phosphorylates both nucleosides and one or more viral proteins. Its phosphorylation of the nucleoside analog ganciclovir, which is structurally similar to acyclovir, confers sensitivity of CMV replication to that drug. The drug maribavir resembles other nucleoside analogs in structure but is actually an inhibitor of UL97 kinase itself. The inhibition of UL97 protein kinase activity interferes with the packaging of viral DNA into preassembled virus capsids. Apart from mutations in UL97 itself, mutations in several other viral genes have been connected with resistance to maribavir.

Cytomegalovirus is also inhibited by an oligonucleotide termed *fomivirsen*. The sequence of fomivirsen is complementary to a certain viral gene that is transcribed in an early stage of viral multiplication. Like other oligonucleotides, fomivirsen is pharmacokinetically challenged; however, this problem can be circumvented by local application in eye infections caused by the virus [13].

With influenza virus, the hemagglutinin surface protein mediates attachment of a parental virion to neuraminic acid moieties on its target cell. When newly formed progeny virions bud from the cell, however, they will again stick to neuraminic acid; the viral neuraminidase enzyme, which is also located in the virus envelope, will at this stage enzymatically cleave the neuraminic acid residue and free the virions from the originating cell (Figure 11.17). In the presence of neuraminidase blockers such as oseltamivir (Figure 11.23), the virions will remain stuck on the expired host cell and not be able to propagate the infection to neighboring cells.[29]

11.7.4 Interferons

Interferons are cytokines that are produced in response to, among other things, viral infection. When released by a virus-infected cell, they induce in adjacent cells a variety of anti-viral proteins, including an RNAse that will cleave viral

[29]Neuraminic acid occurs not only on cell surfaces but also in extracellular proteins in bronchial mucus, where it may function as a decoy for the virus. The activity of neuraminidase must strike a balance that will permit the virus to adhere long enough for entry but also enable liberation and departure; this explains why its catalytic rate is fairly slow. Paramyxoviruses such as measles virus or mumps virus combine the hemagglutinin and neuraminidase activities on a single protein.

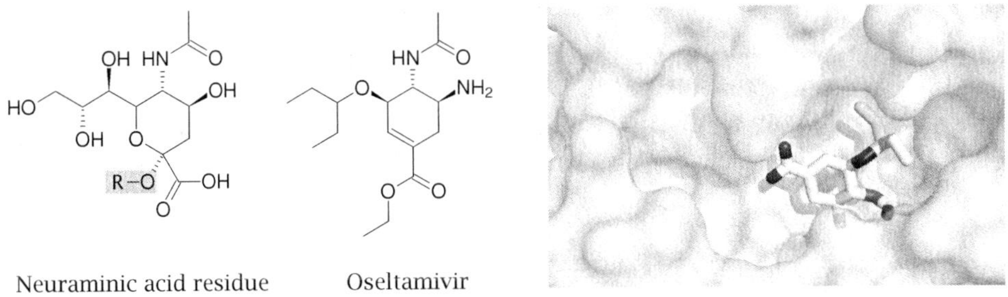

Figure 11.23 Binding of the inhibitor oseltamivir to influenzavirus neuraminidase. The enzyme cleaves neuraminic acid from the free end of a cell surface oligosaccharide, represented by "R" in the structure of neuraminic acid. Oseltamivir is a transition state analog. It only occupies part of the active site; with oligosaccharide substrates, the adjacent groove accommodates the sugar moieties next to neuraminic acid. (Rendered from 1hu4.pdb [35]. A colored version of this figure is shown in the insert in the middle of this book.)

RNA and a protein kinase that will inactivate elongation factor 2, thus shutting down protein synthesis. Interferons therefore have an important role in slowing down the potentially explosive multiplication of viruses before a specific immune reaction can be mounted. Recombinantly expressed interferons are used in chronic viral infections such as hepatitis B and C.

STUDY QUESTIONS

11.1 The plot below shows isobolograms (see Question 2.2) for growth inhibition of the Gram-negative bacterium *Pseudomonas aeruginosa* by carbenicillin, in combination with EDTA (A [14]) and with polymyxin B (B [10]), respectively.

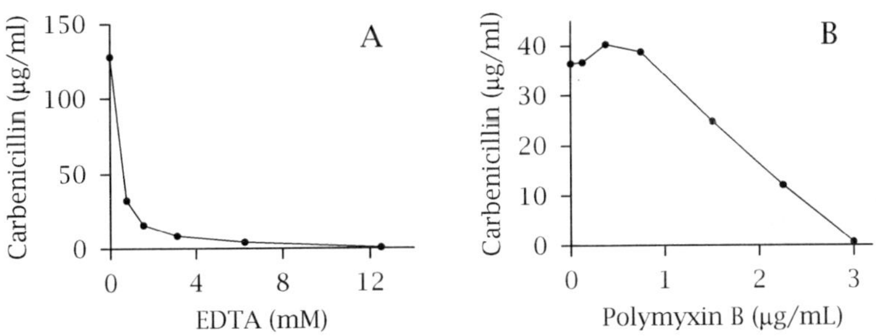

What do the shapes of the curves indicate regarding the synergistic or antagonistic effects of these combinations? Can these effects be understood from what we know about the modes of action of these agents?

11.2 While virtually all antimicrobial compounds are affected by the development of resistance, the rate of such development can vary widely. With two compounds discussed in this chapter, namely, fosfomycin (Section 11.4.2) and flucytosine (Section 11.5), development of resistance under therapy is so rapid that they cannot be used alone but only in combination with other drugs. What is the common motif between fosfomycin and flucytosine that causes resistance to develop much more rapidly than, for example, with penicillin or ketoconazole?

11.3 Figure 11.19 shows amantadine bound inside the M_2 proton channel of influenzavirus. Binding appears to be mediated by the aliphatic carbon moiety of the drug alone, whereas the amino group makes no connection with the protein. Nevertheless, the functionally equivalent drug rimantadine also has an amino group, which is indeed functionally important. Can you explain how the amino groups will help both drugs to reach their destination?

REFERENCES

[1] E. Breukink and B. de Kruijff, "Lipid II as a target for antibiotics," *Nat. Rev. Drug. Discov.* 5 (2006), 321–332.

[2] J. Czub, E. Borowski, and M. Baginski, "Interactions of amphotericin B derivatives with lipid membranes--a molecular dynamics study," *Biochim. Biophys. Acta* 1768 (2007), 2616–2626.

[3] B. D. Davis, L. L. Chen, and P. C. Tai, "Misread protein creates membrane channels: an essential step in the bactericidal action of aminoglycosides," *Proc. Natl. Acad. Sci. U. S. A.* 83 (1986), 6164–6168.

[4] E. De Clercq, "Clinical potential of the acyclic nucleoside phosphonates cidofovir, adefovir, and tenofovir in treatment of DNA virus and retrovirus infections," *Clin. Microbiol. Rev.* 16 (2003), 569–596.

[5] E. De Rossi, J. A. Aínsa, and G. Riccardi, "Role of mycobacterial efflux transporters in drug resistance: an unresolved question," *FEMS Microbiol. Rev.* 30 (2006), 36–52.

[6] J. Delmas et al., "Prediction of the evolution of ceftazidime resistance in extended-spectrum β-lactamase CTX-M-9," *Antimicrob. Agents Chemother.* 50 (2006), 731–738.

[7] J. J. Dwyer et al., "Design of helical, oligomeric HIV-1 fusion inhibitor peptides with potent activity against enfuvirtide-resistant virus," *Proc. Natl. Acad. Sci. U. S. A.* 104 (2007), 12772–12777.

[8] U. Eckstein-Ludwig et al., "Artemisinins target the SERCA of Plasmodium falciparum," *Nature* 424 (2003), 957–961.

[9] D. I. Edwards, "Nitroimidazole drugs--action and resistance mechanisms. I. Mechanisms of action," *J. Antimicrob. Chemother.* 31 (1993), 9–20.

[10] T. C. Eickhoff, "In vitro effects of carbenicillin combined with gentamicin or polymyxin B against Pseudomonas aeruginosa," *Appl Microbiol.* 18 (1969), 469–473.

[11] M. E. Falagas and S. K. Kasiakou, "Colistin: the revival of polymyxins for the management of multidrug-resistant gram-negative bacterial infections," *Clin. Infect. Dis.* 40 (2005), 1333–1341.

[12] M. E. Falagas et al., "Toxicity after prolonged (more than four weeks) administration of intravenous colistin," *BMC Infect. Dis.* 5 (2005).

[13] R. S. Geary, S. P. Henry, and L. R. Grillone, "Fomivirsen: clinical pharmacology and potential drug interactions," *Clin. Pharmacokinet.* 41 (2002), 255–260.

[14] G. F. Gerberick and P. A. Castric, "In vitro susceptibility of Pseudomonas aeruginosa to carbenicillin, glycine, and ethylenediaminetetraacetic acid combinations," *Antimicrob. Agents Chemother.* 17 (1980), 732–735.

[15] D. Hazuda, M. Iwamoto, and L. Wenning, "Emerging pharmacology: inhibitors of human immunodeficiency virus integration," *Annu Rev Pharmacol Toxicol* 49 (2009), 377–394.

[16] R. R. Hiebsch, T. J. Raub, and B. W. Wattenberg, "Primaquine blocks transport by inhibiting the formation of functional transport vesicles. Studies in a cell-free assay of protein transport through the Golgi apparatus," *J. Biol. Chem.* 266 (1991), 20323–20328.

[17] C. Hoffmann et al., "Disclosure of the mycobacterial outer membrane: cryo-electron tomography and vitreous sections reveal the lipid bilayer structure," *Proc. Natl. Acad. Sci. U. S. A.* 105 (2008), 3963–3967.

[18] M. P. Hudock et al., "Inhibition of Trypanosoma cruzi hexokinase by bisphosphonates," *J. Med. Chem.* 49 (2006), 215–223.

[19] J. W. Johnson et al., "Cyclobutanone analogues of beta-lactams revisited: insights into conformational requirements for inhibition of serine- and metallo-beta-lactamases," *J. Am. Chem. Soc.* 132 (2010), 2558–2560.

[20] C. Lam et al., "Membrane-disorganizing property of polymyxin B nonapeptide," *J. Antimicrob. Chemother.* 18 (1986), 9–15.

[21] A. Lee et al., "Interplay between efflux pumps may provide either additive or multiplicative effects on drug resistance," *J. Bacteriol.* 182 (2000), 3142–3150.

[22] I. Letunic and P. Bork, "Interactive Tree Of Life (iTOL): an online tool for phylogenetic tree display and annotation," *Bioinformatics* 23 (2007), 127–128.

[23] Y. Ling et al., "Bisphosphonate inhibitors of Toxoplasma gondi growth: in vitro, QSAR, and in vivo investigations," *J. Med. Chem.* 48 (2005), 3130–3140.

[24] B. Marchand, E. P. Tchesnokov, and M. Götte, "The pyrophosphate analogue foscarnet traps the pre-translocational state of HIV-1 reverse transcriptase in a Brownian ratchet model of polymerase translocation," *J. Biol. Chem.* 282 (2007), 3337–3346.

[25] S. T. Micek, "Alternatives to vancomycin for the treatment of methicillin-resistant Staphylococcus aureus infections," *Clin. Infect. Dis.* 45 Suppl 3 (2007), S184–S190.

[26] V. Molteni et al., "Identification of novel potent bicyclic peptide deformylase inhibitors," *Bioorg. Med. Chem. Lett.* 14 (2004), 1477–1481.

[27] T. Naito et al., "Aminoglycoside antibiotics. IX. 1-N-acyl derivatives of kanamycin A (amikacin analogs)," *J. Antibiot. (Tokyo)* 27 (1974), 851–858.

[28] K. Neupert-Laves and M. Dobler, "The crystal structure of a K^+ complex of valinomycin," *Helv. Chim. Acta* 58 (1975), 432–442.

[29] M. Palmer, "Efflux of cytoplasmically acting antibiotics from gram-negative bacteria: periplasmic substrate capture by multicomponent efflux pumps inferred from their cooperative action with single-component transporters," *J. Bacteriol.* 185 (2003), 5287–5289.

[30] L. E. Pope et al., "Anti-herpes simplex virus activity of n-docosanol correlates with intracellular metabolic conversion of the drug," *J. Lipid Res.* 37 (1996), 2167–2178.

[31] L. E. Pope et al., "The anti-herpes simplex virus activity of n-docosanol includes inhibition of the viral entry process," *Antiviral Res.* 40 (1998), 85–94.

[32] P. Pristovsek and J. Kidricleu, "Solution structure of polymyxins B and E and effect of binding to lipopolysaccharide: an NMR and molecular modeling study," *J. Med. Chem.* 42 (1999), 4604–4613.

[33] W. G. Prowse et al., "Conformation of A82846B, a glycopeptide antibiotic, complexed with its cell wall fragment: an asymmetric homodimer determined using NMR spectroscopy," *Biochemistry* 34 (1995), 9632–9644.

[34] D. Ro et al., "Production of the antimalarial drug precursor artemisinic acid in engineered yeast," *Nature* 440 (2006), 940–943.

[35] R. J. Russell et al., "The structure of H5N1 avian influenza neuraminidase suggests new opportunities for drug design," *Nature* 443 (2006), 45–49.

[36] N. S. Shah et al., "Extensively drug-resistant tuberculosis in the United States, 1993-2007," *JAMA.* 300 (2008), 2153-2160.

[37] M. V. Shilova and C. Dye, "The resurgence of tuberculosis in Russia," *Philos. Trans. R. Soc. Lond. B Biol Sci.* 356 (2001), 1069–1075.

[38] S. M. Simon and G. Blobel, "A protein-conducting channel in the endoplasmic reticulum," *Cell* 65 (1991), 371–380.

[39] S. M. Simon and G. Blobel, "Signal peptides open protein-conducting channels in E. coli," *Cell* 69 (1992), 677–684.

[40] A. Singh and P. J. Rosenthal, "Selection of cysteine protease inhibitor-resistant malaria parasites is accompanied by amplification of falcipain genes and alteration in inhibitor transport," *J. Biol. Chem.* 279 (2004), 35236–35241.

[41] I. K. Srivastava and A. B. Vaidya, "A mechanism for the synergistic antimalarial action of atovaquone and proguanil," *Antimicrob. Agents Chemother.* 43 (1999), 1334–1339.

[42] A. L. Stouffer et al., "Structural basis for the function and inhibition of an influenza virus proton channel," *Nature* 451 (2008), 596–599.

[43] D. Sulton et al., "Clavulanic acid inactivation of SHV-1 and the inhibitor-resistant S130G SHV-1 β-lactamase. Insights into the mechanism of inhibition," *J. Biol. Chem.* 280 (2005), 35528–35536.

[44] P. M. Tarantino et al., "Inhibitors of DNA polymerase III as novel antimicrobial agents against gram-positive eubacteria," *Antimicrob. Agents Chemother.* 43 (1999), 1982–1987.

[45] T. Teerlink, B. de Kruijff, and R. A. Demel, "The action of pimaricin, etruscomycin and amphotericin B on liposomes with varying sterol content," *Biochim. Biophys. Acta* 599 (1980), 484–492.

[46] M. M. Thapar et al., "Pharmacodynamic interactions among atovaquone, proguanil and cycloguanil against Plasmodium falciparum in vitro," *Trans. R. Soc. Trop. Med. Hyg.* 97 (2003), 331–337.

[47] A. Uhlemann et al., "A single amino acid residue can determine the sensitivity of SERCAs to artemisinins," *Nat. Struct. Mol. Biol.* 12 (2005), 628–629.

[48] M. Vaara and T. Vaara, "Sensitization of Gram-negative bacteria to antibiotics and complement by a nontoxic oligopeptide," *Nature* 303 (1983), 526–528.

CHAPTER 12

TUMOR CHEMOTHERAPY

Tumors arise from uncontrolled growth of cells. If this growth remains confined to a single site and limited by a clearly defined anatomical barrier, the tumor is considered benign, and it can usually be cured by surgical removal. Malignant tumors, in contrast, invade and infiltrate their surroundings. In advanced stages, they shed cells that are carried by the blood or lymph streams to remote locations, where they then may form secondary foci of growth, called *metastases.*

In the treatment of most malignant tumors, surgical resection is important, too. It is often followed up with irradiation of the excised tumor's surroundings in order to suppress recurrence of the tumor through renewed growth of any tumor cells that may have been left behind. Chemotherapy frequently complements irradiation and is intended to snuff out the growth of tumor cells that may have settled in distant organs.

In contrast to solid malignant tumors, leukemias and many lymphomas grow diffusely right from the beginning, and chemotherapy is therefore the primary form of treatment.[1] Of the various forms of therapy, only chemotherapy

[1]Most solid tumors are carcinomas, that is, of epithelial origin; the less common sarcomas arise from connective and muscle tissues. Lymphomas, which can be solid or diffuse, arise from lymphatic cells; leukemias grow diffusely and are derived from blood cell precursors.

concerns us here; however, it is worth noting that many antitumor drugs act by causing DNA damage, which is also the mode of action of ionizing radiation.[2]

Malignant tumors, while not as diverse as pathogenic microbes, still have enough biochemical diversity to require a made-to-measure approach to chemotherapy.[3] This diversity is in part inherited from their tissues of origin. Some retained organ-specific characteristics afford opportunities for selective toxicity. An example is the dependence of many breast cancers on female sexual hormones, which makes them susceptible to estrogen or progestin hormone antagonists. Diversity also arises directly from the most important immediate cause of tumor formation, that is, from *somatic mutations*. In some specific cases, one decisive mutation induces formation of a tumor and sustains its growth, and the tumor may then be treated by targeting this single mutated protein. More commonly, however, the tumor becomes manifest only after it has already acquired a multitude of mutations, which will cause tumors originating within the same tissue to differ from one another. Typically, some of these mutations have increased the mutation rate itself or caused other kinds of genetic instability; this will cause cells originating within the same tumor to diverge from one another. When chemotherapy is applied, this genetic variability favors the emergence of cell clones with partial or complete drug resistance.

As was discussed in Chapter 11 regarding antimicrobial chemotherapy, a useful strategy to counter the emergence of drug resistance is the simultaneous application of drugs with different targets. In some forms of lymphoma and testicle carcinoma, the optimization of combination therapy protocols has yielded remarkable improvements in cure rates, to the point where even advanced cases are cured 80–90% of the time. However, some other types of tumors have largely withstood such efforts, and sometimes chemotherapy can achieve only marginal benefits.

Keeping in mind that tumors are diverse, we will first consider some general principles that are relevant to tumor chemotherapy and then consider instructive examples of antitumor drugs.

12.1 SOME PRINCIPLES OF TUMOR BIOLOGY

Melanoma is a highly malignant tumor derived from the pigment cells of the skin. It often develops from certain types of pigment spots (called *dysplastic naevi*) that therefore constitute a *precancerous lesion*. In its early stage, a melanoma will only grow within the epithelial layer of the skin, without invading the connective tissue underneath. Such early tumor stages are referred to as *carcinoma in situ*, that is, cancer confined within the normal tissue demarcations.

When the tumor breaks the barrier to the connective tissue beneath, it marks the transition to invasiveness, and is often followed by the formation of

[2]The term *radiomimetic* is sometimes applied to this type of drug action.

[3]Careful and detailed individual planning is essential with all forms of tumor therapy and has been widely practiced for a long time. The recent hijacking of the term "personalized medicine," which now supposedly refers to therapeutic diversification according to molecular-biological test results only, betrays a profound ignorance of real medicine on the part of the hijackers.

metastases in distant parts of the body. Concomitantly with the progression of malignant behavior, organ-specific features in individual cells and in tissue organization tend to get lost. In the case of melanoma, there are amelanotic variants that have lost the otherwise distinct and diagnostically helpful feature of pigment production. In advanced cases, it is not uncommon to see pigmented and nonpigmented metastases in the same patient.

As one would expect, prognosis changes greatly as the tumor progresses along this path. Surgical removal is curative at the precancerous and carcinoma in situ stages, but invasive melanoma is very aggressive and difficult to treat. When removing a melanoma by surgery, it is very important to apply the cut at a safe distance from the tumor, because otherwise there is a risk of tumor cells remaining or being carried into the bloodstream and then settling in other organs to form metastatic tumors.

The behavior of melanoma cells stands in striking contrast to that of normal skin cells. A cut or puncture through healthy skin will virtually never give rise to a metastatic lump of skin cells in the lung or elsewhere, although very likely healthy skin cells can be pried loose and disseminated by trauma just like tumor cells can. Contemplating this difference is a good starting point to understand some of the key principles of tumor formation and chemotherapy.

12.1.1 Disinhibited cell growth

In contrast to the fairly unrestrained growth of single-celled organisms, the cells of higher organisms such as ourselves are thoroughly hog-tied, and each cell is given very explicit and stringent rules about when to grow, what shape to take, and how long to live. Law-abiding cells will proliferate only while they receive sufficient and specific stimulation by diffusible growth factors, specific matrix macromolecules, and cell-to-cell adhesion molecules; most of these stimuli are received through specific cell surface receptors. Deprived of these stimuli, a well-behaved cell will promptly commit harakiri through a process known as *apoptosis* (Figure 12.1). Further to our example above, it is through this latter provision that the growth of displaced normal skin cells into disconnected, metastatic lumps of skin tissue is avoided.[4]

It follows, then, that in tumorous cells—both benign and malignant—the requirement for one or more specific growth stimuli, or the responsiveness to some growth-inhibiting signals must have been abolished. Tumor cells often express growth factor receptors that have been mutated to be constitutively active, that is, active even without binding of their cognate growth factors. Alternatively, a mutation toward constitutive activity may have affected some downstream element that relays the signal from the growth factor receptor on the cell surface to the intranuclear triggers of proliferation.

Reduced dependence on extrinsic growth stimulation may also result from mutations that change the level of expression of a protein rather than its struc-

[4]Apoptosis that is triggered through loss of cell-cell and cell-matrix contact is referred to as *anoikis* (Greek, for homelessness). Some cells such as macrophages, whose function requires them to travel through different tissues, are not subject to anoikis.

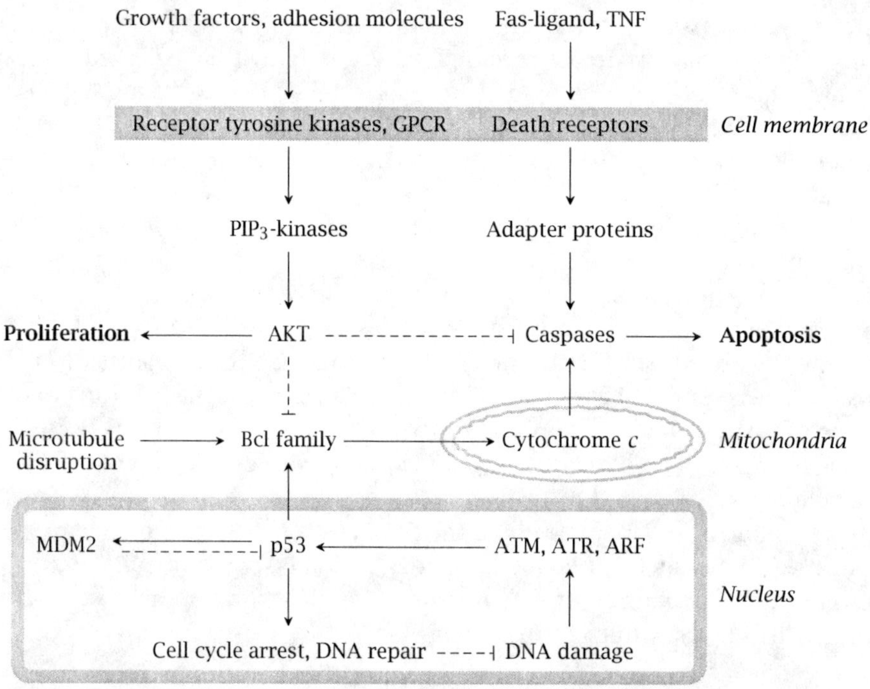

Figure 12.1 Outline of cellular pathways that control proliferation and apoptosis. Growth factors and adhesion molecules stimulate the PIP_3-kinase pathway and thereby stimulate proliferation and inhibit apoptosis. The extrinsic pathway to apoptosis is triggered by proteins such as Fas-ligand and tumor necrosis factor (TNF) and activates *caspases*, a family of cysteine proteases that cleave their numerous substrates at aspartate residues, and thereby destroy the cell. In the intrinsic pathway of apoptosis, DNA damage leads to the activation of protein kinases such as ATM and ATR, which in turn activate the transcriptional regulator p53. Subsequently, increased expression of activating regulatory proteins in the Bcl family leads to the permeabilization of mitochondria by Bax, another Bcl family member. Cytochrome *c* released from the mitochondria then again activates caspases. Mutations in many of the proteins shown here have been implicated in the induction or progression of cancer. Note that this figure does *not* include all signaling pathways relevant to malignancy—it is merely a sketch, intended to provide some context for the drugs discussed in this chapter.

ture. Dysregulation of gene expression may, for example, result in the *autocrine* secretion of a required growth factor by the tumor cell itself. Changes in the level of gene expression may result from *translocation*, that is through reciprocal exchange of fragments between two chromosomes, which will place the genes close to the fault lines of translocation into a new regulatory context. Indeed, chromosomal translocations are involved in the causation of many solid tumors and leukemias. As the examples of chronic myeloid leukemia and of promyeolocyte leukemia show, translocation may also result in the formation chimeric genes that promote growth factor-independent cell proliferation.

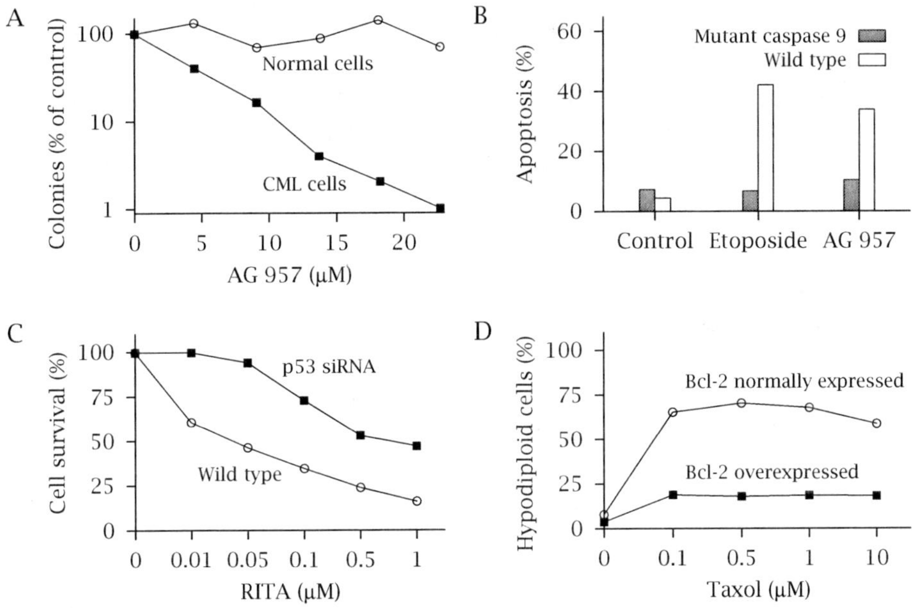

Figure 12.2 Sensitivity of tumor cells to apoptosis, and experimental models of inhibited apoptosis. **A**: Apoptotic response of white blood cells isolated from a leukemia (CML) patient and of normal control cells to the tyrosine kinase inhibitor AG 957. **B**: Expression of a dominant-negative mutant of caspase 9 in a leukemic cell line (K562) inhibits apoptosis in response to the topoisomerase II inhibitor etoposide and to AG-957. **C**: The experimental compound RITA [2,5-bis-(5-hydroxymethyl-2-thienyl)-furan] inhibits the binding of the inhibitory protein MDM2 to p53, thus increasing the proapoptotic activity of p53. Inhibiting the expression of p53 with a specific siRNA in a model cell line reduces apoptosis. **D**: The drug taxol disrupts microtubules, which then release an activating regulator of the Bcl protein family. Overexpression of the inhibitory regulator protein Bcl-2 reduces apoptosis, quantified here by FACS counting of "hypodiploid cells," which are fragments of nuclei released from apoptotic cells. (Figure panels redrawn from Refs. [28,17,18].)

12.1.2 Genetic instability and dysregulation of apoptosis

If we consider the huge number of cells that arise within a human body during its lifetime, which collectively must carry an enormous number of potentially dangerous mutations, we must marvel at the fact that benign or malignant tumors are not more common; after all, most of us do not experience cancer at all, or only late in our lives. This rarity of cancer is largely due to cellular mechanisms of tumor suppression. Various enzyme systems continually scan the DNA for mutagenic damage and attempt to repair it. The importance of DNA repair in the suppression of tumors is evident from the fact that *ataxia telangiectasia*, a hereditary defect in the ATM protein kinase that activates DNA repair (see Figure 12.1), greatly increases the risk of developing tumors. DNA repair is also linked to the regulation of cell proliferation and survival. Activation of DNA repair will stall cell division until repair is completed; activation beyond a certain threshold will trigger apoptosis, that is, the cell will undergo voluntary death.[5]

Apart from response to genetic damage, the proliferation of normal somatic cells is also limited by the telomerase mechanism (see Section 13.1.3) and by what is called *cellular senescence*, which will invoke apoptosis after a certain number of somatic cell divisions. It is clear, then, that apoptosis itself is the major effector mechanism that restrains the proliferation of cells, and that inside tumor cells apoptosis must be suppressed in order to permit their unrestrained growth. This conclusion is borne out by the observation that p53, a *tumor suppressor* protein and transcription factor that plays a central role in activating both DNA repair and apoptosis in response to intracellular triggers, is affected by mutations in as many as 50% of all malignant tumors.[6]

Yet, one single mutation does not a tumor make. Even a complete genetic knockout of p53 in mice does not result in the induction of tumors; once induced, however, tumors in p53-deficient mice undergo accelerated progression to malignancy [19]. Therefore, if we use the traditional distinction of tumor *induction* and tumor *progression*, mutations of p53 belong in the latter category, whereas tumor-inducing mutations are typically related to independence from extrinsic growth stimulation. Moreover, the fact that the p53 knockout mice develop normally tells us that the *extrinsic* pathway to apoptosis, which is triggered from *death receptors* on the cell surface (see Figure 12.1), remains functional without p53, since it plays an essential role in embryonic development. In addition to mutations of p53, tumor cells often acquire mutations affecting the extrinsic apoptosis pathway. Figure 12.2 summarizes some in vitro experiments that illustrate the inhibition of apoptosis by several such mutations.

Even after acquiring the abilities to detach, invade, and multiply independently from extrinsic growth regulation, a tumor cell still depends on the host organism for nutrient supply. Without the formation of new blood vessels, or *angiogenesis*, a growing lump of tumor cells would soon run into limitations of nutrient and oxygen supply. Accordingly, tumor cells frequently acquire mutations that cause the secretion of angiogenetic growth factors.[7]

It becomes clear, then, that a full-fledged tumor cell must be armed with several strategic mutations, which are acquired successively as the tumor advances through the stages toward complete malignancy. For this combination of traits to arise with any degree of likelihood at all, *the mutation rate itself must be increased*, which in turn is achieved by mutations in one or more genetic repair systems. The accompanying increased level of genetic damage should

[5]This latter reaction leads to the wholesale destruction of bone marrow cells after whole body irradiation, which can be used to condition recipients of bone marrow transplants.

[6]It has been proposed that inactive mutants of p53 could be reverted to activity by specific drugs, and an in vitro model study on one mutant has been carried out [5]. While this approach is ingenious and elegant, it remains to be seen whether it can succeed in practice.

[7]Cell degeneration due to shortage of blood supply is indeed often seen in the center of larger primary tumors. The irregular vascularization of tumor tissue may hamper the homogeneous distribution of anticancer drugs. On the other hand, tumor microvasculature tends to be leaky toward macromolecules, which can be exploited for selective drug delivery (see Section 14.3.1).

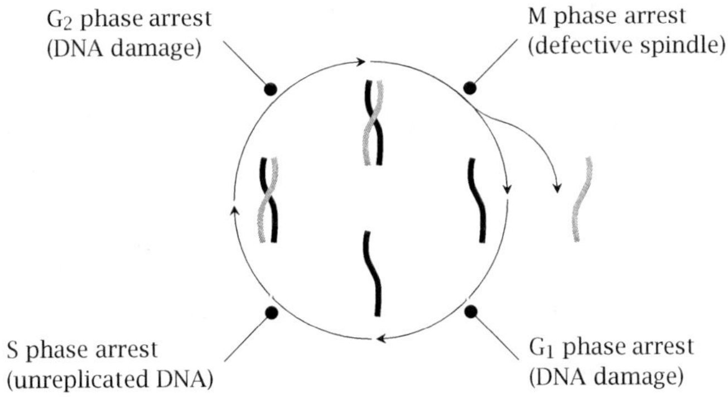

Figure 12.3 The cell cycle and its checkpoints. Cell proliferation undergoes coordinated phases. A new cell that has arisen by mitosis—referred to as M phase within the cell cycle—prepares for the next division in the G_1 phase. It enters the S phase, in which the DNA is replicated, that is, a second double-stranded DNA copy is synthesized for each chromosome. After another short resting (G_2) phase, mitosis repeats. In each of the major phases, specific checkpoint proteins are activated that will halt the progress of the cycle if a specific form of damage is detected. Depending on the extent of damage, mitosis may resume after DNA repair, or apoptosis will be triggered.

normally lead to apoptosis, but it will be tolerated once the activation threshold of apoptosis has been raised, again by mutation. In this way, mutations that compromise genetic stability and apoptosis, respectively, encourage and reinforce one another.

12.1.3 Dysregulation of cell division

Cell proliferation goes through several stages that together are referred to as the *cell cycle* (Figure 12.3). Anticancer drugs that inhibit cell proliferation as such, regardless of cell type, may act during specific phases of the cell cycle or may be active throughout the cycle.

After emerging from mitosis, the newly formed daughter cells enter into a resting state, the G_0 phase.[8] The G_0 phase may be of very short duration in rapidly proliferating tissues; on the other hand, it may last a lifetime in cells that have reached their terminal point of differentiation. It tends to be shorter in tumors than in related normal tissues, although cells in the bone marrow and many mucous membranes have very high rates of proliferation and therefore very short G_0 phases as well. Throughout the G_0 phase and the subsequent G_1 phase, no DNA replication takes place, so that there is only one double-stranded DNA molecule per chromosome. DNA replication occurs in the S (synthesis) phase, which is followed by another brief resting phase, the G_2 phase. Mitosis occurs in the M phase, and the cycle repeats.

Each transition from one phase to the next is controlled by regulatory proteins called *cyclins*, which activate cognate protein kinases. These cyclin-dependent kinases are inhibited by proteins such as p16 or, indirectly, by the

[8]The G in G_0 represents "gap." This phase is filled, of course, with all kinds of interesting cell-specific activities; it is a gap phase only with respect to cell proliferation.

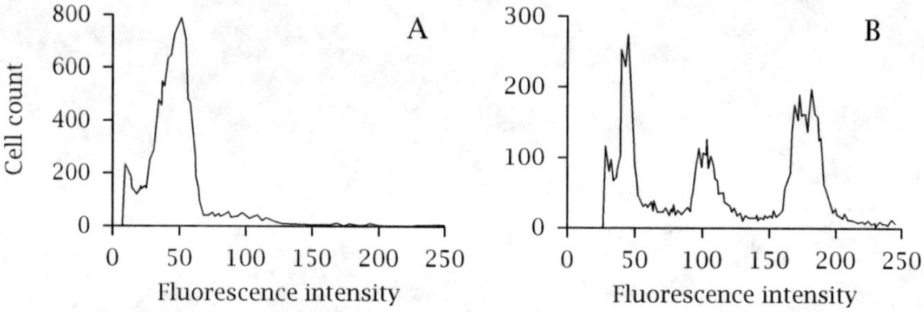

Figure 12.4 Progressive aneuploidy in cells of a recurring tumor of connective tissue (sarcoma), detected by flow cytometry. Cell nuclei were isolated and incubated with a fluorescent dye such as propidium iodide, which emits fluorescence only after intercalation into DNA. The stained nuclei were then passed through a laser beam one by one. Each nucleus causes one fluorescence pulse, the intensity of which is proportional to its DNA content. **A:** In the primary tumor, the DNA content per cell shows one major peak, corresponding to the diploid chromosome complement. **B:** In the recurring tumor, there are multiple peaks, indicating aneuploidy and clonal divergence. (Figure redrawn from original data in Ref. [24].)

retinoblastoma (Rb) protein. Inactivation of p16, Rb, or related proteins through mutation will increase the rate of proliferation; such mutations are found in many tumors. Mutations of Rb in the germline or early in embryonic development cause retinoblastoma, a tumor of the eye that occurs in early life and from which the Rb protein and the related retinoblastoma regulatory pathway derive their names. Similarly, gain of function mutations in cyclin proteins will deregulate proliferation and are found in many different tumors, too [33].

At the conclusion of the G_1 phase and again at the end of the G_2 phase, genome integrity checkpoints are reached, at which the current level of DNA damage is evaluated in a kind of molecular triage. With low levels of damage, the cycle proceeds. Higher levels of damage will pause the cycle and activate DNA repair, whereas very high levels of damage will trigger apoptosis. If mitosis is allowed to proceed, disaster may still strike if the segregation of individual chromosomes fails. Unchecked, this would give rise to aneuploid daughter cells. However, there are checkpoints in mitosis as well, which will abort mitosis and again activate apoptosis.

Mutational inactivation of the checkpoint proteins is another mechanism that permits tumor cells to accumulate higher loads of DNA damage and therefore increase the genetic instability and variability. Inactivation of mitosis checkpoints will permit the propagation of aneuploidy (Figure 12.4), which will accelerate genetic divergence of the tumor and, through gene dosage effects, may further growth dysregulation and drug resistance.

The increased genetic mutability of tumor cells is manifest both in a high number of point mutations and in gross genetic changes such as chromosome deletions, duplications, translocations, and nondisjunctions, to the point that deviations from the diploid complement of chromosomes become the norm and can be used as a diagnostic criterion of malignancy. Genetic variability will lead to clonal divergence and facilitate the emergence of cell clones resistant to anticancer drugs, which may be further hardened by selection during drug therapy.

12.1.4 Most antitumor drugs induce apoptosis

Most traditional antitumor drugs act either through inflicting outright damage on DNA or by stalling DNA replication and mitosis, which will then trigger apoptosis through the mechanisms outlined above. Having just convinced ourselves that apoptosis is compromised in tumor cells, this seems paradoxical. After all, anticancer drugs can be useful only if they are more toxic to tumor cells than to normal ones, which implies that on the whole *tumor cells must be more prone to apoptosis than normal cells.* How can this conundrum be resolved?

There are only partial answers to this question. Firstly, malignant cells that have left their native tissue environment are deprived of the growth stimulation that normal, well-behaved cells enjoy. To the extent that they remain dependent on extrinsic growth signals, this should expose them to increased apoptotic stimulus. Secondly, deficient cell cycle checkpoints may allow mitoses to progress at full speed into disaster ("mitotic catastrophe"). Finally, and more speculatively, apoptosis may be promoted by competition of tumor cells among themselves. The loss of proper growth regulation within a tumor must give rise to a resource-starved, chaotic, and lawless environment. Individual cells competing to survive here will engage in an evolutionary race to the bottom, accumulating as many growth-enhancing mutations as their remaining capacity for apoptosis will bear. Any leeway gained through mutations that inhibit apoptosis will promptly be consumed by a new wave of genetic derangement, until the surviving "fittest" cells will again teeter on the brink of apoptosis.

Even if we don't yet fully understand the causes of increased susceptibility to apoptosis in tumor cells, it thus turns out that the seemingly crude and unsophisticated action mode of most traditional anticancer drugs is indeed aimed remarkably well at a fundamental feature of malignancy. The focus on basic cellular targets, as opposed to tissue- or organ-specific ones, makes such drugs broadly applicable, and it also denies the tumor cells any gain of resistance from their foundational mutations that are related to tissue-specific growth signaling. On the other hand, their modes of action render these drugs highly toxic for noncancerous cells, too. Hence, drugs with a more targeted mode of action and lower degree of toxicity are an attractive proposition, even if they may be applicable only with very specific types of tumors.

12.1.5 Tumor stem cells

The differentiated cells that are characteristic of each tissue are derived from stem cells that make up only a small fraction of the total cell population. A stem cell may undergo either symmetric mitosis, which simply produces two stem cells, or differential mitosis, which yields one stem cell and one partially or fully differentiated, organ-specific cell that may have reduced or altogether lacking proliferative potential. It appears that stem cells also occur within tumors, and while they again make up only a small fraction of the total cell mass, they may be important in the recurrence of tumors and the formation of metastases; the characterization of drug susceptibility and resistance of these stem cells may therefore be a useful exercise [11]. One must keep in mind, however, that genetic instability and the general loss of proper growth regulation may diminish the central role of stem cells in proliferation; the distinction between stem cells and differentiated cells may become blurred in advanced stages of malignancy.

12.2 CELL-TYPE-SPECIFIC ANTITUMOR DRUGS

Specific tumors may retain from their respective tissues of origin some physiological traits that provide targets for selective chemotherapy. The most common case is the dependence on growth stimulation by specific hormones. Many breast cancers, for example, remain dependent on estrogens or progestins for growth. Estrogen receptor antagonists such as tamoxifen (Figure 12.5), or progestin receptor antagonists such as mifepristone (see Figure 7.9) will therefore stop or delay growth of these cells. Of course, these drugs will affect other hormone-dependent tissues as well. Mifepristone, for example, also disrupts the function of the placenta in pregnancy and is used for early abortion.

Among all classes of steroid hormones, estrogens are unique in possessing an aromatic A ring within the sterol scaffold. Formation of this aromatic ring from a nonaromatic precursor is catalyzed by the enzyme *aromatase*, which belongs to the cytochrome P450 family. Inhibition of aromatase depletes estrogens and can be used against estrogen-dependent breast cancers as well. There are reversible and irreversible aromatase inhibitors. Several of the latter have reactive double bonds or leaving groups attached to the B ring, suggesting reaction with a nucleophile in the active site. Exemestane is a covalent inhibitor, and letrozole is a noncovalent one (Figure 12.6). Letrozole bears some resemblance to the triazole family of antifungal drugs, which inhibit a fungal cytochrome P450 sterol demethylase enzyme (Section 11.5). The substrates of aromatase are androgen hormones; for example, estradiol is produced from testosterone. Androgens are required by prostate cancers, and androgen antagonists accordingly are used in their treatment.

Another example of cell type-specific chemotherapy is the use of mitotane in cancer of the cortex of the adrenal glands (Figure 12.7). Mitotane is a prodrug and is activated by 11-β-hydroxylase.[9] This enzyme, a member of the cytochrome P450 family, is located in the mitochondria of the adrenal gland cortical cells and participates in the synthesis of cortisol (Section 7.4.1). Activation of mitotane by 11-β-hydroxylase yields an acyl chloride, which in turn

[9]The enzyme hydroxylates the carbon 11 of the sterol scaffold. Other substrates include metabolites of the insecticide dichlorodiphenyltrichloroethane (DDT), which is very similar to mitotane in structure; this may again result in toxicity.

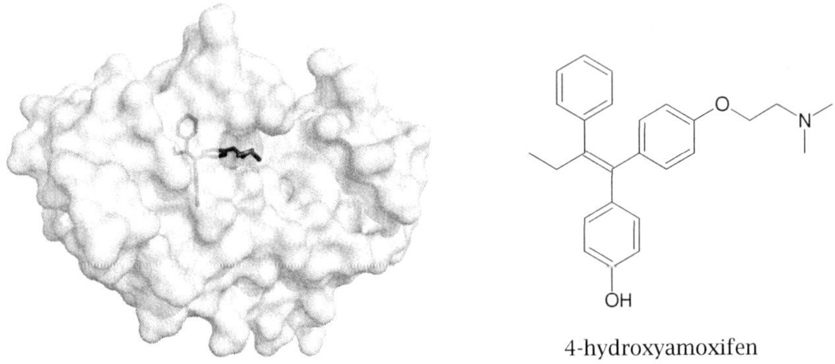

4-hydroxyamoxifen

Figure 12.5 Structure of the estrogen receptor antagonist 4-hydroxytamoxifen, lodged inside estrogen-related receptor γ. The formula of the compound, which is an active metabolite of the drug tamoxifen, is rendered in an orientation similar to that in the molecular structure. (Structure rendered from 2p7z.pdb [1].)

reacts with macromolecules in the mitochondria and induces irreversible cell damage.

Hairy cell leukemia is a rare form of lymphatic leukemia.[10] Proliferation of the malignant cells in this disease is very efficiently inhibited by interferon-α. This is one of the few cases so far in which a purely "biological" tumor therapy actually works.

Tissue-specific toxicity may also result from selective cellular uptake. The antibiotic streptozotocin contains an amino-sugar moiety and enters cells through the GLUT2 glucose transporter [26]. This transporter is highly expressed in pancreatic islet β cells, and the drug can therefore be used to treat tumors derived from this cell type. It is also used to destroy β cells in experimental animal models of type 1 diabetes. Streptozotocin contains a nitrosourea group, which alkylates DNA. The structure of streptozotocin and the reaction mechanism of the nitrosourea group are shown in Figure 12.15.

Among all organs, the thyroid gland alone utilizes large amounts of iodide, which it incorporates into the hormones tri- and tetraiodothyronine (see Section 7.3). Benign tumors of this gland, and many malignant ones as well, retain the capacity to accumulate iodide by active transport. Accumulation of radioactive iodine isotopes can be used both diagnostically and therapeutically. Therapy is performed with ^{131}I, which is a β-emitting radioisotope with a half-life of approximately 8 days. While the thyroid gland is fairly superficially located and therefore, in principle, within easy reach of conventional radiation therapy, the advantage of ^{131}I is that it will also reach metastases in remote locations, as long as they continue to accumulate iodide. Apart from its mode of targeting, the therapeutic effect of ^{131}I resembles that of conventional exterior irradiation.

[10]Leukemias are broadly classified according to the lines of blood cell precursors they arise from. Lymphatic leukemias, which derive from precursor cells of lymphocytes, and myeloid leukemias, which originate from precursors of granulocytes and monocytes, are most common. The finer points of leukemia classification would have any medieval scholastic overcome with joy but are not considered here, although they are certainly important for optimization of therapy.

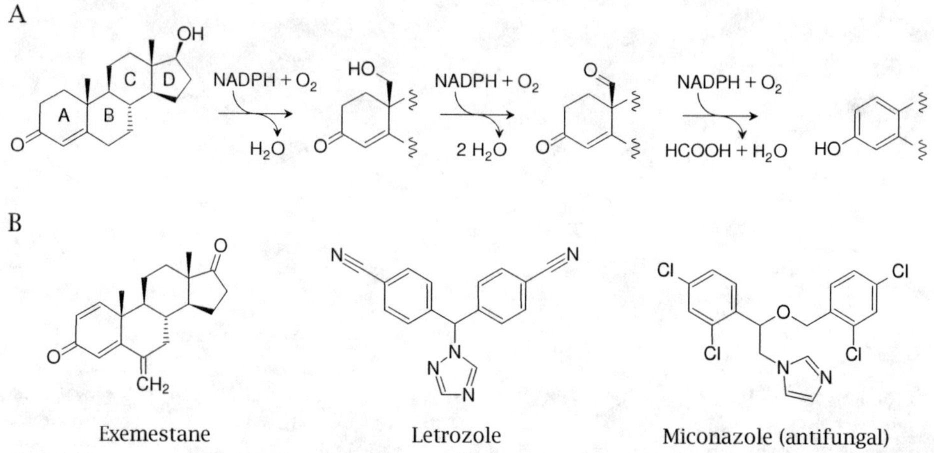

Figure 12.6 Aromatase as an antitumor target. The enzyme converts testosterone to estradiol, which is required for growth by many breast cancers. **A:** The aromatase reaction. The methyl group between rings A and B is converted in two steps to an aldehyde. The keto group in ring A then becomes its enol tautomer, which creates a second double bond in the ring; the third one is introduced concomitantly with oxidative cleavage of the exocyclic aldehyde as formic acid. **B:** Aromatase inhibitors for cancer therapy. Exemestane is a steroidal, covalent inhibitor. Its exocyclic carbon–carbon double bond likely reacts with a nucleophile in the active site. Letrozole is a nonsteroidal, noncovalent inhibitor. It is somewhat similar in structure to the antifungal drug miconazole, an inhibitor of 14-α sterol demethylase, which, like aromatase, belongs to the cytochrome P450 family. (Figure 4.5 shows the related drug ketoconazole bound to a cytochrome P450 enzyme.)

While high dosages of ^{131}I are used to cure thyroid cancer, lower amounts of the isotope can actually *induce* it. ^{131}I is a major product of ^{235}U nuclear fission and as such forms in nuclear reactors. The amounts of ^{131}I released by the Chernobyl reactor meltdown in 1986 caused a significant number of new cases of thyroid cancer [31]. Generally speaking, secondary cancers or leukemias can occur after treatment with other radioactive isotopes, radiotherapy, or DNA-alkylating ("radiomimetic") agents. With overly aggressive tumor therapy, the risk of secondary malignancy may outweigh the gains in cure rates of the primary one.

12.3 DRUGS THAT TARGET SPECIFIC ONCOPROTEINS

Some tumors and leukemias are induced by single, specific *oncogenes* that are translated to their corresponding *oncoproteins.* The first such oncogenes were discovered in the genomes of retroviruses that cause malignancies in animals; examples are the *v-src* gene in Rous sarcoma virus and the *v-abl* gene in Abelson leukemia virus. These two oncogenes encode mutated, dysregulated protein tyrosine kinases that interfere in the regulation of cellular proliferation.

Similar deviant protein kinases can arise through somatic mutations or chromosomal translocations. One example is the *bcr-abl* chimeric tyrosine kinase

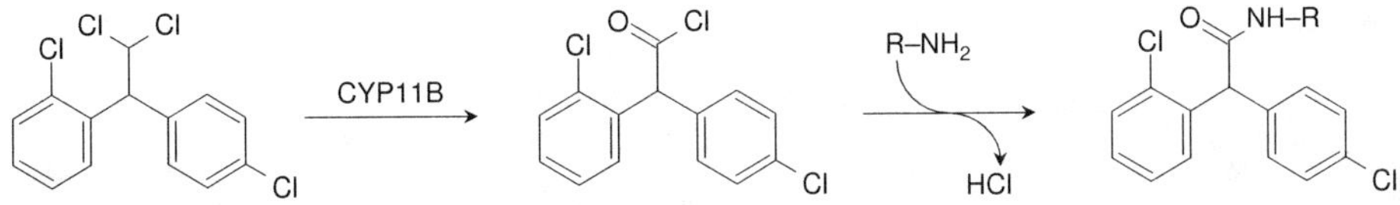

Figure 12.7 Mitotane is converted to an acyl chloride by the cytochrome P450 enzyme 11B (CYP11B). The acyl chloride then reacts with amino groups or other nucleophiles in proteins or nucleic acids. CYP11B is highly expressed in the mitochondria within the cortex of the adrenal gland, rendering this tissue and the tumors derived from it selectively susceptible to mitotane.

[10], which results from a specific reciprocal translocation between chromosomes 9 and 22 that gives rise to the so-called Philadelphia chromosome.[11]

If this translocation occurs in bone marrow stem cells, it gives rise to chronic myeloid leukemia (CML), a *myeloproliferative disease* in which a spectrum of leukocyte subtypes is produced at a dysregulated, enhanced pace. This condition is similar to a benign tumor of a solid organ. After several years, this benign behavior will give way to a *blast crisis* that behaves like an acute leukemia, with the usual signs of malignancy.[12] In the myeloproliferative stage of CML, it is possible to control the disease solely through pharmacological inhibition of the *bcr-abl* kinase. The first effective inhibitor of *bcr-abl* kinase, and indeed the first of all antitumor tyrosine kinase inhibitors, was imatinib. The drug competitively blocks the ATP binding site of the enzyme (Figure 15.7). As with antimicrobial drugs, selection of resistant mutants may occur under therapy. Again as with antimicrobials, the structure of the drug molecule can be varied to combat this mode of resistance. The inhibitor dasatinib, for example, remains active against several imatinib-resistant mutants of the kinase [30].

The target site of imatinib is located within a functional domain that is conserved and shared among a large number of protein tyrosine kinases, many of which are involved in some way in the regulation of cell growth. In addition to *abl*, imatinib acts on at least two other kinases, including the platelet-derived growth factor receptor, which is mutated toward constitutive activity in many epithelial tumors. Moreover, *bcr-abl* or other constitutively active variants of the *abl* kinase also occur in tumors other than CML, and imatinib is useful in these tumors, too [20]. Several more tyrosine kinase inhibitors have now been approved for clinical use or are in advanced clinical testing [3]. Their targets include both intracellular and cell-membrane associated kinases. Among the latter, there are several growth factor receptors that are mutated in many types of carcinoma.

While the binding sites of imatinib and related drugs are on the cytosolic side of the membrane, the extracellular receptor domains are accessible to antibodies. Several monoclonal antibodies that inhibit growth factor receptors have been introduced into clinical practice and have in some cases proven quite suc-

[11] The *abl* gene involved here is the same that also occurs as the virally transmitted variant in the aforementioned Abelson leukemia virus.

[12] The term "blast" has nothing to do with explosives but instead refers to the similarity of the leukemic cells to immature leukocyte precursors such as myeloblasts.

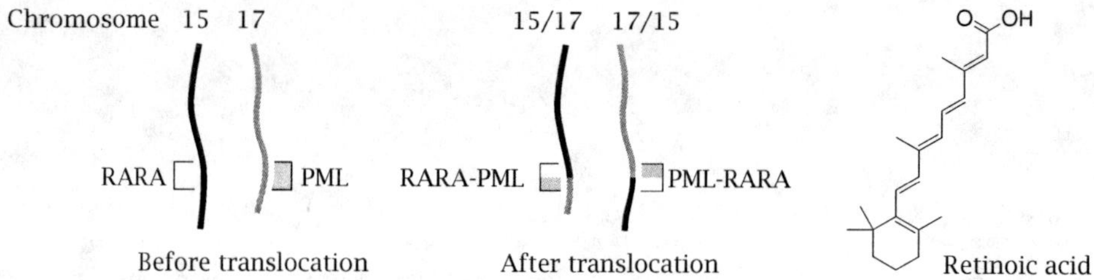

Figure 12.8 Causation of promyelocytic leukemia by reciprocal translocation of chromosomes 15 and 17. The two genes PML and RARA span the fault line of the translocation; the latter encodes the retinoic acid receptor α. Translocation generates the PML-RARA oncoprotein, an in-frame fusion between the N-terminal portion of the PML gene product and the C-terminal part of the retinoic acid receptor. This molecule dimerizes and causes deviant transcriptional regulation, leading to uncontrolled cell proliferation.

cessful. An example is trastuzumab, a monoclonal antibody directed against an epithelial growth factor receptor termed HER2/neu or ErbB-2. Overexpression of this receptor is frequent in breast cancer, and these cases are now commonly treated with the antibody. The same receptor is also targeted by the kinase inhibitor lapatinib, which has shown clinical benefits [25] but is not yet routinely used.

Like CML, promyelocytic leukemia is caused by specific chromosomal translocation, this time involving chromosomes 15 and 17 (Figure 12.8). The rift of this translocation goes right through the gene for α-retinoic acid receptor (RARA), which acts as a transcriptional regulator in the final differentiation of the promyelocytes to granulocytes. As a result of translocation, RARA is expressed in two deviant forms. The one that retains the C-terminal end of the retinoic acid receptor is denoted as PML-RARA. It no longer properly fulfills the function of active RARA, but it dimerizes and binds to DNA on its own, thus interfering with proper gene regulation. It also recruits and thereby may deplete the co-receptor retinoid X receptor (RXR), which is needed by the remaining proper RARA molecules that are expressed from the second, intact copy of chromosome 15.

The disruption of RARA signaling by the structurally and functionally deviant receptors leads to an arrest of cell differentiation at the promyelocyte stage. Promyelocytes proliferate, while granulocytes don't; arrest of differentiation therefore leads to the buildup of a proliferating yet useless and even harmful cell population. However, it is possible to restore differentiation in the aberrant cells by the application of retinoic acid in larger than physiological amounts. Interestingly, binding of retinoic acid to PML-RARA promotes its degradation by proteolysis; this reportedly leads to the destruction of an actively proliferating subpopulation of the leukemic cells that is important in sustaining the disease [23].

Resistance is a common occurrence with all of the above-mentioned cell type- or oncoprotein-specific drugs and antibodies. Some cells in the tumor may have lost the requisite organ-specific features to begin with, or resistant clones may emerge under therapy. These drugs therefore usually have to be combined or followed up with one of the general, non-cell-specific cytotoxic therapies discussed below.

12.4 CYTOTOXIC ANTITUMOR DRUGS

Most conventional anticancer drugs inhibit cell proliferation by damaging cellular DNA or by inhibiting cell division. This may seem to be a crude mode of action, and it is; application of these drugs is accompanied by the widely known side effects of tumor chemotherapy such as depressed blood cell counts, immunosuppression, and hair loss. These side effects result from the inhibited proliferation of normal tissues that have high rates of mitosis, often higher than those of the tumors being treated. The selective toxicity of these drugs for tumor cells is based not so much on rates of mitosis as it is due to the tumor cells' greater tendency to undergo apoptosis (Section 12.1.4).

Antiproliferative drugs may act at different stages of the cell cycle. We can distinguish the following major functional classes: (1) antimetabolites, (2) inhibitors of DNA topoisomerase, (3) inhibitors of mitosis, and (4) DNA-damaging agents.

Antimetabolites inhibit the synthesis of DNA or of its precursors and therefore affect the S phase of the cell cycle, as do the topoisomerase inhibitors. Mitosis inhibitors interfere with the assembly of the mitotic spindle that brings about the physical separation of the chromatides and act in the M phase of the cycle.

DNA-damaging agents cause covalent modification of bases, single or double strand breaks, or crosslinks within the cellular DNA in a cumulative fashion throughout the cell cycle. The lethal effect occurs when the cell cycle runs up against a DNA integrity checkpoint, at which time a high level of damage will trigger apoptosis.

All these drugs cause the typical, well known short-term "chemo" side effects. Because of their strong mutagenic action, the DNA-damaging agents cause the most significant long-term damage, while antimetabolites and antimitotic agents are less toxic. Nevertheless, DNA-damaging agents are an indispensable element of combination therapy, because they act throughout the cell cycle and therefore better reach those tumor cells that are dormant at the time of therapy.

12.4.1 Antimetabolites

Many antimetabolites in clinical use are analogues of purine or pyrimidine bases or nucleosides. We will consider a few representative examples.

5-Fluorouracil. The fluorine atom found in 5-fluorouracil (5-FU) is slightly larger than the hydrogen found in uracil but smaller than the methyl group that occurs in the same position in thymine. 5-FU mimics both thymine and uracil in various enzymatic reactions. Activation to 5-fluorodeoxyuridine monophosphate (5-FdUMP) is initiated by enzymes in the pyrimidine salvage pathway, whose normal function it is to recover bases released in the degradation of nucleic acids (Figure 12.9A). 5-FdUMP irreversibly inhibits thymidylate synthase, which robs the cell of its biosynthetic capacity for thymidine and therefore inhibits DNA synthesis.

The catalytic mechanism of thymidylate synthase is depicted in Figure 12.9B. Thymidylate synthase requires *N,N'*-methylenetetrahydrofolic acid as a cosubstrate. The reaction is initiated by a cysteine residue in the active site of the enzyme and goes through an intermediate stage in which the enzyme, the substrate (dUMP), and the cosubstrate are all covalently connected. This adduct is resolved in the following step, which involves abstraction of a proton from position 5 of the uracil ring by a basic residue in the active site. The trick with 5-FU is that this abstraction cannot happen, since position 5 is occupied by fluorine, which holds on very tightly to its bond electrons. Therefore, the enzyme remains locked up in this covalent intermediate and is irreversibly inactivated.

Another intriguing consequence of the fluorine substitution of uracil is that it favors the iminol tautomer of the uracil ring, which has base-pairing properties that resemble those of cytosine (Figure 12.9C). Synthetic incorporation of 5-FU—subsequent to further phosphorylation of 5-FdUMP to 5-FdUTP—into DNA therefore induces mutations due to misincorporation of guanine instead of adenine.

The bromine and iodine analogues of 5-FU are more efficiently incorporated into DNA, since their larger substituents are similar in size to a methyl group and they therefore resemble thymine more closely. The iodo analogue—in its deoxyriboside form, dubbed *idoxuridine*—is used in cancer therapy, whereas 5-bromodeoxyuridine has been widely used as a mutagen in experimental genetic research.

Dihydrofolate reductase inhibitors. In the discussion of 5-fluorouracil, we saw that *N,N'*-methylenetetrahydrofolate (methylene-THF) functions as a cosubstrate in the synthesis of thymidylate. It also donates methyl groups in the synthesis of purine bases. Its crucial role in the synthesis of DNA precursors makes methylene-THF a good target for cytotoxic chemotherapy.

After methylene-THF donates a methyl group in the thymidylate synthase reaction, the remainder is regenerated in two steps. The first step is the reduction of dihydrofolate to tetrahydrofolate by dihydrofolate reductase. This enzyme is inhibited by methotrexate (Figure 12.10). Unlike 5-fluorouracil or cytosine arabinoside (see text below), methotrexate does not directly interact with the DNA and thus is not mutagenic or carcinogenic. It therefore is more suitable for long-term application and is used to induce immunosuppression in autoimmune diseases such as multiple sclerosis, myasthenia gravis, Crohn's disease, or lupus erythematosus. The inhibitor pemetrexed competes not only at the dihydrofolate reductase enzyme but also at enzymes in purine biosynthesis [9]. It is therefore less affected by DHFR gene amplification, which is a common tumor cell resistance mechanism.

You may recall that dihydrofolate reductase inhibitors are also useful in the chemotherapy of infections (see Figure 11.2). Most inhibitors are used only in

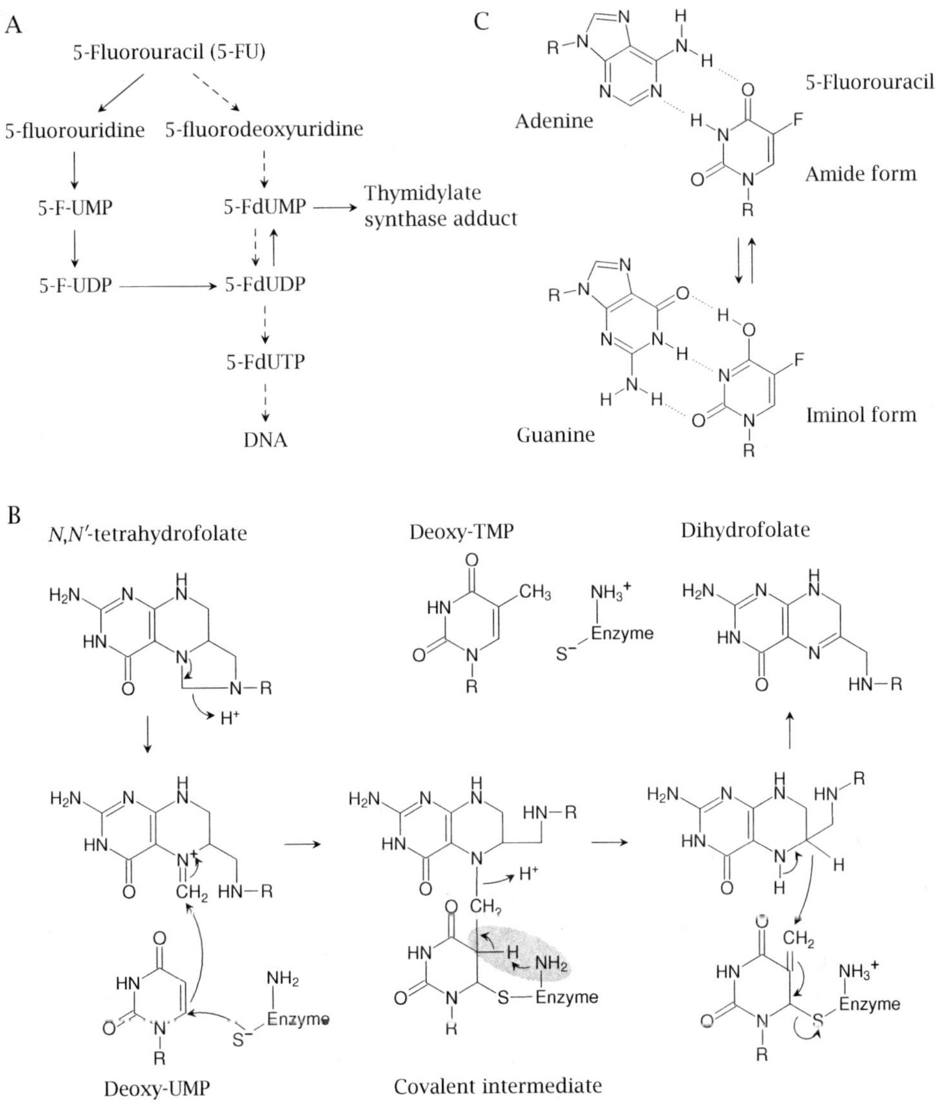

Figure 12.9 Mode of action of the antimetabolites 5-fluorouracil (5-FU) and 5-fluorodeoxyuridine. **A:** Metabolic activation. 5-FU mimics uracil (solid arrows) as well as thymine (dashed arrows), both in the initial activation through the salvage pathway and in subsequent reactions. One of the initial activation products, 5-fluorodeoxyuridine, is also used as a drug itself. The key active metabolite of 5-FU is 5-fluoro-dUMP, which mimics 5-dUMP as a substrate of thymidine synthase and irreversibly inhibits the enzyme (see B). 5-Fluoro-dUTP mimics dTTP and may be incorporated into DNA, which may cause point mutations (see C). **B:** Catalytic mechanism of thymidylate synthase. Halfway through the reaction, the enzyme, the substrate, and the cosubstrate *N,N′*-methylenetetrahydrofolate are all covalently connected. Resolution of this covalent intermediate involves abstraction of a proton from position 5 of uracil (highlighted). Fluorine in position 5 will resist abstraction, and therefore the enzyme will remain covalently trapped. **C:** The fluorine substituent favors the iminol tautomer of the uracil ring, which results in mispairing with guanine instead of adenine during DNA synthesis.

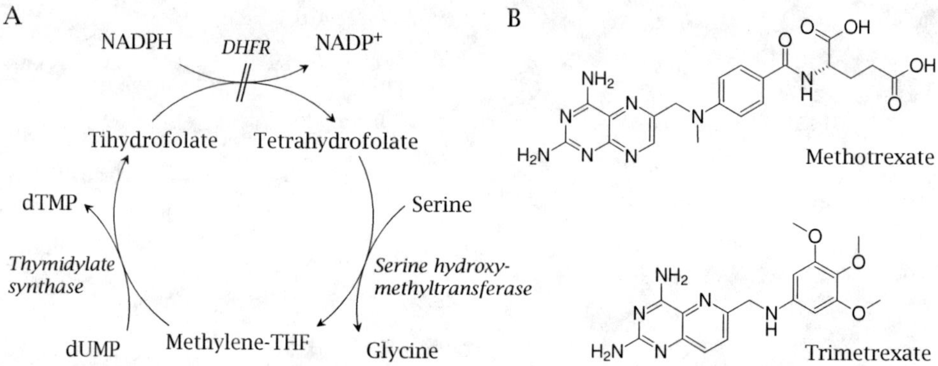

Figure 12.10 Inhibitors of dihydrofolate reductase (DHFR). **A:** Blockade of DHFR indirectly blocks thymidylate synthesis. **B:** Structures of the DHFR inhibitors methotrexate and trimetrexate.

either antitumor or antimicrobial chemotherapy, but not both. However, the drug trimetrexate (Figure 12.10) can be used for tumor therapy as well as some relatively recalcitrant infectious agents, such as *Pneumocystis carinii*, which is a fungus that causes severe lung infections in immunocompromised patients, and *Trypanosoma cruzi*, the causative agent of Chagas disease.

A deficiency of folic acid causes *megaloblastic* anemia. In this condition, the number of blood cells is reduced, while the size of individual blood cells and their precursor cells in the bone marrow is increased.[13] Similar changes to the blood cells also occur under treatment with folate antimetabolites.

Asparaginase. The amino acid asparagine is a precursor in the biosynthesis of pyrimidine and purine bases. Asparagine itself is not an essential amino acid; a cell finding itself short of it will respond by induction of asparagine synthetase, which synthesizes asparagine from aspartic acid and glutamine. The malignant cells in some forms of leukemia and lymphoma lack the ability to adequately induce asparagine synthetase, however. They are therefore susceptible to asparagine depletion, which can be induced therapeutically by applying the enzyme L-asparaginase. The enzyme is most commonly isolated from *Escherichia coli.*

In healthy patients, repeated parenteral application of a bacterial enzyme should before long result in the formation of antibodies, which will inactivate the enzyme and may even induce hypersensitivity. This problem is mitigated in leukemic patients, in whom, due to both the disease itself and the accompanying cytotoxic treatment (asparaginase is always combined with other drugs), immune system are significantly suppressed. However, immune reactions do occur, and even if no serious complications arise, the antibodies will inhibit the enzyme and promote its rapid clearance from the circulation. In this case,

[13]This is caused by DNA synthesis lagging behind the synthesis of other cellular components. The same clinical picture is also observed with a deficiency of vitamin B_{12} (cobalamine), which is required as a cofactor for methyl group transfer in the regeneration of *N,N'*-methylenetetrahydrofolate.

asparaginase isolated from another bacterium, *Erwinia carotovora*, can be used to continue the treatment.

One strategy for preventing immune reactions to asparaginase consists in its covalent modification with polyethylene glycol (PEG). Intriguingly, in some patients, even ones not previously exposed to PEG-modified therapeutics, antibodies may be present that react with PEG itself and promote inactivation of the enzyme [2].

Cytosine arabinoside. A nucleotide antimetabolite that differs from the regular nucleotide in the sugar rather than in the base is cytosine arabinoside (araC; Figure 12.11). In this molecule, the OH group in position 2 of the sugar ring points into the direction opposite to that in ribose. AraC gets incorporated into DNA but then interferes with further DNA synthesis, since the extra OH group sterically hinders the proper alignment of the cytosine base. This may affect different DNA polymerases to different degrees [6], and—in contrast, for example, to the dideoxy nucleotide analogs used in antiviral chemotherapy—the block is not absolute. If DNA synthesis continues, the resulting strand is prone to cleavage by DNA topoisomerase II at the site of the incorporated araC. This results in the accumulation of double-strand breaks and may contribute significantly to the cytotoxicity of araC [7].

Cytosine arabinoside is widely used in the treatment of leukemia, and both its mode of action and the mechanisms of cellular resistance to it have been extensively studied. Cellular concentrations of araC itself are lowered by reduced expression of the uptake transporter ENT or by increased expression of multidrug resistance transporters such as P-glycoprotein, respectively. Activation of araC to araCTP is prevented by increased expression of 5-nucleotidase, which reverts the initial phosphorylation, or of amidases that degrade araC or araCMP. Increased expression of DNA polymerase α, which may alleviate the partial block of DNA replication, and decreased expression of topoisomerase, which may reduce the formation of strand breaks at sites of araC incorporation, also cause resistance to araC [12].

12.4.2 DNA topoisomerase inhibitors

As discussed in Section 11.4.4, unwinding and unpacking by topoisomerases is necessary to render the usually heavily coiled and supercoiled DNA amenable to transcription and replication, and complete inhibition of topoisomerases I or II cannot fail to kill the cell outright. This applies, of course, to both malignant cells and healthy ones; therefore, therapeutic inhibition of topoisomerase cannot be complete. The preferential effect of topoisomerase inhibitors on malignant cells must again be ascribed to their greater readiness to plunge into apoptosis in response to any interference with the DNA synthesis process.

Inhibitors for both topoisomerase I and II are available. Figure 12.12 illustrates the mode of action of topotecan, an inhibitor of topoisomerase I, which cleaves and religates single strands. Topoisomerase II cleaves and rejoins double strands, and if the rejoining step is inhibited more strongly than the initial cleavage step, this will result in the accumulation of DNA double-strand breaks.

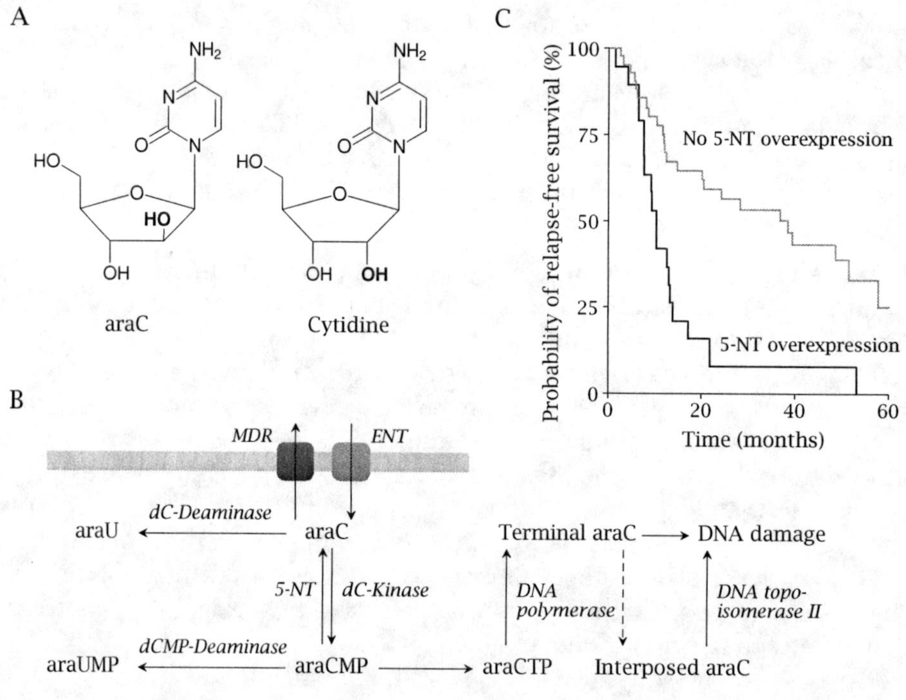

Figure 12.11 Cytosine arabinoside (araC). **A:** Structure. The arabinose ring carries a hydroxy group at position 2 that points "up" instead of "down" as it does in ribose. **B:** Activation, mode of action, and resistance. AraC enters the cell by facilitated diffusion through the equilibrative nucleoside transporter (ENT) and is extruded by multidrug resistance transporters (MDRs) such as P-glycoprotein. It is activated and degraded like deoxycytidine, and as araCTP becomes a substrate for DNA polymerase. Incorporation of araC into DNA obstructs further DNA synthesis. If synthesis proceeds past this obstacle, interposed araC residues promote DNA strand cleavage by topoisomerase II. Increased expression of MDR or of enzymes that divert araC from activation to araCTP, such as 5-nucleotidase (5-NT), cause tumor cell resistance. **C:** AraC is a key drug in the treatment of acute myeloic leukemia (AML). Resistance to araC due to increased expression of 5-NT reduces the duration of relapse-free survival in AML patients. (Figure redrawn from original data in Ref. [12].)

Belated, faulty rejoining of such breaks can in turn give rise to chromosome translocations. As illustrated in some of the examples above, chromosome translocations may cause leukemias, and an association has indeed been observed between the application of topoisomerase inhibitors and the occurrence of leukemias after completion of the treatment.

12.4.3 Proteasome inhibitors

The drug bortezomib (Figure 12.13) inhibits the protease activity of proteasomes. Since proteasomes degrade a large number of proteins, the consequences of such inhibition are complex. Reduced degradation of proapoptotic proteins is assumed to be responsible for the antitumor effect of bortezomib. The drug is clinically used with plasmocytoma, which is a malignancy derived from B-lymphocytes, and in various solid tumors as well.

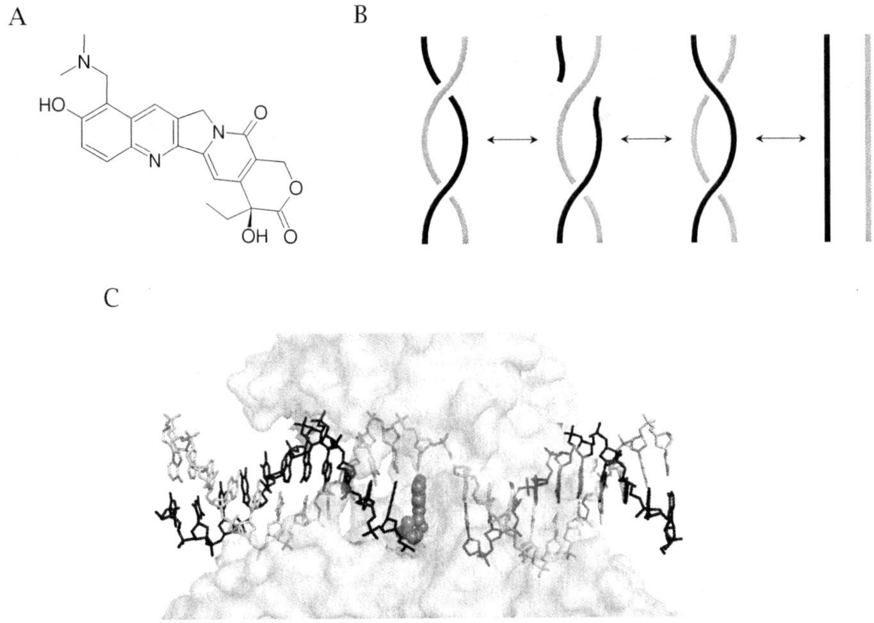

Figure 12.12 Topotecan, an inhibitor of DNA topoisomerase I. **A**: Structure. **B**: Function of DNA topoisomerase I. The enzyme breaks one strand of a DNA molecule, swivels the ends around the other strand, and then religates them. It thus adds coils to the double-stranded DNA or removes them. **C**: Structure of topotecan (shown as spheres), bound to topoisomerase I and DNA. The drug binds to the enzyme such that it inserts itself between the free ends of the cleaved DNA strand and thereby prevents religation. (Rendered from 1k4t.pdb [27].)

12.4.4 Inhibitors of mitosis

Most drugs that inhibit mitosis interfere with the assembly of microtubules. The latter consist of tubulin $\alpha\beta$-dimers, assembled in a regular, helical fashion into a hollow fiber (Figure 12.14). These tubules are being formed and destroyed dynamically as needed by the cell; polymerization is GTP-dependent. They are important in molecular transport and motion; tubulin serves as the track to the locomotives dynein and kinesin, much like actin serves as the track for myosin.[14] However, tubulin, kinesin, and dynein participate in other processes of cellular motility than actin and myosin do. A prominent function of tubulin and its associated motor proteins is the segregation of the chromosomes in mitosis; the mitotic spindle is the tubulin framework assembled for this purpose. Formation of the spindle is inhibited by various plant alkaloids. Commonly used drugs include taxanes such as paclitaxel and docetaxel, which occur naturally in yew trees, and *Vinca rosea* (rosy periwinkle) alkaloids such as vinblastine (Figure 12.14). This drug molecule fits between the α- and β-tubulin subunits and subtly alters their conformation. The drug-bound $\alpha\beta$-dimer will still associate with the growing end of a nascent microtubule; however, once

[14]Dynein and kinesin both move along microtubules but do so in opposite directions.

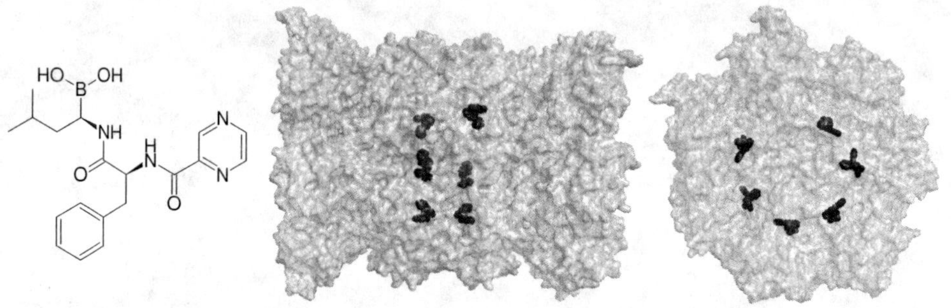

Figure 12.13 Structure and target binding of bortezomib. Proteasomes are drum-shaped, large protease complexes. They unfold and cleave intracellular proteins destined for degradation. Bortezomib binds and inhibits the protease subunits of proteasomes, thereby causing cellular dysregulation and apoptosis. (Structure rendered from 2f16.pdb [15].)

in place, it inhibits the addition of further dimers. The tubule thus becomes capped and trapped.

When saturated with vinblastine, tubulin dimers can be observed polymerizing into atypical helical shapes; this effect, however, may be limited to the high drug concentrations that can be employed in cell culture and may not be relevant in vivo.[15]

12.4.5 DNA-alkylating agents

Antiproliferative drugs that act throughout all phases of the cell cycle are mostly DNA-alkylating agents (Figure 12.15). A straightforward example is mechlorethamine, which contains an *N*-mustard or dichloroethylamine reactive moiety. The chloroethyl groups of this structure form aziridine intermediates that react with nucleophiles. This same reactive group also occurs with the irreversible *α*-adrenergic blocker phenoxybenzamine (see Section 2.2.4). In contrast to that drug, however, mechlorethamine does not possess any molecular feature designed to steer it to one particular target. Therefore, most mechlorethamine molecules will not react with DNA but instead with some other nucleophiles hopping about in the cell, in particular glutathione or other sulfhydryl compounds. However, those molecules that happen to react with DNA are the ones that matter. The two chloroethyl groups of mechlorethamine enable it to crosslink two sites in the DNA. Such crosslinks are particularly effective if they bridge both strands. If only one strand is damaged, DNA repair may use the remaining intact strand as a template for repairing the damaged one. Damage to both strands prevents this and increases the likelihood of irreversible genetic damage.

Mechlorethamine enters cells through active transport by a choline transporter. The drug melphalan, which contains the same reactive group as me-

[15]The drug colchicine, which occurs naturally in autumn crocus, binds to tubulin in the vicinity of the vinblastine binding site and has virtually the same effect on tubulin polymerization. However, it is used in the treatment of gout rather than cancer (see Section 10.2.2). The literature does not offer a rational explanation for the preferment of either drug in its respective application.

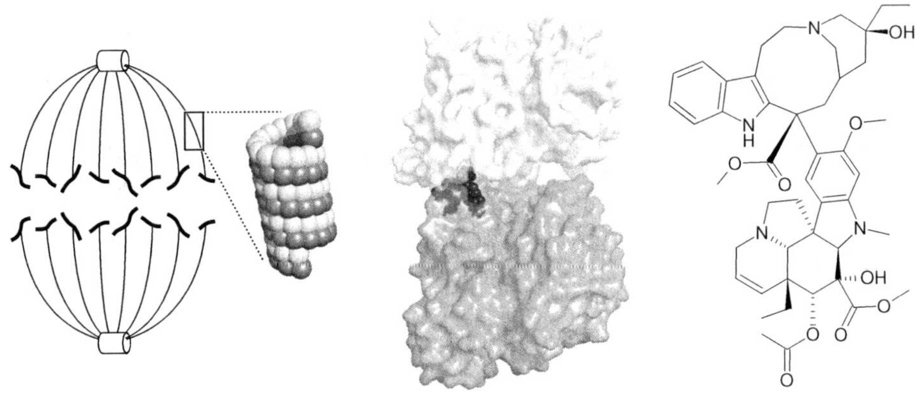

Figure 12.14 Structure and mode of action of vinblastine. The mitotic spindle separates the chromosomes during the metaphase of mitosis. It consists of microtubules, which are highly regular, helical assemblies of tubulin $\alpha\beta$-dimers. Vinblastine binds between α-tubulin (white) and β-tubulin (gray) and distorts the geometry of the dimer and by extension that of the polymeric assembly. Structure rendered from 1z2b [14].

chlorethamine, also contains a phenylalanine moiety, which makes it a substrate for some amino acid transporters, including LAT1 [4]. This transporter is overexpressed in many cancers, which augments the effectiveness of the drug [21].

One commonly used drug containing the dichloroethylamine moiety is cyclophosphamide. This drug is metabolized extensively and gives rise to several active metabolites, the exact contribution of which to the overall therapeutic effect is not clear. Metabolism is initiated by a cytochrome P450 enzyme (CYP2B) in the liver and continues with several enzymatic and nonenzymatic steps. One of the metabolites of cyclophosphamide is acrolein, which also occurs in tobacco smoke and forms in the frying pan. Acrolein can form tricyclic adducts with guanine that prevent the incorporation of any base opposite to them, thus interfering with DNA synthesis and repair.

The drug 1,3-bis(2-chloroethyl)-1-nitrosourea (BCNU, carmustine) combines two chloroethyl moieties with a nitrosourea group. The latter undergoes rapid spontaneous decomposition, releasing two reactive products (Figure 12.15C). The rapid decay limits the tissue penetration of BCNU; this is discussed further in Section 14.4.1.

Cisplatin is not strictly an alkylant but also acts by introducing covalent modifications into the DNA. In aqueous solution, the chloride in the complex is substituted by water. The new complex readily reacts with the nitrogens in purine rings, particularly the N7 positions of guanine bases. This leads to intra- or interstrand crosslinks. Cisplatin enters cells through a specific copper ion transporter [16] and is extruded by another one [22]; mutations that change the level of expression of these transporters are involved in tumor cell resistance to this drug.

While the structures and reactive groups differ between alkylating agents, one commonality is a preference to react with the N7 of guanine bases in the

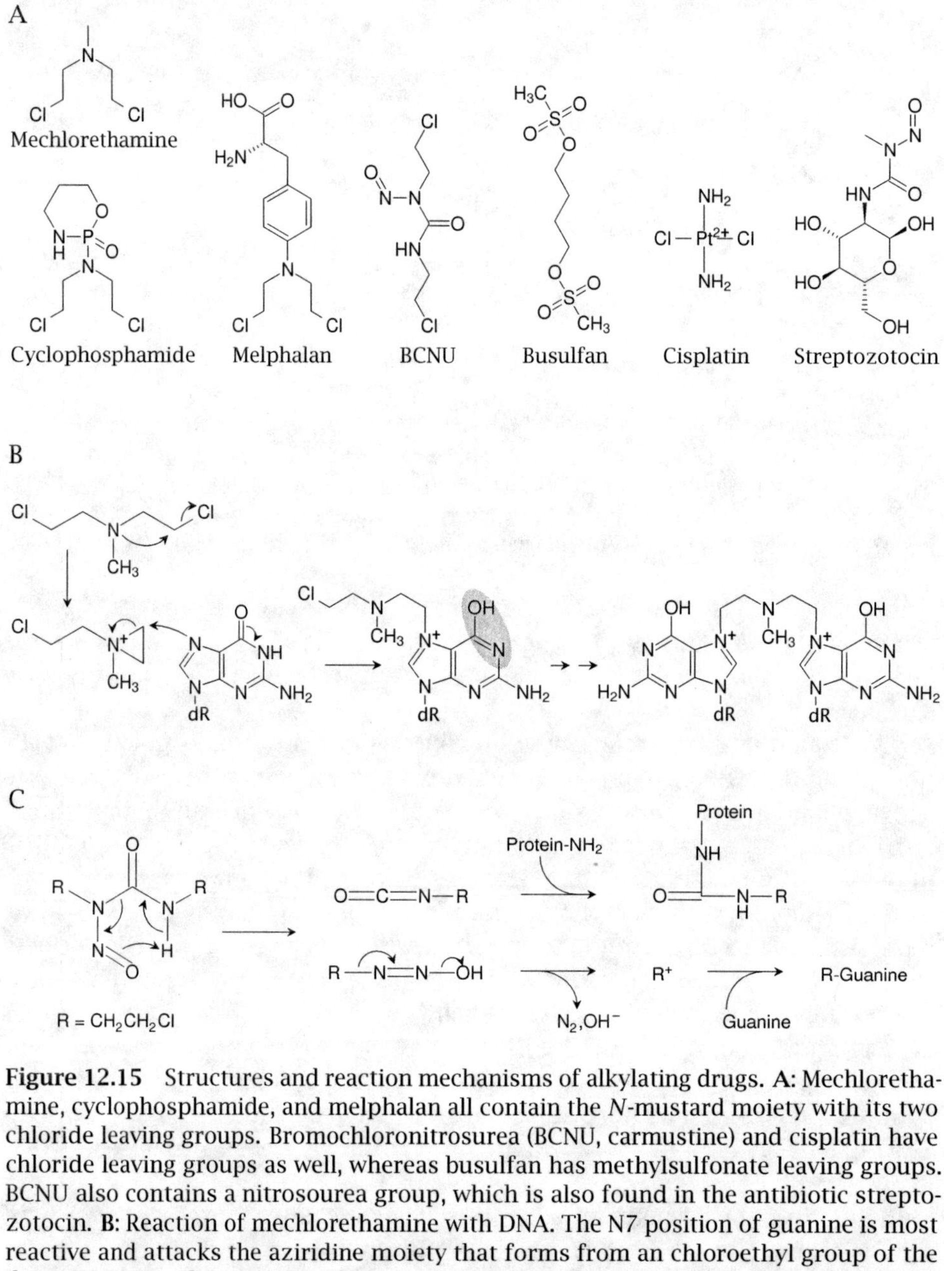

Figure 12.15 Structures and reaction mechanisms of alkylating drugs. **A:** Mechlorethamine, cyclophosphamide, and melphalan all contain the *N*-mustard moiety with its two chloride leaving groups. Bromochloronitrosurea (BCNU, carmustine) and cisplatin have chloride leaving groups as well, whereas busulfan has methylsulfonate leaving groups. BCNU also contains a nitrosourea group, which is also found in the antibiotic streptozotocin. **B:** Reaction of mechlorethamine with DNA. The N7 position of guanine is most reactive and attacks the aziridine moiety that forms from an chloroethyl group of the drug. Reaction of two guanine bases with the same drug molecule leads to crosslinks in the DNA. In the reaction product, the iminol tautomer is favored (highlighted in the first reaction step), which changes its base-pairing preference from cytosine to thymine. **C:** Activation of BCNU and subsequent reactions. The nitrosurea undergoes spontaneous decomposition, yielding an isocyanate compound that reacts with primary amines (top) and a diazo compound that decays further into a carbocation, which in turn reacts with any kind of nucleophiles, including bases in DNA. Each of the two fragments of BCNU retains one chloroethyl moiety, which can subsequently react as shown above for mechlorethamine.

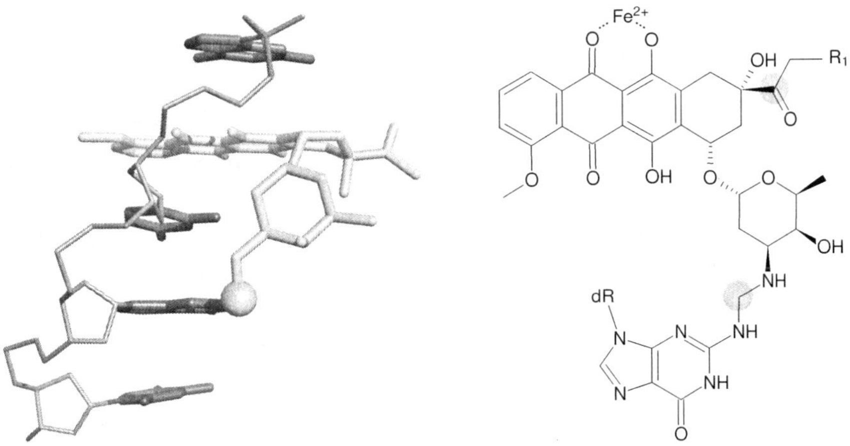

Figure 12.16 Reaction of the anthracycline antibiotics daunorubicin (R_1 = H) and doxorubicin (R_1 = OH) with DNA. The planar ring structure of the drug intercalates between two base pairs of a DNA double strand (only one DNA strand is shown here). The amino group on the daunosamine sugar moiety of the drug then becomes joined to the N2 amino group of a guanine base by a methylene bridge (highlighted). The carbon that forms this bridge is derived from formaldehyde, which is released from cellular precursors or, in doxorubicin, from the hydroxyacetyl side chain on the drug itself (also highlighted) through a series of redox reactions that occur between glutathione, iron and the drug's quinoid rings [8,29]. (Structure rendered from 2d34.pdb [32].)

DNA. Why is that so? The nitrogens in the five-membered rings of guanine and adenine are not directly involved in basepairing with the complementary DNA strand and are therefore more easily accessible than most of the other nucleophilic groups; this leaves the N7 atoms in adenine and guanine. Why would the one in guanine be more reactive? The guanine ring has a less aromatic character than the adenine ring. The π electrons of the ring nitrogens are therefore not as completely delocalized, which makes them stronger nucleophiles. However, the preference is not absolute, and alkylation of adenine and the pyrimidine bases occurs as well.

An interesting consequence of guanine N7-alkylation is the increased propensity of the guanine ring to adopt the tautomeric iminol form (Figure 12.15B). In the iminol form, the orientation of hydrogen bond donors and acceptors is reversed and now resembles that of adenine, thus causing the modified guanine to basepair with thymine instead of cytosine. This effect contributes to the mutagenic effect of guanine alkylation.

A particularly sophisticated group of alkylating agents are the anthracycline antibiotics, exemplified by doxorubicin (Figure 12.16). This drug molecule has a large, flat polycyclic moiety, which will intercalate between the stacked base pairs of DNA. This ring also chelates iron (Fe^{3+}), which is then reduced by glutathione. The reduced iron, in turn, reduces the quinoid drug molecule, which can lead to the formation of reactive oxygen species and of formaldehyde. The formaldehyde crosslinks the amino-sugar moiety of the drug to a guanine base in the DNA, whereas the reactive oxygen species can induce further damage to the DNA and to other constituents of the cell. In particular, oxydative damage to mitochondrial lipids has been linked to the cardiotoxicity of the drug, which limits the dosage that can be used. Techniques that aim to reduce such side effects are discussed in Section 14.3.1.

The covalent DNA adducts of doxorubicin inhibit replication of DNA and are mutagenic. In addition, doxorubicin also inhibits topoisomerase II; the relative contributions of these and other mechanisms to the overall cytotoxicity of the drug are a matter of debate [13].[16]

STUDY QUESTIONS

12.1 The structure below shows pentostatin, an inhibitor of adenosine deaminase (ADA).

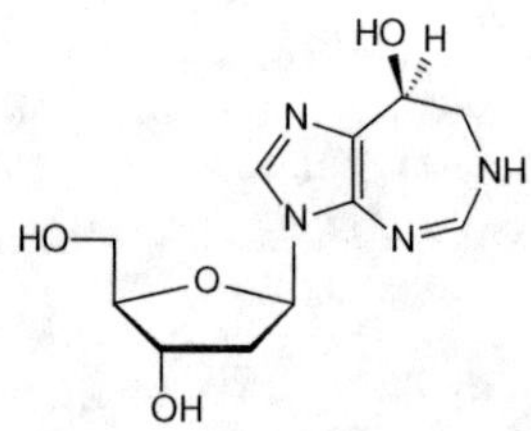

Questions: (1) Explain how this inhibitor inhibits the enzyme. (Hint: Consider the most likely mechanism of the reaction catalyzed by ADA.) (2) Explain how inhibition of ADA can be useful in the treatment of tumors, and what types of tumors are most likely to be susceptible. (You may want to consult Section 10.1.4 for this question.)

12.2 Tamoxifen (Section 12.2) exerts antagonistic effects on the estrogen receptor in the breast gland and the endometrium, whereas in the bone it acts as an estrogen receptor agonist. In contrast, the drug fulvestrant acts as an antagonist in all tissues. Fulvestrant is used in patients with breast cancer who do not or no longer respond favorably to tamoxifen. Can you suggest an explanation as to why fulvestrant has a more robust antitumor activity than tamoxifen?

12.3 Combination of drugs with different targets is a key principle in the chemotherapy of most tumors. However, quite a few antitumor drugs—e.g. 5-FU, araC, cyclophosphamide and doxorubicin—appear to have more than one mode of action, and there often is some argument about the "real target". Is there a connection between these two observations?

REFERENCES

[1] M. C. Abad et al., "Structural determination of estrogen-related receptor gamma in the presence of phenol derivative compounds," *J. Steroid Biochem. Mol. Biol.* 108 (2008), 44-54.

[2] J. K. Armstrong et al., "Antibody against poly(ethylene glycol) adversely affects PEG-asparaginase therapy in acute lymphoblastic leukemia patients," *Cancer* 110 (2007), 103-111.

[3] A. Arora and E. M. Scholar, "Role of tyrosine kinase inhibitors in cancer therapy," *J. Pharmacol. Exp. Ther.* 315 (2005), 971-979.

[16] The amplification of the DNA toxicity of intercalation through covalent reaction is reminiscent of benzopyrene, even though in that case the chemistry of the covalent linkage is different (see Figure 4.8).

[4] A. Begleiter et al., "Evidence for active transport of melphalan by two amino acid carriers in L5178Y lymphoblasts in vitro," *Cancer Res.* 39 (1979), 353–359.

[5] F. M. Boeckler et al., "Targeted rescue of a destabilized mutant of p53 by an in silico screened drug," *Proc. Natl. Acad. Sci. U. S. A.* 105 (2008), 10360–10365.

[6] G. M. Carbone, C. V. Catapano, and D. J. Fernandes, "Imbalanced DNA synthesis induced by cytosine arabinoside and fludarabine in human leukemia cells," *Biochem. Pharmacol.* 62 (2001), 101–110.

[7] S. D. Cline and N. Osheroff, "Cytosine arabinoside lesions are position-specific topoisomerase II poisons and stimulate DNA cleavage mediated by the human type II enzymes," *J. Biol. Chem.* 274 (1999), 29740–29743.

[8] C. Cullinane et al., "Formation of adriamycin--DNA adducts in vitro," *Nucleic Acids Res.* 22 (1994), 2296–2303.

[9] N. J. Curtin and A. N. Hughes, "Pemetrexed disodium, a novel antifolate with multiple targets," *Lancet Oncol.* 2 (2001), 298–306.

[10] G. Q. Daley, R. A. V. Etten, and D. Baltimore, "Induction of chronic myelogenous leukemia in mice by the P210bcr/abl gene of the Philadelphia chromosome," *Science* 247 (1990), 824–830.

[11] M. Dean, T. Fojo, and S. Bates, "Tumour stem cells and drug resistance," *Nat. Rev. Cancer* 5 (2005), 275–284.

[12] C. M. Galmarini et al., "In vivo mechanisms of resistance to cytarabine in acute myeloid leukaemia," *Br. J. Haematol.* 117 (2002), 860–868.

[13] D. A. Gewirtz, "A critical evaluation of the mechanisms of action proposed for the antitumor effects of the anthracycline antibiotics adriamycin and daunorubicin," *Biochem. Pharmacol.* 57 (1999), 727–741.

[14] B. Gigant et al., "Structural basis for the regulation of tubulin by vinblastine," *Nature* 435 (2005), 519–522.

[15] M. Groll et al., "Crystal structure of the boronic acid-based proteasome inhibitor bortezomib in complex with the yeast 20S proteasome," *Structure* 14 (2006), 451–456.

[16] S. Ishida et al., "Uptake of the anticancer drug cisplatin mediated by the copper transporter Ctr1 in yeast and mammals," *Proc. Natl. Acad. Sci. U. S. A.* 99 (2002), 14298–14302.

[17] N. Issaeva et al., "Small molecule RITA binds to p53, blocks p53-HDM-2 interaction and activates p53 function in tumors," *Nat. Med.* 10 (2004), 1321–1328.

[18] K. Janssen et al., "Apaf-1 and caspase-9 deficiency prevents apoptosis in a Bax-controlled pathway and promotes clonogenic survival during paclitaxel treatment," *Blood* 110 (2007), 3662–3672.

[19] C. J. Kemp et al., "Reduction of p53 gene dosage does not increase initiation or promotion but enhances malignant progression of chemically induced skin tumors," *Cell* 74 (1993), 813–822.

[20] C. E. de Kogel and J. H. M. Schellens, "Imatinib," *Oncologist* 12 (2007), 1390–1394.

[21] J. Lin et al., "L-type amino acid transporter-1 overexpression and melphalan sensitivity in Barrett's adenocarcinoma," *Neoplasia* 6 (2004), 74–84.

[22] K. Nakayama et al., "Copper-transporting P-type adenosine triphosphatase (ATP7B) as a cisplatin based chemoresistance marker in ovarian carcinoma: comparative analysis with expression of MDR1, MRP1, MRP2, LRP and BCRP," *Int. J. Cancer* 101 (2002), 488–495.

[23] R. Nasr et al., "Eradication of acute promyelocytic leukemia-initiating cells through PML-RARA degradation," *Nat. Med.* 14 (2008), 1333–1342.

[24] M. L. Ostrowski et al., "Malignant chondroblastoma presenting as a recurrent pelvic tumor with DNA aneuploidy and p53 mutation as supportive evidence of malignancy," *Skeletal Radiol.* 28 (1999), 644–650.

[25] S. K. Pal and M. Pegram, "HER2 targeted therapy in breast cancer...beyond Herceptin," *Rev. Endocr. Metab. Disord.* 8 (2007), 269–277.

[26] W. J. Schnedl et al., "STZ transport and cytotoxicity. Specific enhancement in GLUT2-expressing cells," *Diabetes* 43 (1994), 1326–1333.

[27] B. L. Staker et al., "The mechanism of topoisomerase I poisoning by a camptothecin analog," *Proc. Natl. Acad. Sci. U. S. A.* 99 (2002), 15387–15392.

[28] P. A. Svingen et al., "Effects of the bcr/abl kinase inhibitors AG957 and NSC 680410 on chronic myelogenous leukemia cells in vitro," *Clin. Cancer Res.* 6 (2000), 237–249.

[29] D. J. Taatjes et al., "Redox pathway leading to the alkylation of DNA by the anthracycline, antitumor drugs adriamycin and daunomycin," *J. Med. Chem.* 40 (1997), 1276–1286.

[30] M. Talpaz et al., "Dasatinib in imatinib-resistant Philadelphia chromosome-positive leukemias," *N. Engl. J. Med.* 354 (2006), 2531–2541.

[31] M. D. Tronko et al., "A cohort study of thyroid cancer and other thyroid diseases after the chornobyl accident: thyroid cancer in Ukraine detected during first screening," *J. Natl. Cancer Inst.* 98 (2006), 897–903.

[32] A. H. Wang et al., "Formaldehyde cross-links daunorubicin and DNA efficiently: HPLC and X-ray diffraction studies," *Biochemistry* 30 (1991), 3812–3815.

[33] I. B. Weinstein, "Disorders in cell circuitry during multistage carcinogenesis: the role of homeostasis," *Carcinogenesis* 21 (2000), 857–864.

CHAPTER 13

RIBONUCLEIC ACIDS AS DRUG TARGETS AND DRUGS

by Thorsten Dieckmann

Not too long ago, ribonucleic acids (RNAs) were seen as the molecular equivalent of carbon paper: Their main role seemed to be to serve as intermediate between genetic information (DNA) and function (proteins). Research over the past few decades has revealed that this view was far from reality. RNA has been shown to play many pivotal roles in cellular processes, ranging from metabolic control and the regulation of protein expression to cellular development and the identification and destruction of foreign genetic information [13]. Even the long-established structural or accessory role of RNA in the ribosome, the cellular protein synthesis machinery, is now seen in a different light. Ribosomal RNA rather than the ribosomal proteins is primarily responsible for the catalysis of peptide bond formation and for most steps of translation, while the ribosomal proteins fulfill accessory functions. This newly established significance of RNA has led to increased interest in RNA as a potential drug target, as well as in its use as a therapeutic agent itself.

Drugs that target RNA are not a recent invention and have been in use since long before their detailed mechanisms of action were discovered. Several antibiotics that target the prokaryotic ribosome interact with its RNA components rather than ribosomal proteins and interfere with protein synthesis by directly blocking the catalytic center or the substrate binding sites. Beyond these well-established applications, RNA has more recently become a prime target for the development of antiviral and anticancer therapeutics.

13.1 RNA AS DRUG TARGET

The first aminoglycoside antibiotic agent, streptomycin, was discovered during the systematic screening of bacterial culture supernatants for inhibitory effects against *Mycobacterium tuberculosis* in 1944. Since then, many new aminoglycoside antibiotics have been isolated from various actinomycetes. Unfortunately, the number of resistance mechanisms expressed by pathogenic microorganisms has increased at the same time and provides a major challenge to the use and development of antibiotics in general.

The following section first discusses the interaction of aminoglycosides with ribosomal RNA as an example of a well-established RNA target. This is followed by a brief discussion of telomerase and of the HIV virus as examples of novel targets for the development of RNA targeting drugs.

13.1.1 Targeting the prokaryotic ribosome

Owing to the structural differences between prokaryotic and eukaryotic ribosomes, the cellular protein factory presents an attractive target for antibacterial drugs [30]. The ribosome contains multiple functional sites that perform different steps in translation. The aminoacyl site accepts the tRNA molecule loaded with the next amino acid residue to be incorporated. The growing peptide chain, which at this point is still attached to the tRNA molecule used in the preceding round, is then transferred to the new amino acid by the ribosome's peptidyl transferase activity. The growing peptide leaves the ribosome through the exit tunnel.

Antibiotics that interfere with each of these different functions have been instrumental in their experimental elucidation. We will concentrate here on two types of drugs, which specifically target the RNA components of the ribosome and thus provide ideal case studies for how specificity and function are achieved with RNA as a drug target [42,39]. The first model compound is paromomycin, which belongs to the class of aminoglycoside antibiotics and blocks the ribosomal aminoacyl-accepting site. Chloramphenicol serves as an example for a peptidyl transfer inhibitor. Figure 13.1 summarizes some ribosome-targeting drugs and their specific molecular sites of action.

Structure of aminoglycosides. The class of aminoglycosides [23] includes many hydrophilic molecules, which all contain a central aminocyclitol that is linked to one or two amino sugars via pseudo-glycosidic bonds (Figure 13.2). Most of the clinically important aminoglycosides have 2-deoxystreptamine as their aminocyclitol moiety and are either monosubstituted at position 4 or disubstituted at positions 4 and 5 or 4 and 6. However, there are several important aminoglycosides that do not follow this pattern. One such atypical aminoglycoside is streptomycin.

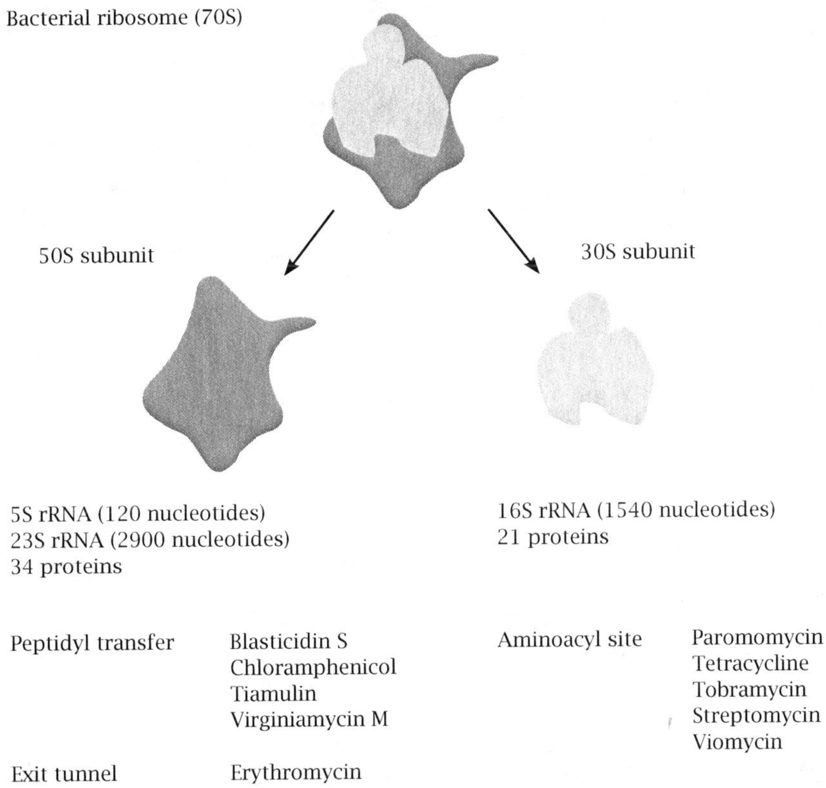

Figure 13.1 Global structure of the bacterial ribosome and target sites of several important antibiotics that inhibit ribosomal protein synthesis.

Molecular interactions between ribosomal RNA and aminoglycosides, and their effect on translation. The molecular details of aminoglycoside interactions with the bacterial ribosome were initially characterized by NMR spectroscopy in 1996 using a 27-nucleotide model RNA in complex with paromomycin (Figure 13.3). The determination of crystal structures of the complete ribosome and of several ribosome–aminoglycoside complexes confirmed most of the earlier results and provided additional detail regarding how they achieve their effect on translation. Two examples of this important class of antibiotics, paromomycin and streptomycin, are discussed below. The former is a member of the family of typical aminoglycosides; the latter belongs to the atypical aminoglycosides.

Typical aminoglycosides: paromomycin. Paromomycin interacts with the bacterial ribosome at its aminoacyl site (A-site) by binding to the major groove of helix 44 in the 30S ribosomal subunit. Ring I is stacked on the base of G1491, and numerous specific hydrogen bonds are formed between the antibiotic and bases as well as phosphate groups in the ribosomal RNA (Figure 13.3). The binding of paromomycin causes a profound change in the structure of the ribosomal RNA that directly interferes with the fidelity of aminoacyl-tRNA selection.

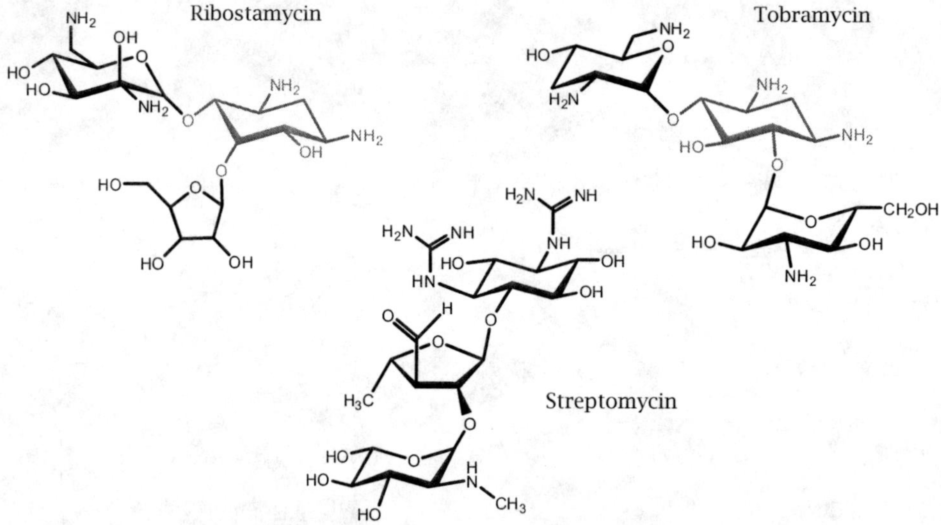

Figure 13.2 Structures of representative aminoglycoside antibiotics. Ribostamycin is a typical aminoglycoside of the 4,5-disubstituted deoxystreptamine group. Tobramycin represents the typical aminoglycosides with 4,6-disubstituted deoxystreptamine. It is clinically important in the treatment of *Pseudomonas aeruginosa* infections. In the structures of ribostamycin and tobramycin, the aminocyclitol moiety is highlighted. Streptomycin is an atypical aminoglycoside.

Comparison of the 30S ribosomal subunit crystal structures in the presence and absence of the drug shows that A1492 and A1493 are pushed away from their normal positions and flipped out of helix 44 (Figure 13.3). These two nucleotides play a crucial role in translation by forming a hydrogen bond network with the first two bases of both the mRNA codon and the anticodon of the tRNA. These interactions allow the ribosome to discriminate between cognate and noncognate mRNA-tRNA pairings. Binding of a correct (cognate) tRNA results in the two adenines being flipped out of the helix, causing a net increase of the affinity for the cognate tRNA. Binding of paromomycin as well as several other aminoglycosides essentially mimics this effect even in the presence of a noncognate tRNA. This will increase the affinity for tRNA binding and decrease the rate of A site tRNA dissociation. The ribosome therefore can no longer discriminate between correct and incorrect tRNAs, which explains the miscoding effect of this class of antibiotics.

Atypical aminoglycosides: streptomycin. Streptomycin belongs to the so-called atypical aminoglycosides. Because of the particularly rapid appearance of bacterial resistance, streptomycin was replaced in most applications by other drugs within a few years after its discovery. However, it has more recently found renewed applications in the treatment of multidrug resistant strains of *Mycobacterium tuberculosis.*

Streptomycin also targets the A-site of the bacterial ribosome; however, its binding site is not identical but adjacent to that of the typical aminoglycosides like paromomycin. The ribosome-bound antibiotic makes contact with four

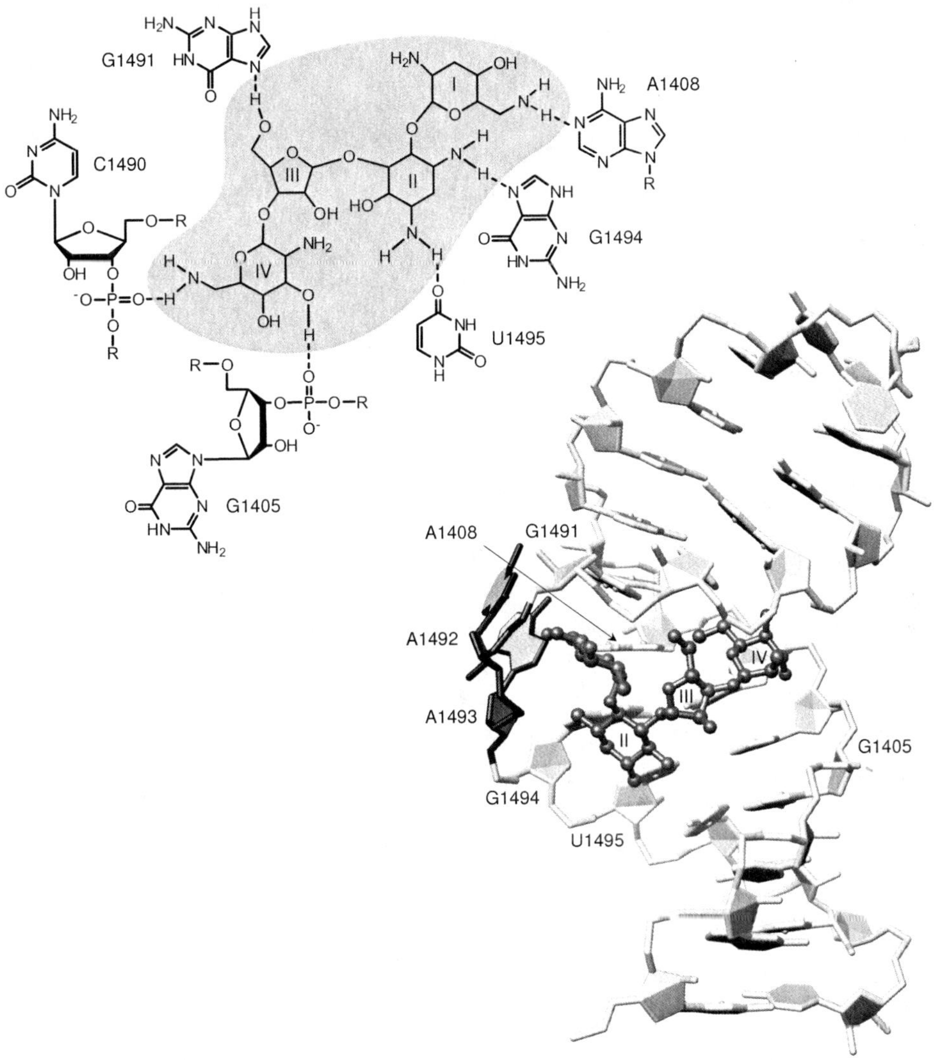

Figure 13.3 Paromomycin and its interactions with the ribosomal A-site RNA. Shown are the chemical structure of paromomycin (gray background) and its hydrogen bond interaction network with components of the ribosomal RNA (top) and the three-dimensional structure of the RNA-paromomycin complex as determined by NMR spectroscopy (bottom). A colored version of this figure is shown in the insert in the middle of this book.

domains of the ribosomal 16S RNA. These include helices 1, 27, 28 and 44. Moreover, the molecule also interacts with a lysine side chain from the ribosomal protein S12. This complex network of interactions is in sharp contrast to paromomycin and the typical aminoglycosides, which interact solely with rRNA elements from helix 44.

Streptomycin causes its antibacterial effect by interfering with the ribosomal proofreading, which leads to increased miscoding of the newly synthesized

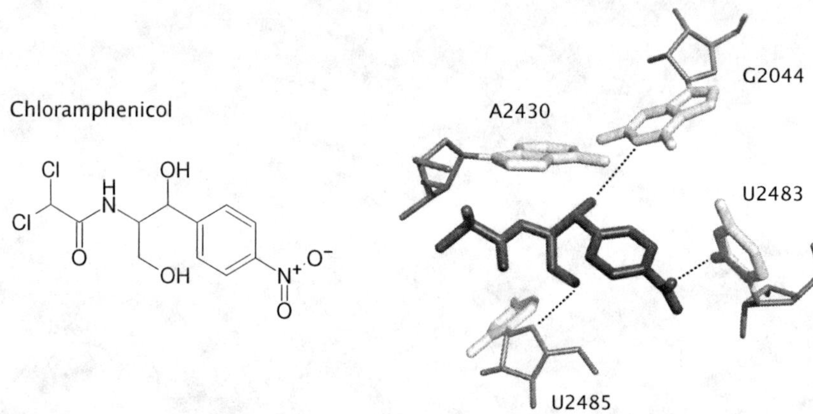

Figure 13.4 Molecular interactions of chloramphenicol within the peptidyl transferase catalytic site of the bacterial ribosome. The orientation of chloramphenicol in the molecular structure is similar to the structural formula on the left. Nucleotide numbers refer to position within the A chain of the *Deinococcus radiophilus* rRNA. Molecular structure rendered from 1K01.pdb [34].

proteins. On the molecular level, this is achieved by affecting the dynamic equilibrium between two conformations of rRNA helix 27, locking it into a high-affinity state for binding of both cognate and noncognate tRNAs.

Chloramphenicol. Chloramphenicol is a bacteriostatic antibiotic that was introduced into clinical application in 1949. The drug was originally discovered as a natural product of the soil bacterium *Streptomyces venezuelae*. It became the first antibiotic to be manufactured synthetically on a large scale. Together with the antibiotics of the tetracycline family, it is considered the classical broad-spectrum antibiotic, which means that it is clinically effective against both Gram-positive and Gram-negative bacteria, both aerobic and anaerobic ones. This broad spectrum of activity can at least in part be attributed to the small size of the drug molecule and nicely illustrates the advantages of small molecule drug molecules versus larger ones.

The mode of action of chloramphenicol differs significantly from the aminoglycoside antibiotics. Rather than interfering with the accuracy of translation, chloramphenicol directly inhibits peptide bond formation by binding to and interacting with key nucleotides of the 23S ribosomal RNA in the peptidyl-transfer center of the ribosome (Figure 13.4). Binding of the drug directly interferes with substrate binding. The small size of the drug molecule allows easy access to the target site deep inside the ribosome. In combination with its lipophilic character, it also facilitates access to all tissues of the body, including the brain. Even the only somewhat larger, yet more polar, aminoglycosides have already a much more limited tissue distribution and activity spectrum. This effect of drug size and polarity becomes even more problematic for macromolecule-based drugs, such as the nucleic acids and derivatives thereof discussed later in this chapter.

13.1.2 Bacterial resistance as a challenge to drug development

Antibacterial agents that target RNA such as the ones discussed above are quite efficient and highly specific to their prokaryotic targets. However, they have several weaknesses that can lead to the rapid development of resistant bacterial strains. The widespread use of antibiotics has led to the appearance of multidrug resistant strains of several pathogenic bacteria. An infection with such pathogens will not respond to standard treatments and can be life threatening. In most cases, the only treatment options are antibiotic drugs of "last resort" such as vancomycin or streptogramins, often administered as a combination therapy involving cocktails of multiple antibiotic agents.

Resistance mechanisms. Because ribosomes are located inside the bacterial cell, antibiotics that target them can be effective only if they enter the cell in sufficient amounts. Generally speaking, there are three major pathways through which bacterial resistance to intracellularly acting antibiotics can appear [42,39]: (1) reduction of the intracellular concentration of the molecule, (2) modification of its target, and (3) modification of the antibiotic itself.

Under physiological conditions, aminoglycosides are positively charged and thus cannot easily travel through biological membranes by simple diffusion. With Gram-negative bacteria, the mechanism for aminoglycoside uptake is thought to start with adsorption by electrostatic interactions between the cationic drug and the negatively charged lipopolysaccharides of the outer membrane. Once the outer membrane has been traversed, uptake across the inner membrane depends on the membrane potential, which renders the cytosol negative relative to the extracellular space and is generated by the respiratory chain.

The importance of the membrane potential for antibiotic action can be seen in the fact that obligatory anaerobic bacteria, which do not have a respiratory chain to maintain a membrane potential, are intrinsically resistant to aminoglycosides. Facultatively anaerobic bacteria, which perform respiration and have a polarized membrane in the presence of O_2 but can survive on anaerobic metabolism in its absence, can acquire partial or complete resistance by modifications of the respiratory chain, for example, ATP synthase mutations, that will dissipate the membrane potential. Other resistance pathways that affect drug uptake are modifications of the membrane components that decrease the density of negative charges at the membrane surface. Finally, bacteria also utilize active energy-dependent transport mechanisms to reduce the antibiotic concentration in the cell plasma. This pathway is particularly important in resistant strains that overexpress multidrug transporter proteins (see Section 11.4.1).

An alternate pathway to resistance is the modification of the drug target, in the case of aminoglycosides the ribosome. The two major mechanisms that fall into this group are methylation of the 16S rRNA and ribosomal mutations. The genetic recipes for these changes are already available in nature because the organisms that produce aminoglycosides, mostly members of the actinomycetes, need to protect themselves from their action. The genes encoding

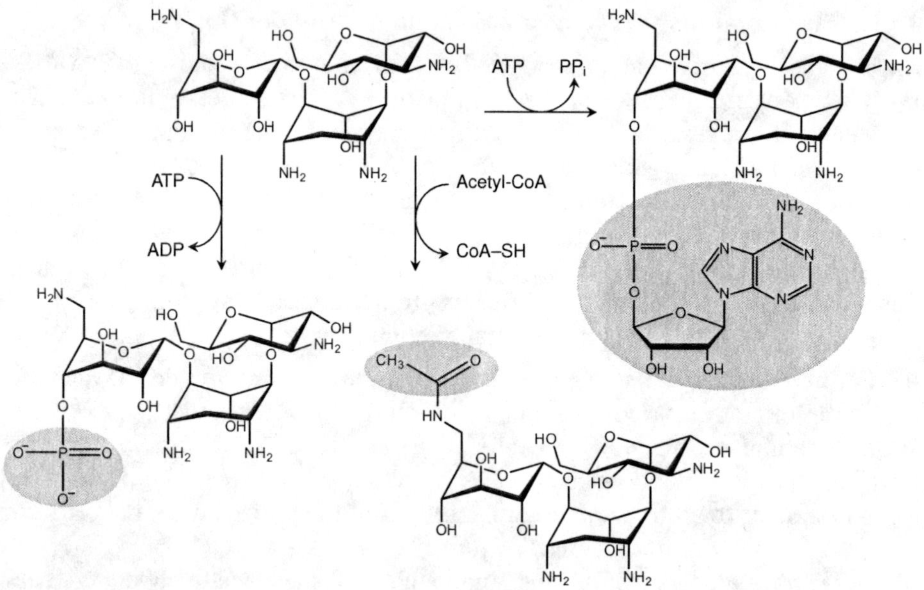

Figure 13.5 Bacterial resistance to aminoglycosides through enzymatic modification of the drug molecules. Kanamycin A is shown as an example. From left to right: Phosphorylation is catalyzed by aminoglycoside *O*-phophotransferases. Acetylation is catalyzed by aminoglycoside *N*-acetyltransferases. Adenylation is catalyzed by aminoglycoside *O*-adenylyltransferases.

these resistance mechanism may be transferred to other bacterial species on plasmids or bacteriophage genomes.

Finally, the bacteria can utilize enzymes to modify and thus inactivate the drug molecule itself by lowering its affinity for the target RNA. Three major classes of modifying enzymes have been described for the aminoglycosides (Figure 13.5): (1) aminoglycoside phosphotransferases, which transfer the γ-phosphate from ATP to one of the hydroxyl groups of the antibiotic; (2) acetyltransferases, which transfer acetyl groups from acetyl-CoA to one of the amino groups in typical aminoglycosides; and (3) aminoglycoside adenylyltransferases, which use Mg-ATP to produce *O*-adenylylated aminoglycosides.

The large variety of pathways that can lead to resistance against aminoglycosides outlined above highlights the difficulties when trying to overcome antibiotic resistance. Most commonly, acquired aminoglycoside resistance is due to the production of enzymes that modify the drug and interfere with its interaction with ribosomal RNA. The structural genes for these enzymes are usually encoded on transferable genetic elements such as plasmids and transposons and can fairly easily be disseminated to other bacteria. Given enough selective pressure, that is, the presence of antibiotics at inhibitory but not lethal concentrations, resistance will quickly spread throughout a bacterial population. Hence it is not surprising that the worst cases of multidrug-resistant bacterial strains are frequently found in hospitals where antibiotic use is common and patients with compromised immune systems provide ideal hosts.

13.1.3 New RNA targets for drug design

The development of bacterial resistance against antibiotics as outlined for the aminoglycosides above, and the central role that RNA plays in viral infection and the origin of various cancers has prompted research efforts to utilize RNA and RNA-protein interactions as novel drug targets. Even though most of these efforts are still in the experimental stage, three representative examples will be briefly discussed in the following section.

Telomerase. Telomeres are specialized structures located at the ends of linear chromosomes that are crucial for the maintenance of their integrity during cell division [4,5]. In mammalian cells they consist of tandem repeats of six nucleotides (TTAGGG) that are repeated along a stretch of 5,000–30,000 base pairs. The length of these repeat structures has been correlated with cellular aging and the induction of cellular senescence and cell death. Telomeres are maintained by a specialized enzyme, telomerase, which utilizes an RNA template to extend the length of the telomere repeats [42,39]. Telomerase is a ribonucleoprotein complex that consists of a catalytic protein subunit, the human telomerase reverse transcriptase (hTERT), and the human telomerase RNA template (hTR). The observation that telomerase activity is upregulated in many rapidly dividing cancer cells has led to an intense effort to explore the potential of telomerase inhibitors as antitumor agents.

The RNA component of telomerase (Figure 13.6) is an essential functional unit of the enzyme and as such represents a promising target for the development of inhibitors. Two different strategies have been explored in recent studies. One approach utilizes short DNA and RNA molecules with unusual (non-natural) linkages between bases/nucleotides (Figure 13.7). These molecules have sequences that are complementary to parts of the hTR sequence and utilize the antisense approach in order to inhibit telomerase activity (Figure 13.6). A modification of this approach utilizes an antisense DNA oligomer with an attached 2′,5′-oligoadenylate moiety. The latter has been shown to induce cell death in ovarian cancer cells, while having no effect on normal ovarian cells [22].

Oligonucleotides with modified backbones. The major roadblock for a medical application of drugs based on these principles is the delivery of the molecules into the target cells. Even though the modified antisense molecules are very stable in a biological environment, they are large and typically highly charged and thus cannot easily cross the cell membrane. In the "proof of principle" studies mentioned above, the cells were transfected multiple times with cationic lipids in order to deliver the antisense agents to the target cells. Antitumor drugs usually must be able to reach their target in all organs and tissues, which is much more difficult to accomplish. Therefore, future clinical applications of this strategy will critically depend on improvements in drug delivery technology. Where such pharmacokinetic problems can be bypassed or overcome, therapeutic success is feasible. This is illustrated by the success of fomivirsen (Vitravene™), a DNA-based antisense drug for the treatment of cytomegalovirus-induced retinitis. This drug is applied at high concentration

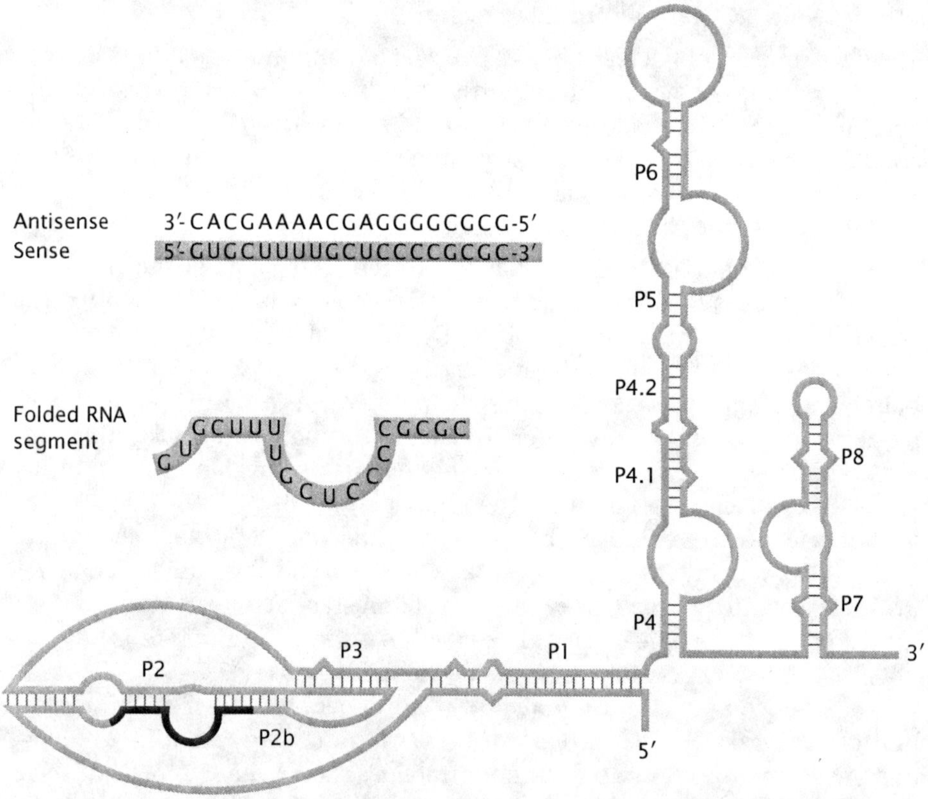

Figure 13.6 RNA component of human telomerase (hTR). The secondary structure is shown as a gray ribbon and labeled with standard domain numbers. A possible target site for antisense therapy is highlighted. In order to maximize affinity, an antisense nucleotide should hybridize with all bases in the target sequence, including those that are not base-paired within the native RNA molecule.

and in close vicinity to its target by direct injection into the *corpus vitreum* of the eyeball [42,39].

Fomivirsen is protected from enzymatic degradation by its phosphorothioate backbone (Figure 13.7A). Another phosphorothioate antisense oligonucleotide targets B-cell lymphoma protein 2 (Bcl2), which is an important regulator in programmed cell death (see Figure 12.1; [26]). This drug, named *oblimsen*, is one of several oligonucleotides destined for systemic application that are currently in phase III clinical trials.

Some additional aspects of nucleic acid delivery to target tissues are discussed in Section 14.6.

Targeting RNAs with small molecules. A more conventional drug discovery approach that also targets hTR employed a virtual screening of small molecule libraries, followed by NMR-based characterization of the most promising drug candidates [27]. In this study, the P2b stem-loop of hTR (Figure 13.6) was targeted. This region of the RNA is thought to act as a crucial molecular switch

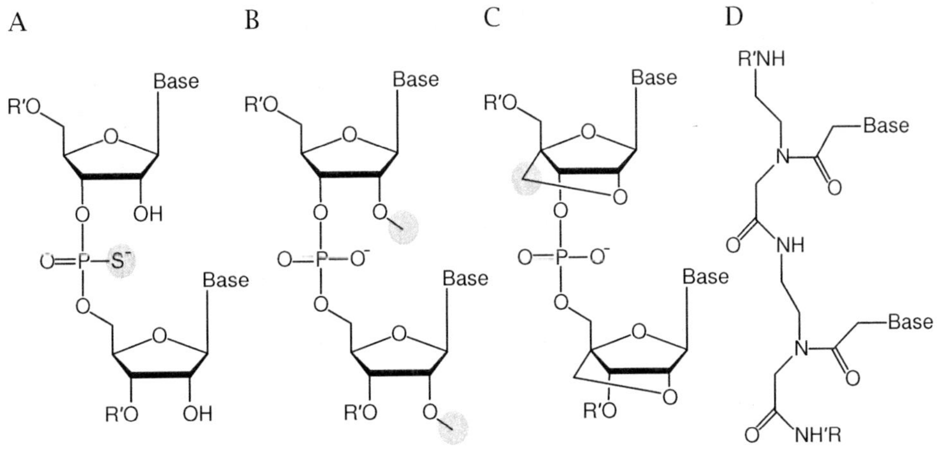

Figure 13.7 Unusual nucleotide linkages used in oligonucleotides used as therapeutic agents. **A**: Phosphorothioates, **B**: 2′-*O*-Methyl RNA, **C**: Locked nucleic acids (LNA), **D**: Peptide nucleic acids (PNA). Modified ribosephosphate or deoxyribosephosphate backbones are protected from enzymatic degradation. Peptide nucleic acids additionally render the molecule less polar and improve its permeation across membranes.

that controls access to the template portion of hTR. The study identified several small molecules that showed micromolar dissociation constants and a strong preference for binding to the hTR site over competitor RNAs, for example, the ribosomal A-site.

These results indicate that it may be possible to find small molecule inhibitors of telomerase that bind and inhibit its RNA portion. This strategy should be much less affected by the drug delivery problem that complicates the antisense approaches outlined above. It should be noted that at this time it is not clear how effective the small molecules will be at blocking RNA structural transitions in the switch region and thus at modulating telomerase function.

13.1.4 Viral RNA targets

RNA and interactions between RNA and proteins play a crucial role in the life cycle of (+)- and (−)-stranded RNA viruses and retroviruses. As such they are attractive targets for anti-viral therapy, and significant efforts are under way to identify suitable target systems and drug candidates. The human immunodeficiency virus (HIV), the causative agent of acquired immunodeficiency syndrome (AIDS), is a particularly well characterized retrovirus and can serve as an example. Much effort has gone into the development of antiviral drugs that can delay or stop the development of AIDS after an HIV infection. All currently available anti-HIV drugs target protein components of HIV, for example, the HIV protease or the integrase, which mediates the integration of the viral dsDNA into the host genome. However, the genomic RNA of the virus also contains several highly conserved elements that are crucial for function. One such element is the transactivation responsive region (TAR) at the 5′-end of all nascent HIV-1 transcripts. This 59 base stem-loop structure is crucial for viral gene expression

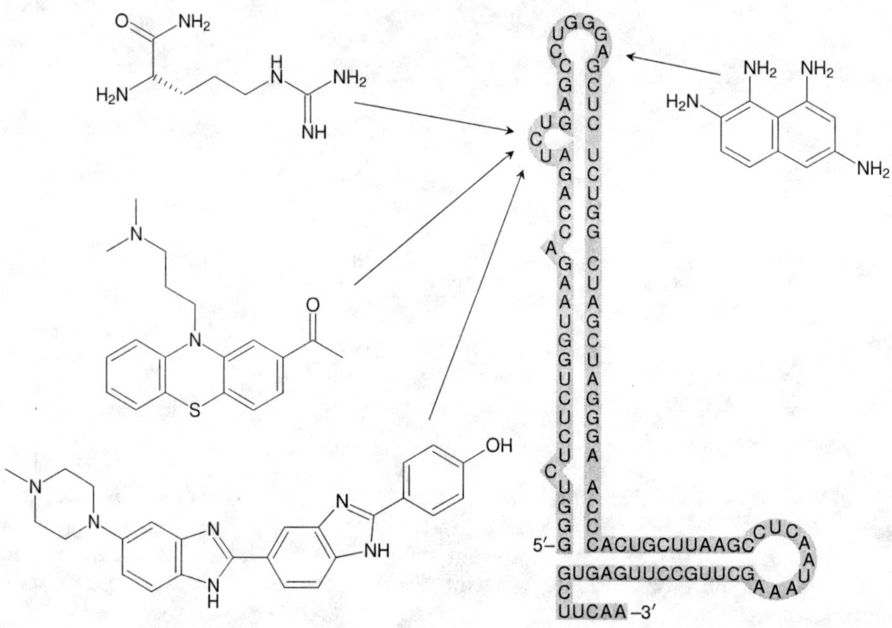

Figure 13.8 Secondary structure schematic of the HIV transactivation RNA (TAR) and structures as well as binding sites of small molecules that have demonstrated a preference for binding to this crucial RNA.

and is absent in uninfected human cells (Figure 13.8). TAR interacts with a viral protein (tat) and causes a dramatic increase in the number of viral transcripts. Thus, specific ligands that bind to TAR and interrupt its interaction with tat have the potential to be effective HIV-1 inhibitors.

Two strategies have been pursued in order to identify suitable ligands. The first utilizes small molecules and the second peptide-based ligands [21]. In these studies, both in silico screening and combinatorial libraries were applied. All of these ligands bind at the hairpin loop or the three-nucleotide bulge region of the TAR stem-loop structure (Figure 13.8). The most active of these small compounds, CGP 40336A, has IC_{50} values of 22 nM and 1.2 μM in in vitro and cellular assays, respectively. The difference is due predominantly to the differences in drug delivery; while the drug has direct access to the RNA in the in vitro assay, it needs to cross the cell membrane in the cellular assay. Among the peptide based ligands, the nonapeptide CGP 64222 has been shown to inhibit tat/TAR interactions in cellular assays at concentrations of approximately 20 μM. It was also demonstrated that it blocks HIV-1 replication in primary human lymphocytes.

These preliminary studies indicate that viral RNA can indeed be a viable target for drug design. However, to date none of these strategies have resulted in clinical applications. The main bottlenecks in the development of this type of antiviral drugs are the insufficient knowledge of the cellular targets and pharmacological effects of the compounds that have been identified so far. In

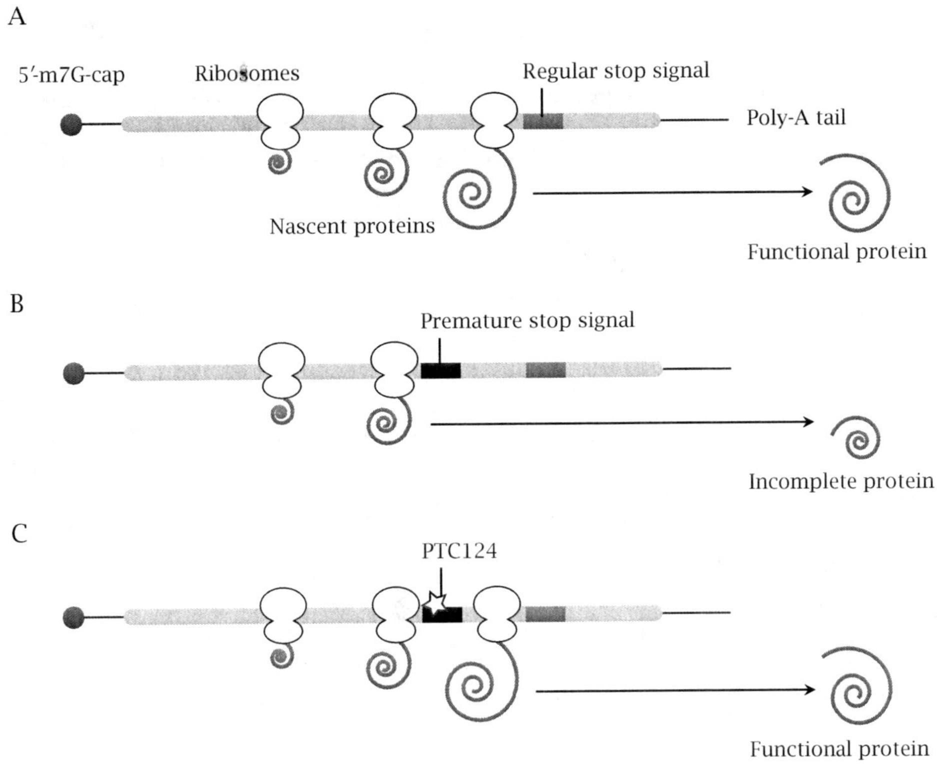

Figure 13.9 Schematic of the effect of premature termination signals and their treatment with PTC124. **A**: The normal mRNA leads to the production of full-length, active protein. **B**: A premature stop signal causes the ribosomes to terminate translation before reaching the end of the gene, resulting in truncated, inactive protein. **C**: PTC124 interacts with the mRNA/ribosome and promotes read-through, thereby allowing the production of full-length, active protein.

addition, the aforementioned differences in IC_{50} values from in vivo and in vitro assays indicates potential drug delivery problems due to low solubility, exclusion by cell membranes, and with peptides drugs also the susceptibility to degradation by proteases.

13.1.5 Genetic disorders caused by nonsense mutations

Genetic disorders are usually heterogeneous. While the disrupted gene is often the same in all patients with a given disease, the particular mutations within this gene are not. In some patients, the mutation may result in substitution of a functionally essential amino acid residue, while in others the inactivation is due to a so-called *nonsense mutation*, that is, a mutant UAA, UAG, or UGA stop codon within the coding region of the messenger RNA. A nonsense mutation will cause the premature termination of mRNA translation, which results in a truncated, non-functional peptide product (Figure 13.9). Premature translational termination may also cause destabilization of the mRNA; this is referred to as *nonsense mediated decay* [24].

Nonsense mutations cause many individual cases in cystic fibrosis and Duchenne muscular dystrophy [2].[1] Clearly, a causal therapy with conventional drugs, that is, with drugs that bind to a protein target, is not possible in patients who don't express the target protein to begin with. Therefore, until recently, the only hope for a causal treatment rested on the development of gene therapy methods for replacing the deficient gene.

However, an alternate approach has now emerged. Small molecules have been identified that bind to the complex of the mRNA and the nascent protein, and which induce the ribosome to read through the premature stop codon, presumably by accepting a mismatched tRNA in its position (Figure 13.9). These drug molecules were screened from libraries using an essay that employed a reporter gene, firefly luciferase, into which a premature stop codon had been inserted. Molecules that promoted translational read-through and thereby induced the expression of intact luciferase could then be identified by the restoration of chemiluminescence. One of these molecules, PTC124 (ataluren), is highly effective with both luciferase and dystrophin, the gene that is deficient in Duchenne muscular dystrophy [43]. While its mode of action is not yet understood in molecular detail, the drug is already in clinical testing for the treatment of both cystic fibrosis and Duchenne muscular dystrophy [36,10,14]. If this class of drugs proves to be clinically effective and safe, it should become valuable in treating genetic defects at the translational level.

13.2 RNA AS A THERAPEUTIC AGENT

In addition to their potential as drug targets, ribonucleic acids have also been the focus of intense research toward their application as drugs themselves. These studies make use of either the RNA interference (RNAi) principle or of the catalytic and ligand binding properties of specifically engineered RNA molecules. As with other biopharmaceutical drugs, the production of RNA-based drugs involves significantly higher costs when compared to small molecule drugs. The following section provides a brief overview of recent studies that explore the application of RNAi in the treatment of neurological disorders, and the application of ribozymes and RNA aptamers as therapeutic agents.

13.2.1 RNA interference

The discovery of highly specific genetic interference by double-stranded RNA (RNAi) in 1998 has provided a major new tool for cell biology. Its main impact in pharmacological research has been in target validation in vivo. However, the therapeutic potential of highly specific suppression of protein synthesis at the

[1] Cystic fibrosis results from the disruption of the cystic fibrosis transmembrane conductance regulator (CFTR) protein, which is a chloride transporter. Chloride secretion becomes limiting for overall fluid secretion across mucous membranes, which ultimately causes the symptoms. Duchenne muscular dystrophy is due to the inactivation of the protein dystrophin, which is necessary for anchoring actomyosin filaments to the cell membrane in muscle cells.

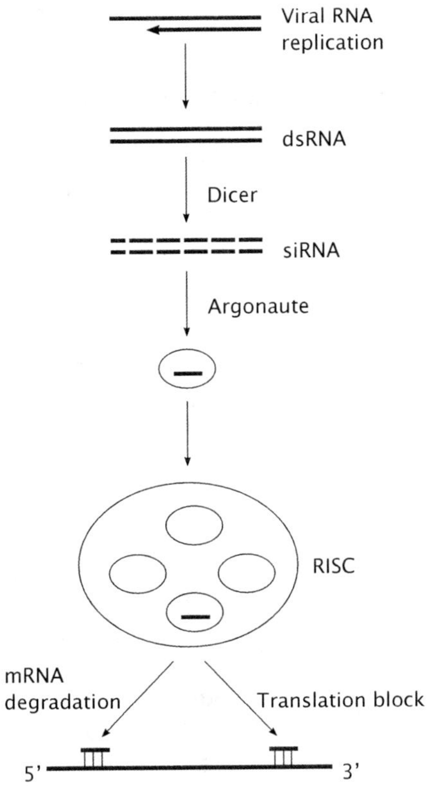

Figure 13.10 Schematic of the RNA interference pathway and its effect on protein expression. Cleavage of double-stranded RNA (dsRNA) yields small interfering RNA (siRNA). Single-stranded siRNA fragments bind to the multi-protein *RNA-induced silencing complex* (RISC), which then attaches to complementary mRNA molecules, blocking their translation and initiating their degradation.

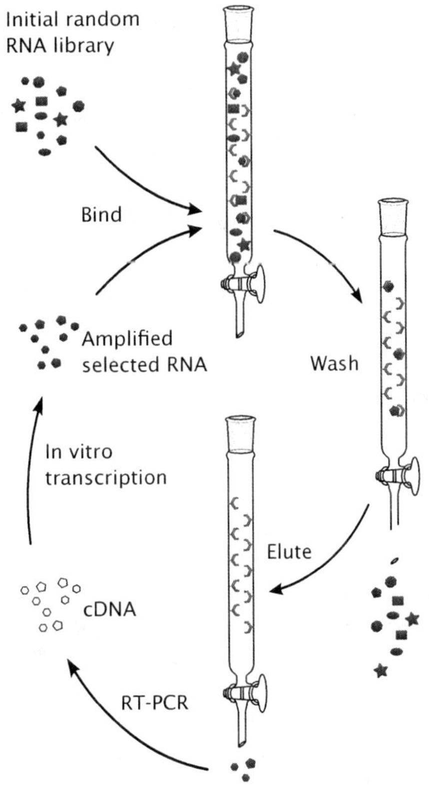

Figure 13.11 The SELEX process for the generation of RNA aptamers. The target ligand is bound to a solid phase. A library of random RNA sequences is incubated with the immobilized ligand, and unbound RNAs are washed away. Ligand-bound RNAs are eluted and amplified through RT-PCR and in vitro transcription, and the process repeated using more stringent conditions for binding and washing.

messenger RNA level was noticed and many studies are underway, some of which have reached the stage of trials in animal models [42,39].

The natural RNA interference pathway leads to posttranscriptional gene silencing via specific, small RNAs, either the double-stranded short interfering RNAs (siRNA) or micro RNAs (miRNA). These two types of interfering RNA molecules differ by their cellular origin, but ultimately they enter the same cellular pathway that eventually leads to mRNA cleavage (Figure 13.10). siRNAs are promising therapeutic agents in situations where it is necessary to reduce the expression of a specific protein in order to address a particular pathological condition. Typically siRNAs will not completely abolish the production of their target protein, but rather will reduce its expression by 30–50%.

As with other nucleic acid-based drugs, the major challenge for these applications lies in effective drug delivery. Both viral delivery and direct injection of naked siRNAs have been successfully applied in animal models of various human diseases or pathological conditions. These include depression (molecular target: serotonin transporter), pain (δ-opioid receptor, P2X3 ATP-gated cation channel, and NMDA receptor 2B), and Parkinson's disease (dopamine transporter), as well as Huntington's disease and amyotrophic lateral sclerosis (ALS) [33].

These studies demonstrate RNA interference has the potential to provide effective treatments for a number of neurological disorders. However, major challenges remain, including efficient delivery of the required amounts of siRNA, which can be hundreds of micrograms per day over extended treatment times as observed in animal models.

13.2.2 RNA aptamers

Aptamers are synthetic nucleic acids that bind to arbitrary other molecules. They are obtained by in vitro selection from very large random sequence libraries, typically containing 10^{12}–10^{14} unique sequences, through a process termed *systematic evolution of ligands by exponential enrichment*, or SELEX for short (Figure 13.11; [9,35]). Since the inception of this technique in the 1990s, aptamers have been selected for many different targets, ranging from small molecules to proteins and other nucleic acids. The pharmaceutical potential of this methodology was realized early on, but the same problems that accompany the widespread use of RNAi strategies, namely, drug delivery and cost, apply to aptamer drugs also. However, a careful choice of the target in combination with local delivery by injection can be successful, as the recent FDA approval of Macugen® (pegaptanib, by Pfizer and Eyetech) illustrates. Macugen is an aptamer drug that targets and inhibits vascular endothelial growth factor (VEGF) and is approved for the treatment of "wet" age-related macular degeneration [42,39].

13.2.3 Ribozymes

The discovery of natural ribozymes—RNA molecules that act as catalysts—was a turning point in nucleic acid science. Most natural ribozymes are catalysts of nucleolysis reactions; that is, they accelerate the cleavage of specific RNA sequences. The very high specificity of these reactions—ribozymes can target a single RNA sequence element in a cellular assay—makes ribozymes a promising tool for pharmaceutical applications [42,39]. The general principle behind these potential applications is the targeting of messenger RNAs for proteins that are related to a disease state. The destruction of the mRNAs leads to a depletion of the protein product and can thus positively affect the condition of the patient. A typical example for this type of drug candidate are ribozymes that target human epidermal growth factor receptor type 2 (HER2/neu), which are aimed at the treatment of breast and ovarian cancer.

As with all nucleic acid-based drugs discussed above, the major roadblocks for the development of efficient drugs based on ribozymes are drug delivery and cost as well as stability. While the latter problem can be overcome via the use of modified backbone chemistries, the former still pose a considerable challenge, and much of the effort for the development of ribozyme pharmaceuticals has been shifted to RNAi-based approaches that utilize significantly smaller RNA constructs.

STUDY QUESTIONS

13.1 Antibiotics that act by targeting the ribosome in bacteria have been a major game changer in the therapy of bacterial infections. What are the molecular features that differentiate the prokaryotic from the eukaryotic ribosome and that allow this level of specificity?

13.2 In addition to the examples given in this chapter, RNA–protein interactions are a major target for the development of new anticancer drugs. Several of these approaches involve signal sequences in the untranslated 5′ and 3′ regions of messenger RNAs. What kind of signals are targeted, and in which cellular pathways are these interactions found?

13.3 Micro-RNAs (miRNAs) are important signal and control markers that regulate cellular cycles and metabolism. What kind of roles do miRNAs play in cancer biology, and how can they potentially be targeted?

13.4 While RNA itself has the potential to be a very specific and effective drug in targeting protein expression related diseases, there are significant challenges with respect to the development of broadly applicable treatment strategies. What are the main challenges, and how can they potentially be overcome?

REFERENCES

[1] M. Barinaga, "The telomerase picture fills in," *Science* 276 (1997), 528–529.

[2] M. Bhuvanagiri et al., "NMD: RNA biology meets human genetic medicine," *Biochem. J.* 430 (2010), 365–377.

[3] E. H. Blackburn, "The telomere and telomerase: Nucleic acid-protein complexes acting in a telomere homeostasis system. A review," *Biochemistry (Mosc.)* 62 (1997), 1196–1201.

[4] E. H. Blackburn, "Telomerase," *Annu. Rev. Biochem.* 53 (1992), 113–129.

[5] E. H. Blackburn, "Telomeres: no end in sight," *Cell* 77 (1994), 621–623.

[6] A. Capone, "Intravitreous pegaptanib sodium (Macugen) in patients with age-related macular degeneration (AMD): Safety and pharmacokinetics," *Invest. Ophthalmol. Vis. Sci.* 46 (2005), E-Abstract 2362.

[7] R. K. Chandrakanth, S. Raju, and S. A. Patil, "Aminoglycoside-resistance mechanisms in multidrug-resistant Staphylococcus aureus clinical isolates," *Curr. Microbiol.* 56 (2008), 558–62.

[8] Y. Chen, G. Cheng, and R. I. Mahato, "RNAi for treating hepatitis B viral infection," *Pharm. Res.* 25 (2008), 72–86.

[9] A. D. Ellington and J. W. Szostak, "In vitro selection of RNA molecules that bind specific ligands," *Nature* 346 (1990), 818–822.

[10] R. S. Finkel, "Read-through strategies for suppression of nonsense mutations in Duchenne/Becker muscular dystrophy: aminoglycosides and ataluren (PTC124)," *J. Child Neurol.* 25 (2010), 1158–1164.

[11] S. Fox, "Fomivirsen for CMV retinitis, new DTaP vaccine," *Infect. Med.* 15 (1998), 590–592.

[12] R. Geary, S. Henry, and L. Grillone, "Fomivirsen: clinical pharmacology and potential drug interactions," *Clin. Pharmacokinet.* 41 (2002), p. 255.

[13] R. F. Gesteland, T. Cech, and J. F. Atkins, *The RNA world,* 2nd ed. Cold Spring Harbor monograph series 37, Cold Spring Harbor Laboratory Press, 1999.

[14] J. L. Goodier and J. Mayer, "PTC124 for cystic fibrosis," *Lancet* 373 (2009), 1426; author reply 1426–1426; author reply 1427.

[15] L. R. Grillone and R. Lanz, "Fomivirsen," *Drugs of Today* 37 (2001), 245–255.

[16] D. Haussecker, "The business of RNAi therapeutics," *Hum. Gene Ther.* 19 (2008), 451–62.

[17] S. Jana and J. K. Deb, "Molecular understanding of aminoglycoside action and resistance," *Appl. Microbiol. Biotechnol.* 70 (2006), 140–150.

[18] S. Jana et al., "RNA interference: potential therapeutic targets," *Appl. Microbiol. Biotechnol.* 65 (2004), 649–657.

[19] M. Ju et al., "Macugen Efficacy Study in Corneal Neovascularization Model," *Invest. Ophthalmol. Vis. Sci.* 46 (2005), p. 3025.

[20] D. Kim and J. Rossi, "RNAi mechanisms and applications," *BioTechniques* 44 (2008), 613–6.

[21] A. Krebs et al., "Targeting the HIV trans-activation responsive region - Approaches towards RNA-binding drugs," *ChemBioChem* 4 (2003), 972–978.

[22] D. M. Kushner et al., "2-5A antisense directed against telomerase RNA produces apoptosis in ovarian cancer cells," *Gynecol. Oncol.* 76 (2000), 183–192.

[23] S. Magnet and J. S. Blanchard, "Molecular insights into aminoglycoside action and resistance," *Chem. Rev.* 105 (2005), 477–497.

[24] L. E. Maquat, "Nonsense-mediated mRNA decay: splicing, translation and mRNP dynamics," *Nat. Rev. Mol. Cell Biol.* 5 (2004), 89–99.

[25] T. Nguyen et al., "RNAi therapeutics: an update on delivery," *Curr. Opin. Mol. Ther.* 10 (2008), 158–67.

[26] J. B. Opalinska and A. M. Gewirtz, "Nucleic-acid therapeutics: basic principles and recent applications," *Nat. Rev. Drug. Discov.* 1 (2002), 503–514.

[27] I. G. Pinto et al., "Discovery of Ligands for a Novel Target, the Human Telomerase RNA, Based on Flexible-Target Virtual Screening and NMR," *J. Med. Chem.* 51 (2008), 7205–7215.

[28] J. Poehlsgaard and S. Douthwaite, "The bacterial ribosome as a target for antibiotics," *Nat. Rev. Microbiol.* 3 (2005), 870–881.

[29] C. Potera, "First antisense NDA filed - Fomivirsen goes before the agency for product review," *Genet. Eng. News* 18 (1998), p. 1.

[30] M. I. Recht, S. Douthwaite, and J. D. Puglisi, "Basis for prokaryotic specificity of action of aminoglycoside antibiotics," *EMBO J.* 18 (1999), 3133–3138.

[31] E. M. Rezler, D. J. Bearss, and L. H. Hurley, "Telomere inhibition and telomere disruption as processes for drug targeting," *Annu. Rev. Pharmacol. Toxicol.* 43 (2003), 359–379.

[32] E. M. Rezler, D. J. Bearss, and L. H. Hurley, "Telomeres and telomerases as drug targets," *Curr. Opin. Pharmacol.* 2 (2002), 415–423.

[33] D. W. Y. Sah, "Therapeutic potential of RNA interference for neurological disorders," *Life Sci.* 79 (2006), 1773–1780.

[34] F. Schlünzen et al., "Structural basis for the interaction of antibiotics with the peptidyl transferase centre in eubacteria," *Nature* 413 (2001), 814–821.

[35] D. Schneider, L. Gold, and T. Platt, "Selective enrichment of RNA species for tight binding to Escherichia coli Rho factor," *FASEB J.* 7 (1993), 201–207.

[36] I. Sermet-Gaudelus et al., "Ataluren (PTC124) Induces CFTR Protein Expression and Activity in Children with Nonsense Mutation Cystic Fibrosis," *Am. J. Respir. Crit. Care Med.* (2010).

[37] L. Singerman, "Findings of the Phase 2 Trial of the Safety and Efficacy of Pegaptanib Sodium (Macugen&x2122;) in Patients With Diabetic Macular Edema," *Invest. Ophthalmol. Vis. Sci.* 46 (2005), p. 4674.

[38] M. D. de Smet, C. Meenken, and G. J. van den Horn, "Fomivirsen - a phosphorothioate oligonucleotide for the treatment of CMV retinitis," *Ocular Immunology and Inflammation* 7 (1999), 189–198.

[39] B. A. Sullenger and E. Gilboa, "Emerging clinical applications of RNA," *Nature* 418 (2002), 252–258.

[40] T. Tenson and A. Mankin, "Antibiotics and the ribosome," *Mol. Microbiol.* 59 (2006), 1664–1677.

[41] C. A. Theimer and J. Feigon, "Structure and function of telomerase RNA," *Curr. Opin. Struct. Biol.* 16 (2006), 307–318.

[42] N. G. Walter and D. R. Engelke, "Ribozymes: catalytic RNAs that cut things, make things, and do odd and useful jobs," *Biologist (London)* 49 (2002), 199–203.

[43] E. M. Welch et al., "PTC124 targets genetic disorders caused by nonsense mutations," *Nature* 447 (2007), 87–91.

CHAPTER 14

DRUG DELIVERY

by Alice Chan

Many drugs that have potentially useful therapeutic activity fall short in practice for various reasons. They may fail to reach their targets as a result of inefficient absorption or rapid metabolic inactivation, or because their target sites are poorly accessible, such as those behind the blood–brain barrier. Other drugs, while able to reach their targets, may have unacceptably low therapeutic indices.

The diverse techniques, materials, and devices that have been invented to overcome such problems by controlling the time, rate, and destination of drug release are subsumed under the name *drug delivery*. This is a varied and rapidly growing field; here, we will consider a few selected examples to illustrate the possibilities. We will see that drug delivery methods may provide benefits such as (1) improved drug solubility, absorption, and distribution, (2) targeted delivery, (3) kinetically controlled release, and (4) reduced drug toxicity [32,1].

14.1 IMPROVING INTESTINAL DRUG ABSORPTION

Section 3.3 presented the different routes of drug application. Drugs that are applied orally meet their first challenge when exposed to hydrochloric acid in the stomach. Of those that are stable enough to reach the intestine, only the ones with suitable solubility and polarity will be able to traverse the gut

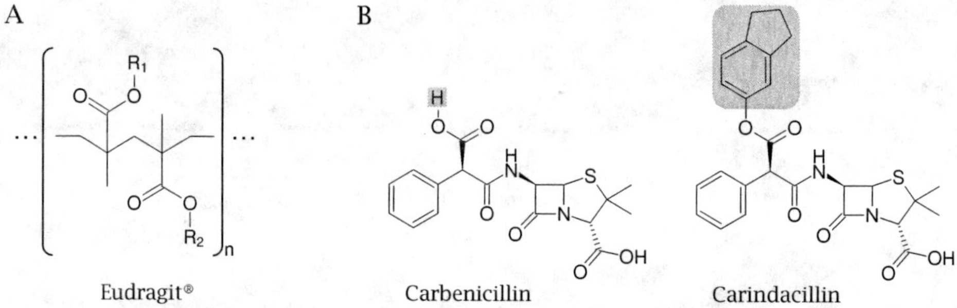

Figure 14.1 Protecting drugs from degradation by gastric acid through polymer coating or prodrug formation. **A:** General structure of methacrylic acid copolymers (Eudragit®), a class of polymers used for coating drugs. Examples: R_1 = H, R_2 = CH_3: Eudragit L; R_1 = CH_3, R_2 = CH_2CH_3: Eudragit L100-55; R_1 = R_2 = CH_3: Eudragit NE30D. **B:** Carindacillin is an orally applicable prodrug of carbenicillin. The ester has a greater acid resistance than the parent carboxylic acid and also crosses the intestinal epithelium more easily.

epithelium, and among these, only the ones that survive the first-pass effect will reach the systemic circulation.

Notwithstanding all these hurdles, for patients' convenience and compliance, oral application is usually preferable. Much effort has therefore been dedicated to improving the stability of drugs within and their absorption from the digestive tract.

14.1.1 Protecting drugs from degradation

Coating oral drugs with polymers allows us to control the location of release and absorption within the digestive system. For instance, methacrylic acid copolymers are resistant to gastric acid because of their low solubility at low pH (Figure 14.1A). Hence, a drug coated with this copolymer will not be released in the stomach and will thereby be protected from the gastric acid. When it encounters the slightly alkaline milieu in the small intestine, the polymer dissolves, and the drug enclosed within is released.

The earliest example of this type of protective copolymer dates back to 1953 and was marketed as Eudragit®. Subsequently introduced variants of Eudragit have different ratios of carboxylic acid groups to ester groups, which give them different solubility profiles, and provide varying rates of drug release at different pHs [12].

Another approach to avoid undesired drug degradation is to make the drug itself more resistant to acidic or enzymatic breakdown, that is, to create a prodrug.[1] Carbenicillin, a penicillin, cannot be applied orally, since it is highly susceptible to gastric acid; at pH 2.0 and 37°C, the drug is completely hydrolyzed within less than one hour. In contrast, if the carboxylic acid group on carbenicillin is converted to its 5-indanyl ester, the resulting compound is

[1]Some authors do not consider prodrugs a "proper" drug delivery strategy. However, since prodrugs and other methods of drug delivery address fundamentally the same problems, we will consider prodrugs here, too.

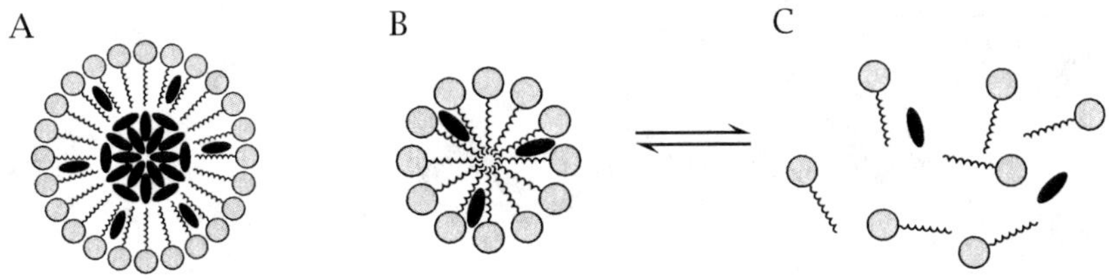

Figure 14.2 Solubilization of drugs with surfactants. Surfactants (or detergents; gray) are amphiphilic molecules that in aqueous solution aggregate into micelles, which can accommodate aggregates (**A**) or individual molecules (**B**) of hydrophobic drugs (black) within. Most surfactants rapidly equilibrate between the micellar state and the monomeric state. When the surfactant is diluted to below its critical micellar concentration, the micelles will rapidly dissipate, and the drugs contained within will be released (**C**).

stable under the same conditions for 23 hours. In addition, esterification of the carboxylic acid group also enhances the absorption. The prodrug, carindacillin, can thus be administered orally. Once absorbed, carindacillin is hydrolyzed in the liver to release carbenicillin [47,20,39] (Figure 14.1B).

14.1.2 Improving drug solubility

The solubility of a drug is one of the key factors in determining its pharmacokinetics with any route of administration. Drug molecules with many ionic groups may become insoluble at certain pH values and form precipitates in our body fluids, whereas drug molecules with few or no polar moieties will be very hydrophobic and not be water-soluble to begin with. Either way, the lack of solubility will hamper absorption and distribution. The solubility of drugs may be improved either through the use of surfactants or through modifications of functional groups on the drug molecules themselves.

Surfactants or detergents are amphiphilic molecules that solubilize drugs in much the same way as they solubilize fat residues in dish washing. When a surfactant is added to a suspension of a hydrophopic drug in water, it will disperse the "oily" or "fatty" drug into very small droplets or micelles, which are covered and stabilized by a monolayer of surfactant molecules; at a sufficiently high ratio of surfactant to cargo drug, there may be only a few drug molecules contained in each micelle. This facilitates the rapid dissipation and distribution of the drug once the emulsion is diluted into the body fluids (Figure 14.2).

Examples of drug preparations containing surfactants are oil-soluble vitamin D_3 (cholecalciferol; see Figure 7.14), which is admixed with the surfactant Tween-80 [35], and the antifungal drug amphotericin B admixed with the surfactant deoxycholate; both are applied parenterally. Deoxycholate forms very small micelles and has a high critical micellar concentration. It therefore dissipates very quickly upon dilution after parenteral application and releases the cargo drug all at once. Liposomes, which are more stable than surfactant micelles, will release the drug in a more controlled fashion, with can help to control toxicity. The use of deoxycholate and of liposomes for amphotericin B delivery is discussed in Section 14.5.

Box 14.1 Capsules and pellets: Of crime and its role in the progress of drug therapy

Protection of drugs with copolymers can be applied in various forms. The simplest form is a capsule that is filled with powdered drug preparations. A more involved procedure is to coat small drug particles individually and then press them into pellets.

Two decades ago, a Tylenol™ (acetaminophen) pill was a capsule; however, nowadays, it is usually a pellet. This change in drug formulation was motivated by the following events:

In September 1982, in a suburb of Chicago, a 12-year-old girl was found dead after taking a capsule of Extra Strength Tylenol. A few hours later, a young post office worker in another suburban area collapsed and died, also after taking some Tylenol capsules. Subsequently, the worker's wife and brother took pills from the same bottle and also died within a few minutes.

Forensic investigations revealed that the capsules had been opened and filled with potassium cyanide. A man named James Lewis was arrested and convicted for sending extortion letters to the drug's manufacturer, Johnson & Johnson, demanding payment of one million dollars in return for discontinuation of the manipulations; however, he was not found guilty of the murders themselves. The investigation has recently been renewed, but no new evidence has been released to the public.

To prevent any future tampering, efforts were made in the industry to replace capsules by polymer-coated Tylenol tablets. Research directed toward tamper-resistant packages was also intensified. In 1983, the US congress passed the Federal Anti-Tampering Act, issued by the US Food and Drug Administration, which made tampering with a consumer product package a crime [29,16,18].

Absorption of an orally applied drug requires the drug to pass through all membranes in the intestinal epithelium, most often by diffusion. If a drug is more lipophilic, it can permeate across the lipid bilayer more easily, whereas polar groups inhibit permeation (see Section 3.1). For instance, the angiotensin receptor inhibitor EXP7711 has poor epithelial permeability, because its carboxylic acid group is ionized at the physiological pH (Figure 14.3). However, if the acid group is replaced by other functional groups without an acidic proton, the potency of the drug decreases significantly [21]. To enhance the absorption of the compound and still maintain a high binding affinity at the target site, a tetrazole group was used to replace the carboxylic acid group. Like the latter, tetrazole is acidic and has a planar structure, but it is 10 times more lipophilic than COOH [47]. The pK_a values of both tetrazoles and carboxylic acids are similar ($pK_a \sim 4$); therefore, both groups are ionized at physiological pH. Tetrazoles cross the epithelium more readily because the negative charge on tetrazole is more delocalized. The lower charge density means that the ion's hydration shell is not as tightly bound and can be more readily be stripped off as the ion partitions into the lipid membrane.

Functional groups that have similar physical or chemical properties, and also have similar biological activities, like carboxylic acid and tetrazole in this example, are called *bioisosteres* [28,54]. Note, however, that the replacement of carboxylic acid by tetrazole may not work with other drugs.

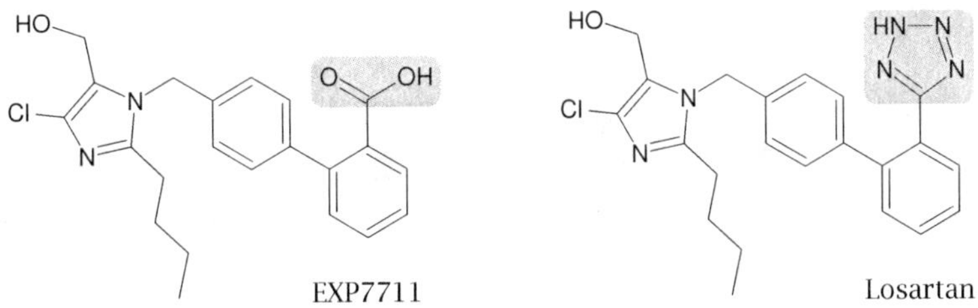

Figure 14.3 Structures of the angiotensin receptor inhibitors EXP7711 and losartan. The acidic carboxylate group of EXP7711 inhibits intestinal absorption, yet an acidic group is necessary to achieve a high IC_{50}. Like carboxylate, tetrazole is acidic and has a planar structure, but it is 10 times more lipophilic than COOH [47]. Replacement of the carboxylate in EXP7711 with tetrazole gives losartan, which is absorbed much more readily.

Replacing carboxylic acid groups with other functional groups is often useful in improving drug absorption, as illustrated by the esterification of ampicillin to bacampicillin (see Figure 3.3) and the esterification of carbenicillin to carindacillin (above). Esterification is just one of the numerous functional group transformations that may enhance the absorption of a drug. For any given drug, the best strategy will have to be determined by trial and error.

Physicochemical properties and solubility. Molecules in a crystal are arranged regularly and held together by intermolecular forces. During solubilization in a solvent, these forces have to be overcome in order to release the molecules from the crystal lattice. The strength of the intermolecular forces changes when the molecules align in a different way. Hence, different patterns of crystal packing may cause significant differences in solubility.

One of the most dramatic examples to demonstrate the impact of a drug's crystal form on bioavailability is the HIV protease inhibitor ritonavir [4]. This drug is marketed in semisolid and liquid forms. Both formulations contain an ethanol/water mixture because the solid form is poorly soluble and poorly absorbed. In 1998, the semi-solid preparation was withdrawn, when it was discovered that over time it gave rise to a previously unknown crystal form that interfered with absorption of the drug. This form, now known as form II, is thermodynamically more stable, and is thus much less soluble, than the previously known crystal form I. The product was reformulated and reintroduced in 1999 as soft gel capsules containing ritonavir in dispersed amorphous form, i.e. with the molecules arranged within the particles in an irregular manner [38].

Another aspect that affects solubility of a drug is its chirality. Undergraduate organic chemistry tells us that enantiomers have identical physical properties. Indeed, the solubilities of two enantiomers are identical in achiral solvent. However, the solubility of pure enantiomers may differ from that of a racemic mixture of the same compound; this is because the crystal packing of the racemic mixture will differ from that of the pure enantiomers.

For instance, the antitumor drug 1,2-di(4-piperazine-2,6-dione)propane is 5 times more soluble in its enantiomerically pure form, which thus is taken up across the skin more readily than the racemic drug [48]. However, whether racemate and pure enantiomers differ in their pharmacokinetics may vary enormously from one drug to another. As long as the pharmacokinetics of the racemate are good enough and neither enantiomer has serious side effects, a drug will usually be marketed as the racemate for economical reasons.[2]

14.1.3 Avoiding drug absorption

In certain cases, drug absorption is not necessarily favorable. For example, in the treatment of intestinal infections, we want to maximize the drug concentration within the intestinal tract itself. In the early days of antimicrobial chemotherapy, the sulfonamide sulfathiazole was commonly used for such treatment. However, when sulfathiazole is absorbed, it is enzymatically acetylated; the resulting metabolite is poorly soluble in the blood plasma. Grave consequences may result when the acetylated metabolite precipitates within and blocks the kidney tubules [47].

To avoid the absorption of sulfathiazole, it can be replaced with the prodrug succinylsulfathiazole (Figure 14.4). The carboxylic acid group of this compound is ionized in the intestine, and is therefore less readily absorbed. Enzymatic hydrolysis generates active sulfathiazole slowly in the intestine, which will reduce the drug concentration and prevent the systemic side effects. Interestingly, the activating enzyme does not come from the human body but from the bacterial flora inside the colon [51]. Other therapies involving colon-specific drug delivery also rely on enzymes from the colon flora. For example, the prodrug balsalazide is converted to the active anti-inflammatory agent mesalazine (also known as 5-aminosalicylic acid) through enzymatic azoreduction in the large intestine [23].

14.2 IMPROVING DRUG DISTRIBUTION

The degree of difficulty associated with the distribution of a drug from the bloodstream to its target depends on the location of the target. A notoriously difficult location is within the central nervous system, that is, behind the blood-brain barrier (BBB).

14.2.1 Dopamine and its prodrugs

In Parkinson's disease, the *substantia nigra* in the brain stem degenerates, causing the dopamine level in the brain to drop; this causes the characteristic neurological deficits. Accordingly, a key objective of pharmacological treatment is to restore the brain dopamine level. Dopamine itself cannot permeate the BBB and thus cannot be used as a drug. Fortunately, its metabolic precursor, L-DOPA or levodopa, manages to get across. Levodopa is converted to dopamine inside the brain by dopa decarboxylase (also called aromatic L-amino acid decarboxylase). Thus, levodopa can be used as a prodrug for dopamine in Parkinson's disease.

[2]Ibuprofen is generally marketed as racemate, in which the *R* enantiomer is the active form but *S* is inactive. An isomerase, 2-arylpropionyl-CoA epimerase, converts the *R* form to the active *S* enantiomer.

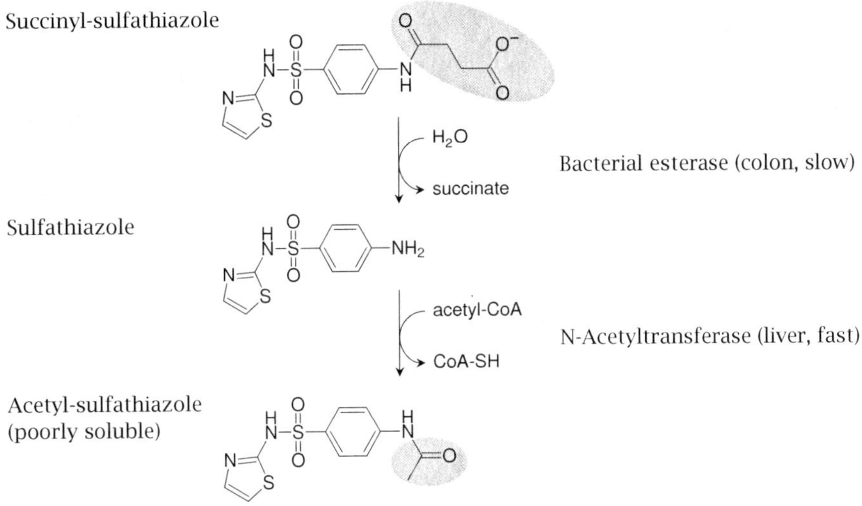

Figure 14.4 Succinylsulfathiazole, a prodrug designed for *reduced* absorption. When sulfathiazole is absorbed, it is converted to a poorly soluble metabolite through *N*-acetylation. The acetylated compound may block the kidney tubules and cause fatal consequences. Succinylsulfathiazole is less toxic because the compound is ionized and thus not taken up efficiently. Once the drug reaches the large intestine, it is hydrolyzed by bacterial esterases. The slow rate of hydrolysis means that systemic sulfathiazole concentrations will remain low throughout.

Levodopa is even more polar than dopamine.[3] Therefore, you may wonder why the blood-brain barrier is permeable to levodopa but not dopamine, because normally the more polar compounds are less likely to traverse membrane barriers (see Section 3.2). Now, if you look more closely in Figure 14.5, levodopa has a phenylalanine group "embedded" in the molecule. This structural similarity allows levodopa to be transported by the neutral amino acid transporter (LAT1), a carrier protein expressed at the BBB that transports phenylalanine as well as several other amino acids.

When levodopa is administered alone, nausea and vomiting are common side effects. That is because dopa carboxylase does not just exist in the brain, but is also present at even higher levels in the liver, heart, lungs, and kidneys. If levodopa is administered orally, more than 95% is converted to dopamine in the periphery and probably less than 1% enters the brain; the dopamine level in the bloodstream is hence greatly increased [8].[4] Dopamine is also the precursor of norepinephrine and epinephrine, which stimulate adrenergic receptors. Peripheral production of dopamine can therefore induce excessive

[3]The logP values (see Section 3.2.1), as estimated by the software program XlogP 3 [11], are -1 and -2.7 for dopamine and levodopa, respectively. This difference indicates that levodopa prefers water about 50 times more strongly than dopamine. XlogP values of many compounds can be found in the PubChem database.

[4]Nausea is triggered in the *area postrema*, a small, specialized group of neurons in the brain stem that senses noxious stimuli and is *not* effectively hidden by the BBB; hence the effect of peripherally converted dopamine.

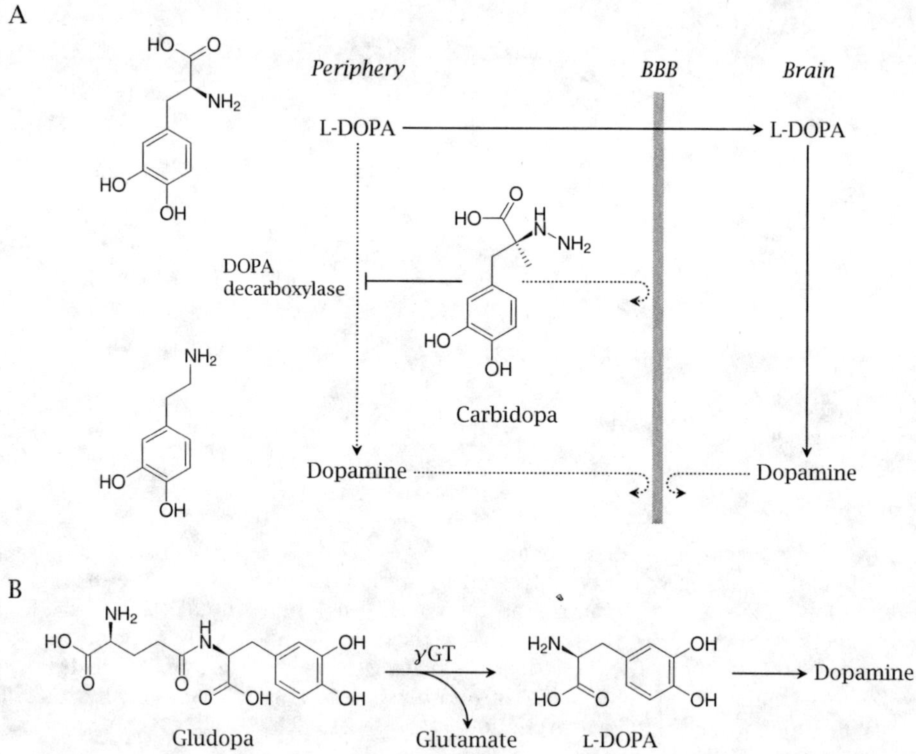

Figure 14.5 Dopamine and its prodrugs. **A:** Dopamine does not cross the BBB. In the treatment of Parkinson's disease, its metabolic precursor, levodopa (L-DOPA), is used as a prodrug. Levodopa crosses the BBB by active transport and is then converted to dopamine by DOPA decarboxylase. Levodopa is combined with carbidopa or benserazide, which inhibit DOPA decarboxylase in the periphery. Like dopamine, these compounds do not cross the BBB and thus do not interfere with dopamine metabolism in the brain. **B:** The kidneys contain a high activity of L-γ-glutamyl-transpeptidase (γGT). This is exploited for the selective release of levodopa from gludopa. The levodopa is then converted to dopamine, which improves kidney perfusion.

adrenergic stimulation, leading to vasoconstriction, increased heart rate and blood pressure.

To prevent peripheral conversion of levodopa to dopamine and to improve its delivery to the brain, DOPA decarboxylase inhibitors such as carbidopa or benserazide are used in combination. These drugs are unable to permeate the BBB; thus, the conversion of levodopa is inhibited outside but not inside the brain. As a result, the oral dose of levodopa can be reduced by more than 75% [52], and the side effects are diminished accordingly.

Aside from its application in Parkinson's disease, dopamine can also be used in the periphery, for example, to increase renal blood flow. In this case, no particular problem of drug distribution occurs, and dopamine can be applied directly; however, this will still induce the side effects described above. To reduce systemic side effects, the prodrug L-γ-glutamyl-L-dopa (gludopa) is used to target the kidney selectively. In the kidneys, there is a particularly high concentration of the enzyme γ-glutamyltransferase (γ-GT), which catalyzes the release of L-DOPA from gludopa [52]. L-DOPA is then locally transformed to dopamine by DOPA decarboxylase (Figure 14.5). In rats, the dopamine level in the kidneys attained by gludopa was found to be 5 times higher than with an equivalent dose of levodopa [58].

14.2.2 A prodrug with a little twist

In contrast to polar molecules such as dopamine, hydrophobic molecules can traverse the BBB very easily. However, these molecules go back to the periphery equally well; therefore, in order to maintain a sufficient level inside the brain, the concentration in the periphery will have to be kept up as well, which may be undesirable.

As an example of the foregoing, consider estradiol, an estrogen that is used to treat menopausal vasomotor symptoms such as "hot flashes" [5]. Estradiol is lipophilic; its esters, such as estradiol valerate, can be even more hydrophobic and will be cleaved by esterases that occur both in the brain and elsewhere. While it thus is easy for estrogen to reach its receptors inside the brain, a continuously maintained high concentration in the periphery is known to put women at increased risk for breast cancer.

A possible way out of this dilemma would be to develop a prodrug that enters the brain and remains inside for extended periods of time, even after the level in the periphery has returned to near zero. An estradiol prodrug has been devised that achieves this effect by way of a functional group that is initially uncharged, but through a metabolic reaction is turned into a cation, rendering the entire compound membrane-impermeant (Figure 14.6). This functional group is 1,4-dihydro-*N*-methylnicotinic acid (dihydrotrigonelline). Its oxidation is facile,[5] and as such, the conversion of the lipophilic prodrug to its charged metabolite will occur on both sides of the blood–brain barrier. However, the molecules converted within the brain will remain trapped there and slowly release estradiol through esterase cleavage [23]. In contrast, outside the brain the charge introduced by the metabolic conversion will promote the renal elimination of the prodrug metabolite, so that the level of free estradiol there will be much lower than in the brain.

The ionized methylnicotinic acid that also results from esterase cleavage is polar but is readily eliminated from the brain by an active transport system for small organic ions. Overall, a preferential delivery and release of estradiol inside the brain is achieved, in which the half-life of the quaternary intermediate inside the brain is several times longer than in the periphery [6]. Phase I and II clinical evaluations in volunteers indicated that, following both buccal and intravenous application, estradiol 17-dihydrotrigonelline achieved more effective brain targeting than estradiol valerate, a compound with similar lipophilicity and BBB permeability [6].

[5]The conversion of dihydrotrigonelline to the pyridinium form resembles that of NADH to NAD^+, but it is not clear which, if any, enzymes catalyze this conversion.

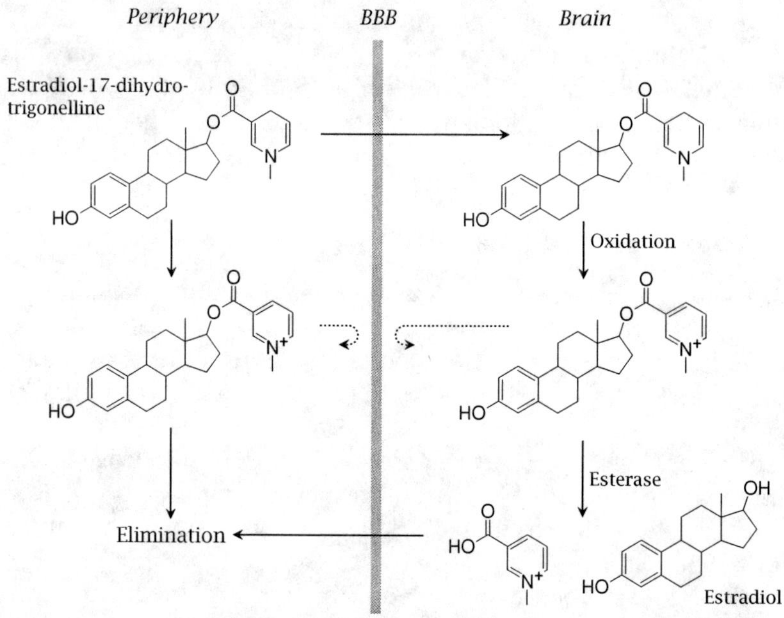

Figure 14.6 Trapping of an estradiol prodrug inside the brain. The 1,4-dihydro-*N*-methylnicotinic acid ester of estradiol can enter the brain by diffusion. Oxidation of 1,4-dihydro-*N*-methylnicotinic acid occurs in both the periphery and the brain. The charge introduced by oxidation will accelerate elimination in the periphery but inhibit elimination in the brain, since the charged molecule is no longer able to cross the BBB. Estradiol is slowly released from the trapped prodrug by esterases.

Like estradiol 17-dihydrotrigonelline, the combinations of levodopa with DOPA decarboxylase inhibitors achieve preferential release of the final drug inside the brain. However, the elegance of the dihydrotrigonelline approach lies in the fact that this preference is conferred entirely by the conjugated carrier group; it is therefore potentially applicable to many other cargo drugs.

Both methods also illustrate the potential of selective drug delivery. Further methods addressing this problem are discussed in the next section.

14.3 TARGETED DRUG DELIVERY

After a drug is absorbed and distributed into the bloodstream, the goal is to maximize the therapeutic action and to minimize the side effects. This is easy to achieve with drugs that have inherently high therapeutic indices, such as penicillin; several grams of penicillin may be used per day on a single patient without causing any serious side effects. However, most drugs have lower therapeutic indices than penicillin; many of the cytotoxic drugs used in cancer therapy are cases in point. In these cases, it would be highly desirable to have the drug selectively, or at least preferentially, delivered to the cells or tissues containing the targets. The drug's therapeutic index would in effect be multiplied by the degree of enrichment within the diseased tissue.

14.3.1 Liposomes as drug carriers and the EPR effect

Around 1990, researchers found that serum albumin that was labeled with Evans Blue preferentially accumulated in tumor tissues after intravenous injection. Larger proteins such as transferrin (90 kDa) and IgG (160 kDa) behaved similarly. However, small proteins such as ovomucoid (29 kDa) and neocarzinostatin (12 kDa) did not accumulate in tumors. This is due to the fact capillaries in tumors, as well as in tissues affected by inflammation, are more porous and leaky. When a tumor metastasis reaches a size of 2–3 mm, angiogenesis is induced. A network of blood vessels branches out from existing blood vessels. The newly formed capillaries are defective, in that the endothelial cells are poorly aligned and disorganized, and the gaps between them are much larger (~400 nm) than in normal tissues (~2 nm). As a result, macromolecules that are too large to extravasate in normal tissues will leak out selectively into the tumor tissues (Figure 14.7) and accumulate there [49,36]. In addition, tumor tissues often have an impaired lymph clearance, which will increase the dwell time of the extravasated macromolecules. This selective extravasation and accumulation of macromolecules in tumor tissue is referred to as the *enhanced permeability and retention* (EPR) effect [30].[6]

In principle, the EPR effect will enable selective accumulation of an antitumor drug that is coupled to a macromolecular carrier. However, the extravasation of macromolecules into tumors is a slow process; therefore, the carrier-bound drug must stay in the circulation for a long period of time. While macromolecular carriers will typically be exempt from rapid clearance by the kidneys or metabolic inactivation, they much more readily draw the attention of the immune system than small molecules do. Rapid clearance can ensue when carrier proteins react with antibodies and subsequently undergo phagocytosis. Carrier liposomes may react with the proteins of the serum complement system, which is somewhat similar to the plasmatic coagulation system in that it can be activated by nonphysiological surfaces, including simple phospholipid membranes. This will lead to permeabilization of the liposomes and thus to premature release of any drugs they contain. It will also accelerate the removal of the liposomes from the plasma through phagocytosis. However, the reaction of liposomes with complement proteins can be largely suppressed with suitable surface modification, in particular with polyethylene glycol (Figure 14.8). Liposomes accordingly are gaining traction as a delivery vehicle for anticancer drugs.

Liposomes may consist of natural or semisynthetic phospholipids. In aqueous suspension, most phospholipids arrange themselves to form spherical bilayer membranes, where the fatty acyl tails form a continuous lipophilic phase and the polar headgroups face the exterior and interior aqueous phase. Within these spherical membrane particles, hydrophilic drugs stay inside the aqueous interior, whereas hydrophobic ones reside in the hydrophobic acyl chain

[6]The increased blood flow toward the tumor may promote vascular permeability, the EPR effect and hence the tumor growth. The use of NO scavengers and NOS inhibitors that cause vasoconstriction suppresses the extravasation of Evans blue.

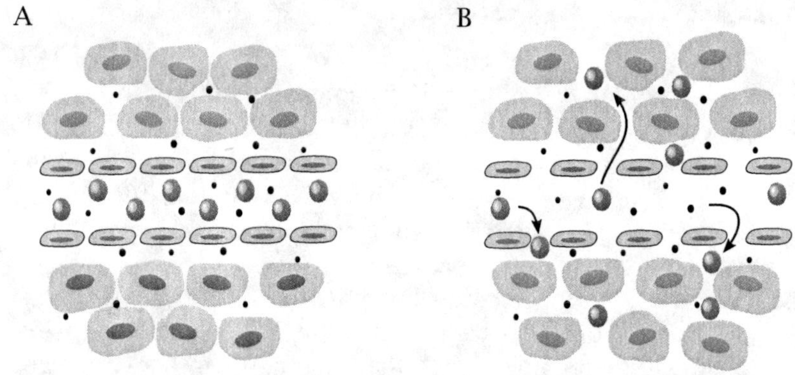

Figure 14.7 The enhanced permeability and retention (EPR) effect. Small molecules can diffuse out of capillaries in both normal tissue (**A**) and cancerous tissue (**B**). In contrast, large molecules and particles can leave the capillaries in tumors only because here the gaps between endothelial cells are enlarged. The lack of lymphatic drainage in tumors further promotes accumulation of macromolecules in these tissues.

region of the phospholipid layer itself [3] (Figure 14.8). The optimal size of liposomes for cancer treatment is around 100 nm. If the size of liposomes exceeds 200 nm, their susceptibility to phagocytosis in the liver and spleen is increased, resulting in rapid clearance. On the other hand, if they are too small, the selectivity of the EPR effect is lowered.

An example of liposomal drug delivery of an anti-cancer drug is doxorubicin (see Figure 12.16). The plasma half-life of free doxorubicin is 5 minutes. When the drug is packaged into liposomes without any particular surface modification, the half life increases to 2–3 hours. It can be lengthened further by derivatizing the surface of the liposomes with polyethylene glycol (PEG), which creates a steric barrier for complement and other proteins that would otherwise interact with the lipid bilayer to induce permeabilization and promote phagocytosis [22,2]. The plasma half-life of doxorubicin contained in PEGylated liposomes (Doxil®) is 8 times longer than that of the drug in unmodified liposomes [27].

In a mouse model of mammary carcinoma, Doxil showed greater efficacy in curing implanted tumors and in lowering the incidence of metastases from these tumors, compared to doxorubicin-loaded liposomes without surface modification [37]. In another animal model, conventional liposomes failed to arrest the growth of a human lung tumor xenograft, while Doxil was effective [59].

14.3.2 Antibody conjugates

Antibodies, in particular monoclonal ones, are increasingly important in tumor therapy. Many antibodies function directly as drugs. An example is the monoclonal antibody trastuzumab, which neutralizes an important growth factor receptor in breast cancer (see Section 12.3).

Alternatively, antibodies may simply act as targeting devices for other types of drugs conjugated to them. In this case, the antigen recognized by the anti-

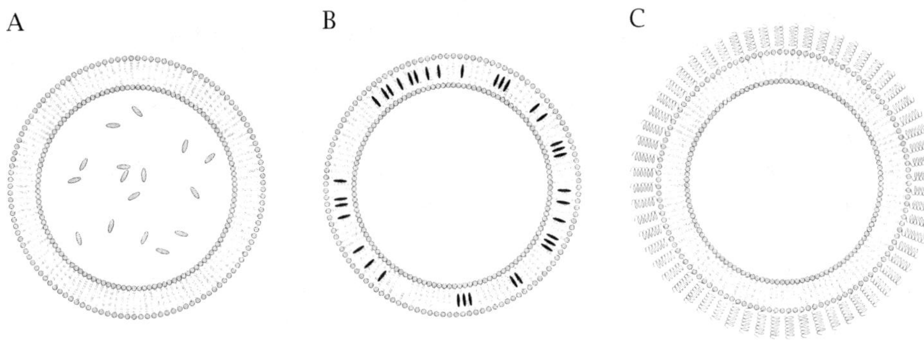

Figure 14.8 Liposomes as drug delivery vehicles. Liposomes are vesicles that consist of natural or synthetic phospholipids. They can vary widely in size, but the ones used in drug delivery are most often between 50 and 200 nm in diameter. Hydrophilic cargo drugs will be enclosed in the lumen (**A**), whereas hydrophobic ones will reside in the bilayer itself (**B**). Liposomes can be protected from disruption by the proteins of the complement system and from phagocytosis by surface derivatization with hydrophilic polymers such as PEG, which will greatly increase their stability in the circulation (**C**).

body need not be functionally crucial to the growth of the tumor by itself; the only requirements are that the antigen be located on the cell surface, and that it be selectively, or at least preferentially, expressed on the tumor cells. Different types of cytotoxic effectors can be targeted to tumor cells using antibodies. We will consider three examples.

Antibody-cytotoxin conjugates. Antibodies that bind to cell surface antigens will often be internalized together with their antigens by the target cell. Any cytotoxic drugs conjugated to the antibody will then be taken along for the ride. To make this useful for drug delivery purposes, the linkage between the antibody and the drug must allow the drug to detach from the antibody, so that it can interact with its intracellular target.

One such conjugate has been used in the treatment of acute myeloid leukemia (AML). This disorder results from the malignant transformation of a type of myeloid stem cell. In about 80% of all AML patients, the cell surface protein CD33 is expressed by the leukemic cells but not by normal hematopoietic stem cells. Therefore, in these patients, conjugated anti-CD33 antibodies can be used to target the leukemic cells. However, the expression level of CD33 in the leukemic cells is low; a highly potent drug is thus required that will deliver a decisive cytotoxic effect towards the cancer cells even at low concentration [26].

This latter requirement is met by calicheamicins, a class of antibiotics obtained from the soil bacterium *Micromonospora echinospora*. Calicheamicins are powerful mutagens and kill cells in low picomolar concentrations [17]. These compounds bind to the minor groove of the DNA and then induce double-strand DNA cleavage [60]. On binding of a calicheamicin molecule, its trisulfide bond is reduced by glutathione. Subsequent rearrangements generate a highly reactive diradical intermediate that causes the DNA damage (Figure 14.9).

The conjugate of one such compound with anti-CD33 antibody has been blessed with the equally informative and melodious name *gemtuzumab ozogamicin*. In the conjugate, the antibody and the calicheamicin are linked by (4-acetylphenoxy)butanoic acid. The ester bond in this linker is stable at extracellular pH and is cleaved only following cellular uptake [42], inside the acidic environment of the phagolysosome (Figure 14.9). Therefore, the drug will be preferentially released inside the leukemic target cells.

While calicheamicins are considered too toxic for therapeutic use when in free form, the conjugate has a more favorable toxicity profile. The drug was approved by the FDA in 2000 for use in acute myeloic leukemia and was marketed under the name Mylotarg by Pfizer. However, it was withdrawn in 2010 after clinical trials raised concerns about its safety and effectiveness.

Another type of effector molecules that can be targeted using antibodies are protein toxins. Diphtheria toxin is highly potent; a single molecule of it suffices to kill a mammalian cell. However, routine immunization against diphtheria is performed using inactivated diphtheria toxin, which may limit the effectiveness of these conjugates. *Pseudomonas aeruginosa* exotoxin A acts by the same mechanism as diphtheria toxin[7] and can replace it in antibody conjugates. Several such conjugates are in clinical trials against various forms of leukemia [46].

Antibody-radionuclide conjugates. Another kind of effector moiety that can be coupled to antibodies are radionuclides that emit either α or β particles. Because α particles travel only a short distance, their energy is deposited with a high *linear energy transfer*, leading to the formation of reactive ion and radical species at high density and, in consequence, to DNA double-strand breaks. As β particles have a longer range, they have a lower linear energy transfer, typically resulting in single-strand DNA breaks [43].

Radionuclides may inherently be a little messier than conventional drugs, since inevitably some of the radioactivity will decay and affect innocent cells before the carrier antibodies have had time to associate with their target antigens. On the other hand, radionuclides do not have to undergo cellular uptake and chemical cleavage from the antibody. Moreover, their radius of action covers more than a single cell; therefore, mutant tumor cells that fail to express the target antigen may still be reached by radionuclides bound to antigen-bearing cells in the vicinity.

An isotope with a useful combination of traits for this application is ^{131}I. As discussed in Section 12.2, the free isotope will accumulate in thyroid gland cells and thus can be used in the treatment of tumors of this organ. However, ^{131}I can also be directed to other targets by conjugating it with a suitable targeting molecule. Under oxidative conditions, iodine covalently reacts with the tyrosine residues of proteins, including antibodies; this reaction resembles the enzymatic iodination of thyroglobulin (see Section 7.3.1).

[7]Both toxins have an enzymatic domain that catalyzes ADP-ribosylation of *diphthamide*, a nonstandard amino acid that arises through posttranslational modification, in elongation factor 2. ADP-ribosylation inactivates this protein and disrupts protein synthesis.

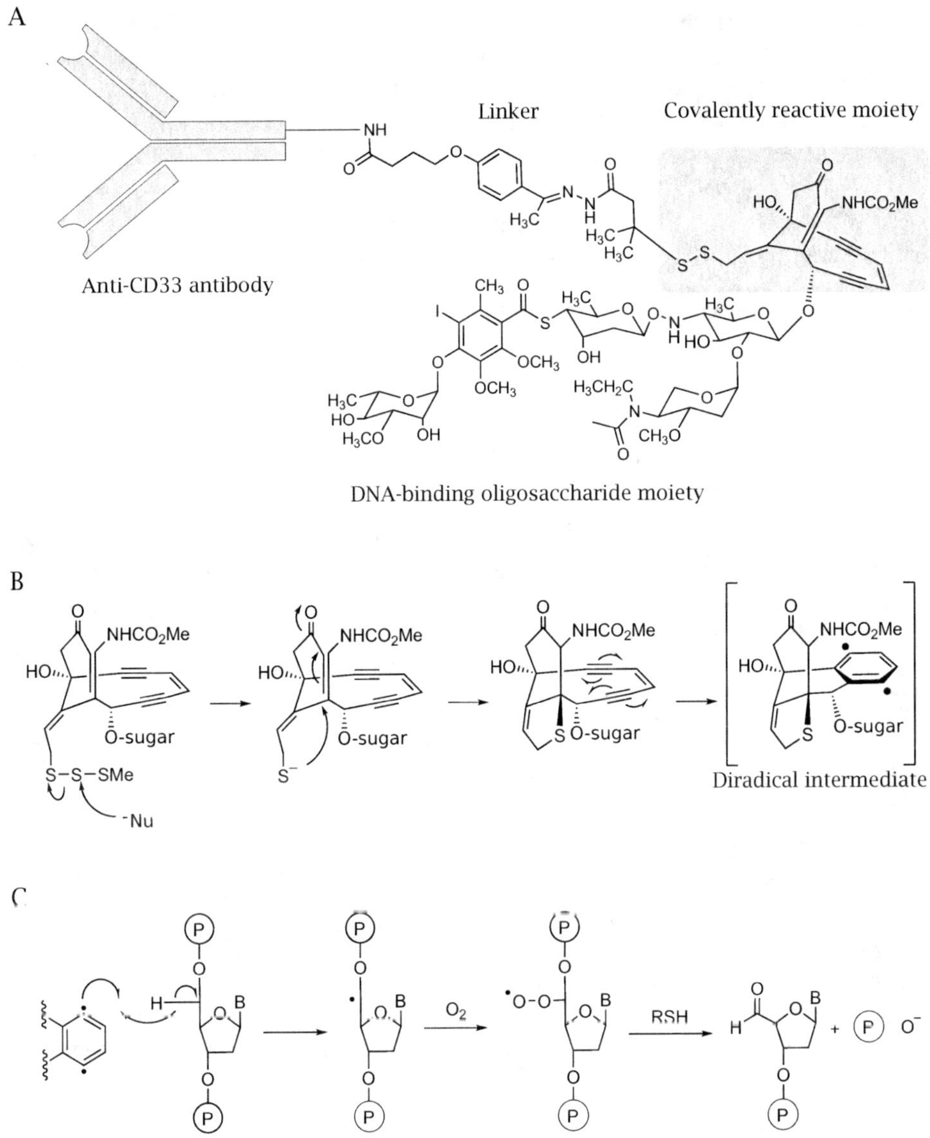

Figure 14.9 Antibody-mediated delivery of calicheamicins, and mechanism of DNA cleavage. **A**: Structure of the conjugate gemtuzumab ozogamicin. A humanized IgG4 anti-CD33 monoclonal antibody and N-acetylcalicheamicin γ-DMH are linked using (4-acetylphenoxy)butanoic acid. Two sugar residues within the oligosaccharide moiety of the drug mediate binding of the minor groove of DNA. Several calicheamicin molecules may be coupled to one antibody molecule. **B**: Activation. Calicheamicins are endiyne antibiotics; that is, they possess an alk**ene** and two alk**ynes**. They also have a trisulfide group (replaced by a disulfide in the antibody conjugate). An intracellular nucleophile (Nu^-), typically glutathione, attacks the trisulfide. Subsequent conjugate addition of the enone and Bergman rearrangement generate a reactive diradical intermediate, which reacts with DNA and induces DNA cleavage. **C**: Mechanism of DNA backbone cleavage by the activated calicheamicin diradical.

A therapeutic conjugate, ^{131}I-tositumomab, contains a monoclonal antibody directed against CD20, a surface antigen that is highly expressed in B-cell lymphoma, a malignancy derived from B lymphocytes. Binding of antibodies to CD20 will promote apoptosis, and antibodies bound to cell surface antigens in general will activate complement, as well as a special type of lymphocyte known as natural killer cells. Therefore, even unlabeled antibodies against CD20 should have a therapeutic effect. However, conjugation with radioactive iodine may extend toxicity to cells that express lower amounts of CD20 or, since the effect of radioactivity affects not only the targeted cell itself but also its vicinity, even to ones that have ceased to express this antigen altogether.

In a similar antibody conjugate, ^{131}I is replaced with ^{90}Y. Possibly because of a less stringent coupling chemistry—the metal ion is attached to the antibody through chelation rather than covalently—diffuse toxicity seems to be somewhat higher with this conjugate [31].

Immunoliposomes. The combination of the EPR effect with antigen recognition can enhance selective drug delivery. Doxorubicin-loaded liposomes can again serve as an example. In a mouse model, the antibody-modified liposomes doubled the survival time in mice and reduced the size of the brain tumor significantly, compared to Doxil without antibodies [19]. Inside a brain tumor, the BBB may be disrupted by abnormal blood vessels, and thus the liposomes are able to pass through the barrier. In another animal study on Lewis lung carcinoma, antibody-modified liposomal doxorubicin showed enhanced accumulation inside the cancer and greater inhibition of the metastatic process than Doxil without antibodies [25].

Humanized antibodies. The selection of a monoclonal antibody starts with the conventional immunization of a mouse. The mouse is then sacrificed, and individual mouse B cells are immortalized through fusion with cells of a suitable immortal cell line before screening for binding to the antigen of interest. The monoclonal hybridoma cells thus created will therefore always produce mouse antibodies. Like other nonhuman proteins, mouse antibodies may be recognized as foreign antigens by the human immune system and, after repeated therapeutic application, induce the formation of neutralizing anti-antibodies.

To reduce the antigenicity of monoclonal antibodies obtained through this process, the functional domains of the mouse antibody that are not involved in antigen binding can be replaced by the homologous domains from human antibodies. This domain swapping process can be more or less fine-grained; "chimeric" antibodies retain relatively more mouse sequences than "humanized" ones, in which only the minimal determinants of antigen complementarity are grafted from the mouse antibody to the human antibody scaffold (Figure 14.10). As expected, humanized antibodies are less immunogenic and are preferred in new development [9].

14.4 KINETICALLY CONTROLLED DRUG RELEASE

If a drug is applied periodically, its plasma concentration will oscillate with an amplitude that is determined by the drug's half-life and the interval of applica-

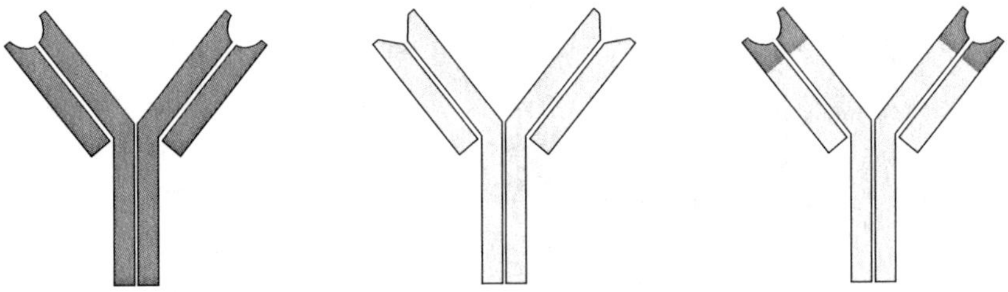

Figure 14.10 Humanized antibodies. Monoclonal antibodies are derived from mice and thus are immunogenic. Their immunogenicity can be substantially reduced by transplanting only their antigen-specific moieties onto the scaffold of a human antibody.

tion. Sometimes, we require more control over the time course of the plasma concentration. The most general solution is to apply the drug through continuous infusion, which is common practice in hospital care but is difficult to use in other settings. Special delivery systems can enable kinetically controlled drug release in outpatients and over extended periods of time.

14.4.1 Implants for long-term delivery

Prostate cancer is one of the most common forms of cancer in males. It has long been known that progression of the cancer is linked to the level of male sex hormones. When the level of testosterone is kept very low, the growth of the hormone-dependent tumor cells is significantly inhibited. This can be achieved using hormone therapy with gonadotropin-releasing hormone (GnRH) analogs such as leuprolide acetate.

GnRH agonists are water-soluble peptides and must be parenterally applied. Their plasma half-life is relatively short—only 3 hours in the case of leuprolide acetate. Even if GnRH and related peptides are extremely potent, injections must occur daily in order to maintain a sufficient concentration.

A more patient-friendly approach is to implant a device subcutaneously that releases the drug at a constant flow rate. Since the absolute concentration of peptide required is very low, a single small cartridge can store enough drug for 12 months [1,41]. The challenge is to ensure a stable, very slow flow out of this cartridge for that length of time. This is achieved with a simple but ingenious and effective trick, referred to as an "osmotic engine" (Figure 14.11). The implanted device is a titanium cylinder, which houses two compartments separated by a piston. One of the two compartments contains the drug, whereas the other contains sodium chloride pellets. The latter compartment is separated from the surrounding tissue by a semipermeable membrane. The osmotic gradient causes water to enter through this membrane and dissolve the sodium chloride. The salt solution expands and displaces the piston, thus slowly expelling leuprolide acetate from the drug compartment. The rate-controlling element in this system is the membrane that confines the sodium chloride pel-

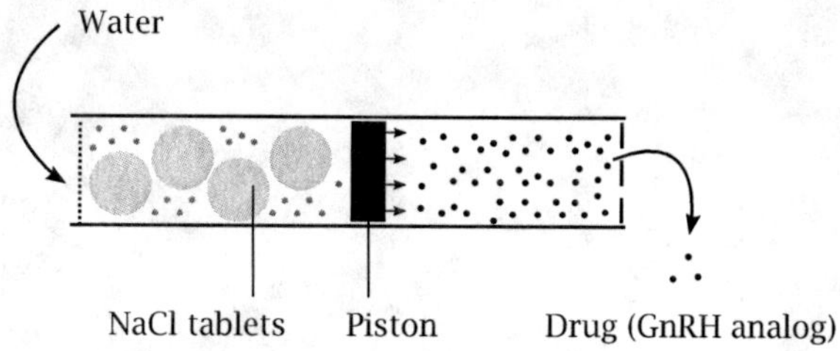

Figure 14.11 The Viadur® implant. Two compartments are separated by a piston, and each is separated from the exterior by a membrane. The outflow of the GnRH analog leuprolide acetate from the drug reservoir is regulated by the movement of the piston, which in turn is driven by the osmotic activity of NaCl in the other chamber. The flow rate is controlled by the permeability of the *diffusion moderator*, that is, the membrane that seals the NaCl-filled chamber.

lets; its permeability could in principle be tweaked to provide for any desired rate of drug release.

Delayed delivery can also be accomplished by enclosing the drug in question within a polymer matrix that undergoes slow metabolic degradation and dissolution. This is used with the cytostatic drug carmustine (BCNU; Figures 14.12 and 12.15) in the treatment of glioblastoma (brain cancer). When surgically excising a glioblastoma, it is impossible to apply the same generous margin of safety as with tumors in other organs, since that would cause grave damage to brain function. Hence, the probability of tumor recurrence is high, and radiotherapy and chemotherapy are applied in a bid to eradicate the remaining cancer cells.

Carmustine crosses the BBB when applied parenterally or intrathecally.[8] However, because of its high chemical reactivity, it travels only a very short distance into the brain tissue—2 mm from the ependymal epithelium, after intrathecal application—and thus delivers most of its cytotoxic impact in the wrong places.

Garfield and coworkers attempted to administer the drug directly into the surgical cavity using a catheter, but the effectiveness of the treatment was apparently hampered by the short half-life of carmustine, which in the systemic circulation is only 12 minutes [7,40]. To overcome the problem of poor delivery and short half-life, carmustine has been embedded into wafers that consist of a biodegradable polymer (Figure 14.12) and are available as Gliadel®. A number of these wafers are placed into the cavity that remains after tumor excision. The polyanhydride polymer layers gradually hydrolyze and dissolve in a few weeks' time, and the enclosed drug is slowly released at a constant rate into the vicinity (Figure 14.12). The products of polymer hydrolysis, carboxyphenoxypropane and decanedioic acid, are eliminated by the kidneys or metabolized by the liver, respectively.

[8]Intrathecal application means injection or infusion into the space that is filled with cerebrospinal fluid and surrounds the brain and spinal cord. The ependymal epithelium lines the liquor-filled cavities (ventricles) in the brain and the central canal of the spinal cord.

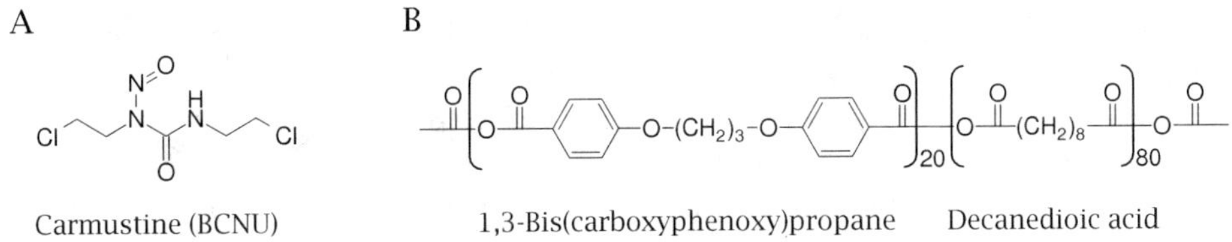

Figure 14.12 Site-selective delivery of carmustine. **A:** Structure. Carmustine can cross the BBB but is highly reactive and therefore consumed within too short a range of penetration into the brain tissue. **B:** Stucture of the polyanhydride polymer in Gliadel® wafers. The ratio of carboxyphenoxypropane and decanedioic acid controls the rate of hydrolysis and thereby drug release from the drug-polymer matrix.

This local delivery approach enables the drug to act on the target site before its rapid metabolism. In addition, the effective concentration of the drug in the resected site can be many times higher than achievable by intravenous administration. Because of its high reactivity, the drug is active only within very close proximity of the wafer inside the brain. Compared to intravenous administration, the use of the wafer avoids the drug traveling through the periphery, which reduces nonspecific toxicity.

Local delivery of carmustine appears to be clinically effective; in clinical studies on glioblastoma patients, the median survival time was increased by 50% in the Gliadel group compared to the placebo group [32].

14.4.2 Insulin delivery

In the cells of most tissues, the uptake and utilization of glucose from the blood depends on the availability of insulin, a peptide hormone secreted by the β cells in the Langerhans islets of the pancreas. The molecule has two chains (the A and B chains) that are connected by disulfide bonds, and contains 51 amino acid residues overall. A lack of insulin activity causes diabetes mellitus. The two major forms of this disease, type 1 and type 2, are discussed in Section 10.3). Substitution of insulin is necessary in all type 1 and in many type 2 patients.

Insulin injection therapy is not easy to get right. Excessive insulin application will result in hypoglycemia, which is a pathologically low level of blood glucose that can induce unconsciousness and even be life-threatening. Insufficient insulin levels will lead to hyperglycemia, which when severe can induce coma, too. Moderate hyperglycemia does not cause problems in the short term, but it will still promote the characteristic long-term deterioration of kidney, nerve, and other organ functions. Ideally, therefore, the time course of the plasma concentration of externally applied insulin should closely match that of the normal pancreatic secretion, in order to keep the glucose level within its narrow physiological range.

Physiologically, the blood insulin concentration peaks after a meal and then returns to a relatively stable basal level after a few hours. The half-life of circulating insulin is only approximately 25 minutes, due to inactivation by specific peptidases. This is much shorter than the duration of the peaks, which therefore are sustained by continuous insulin secretion and cannot be mimicked with single intravenous injections of insulin. The mismatch is even more pronounced with the almost flat level of insulin that prevails between peaks.

In acute therapy and in a hospital setting, the insulin level can be adjusted as needed through continuous intravenous infusion. In outpatients and for sustained therapy, however, we need other methods to approximate the physiological time course.

When insulin is injected subcutaneously, it will wind up in the interstitial space. Insulin then has to "reversely distribute" into the circulation. This uptake is fairly rapid for monomeric insulin, which is small enough to allow diffusion through the pores in the capillary wall. However, at high concentrations, insulin self-associates into hexamers (Figure 14.13), which are too large for rapid capillary uptake. Only the monomers that remain present at equilibrium will enter the bloodstream. If the injection is applied shortly before a meal, the time profile of insulin becoming available in the circulation approximates the duration of the physiological postprandial peak [1].

The problem that remains then is to keep the level of insulin sufficiently high in the periods between peaks. To this end, the rate of capillary uptake must be slowed down further. Preparations that have been conditioned for delayed uptake are referred to as long-acting or *basal* insulins, as opposed to native insulin and other short-acting *bolus* insulins. A combination of basal and bolus insulins, and optionally intermediate-acting ones, can then be used to mimic the overall physiological insulin secretion profile.

Slow-acting insulin preparations. The rate of insulin monomer release from aggregates is influenced by a multitude of effects, which can be exploited in the preparation of long- and intermediate-acting insulins. Insulin is negatively charged at neutral pH, which causes electrostatic mutual repulsion of the monomers and promotes dissociation. Adjusting the pH to lower values reduces repulsion and promotes aggregation, as do positively charged additives such as zinc and, more strongly, protamine.[9] In addition, crystalline insulin dissolves more slowly than amorphous insulin aggregates.

The rate of dissociation can also be changed by specific changes to the insulin molecule. Addition of two positively charged arginine residues to the carboxy terminus of the B chain, in combination with substitution of residue asparagine 21 in chain A by glycine, yields *glargine*, a derivative with stable, slow release kinetics. Covalent attachment of a fatty acyl residue to lysine B29 yields *insulin detemir*, which may form micellar aggregates or bind to other proteins, causing slower uptake into and slower clearance from the blood.

The exact mixture and dosage of slow- and fast-acting insulins has to be adjusted empirically with each patient. Traditionally, when developing an individual treatment plan, emphasis was placed on minimizing the number of daily insulin injections. One limitation of this approach is that patients need to carefully synchronize their meals with his insulin application schedule. Moreover,

[9] Protamine is a small, positively charged nuclear protein found in the testes of rainbow trout.

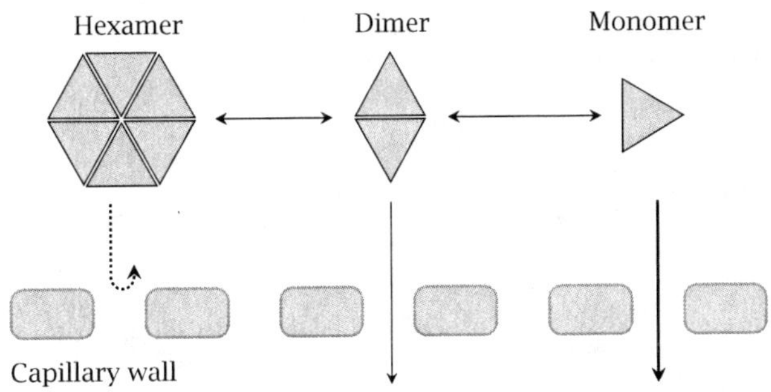

Figure 14.13 Aggregation and capillary wall penetration of insulin. Zinc ions and high insulin concentration promote the formation of hexamers, which don't permeate across the capillary walls. The aggregation equilibrium can be shifted either way by various point mutations, which is exploited in the preparation of both fast- and slow-acting insulins (see text and Ref. [13] for details).

the blood glucose level will seldom be as tightly controlled as is desirable in order to minimize the induction of diabetic long-term complications.

Intensive insulin therapy. Tighter glucose control is the purpose of *intensive insulin therapy*, in which frequent measurements of blood glucose are used to guide the likewise more frequent applications of insulin. One risk inherent in this approach is that hypoglycemia may result when a dose of insulin is applied before the previous ones have been fully taken up into the circulation. In order to minimize this risk, it is desirable to accelerate the capillary uptake beyond the rate achievable with native insulin. Several mutant insulins have been created that aggregate less readily than wild-type insulin and therefore undergo faster capillary uptake. Insulin lispro, in which amino acid residues proline B28 and lysine B29 are switched, and insulin aspart, which contains a mutant aspartate residue at position B28, are in clinical use and reportedly offer a reduced risk of hypoglycemia. Figure 14.14 illustrates how these changes reduce the stability of insulin aggregates.

Insulin pumps. The idea of just-in-time application of insulin leads logically to insulin pumps, which can release insulin continuously, much like the pancreatic islets. Ideally, the flow rate would be automatically controlled without any required user intervention by continuous measurement of the blood glucose level. To avoid undulations in the feedback loop, the delay between the subcutaneous release by the pump and the availability of insulin in the circulation should be as short as possible; therefore, insulin preparations with minimized aggregation will again be preferable.

Other insulin delivery methods have also been developed. Inhalable insulin was available on the market for a short while. However, due to the concerns on the long-term effects of insulin on the lungs and the accuracy in the dosage, demand was lower than expected, and the product line was discontinued [50].

14.5 CONTROLLING DRUG TOXICITY

Drug delivery methods may also reduce toxicity. This is illustrated by the prodrug acyclovir, which is used in the treatment of herpes virus infections.

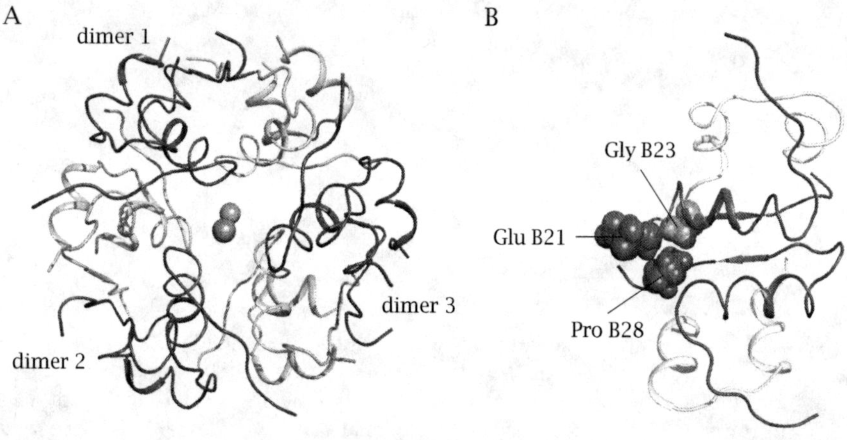

Figure 14.14 Structure of the insulin hexamer. **A:** The hexamer is composed of three dimers and stabilized by two centrally placed zinc ions. **B:** In each dimer, proline 28 of one B chain interacts with glutamate 21 and glycine 23 of the opposite B chain. In insulin lispro, proline B28 is replaced with lysine, which destabilizes the interaction of the two monomers. In insulin aspart, proline 28 is replaced with aspartic acid, which also perturbs this interaction and additionally creates electrostatic repulsion with the opposite glutamate. A colored version of this figure is shown in the insert in the middle of this book.

Acyclovir not only manages to enter the target cells more readily than the phosphorylated compound, which acts as inhibitor of viral DNA polymerase, but it also achieves greater selective toxicity toward herpes virus. This selectivity is due to the fact that the phosphorylation step in the activation of the prodrug is accomplished only by the viral thymidine kinase but not by any of the kinases that occur in uninfected cells (Section 11.7.2). Reducing absorption into the systemic circulation may also lower the toxicity, as exemplified by the prodrug succinylsulfathiazole (Section 14.1.3).

Alternatively, toxicity may be lowered through the use of drug carriers, as we shall see in the following.

14.5.1 Controlling amphotericin B toxicity with liposomes

Amphotericin B is an antibiotic that binds to ergosterol in fungal cell membranes and then permeabilizes them (see Section 11.5). It is active against a broad spectrum of fungal pathogens and one of the mainstays of therapy. However, ergosterol is very similar to cholesterol (Figure 14.15); amphotericin B indeed also binds to cholesterol and permeabilizes mammalian cell membranes and hence is prone to cause serious side effects, particularly affecting the kidneys.

Amphotericin B is very poorly water-soluble and therefore has to be applied with some kind of additive or carrier. The traditional preparation contains the detergent deoxycholate, which will solubilize amphotericin B within mixed micelles. Amphotericin B can also be incorporated into the lipid bilayers of phospholipid liposomes. Interestingly, the liposomal preparations have

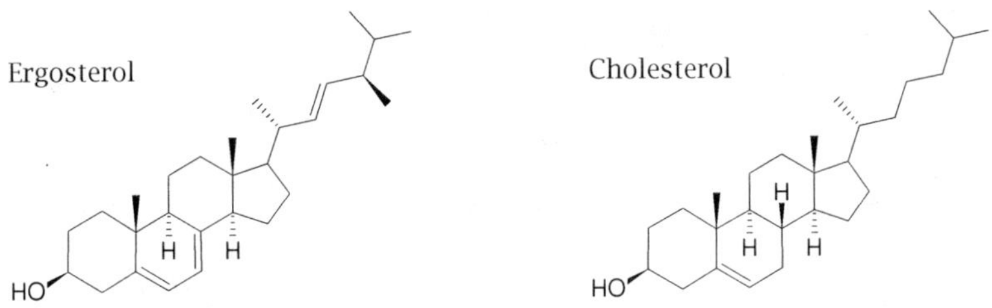

Figure 14.15 Structures of ergosterol and of cholesterol. Cholesterol occurs in mammalian cell membranes and ergosterol in fungal ones. Amphotericin B binds to both; its higher affinity for ergosterol is the basis of its selective toxicity for fungi, whereas its interaction with cholesterol is responsible for its considerable toxicity for human cells. (See also Figure 11.14.)

lower toxicity than the deoxycholate-based one and can be applied at higher dosages than the latter. For example, with one commercial liposomal formulation, AmBisome®, the recommended daily dose is 3–6 mg/kg. In contrast, with the deoxycholate formulation, 0.25–0.5 mg/kg per day is the usual dosage, and the maximum dose cannot be greater than 1.5 mg/kg.

The pharmacokinetics of amphotericin B is incompletely understood, and as a consequence, so is the difference in toxicity of these two preparations. However, deoxycholate has a high critical micellar concentration, which means that on intravenous application and dilution, the detergent molecules will quickly dissipate. This should yield a high local concentration of denuded amphotericin B molecules, which will then likely bind to the nearest targets rather than to those with the highest affinity. The kidneys have a higher rate of perfusion than any other organ and therefore likely receive a disproportionate share of the amphotericin B so deposited, which may account for their high sensitivity.[10]

Liposomes are much more stable than deoxycholate micelles, and the drug therefore will have time to slowly equilibrate with other compartments. Ideally, it would seem, the affinity of the drug for the liposomes should be intermediate between its affinity for the host cell membranes and that for the fungal cell membranes. In an *in vitro* study using liposomes composed of different kinds of phospholipids, those with saturated acyl chains afforded much lower toxicity than those with unsaturated acyl chains [33]. It was not rigorously determined, however, whether this difference was caused by variations in the membranes' affinity for amphotericin B or by other factors.

Another pertinent observation is that amphotericin B associates with both high- and low-density lipoproteins (HDL and LDL). When applied with deoxycholate, it distributes about evenly between them, whereas liposomal preparations deliver preferentially to HDL. The HDL complex is more favorable with respect to both therapeutic action and toxicity [56].

The distribution of the liposomes themselves may also contribute to the lower toxicity and therapeutic effectiveness of liposomal amphotericin B. Ma-

[10]The cytotoxicity of amphotericin B is pH-dependent, and the damage to kidney epithelia may be greater when the urine pH is low [55].

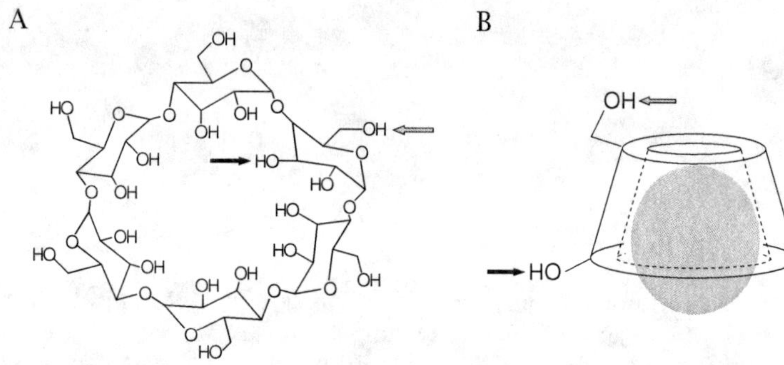

Figure 14.16 Cyclodextrin: Structure and use in drug delivery. **A:** Structure of β-cyclodextrin. **B:** Schematic of a cyclodextrin molecule encasing a hydrophobic drug molecule. The two hydroxy groups shown correspond to those highlighted in (A).

crophages may take up both fungal cells and liposomes, so that the drug may become concentrated near its target cells. The EPR effect, discussed above in the context of tumor therapy, may also be elicited by infections, including fungal ones, and liposomal amphotericin B might thus become enriched at sites of infection.

14.5.2 Cyclodextrins

Cyclodextrins are circular polymers of glucose. The number of glucose residues is 6, 7, and 8 in α-, β-, and γ-cyclodextrin, respectively; all residues are linked by α-1,4-glycosidic bonds. In aqueous solution, cyclodextrins assume a conical shape with a hydrophilic periphery and a hydrophobic central cavity, which can bind many hydrophobic compounds. The inclusion of lipophilic drugs into cyclodextrins not only improves their solubility (Figure 14.16) but can also reduce associated side effects. Here are several examples:

1. Gastrointestinal bleeding is one of the side effects of piroxicam, a cyclooxygenase inhibitor (see Section 9.3). The absorption of piroxicam is hindered by poor solubility. When complexed with β-cyclodextrin, solubility and absorption are improved, resulting in fewer gastrointestinal lesions [15].

2. Prostaglandin E_1 is used in the treatment of atherosclerosis. When it is applied intravenously, it is metabolized rapidly in the lungs, so that it will not reach the capillaries of the systemic circulation. The free drug, therefore, must be applied by intraarterial infusion, which because of the high intraarterial pressure is a somewhat precarious procedure. Complexing prostaglandin E_1 with cyclodextrin slows down its metabolism and permits its intravenous application [15].

3. Nicotine gives an acrid and burning sensation in the mouth. Enclosure of the drug inside β-cyclodextrin prevents it from binding to the receptors. This formulation is used in commercially available chewing gum or sublingual tablets [53].

14.6 DELIVERY OF NUCLEIC ACIDS

Both DNA and RNA can be used as drugs in gene therapy, antisense therapy and for gene silencing. In theory, nucleic acid drugs are more versatile and predictable than any other form of drug therapy, and the possibilities may seem boundless. In practice, however, these drugs are severely limited by pharmacokinetic problems. For specific interaction with their intracellular targets, an oligonucleotide needs a size of at least some 15–20 bases, corresponding to roughly 5–10 kDa. DNA molecules that are used in gene therapy encompass entire genes and are, of course, much larger. The phosphodiester bonds in the DNA and RNA backbones are both polar and susceptible to enzymatic hydrolysis. Therefore, naked, unmodified DNA or RNA cannot achieve systemic distribution and is not able to cross cell membranes by diffusion.

Many strategies have been devised to improve the stability, distribution and cellular uptake of nucleic acids [45,10]. DNA can be protected from enzymatic cleavage by using sulfur to replace one of the nonbridging oxygen atoms in the backbone phosphate [45,21]. This and several other backbone modifications are shown in Figure 13.7.

14.6.1 Viral vectors for DNA delivery

A delivery system for nucleic acids must fulfill a complex set of requirements. First, it has to protect the nucleic acids from degradation and phagocytosis in the bloodstream and the interstitial space. After entering the cell, it must protect the nucleic acid from the acidity and the degradative enzymes inside the endosome. It must then facilitate the escape of the nucleic acids from the endosome and their transport to the nucleus [57,34].

Fortunately, delivery systems for nucleic acids that meet all these criteria can be found in nature; they are called viruses. An intact, unmodified virus will, of course, not cease to function after delivering the viral genome, since the viral genome will direct its own replication and the formation of progeny virus, usually with pathogenic consequences. However, it is possible to genetically modify virus genomes to abolish or subdue their replicative capacity. A replication-incompetent adenovirus-derived vector has been used to deliver intact copies of the p53 gene. This gene is an important cell cycle regulator; it is mutated in 50-70% of human tumors (see Section 12.1.2). The gene transfer vector, called INGN 201, as of 2006 had entered phase III clinical studies for various forms of cancer [14]. A very recent clinical study that also used an adenovirus-derived vector achieved coagulation Factor IX expression in patients with hemophilia B, who are genetically deficient for this protein [44].

If a recombinant virus used in gene therapy retains the ability to replicate, this offers the advantage of amplifying the number of gene copies delivered. Nevertheless, at some point replication will have run its course, and the application of virus will then have to be repeated. Recombinant viruses, however, are not exempt from immune responses, and renewed dosages of virus will be rendered ineffective by the formation of neutralizing antibodies. To obtain a longer-lasting effect, one may use viruses that have the ability to persist intracellularly. Retroviruses, in particular, achieve long-lasting gene expression by inserting copies of their genome into the host cell DNA. This may, however, cause the activation of cellular genes in the vicinity of the integration site, including the activation of oncogenes, and caution must be exercised in the application of such retroviral vectors.

Nonviral vectors for gene therapy possess lower immunogenicity and can more easily be mass-produced. However, their low transfection efficiency remains the obstacle for the development [24].

STUDY QUESTIONS

14.1 Drug carriers that float on top of the gastric chyme (juice) can be used for drug delivery. Suggest a rationale behind this approach. How would you design such a floating carrier system?

14.2 Cationic polymers are used as carriers for the delivery of DNA to cells. What are the advantages of cationic over neutral or anionic carriers?

14.3 Peptide drugs tend to be rapidly degraded after oral ingestion. One way to prevent this may be through the use of protective carriers; another is to modify the peptides themselves in order to reduce their susceptibility to proteolytic cleavage. What structural modifications would be useful for this?

14.4 We have considered the use of antibody–radionuclide conjugates for the targeted delivery of radiation (Section 14.3.2). A commonly used radionuclide is ^{131}I, which has a half-life of 8 days; this requires prolonged isolation and screening of the patient. Radionuclides with shorter half-lives might seem more practical. Would such shorter-lived conjugates also have disadvantages or limitations?

14.5 Can the drugs shown in the scheme below be applied through skin patches?

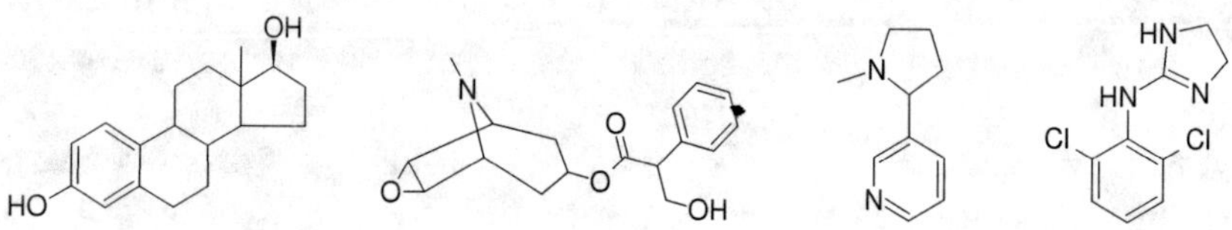

1 2 3 4

REFERENCES

[1] L. Allen, N. Popovich, and H. Ansel, *Ansel's pharmaceutical dosage forms and drug delivery systems,* Lippincott Williams & Wilkins, 2005.

[2] T. M. Allen, "Long-circulating (sterically stabilized) liposomes for targeted drug delivery," *Trends Pharmacol. Sci.* 15 (1994), 215–220.

[3] T. M. Allen, "Controlled drug delivery: designing technologies for the future," K. Park and R. J. Mrsny, ed., ACS Symp. Series vol. 752 (2000), American Chemical Society, 100–108.

[4] J. Bauer et al., "Ritonavir: an extraordinary example of conformational polymorphism," *Pharm. Res.* 18 (2001), 859–866.

[5] N. Bodor and P. Buchwald, "Recent advances in the brain targeting of neuropharmaceuticals by chemical delivery systems," *Adv. Drug Deliv. Rev.* 36 (1999), 229–254.

[6] N. Bodor and P. Buchwald, "Barriers to remember: brain-targeting chemical delivery systems and Alzheimer's disease," *Drug Discov. Today* 7 (2002), 766–774.

[7] D. A. Bota et al., "Interstitial chemotherapy with biodegradable BCNU (Gliadel) wafers in the treatment of malignant gliomas," *Ther. Clin. Risk Manag.* 3 (2007), 707–715.

[8] L. L. Brunton, J. S. Lazo, and K. L. Parker, eds., *Goodman and Gilman's The pharmacological basis of therapeutics,* 11th ed., McGraw Hill, 2005, 1794–1888.

[9] P. Carter, "Improving the efficacy of antibody-based cancer therapies," *Nat. Rev. Cancer* 1 (2001), 118–129.

[10] D. Castanotto and J. J. Rossi, "The promises and pitfalls of RNA-interference-based therapeutics," *Nature* 457 (2009), 426–433.

[11] T. Cheng et al., "Computation of octanol-water partition coefficients by guiding an additive model with knowledge," *J. Chem. Inf. Model.* 47 (2007), 2140–2148.

[12] M. Chourasia and S. K. Jain, "Pharmaceutical approaches to colon targeted drug delivery systems," *J Pharm. Pharm. Sci.* 6 (2003), 33–66.

[13] H. R. Costantino et al., "Therapeutic protein and peptide formulation and delivery," Z. Shahrokh et al., ed., ACS Symp. Series vol. 675 (1997), American Chemical Society, 29–66.

[14] D. Cross and J. K. Burmester, "Gene therapy for cancer treatment: past, present and future," *Clin. Med. Res.* 4 (2006), 218–227.

[15] M. E. Davis and M. E. Brewster, "Cyclodextrin-based pharmaceutics: past, present and future," *Nat. Rev. Drug. Discov.* 3 (2004), 1023–1035.

[16] P. B. Deasy, "Evaluation of drug-containing microcapsules," *J. Microencapsul.* 11 (1994), 487–505.

[17] A. van Duijn-Goedhart et al., "Differential responses of Chinese hamster mutagen sensitive cell lines to low and high concentrations of calicheamicin and neocarzinostatin," *Mutat. Res.* 471 (2000), 95–105.

[18] G. Dunea, "Death over the counter," *Br Med J (Clin Res Ed)* 286 (1983), 211–212.

[19] T. A. ElBayoumi and V. P. Torchilin, "Tumor-specific anti-neucleosome antibody improves therapeutic efficacy of doxorubicin-loaded long-circulating liposomes against primary and metastatic tumor in mice," *Mol. Pharmaceutics* 6 (2009), 246–254.

[20] A. R. English et al., "Carbenicillin indanyl sodium, an orally active derivative of carbenicillin," *Antimicrob. Agents Chemother.* 1 (1972), 185–191.

[21] W. O. Foye and T. L. Lemke, *Foye's principles of medicinal chemistry,* 2nd ed., Lippincott Williams & Wilkins, 2002.

[22] A. A. Gabizon, "Pegylated liposomal doxorubicin: metamorphosis of an old drug into a new form of chemotherapy," *Cancer Invest.* 19 (2001), 424–436.

[23] S. C. Gad, ed., *Drug discovery handbook,* Wiley-Interscience, 2005.

[24] K. Gao and L. Huang, "Nonviral Methods for siRNA Delivery," *Mol. Pharm.* 6 (2009), 651–658.

[25] B. Gupta and V. P. Torchilin, "Monoclonal antibody 2C5-modified doxorubicin-loaded liposomes with significantly enhanced therapeutic activity against intracranial human brain U-87 MG tumor xenografts in nude mice," *Cancer Immunol. Immunother.* 56 (2007), 1215–1223.

[26] P. R. Hamann et al., "Gemtuzumab ozogamicin, a potent and selective anti-CD33 antibody-calicheamicin conjugate for treatment of acute myeloid leukemia," *Bioconjug. Chem.* 13 (2002), 47–58.

[27] J. M. Harris and R. B. Chess, "Effect of pegylation on pharmaceuticals," *Nat. Rev. Drug. Discov.* 2 (2003), 214–221.

[28] R. J. Herr, "5-Substituted-1H-tetrazoles as carboxylic acid isosteres: medicinal chemistry and synthetic methods," *Bioorg. Med. Chem.* 10 (2002), 3379–3393.

[29] B. K. Immel, "A Brief history of the GMPs for pharmaceuticals," *Pharmaceut. Tech.* (2001), 44–53.

[30] A. K. Iyer et al., "Exploiting the enhanced permeability and retention effect for tumor targeting," *Drug Discov. Today* 11 (2006), 812–818.

[31] H. A. Jacene et al., "Comparison of 90Y-ibritumomab tiuxetan and 131I-tositumomab in clinical practice," *J. Nucl. Med.* 48 (2007), 1767–1776.

[32] J. P. Jain, D. Chitkara, and N. Kumar, "Polyanhydrides as localized drug delivery carrier: an update," *Expert Opin. Drug Deliv.* 5 (2008), 889–907.

[33] R. L. Juliano et al., "Mechanism of the selective toxicity of amphotericin B incorporated into liposomes," *Mol. Pharmacol.* 31 (1987), 1–11.

[34] R. Juliano et al., "Mechanisms and strategies for effective delivery of antisense and siRNA oligonucleotides," *Nucleic Acids Res.* 36 (2008), 4158–4171.

[35] E. Kerns and L. Di, *Drug-like properties: concepts, structure design and methods: from ADME to toxicity optimization,* Academic Press, 2008.

[36] F. Kratz et al., "Prodrug strategies in anticancer chemotherapy," *ChemMedChem* 3 (2008), 20–53.

[37] D. D. Lasic and D. Papahadjopoulos, "Liposomes revisited," *Science* 267 (1995), 1275–1276.

[38] D. Law et al., "Physicochemical considerations in the preparation of amorphous ritonavir-poly(ethylene glycol) 8000 solid dispersions," *J. Pharm. Sci.* 90 (2001), 1015–1025.

[39] Y. H. Li et al., "Mechanism of intestinal absorption of an orally active β-lactam prodrug: uptake and transport of carindacillin in Caco-2 cells," *J. Pharmacol. Exp. Ther.* 290 (1999), 958–964.

[40] S. H. Lin and L. R. Kleinberg, "Carmustine wafers: localized delivery of chemotherapeutic agents in CNS malignancies," *Expert Rev. Anticancer Ther.* 8 (2008), 343–359.

[41] L. S. Marks, "Luteinizing hormone-releasing hormone agonists in the treatment of men with prostate cancer: timing, alternatives, and the 1-year implant," *Urology* 62 (2003), 36–42.

[42] P. A. McCarron et al., "Antibody conjugates and therapeutic strategies," *Mol. Interv.* 5 (2005), 368–380.

[43] M. von Mehren, G. P. Adams, and L. M. Weiner, "Monoclonal antibody therapy for cancer," *Annu. Rev. Med.* 54 (2003), 343–369.

[44] A. C. Nathwani et al., "Adenovirus-Associated Virus Vector-Mediated Gene Transfer in Hemophilia B," *N. Engl. J. Med.* (2011), electronically published ahead of print.

[45] J. B. Opalinska and A. M. Gewirtz, "Nucleic-acid therapeutics: basic principles and recent applications," *Nat. Rev. Drug. Discov.* 1 (2002), 503–514.

[46] I. Pastan et al., "Immunotoxin treatment of cancer," *Annu. Rev. Med.* 58 (2007), 221–237.

[47] G. L. Patrick, *An introduction to medicinal chemistry,* 3rd ed., Oxford University Press, 2005.

[48] A. J. Repta, M. J. Baltezor, and P. C. Bansal, "Utilization of an enantiomer as a solution to a pharmaceutical problem: application to solubilization of 1,2-di(4-piperazine-2,6-dione)propane," *J. Pharm. Sci.* 65 (1976), 238–242.

[49] K. Riehemann et al., "Nanomedicine--challenge and perspectives," *Angew. Chem. Int. Ed.* 48 (2009), 872–897.

[50] H. Schachner, "The science behind Exubera®," *Drug Develop. Res.* 69 (2008), 130–137.

[51] R. R. Scheline, "Metabolism of foreign compounds by gastrointestinal microorganisms," *Pharmacol. Rev.* 25 (1973), 451–523.

[52] R. B. Silverman, *The organic chemistry of drug design and drug action,* 2nd ed., Academic Press, 2004.

[53] J. Szejtli, "Past, present, and future of cyclodextrin research," *Pure Appl. Chem.* 76 (2004), 1825–1845.

[54] P. B. Timmermans et al., "Angiotensin II receptor antagonists. From discovery to antihypertensive drugs," *Hypertension* 18 (1991), III136–III142.

[55] I. Walev and S. Bhakdi, "Possible reason for preferential damage to renal tubular epithelial cells evoked by amphotericin B," *Antimicrob. Agents Chemother.* 40 (1996), 1116–1120.

[56] K. M. Wasan and G. Lopez-Berestein, "Modification of amphotericin B's therapeutic index by increasing its association with serum high-density lipoproteins," *Ann. N. Y. Acad. Sci.* 730 (1994), 93–106.

[57] K. A. Whitehead, R. Langer, and D. G. Anderson, "Knocking down barriers: advances in siRNA delivery," *Nat. Rev. Drug. Discov.* 8 (2009), 129–138.

[58] S. Wilk, H. Mizoguchi, and M. Orlowski, "γ-Glutamyl dopa: a kidney-specific dopamine precursor," *J. Pharmacol. Exp. Ther.* 206 (1978), 227–232.

[59] S. S. Williams et al., "Arrest of human lung tumor xenograft growth in severe combined immunodeficient mice using doxorubicin encapsulated in sterically stabilized liposomes," *Cancer Res.* 53 (1993), 3964–3967.

[60] N. Zein et al., "Calicheamicin $\gamma_1{}^{\mathrm{I}}$: an antitumor antibiotic that cleaves double-stranded DNA site specifically," *Science* 240 (1988), 1198–1201.

CHAPTER 15

DRUG DISCOVERY

by John Honek

How does one discover and develop a new drug such as a new antibiotic or a new neuropharmacological agent? This chapter can only give a very cursory overview of the approaches that are utilized; however, it will hopefully give the reader a better perspective on the complex considerations involved in the development of a new pharmacological agent. Additional readings in this area can be found in several textbooks in medicinal chemistry [8,13,24,28,35]. For excellent ongoing series of reviews in drug discovery, the reader is directed to the journals *Nature Reviews Drug Discovery* (Nature Publishing Group) and *Annual Reports in Medicinal Chemistry* (Academic Press).

15.1 TARGET SELECTION AND VALIDATION

Drug development begins with the selection of a target. In a molecule-level strategy, one begins with a disease and evaluates the biochemical pathways and physiological regulations that control its pathogenesis. Then, the next step is to decide on the most suitable target receptor or enzyme within these pathways. As far as possible, the chosen target is *validated* in preliminary experiments. The purpose of validation is to determine whether inhibition or activation of the chosen target will indeed result in the expected physiological effect. If validation succeeds, it justifies the considerable effort that lies ahead; if it fails, the effort can be spared and directed toward another, more promising target.

Depending on the nature of the receptor and the functional response, target validation may be possible in cell culture, or it may require animal experiments. In either case, the functional effect of pharmacological inhibition is often modeled by genetic inactivation of the target. An expedient means for gene inactivation in cell culture is RNA interference (Section 13.2.1). Gene knockouts by homologous recombination can be used with both cell cultures and experimental animals; examples are illustrated in Figure 8.10 and Figure 5.12. The effect of receptor agonists could in principle be mimicked by receptor overexpression (see Figure 2.7), but this approach does not seem to be used much in practice.

The same approach of target selection and validation can be extended to the development of a novel antibacterial compound. Genetic knockout of bacterial proteins can be performed by transposon mutagenesis [29]. If the knockout indicates that an enzyme found in a pathogenic bacterium is *essential*, that is necessary for the viability of the bacterium, then this enzyme might become the target for a drug development program. However, one must bear in mind that a gene may be essential to the bacterium's survival in vitro but not in vivo. For example, bacterial fatty acid biosynthesis is essential in vitro and has been proposed as a target pathway for the development of new antibacterials [32]. However, some bacteria are able to circumvent this inhibition when an extrinsic supply of fatty acids is available, as it is in human plasma [5].

Target validation occurs within the confines of our current understanding, and even if it is deemed successful, the subsequent drug development effort may yet produce unexpected results. Often, what was initially believed to be a single receptor, was found in fact to be an ensemble of various receptor types and subtypes. Striking examples are glutamate receptors, of which there are two major types: (1) the ionotropic receptors or ligand-gated ion channels and (2) the metabotropic or G-protein coupled glutamate receptors (see also Chapters 5 and 6).

The ionotropic glutamate receptors have several subtypes, many of which were discovered by the investigation of an array of chemical compounds that were found to produce different physiological effects on various cell types. These observations were eventually attributed to the existence of several different ionotropic glutamate receptors, which were classified according to their responsiveness to the agonists NMDA, AMPA, and kainic acid (Figure 15.1). Meanwhile, each of these subtypes has been subdivided into even more pharmacologically and molecularly distinct variants.

The metabotropic glutamate receptor class is also divided into three major subtypes with further subclassification, based on sequence analysis, biochemistry, and pharmacology [34]. The major groups can be differentiated with the compounds quisqualic acid, L-AP4, and ACPD (Figure 15.1). Hence, selection of the specific target receptor for agonist or antagonist development is an impor-

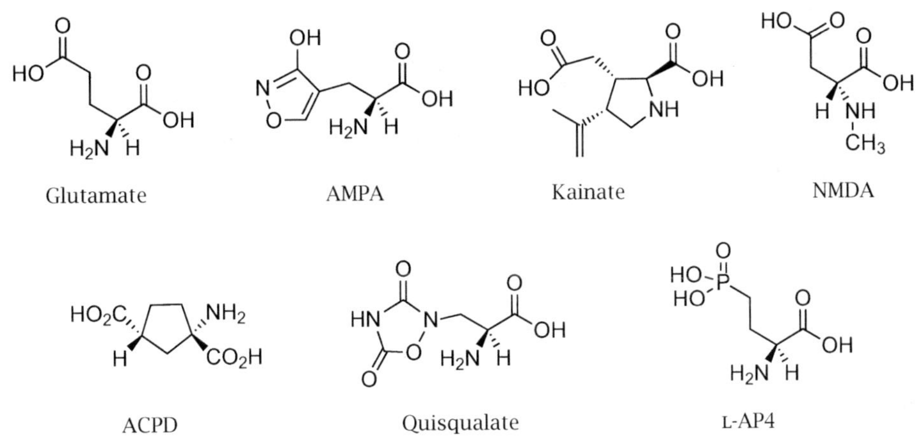

Figure 15.1 Chemical structures of compounds that demonstrate the existence of some of the various glutamate receptor subtypes referred to in the chapter. AMPA [2-amino-3-(5-methyl-3-oxo-1,2-oxazol-4-yl)propanoic acid] and quisqualate activate the same subtype of ionotropic glutamate receptors. Quisqualate also activates group I metabotropic glutamate receptors. NMDA (N-methyl-D-aspartate) and kainate, respectively, activate two other ionotropic receptor subtypes. ACPD (1-aminocyclopentane-1,3-dicarboxylic acid) is an agonist of group I and II metabotropic glutamate receptors. L-AP4 (L-2-amino-4-phosphonobutanoic acid) is an agonist of group III metabotropic receptors.

tant but complex step in a drug discovery program. For example, attempts to develop a selective antagonist for the mGlu4 metabotropic receptor subtype, which belongs to group III, would require the determination of the *receptor profile* of a candidate compound, that is, its effect not only on the mGlu4 receptor but on all other glutamate receptor types and subtypes as well.

The extensive experimentation involved in such screening of new compounds frequently aids in the elucidation of fundamental cellular biochemistry. Of course, this is very beneficial, since the more we know about the cellular process, the more knowledge we have to eventually uncover more selective or even entirely novel pharmacological targets.

15.2 SCREENING OF CANDIDATE COMPOUNDS

Once a biomolecular target has been decided on, assays then need to be developed to screen new compounds for interaction with this target. These assays will typically be performed in stages. In the first round, the goal is to rapidly narrow down a very large number of compounds to a much smaller set of candidates for further evaluation, using simple and robust readouts. These assays can be displacement assays, in which the natural or prototypical ligand is radioactively or fluorescently labeled, and its displacement by a new compound detects that new compound's binding affinity for the macromolecular target.

If the target is an enzyme (an important case; it has been determined that over 320 commercial drugs target over 70 enzymes [25]), the assay will typically be based on the enzyme's activity. Optical assays are particularly well suited for

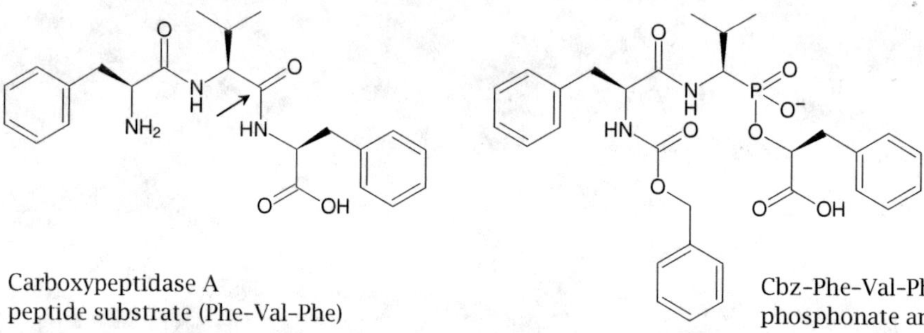

Carboxypeptidase A
peptide substrate (Phe–Val–Phe)

Cbz–Phe–Val–Phe
phosphonate analog

Figure 15.2 Chemical structure of a tripeptide substrate of carboxypeptidase A, and of the inhibitor Cbz-Phe-Val-Phe phosphonate that inhibits the enzyme with femtomolar affinity [12].

adaption to a high-throughput format. For example, a hydrolytic enzyme may release colored or fluorescent products from substrate analogs, and the change in absorbance or fluorescence over time in the assay will indicate that the enzyme is functioning. The readout of the inhibitor screening will then be the suppression of the absorbance or fluorescence change. Inhibition of the enzyme will halt or slow the processing of the substrate. By varying the concentration of the candidate compounds, such assays can be used to determine their inhibitory potency. Initially, however, candidate compounds are examined at just a single, relatively high concentration to determine whether they have any activity at all [9].

Compounds that produce a signal in the initial screen are then subjected to more detailed, specific and quantitative tests. For example, if the compound is an enzyme inhibitor, its potency and mode of inhibition must be characterized. Some inhibitors will competitively inhibit the enzyme, while others may produce noncompetitive or mixed inhibition. Other types of inhibition can be time-dependent, which indicates that the inhibitor needs time to bind productively to the enzyme (k_{on}; seconds to minutes), and with further analysis, the time-dependent release of the inhibitor (k_{off}) from the enzyme target.

Some slow-release inhibitors may exhibit k_{off} rates that vary from seconds to months [9]. An example of an inhibitor that dissociates from its target extremely slowly is the compound Cbz-Phe-Val-Phe phosphonate, which is an analog of a substrate of carboxypeptidase A, Phe-Val-Phe (Figure 15.2; [12]). This compound inhibits carboxypeptidase A in a time-dependent manner, and the half-life of its dissociation from this enzyme has been estimated to be over 8 years! Although the interaction is noncovalent, it will last longer than the lifetime of the enzyme molecule and therefore amounts to irreversible inhibition. While unusual, such behavior may be of pharmacological usefulness.

Properly screening potential drugs against purified receptors and enzymes can be quite challenging. An excellent example has been the investigation of inhibitors of the *Escherichia coli* methionine aminopeptidase. This enzyme functions in the posttranslational processing of proteins. Methionine—or in prokaryotes, *N*-formylmethionine—serves as the initiator amino acid in protein

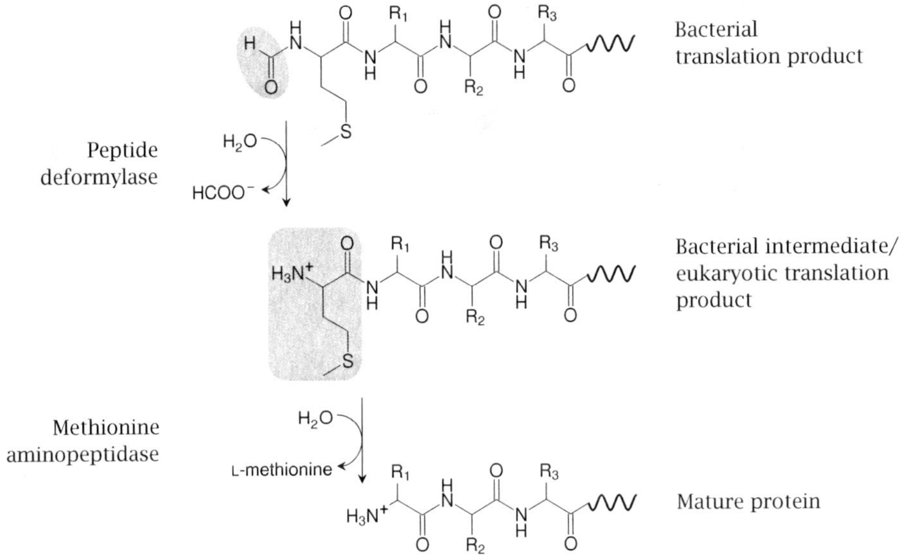

Figure 15.3 Reactions in the posttranslational processing of protein N-termini. *N*-Formylmethionine occurs as the N-terminal amino acid residue in bacterial proteins. Its formyl group is removed by peptide deformylase, and the methionine is subsequently removed by methionine aminopeptidase. The deformylase is a target for antibacterial drugs, whereas methionine aminopeptidase occurs in both bacterial and mammalian cells and thus is a potential target for both antibacterial and antitumor drugs [38].

biosynthesis, and it forms the N-terminus of each newly synthesized protein molecule.[1] Many proteins, however, will become functional only after the N-terminal methionine or *N*-formylmethionine has been removed. In prokaryotes, this removal begins with a peptide deformylase. The resulting methionine is then removed by methionine aminopeptidase, which is also found in eukaryotic cells (Figure 15.3).

The requirement for activation through processing of the N-terminal methionine applies to some essential proteins, so that inhibition of methionine deformylase (in prokaryotes) or of methionine aminopeptidase (in both prokaryotes and eukaryotes) is a lethal event. Therefore, inhibitors of peptide deformylase and methionine aminopeptidase are considered a promising option for the development of novel antibacterial and antitumor agents. Another option are inhibitors that will block protein synthesis altogether by interfering with the incorporation of the initial methionine or *N*-formylmethionine [38].

Methionine aminopeptidases are activated by metal ions, and yet it is has been difficult to determine what the naturally occurring metal is for various sources of the enzyme. For use in enzyme-level screening experiments, the purified enzyme was metallated and thereby activated through the addition of various metal ions. Interestingly, inhibitors that were optimized against

[1]Exceptions to these rules are the occasional occurrence of isoleucine as the first amino acid, and some bacterial species such as *Pseudomonas aeruginosa* that, like eukaryotes, employ non-formylated methionine for initiation [19,30].

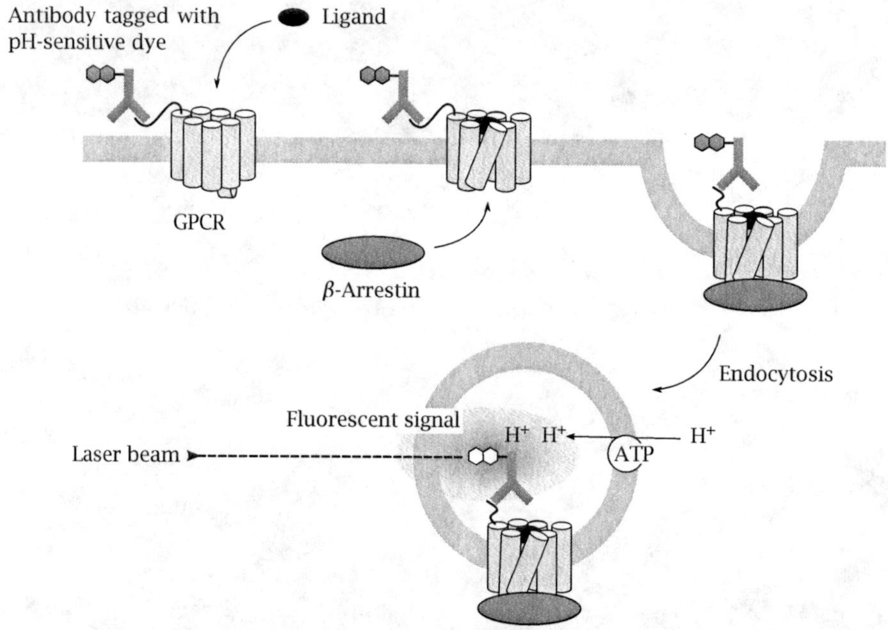

Figure 15.4 The cell-based CypHer 5 assay for GPCR activation [1]. The receptor of interest is overexpressed in cells, with its N-terminus fused to an antigenic peptide that is then labeled with an antibody carrying a pH-sensitive fluorescent dye. When a ligand activates the receptor, β-arrestin binds and triggers receptor endocytosis. When the endosome is acidified, the dye is protonated, making it brightly fluorescent.

enzyme preparations activated with specific metals such as Co^{2+} and Ni^{2+} were found to be inactive against intact *E. coli* cells. In these studies, investigators concluded that their enzyme-level screens were not using the natural metallated form of the enzyme [39]. When inhibitors were screened against the Fe^{2+}-metallated *E. coli* enzyme, the selected compounds showed activity against intact *E. coli* cells.

These observations illustrate how easy it is to obtain misleading results from in vitro screening tests. The fact that enzyme preparations reconstituted with Co^{2+}, Ni^{2+} and Fe^{2+} are all active initially suggested that the nature of the metal is not very critical, yet only one of these preparations turned out to be a valid model of the in vivo-form of the enzyme. Extending this to other enzyme or receptor targets, the purified biomolecules used for the screening process need to be in their most "native" form, such as being properly phosphorylated or posttranslationally modified, containing all their important cofactors, and so on.

While most enzymes can readily be assayed in the form of purified proteins, this is more difficult with other types of receptors such as G protein-coupled receptors and ion channels. A variety of ingenious techniques have been developed to probe these targets within intact cells, and many of these cell-based assays have been commercialized for pharmaceutical screening. For example, cell-based fluorescence assays have been developed to screen for ligands of

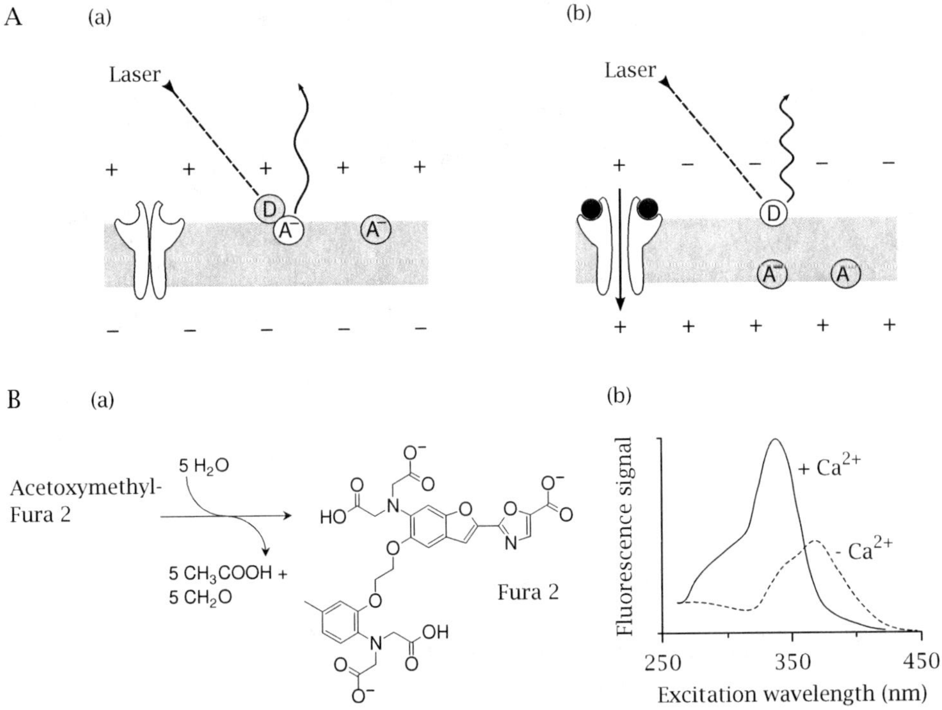

Figure 15.5 Cell-based assays for ion channel activation. **A:** Cell-based FRET assay. The cell membranes are loaded with two fluorescent dyes, which are spectrally matched for fluorescence resonance energy transfer (FRET) but differ in net charge. (a) At the resting potential, the two dyes are close to each other, and excitation of the donor dye causes FRET and emission from the acceptor dye. (b) When a ligand activates the channel and depolarizes the membrane, the dyes are separated, and the donor emits directly. **B:** Fluorescence assay of Ca^{2+} channel activation. (a) The cells are loaded with a resorption ester of the fluorescent dye Fura 2. Cleavage by cellular esterases exposes the dye's two Ca^{2+}-chelating functional groups. (b) When calcium channels are opened, intracellular Ca^{2+} increases and binds to the dye, which leads to an increased fluorescence response upon excitation at 340 nm.

G-protein-coupled receptors (GPCRs) and other potential drugs that regulate GPCRs [21,1]. The assay illustrated in Figure 15.4 relies on the endocytosis of GPCRs, which is triggered by the binding of β-arrestin to the receptor in its activated conformation. Note that, unlike typical enzyme assays, this assay is not restricted to specific receptors or ligands; it can be used with any GPCR that undergoes endocytosis on activation.

Cell-based fluorescence assays have also been developed to screen compounds for activation of ion channels. The channels are again overexpressed in cultured cells. The change in membrane potential that occurs when an agonist opens the channel can be detected using different types of potential-sensitive fluorescent dyes [20,41]. One physical principle is fluorescence resonance energy transfer (FRET; see Section 5.8.3). Here, the cell membranes are loaded with two different fluorescent dyes, which behave as a FRET donor–acceptor

Figure 15.6 Examples of high-throughput assay instrumentation. **A:** IonWorks® Quattro high-throughput electrophysiological screening system. **B:** PatchXpress® automated parallel patch-clamp system. **C:** FLIPR$^{TETRA®}$ high-throughput cell-based assay system. (All reproduced with permission from the manufacturer, MDS Analytical Technologies).

couple and have different net charge. At the resting potential, the two dyes reside on the same face of the membrane. Because of this proximity, donor excitation results in energy transfer to the acceptor dye, which then emits within its own characteristic spectrum. When a ligand activates the channel, the membrane is depolarized, causing one of the dyes to migrate to the other face of the membrane. The dyes are now separated, and upon excitation the donor will emit directly and at shorter wavelengths (Figure 15.5A).

Ca^{2+} channels of various types are important drug targets (see Chapter 6). Since the normal intracellular Ca^{2+} concentration is very low, the opening of Ca^{2+} channels causes a large relative change of the Ca^{2+} level. This change can be detected by calcium-binding fluorescent dyes such as Fura 2 (Figure 15.5B). Such assays are useful not only for drug screening but also in pharmacology and cell biology in general.

While optical readouts scale very well to high-throughput experiments, they don't provide much detail on the mode of action of the tested compounds. More quantitative and specific information can be obtained with electrophysiological assays that characterize the changes in membrane potentials directly. They may be suitable for primary screening but are more involved than fluorescence assays and probably are more often used for secondary screening. Figure 15.6 shows examples of commercial instruments developed for the cell-based fluorescence and electrophysiological assays discussed above.

Even though a great deal of screening and functional characterization can be done in vitro, animal studies are still indispensable in advanced screens for a select group of compounds that show very good to excellent activity in earlier stage biochemical or cell-based assays. They are necessary for rigorous evaluation of toxicity, and for evaluation of compounds that control animal physiology. After all, it is not possible to study physiological effects such as blood pressure, sleep, or epileptic seizures in a petri dish.

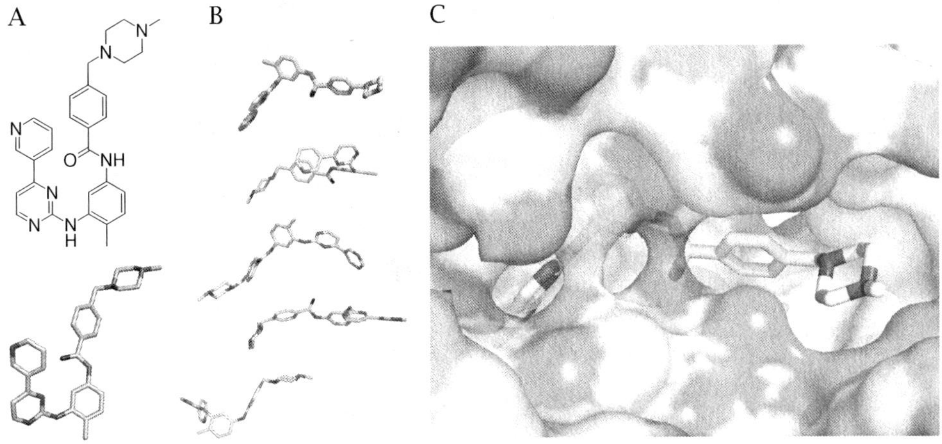

Figure 15.7 An example in silico docking experiment. The protein tyrosine kinase inhibitor imatinib was docked into the active site of the *abl* kinase oncoprotein, using the computer program AutoDock Vina [36]. **A**: Structure of imatinib and the initial conformation that was given to the docking program. **B**: Poses, or ligand orientations produced by the docking program run. The most energetically favorable one is shown at the top. **C**: The most favorable pose shown within the active site of the *abl* kinase. This pose is virtually identical to the one obtained by crystallography [18]. A colored version of this figure is shown in the insert in the middle of this book.

15.3 COMPUTATIONAL SCREENING

The binding site within a macromolecular target can be viewed as a three-dimensional molecular cavity, whose structure can be explored using a series of compounds that share a common molecular scaffold but differ in the selection and spatial arrangement of functional groups attached to it. Chemical moieties such as the hydroxyl, amino, arene, and amide functional groups can be used to map out the useful binding interactions that could be made between the molecule and the target receptor or enzyme.

In order for a drug molecule to bind avidly and specifically to the binding site, it needs not just one but a combination of several high-affinity interactions. If each of these interaction needs to be determined empirically, then the number of combinations of functional groups, and therefore compounds, that must be screened will become very large. In order to reduce this number and to focus the experimental effort on the most promising groups of structures, various theoretical approaches have been developed.

If the target's three-dimensional structure is known or a model can be developed, one approach is to prescreen chemical structures in silico, that is, by computer simulation. Each candidate molecule is computationally positioned into the active site, and the internal flexible bonds in the molecule are rotated in order to *dock* the molecule, that is, to determine the best possible fit to the active site. A series of best docked structures, termed *poses*, are obtained from this approach. These poses are then ranked according to their virtual binding affinities, which are estimated using various energy factors that take into ac-

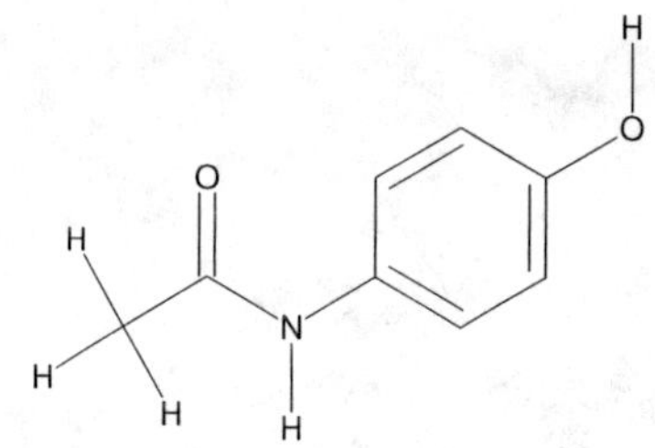

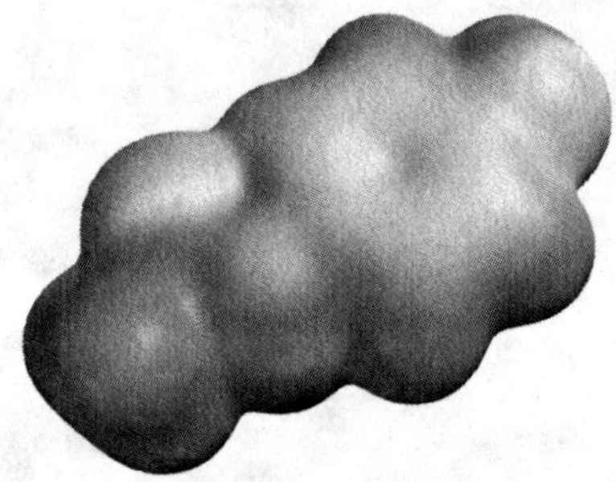

Figure 15.8 Electrostatic potential mapped onto the electronic density for acetaminophen at the RHF/6-31G* level of theory. This view of acetaminophen shows the van der Waals contour of the drug along with an isocontour of charge at the surface of this contour. The covalent structure is shown for comparison. (Software used was Spartan 08, Wavefunction Inc, USA.) A colored version of this figure is shown in the insert in the middle of this book.

count the magnitude of molecular interactions such as sterics, electrostatics, hydrogen bonding, and van der Waals forces [23].

Figure 15.7 illustrates this process for the antitumor drug imatinib, which blocks the active site of the oncoprotein *abl* kinase (see Section 12.3). In this case, the most avidly binding pose is almost identical to that determined by crystallography [18]. However, that is not always the case. One aspect that is missing from this figure is the calculated electrostatic potential that is assigned to each of the atoms in the drug molecule. Such calculations are illustrated, for the drug acetaminophen, in Figure 15.8.

If a detailed three-dimensional structure of the target molecule is not available, it may still be possible to use computational screening through the construction of a *pharmacophore.* According to the International Union of Pure and Applied Chemistry (IUPAC), a pharmacophore is "an ensemble of steric and electronic features that is necessary to ensure the optimal supramolecular interactions with a specific biological target and to trigger (or block) its biological response" [40]. This concept is similar to that of consensus motifs in proteins or nucleic acid molecules: A pharmacophore can be considered the consensus structure of drug molecules that exert a similar pharmacological effect on the target.

From a pharmacophore, a hypothetical, simplified model of the binding site can be constructed. Only key elements of a receptor need to be taken into account. These elements could be the positions of positive or negative charges in the binding site, produced by amino acid residues or tightly bound ions, of hydrogen-bonding residues, as well as hydrophobic and sterically restrictive areas of the active site. As an example, a pharmacophore developed from the bioactivity of a series of related compounds that either do or do not inhibit the enzyme ATP:L-methionine *S*-adenosyltransferase is illustrated in Figure 15.9 [33].

Molecular modeling and computational chemistry can make use of virtual compound libraries, which are either experimentally determined three-

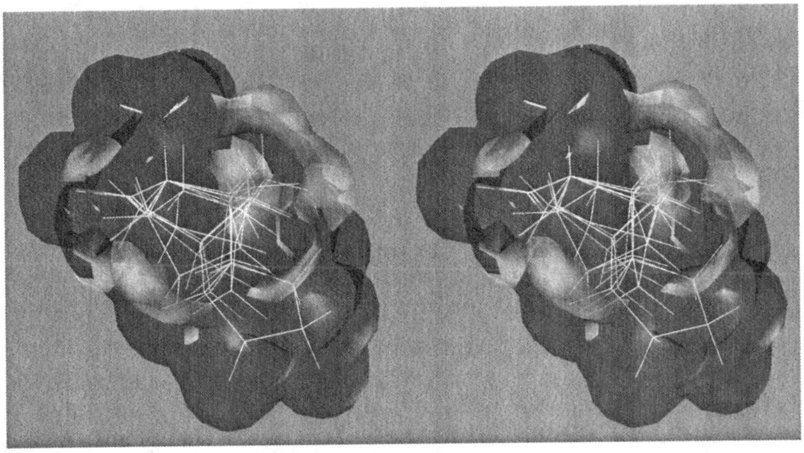

Figure 15.9 Hypothetical pharmacophore for inhibitors of the target enzyme ATP:L-methionine *S*-adenosyltransferase. Six derivatives of 1-aminocyclopentane-1-carboxylic acid were tested for their ability to displace the natural substrate (methionine). White lines indicate the bonds of the aligned compounds. The volume that envelopes all active compounds is outlined in dark gray; the volumes of substituents that prevent binding extend out and are indicated in a lighter tone. (Based on data in [33]. Sybyl8.1 (Tripos, Inc, USA) software was used to produce this cross-eyed stereo figure. A colored version of this figure is shown in the insert in the middle of this book.)

dimensional structures available through the Cambridge crystallographic small molecule database, proprietary structure libraries, or even structures of compounds that have only been drawn on a computer screen and then energy-minimized in silico but that have never actually been synthesized [2]. All these experimentally determined or hypothetical structures can be screened by virtually docking them with the target active site or the pharmacophore.

Recent studies have shown that a number of docking programs can indeed produce fairly accurate poses of molecules bound to the active sites of enzymes. However, it is usually more difficult to accurately rank the compounds that have been docked according to affinity, for the purpose of selecting the best binders from a virtual screen [23]. It has frequently been found that a number of false positives can be produced by virtual docking. This is related to the fact that currently this computational approach often suffers from a shortage of detailed knowledge concerning the intermolecular interactions and their free energies in the active site.

Nevertheless, it should be kept in mind that virtual docking requires fewer resources and is less expensive than large-scale synthetic and screening efforts. Even if not perfect, the best structures identified in virtual screens can usually serve as synthetic starting points for chemical elaboration. Real compounds are then made and tested in the biochemical screens to provide binding data to substantiate or refute the computationally derived hypotheses, as well as to provide checks on the quality of the computational rankings that were determined.

15.4 PHENOTYPIC SCREENING

A problem with screening compounds against purified proteins is that activity observed in such assays may not translate to intact cells, for example, if the compound in question is excluded by the cell membrane or is extruded from the cell by active transport. Conversely, compounds that have no activity on a specific target protein may nevertheless have useful pharmacological activity on the cell, due to activity on other targets, possibly even unknown ones.

In a phenotypic screen, a wider net is cast in order to capture such potentially varied mechanisms of action. Here, the readout consists in a change not in the activity of one specific protein, but rather in the physiological state of the cell as a whole, for example, its motility or metabolic turnover. Such whole cell-based assays are also frequently utilized at an early stage of compound screening [7]. For example, screening of compounds for antitumor activity are routinely performed against panels of human tumor cell lines [27]. These cell proliferation assays allow for the detection of the desired bioactivity—cell growth inhibition or cell death—regardless of the site of action of the compounds being tested. Valuable time may be saved and costly effort avoided by utilizing this approach, since direct activity on the tumor cell is evaluated.

Another example of this approach is found in the screening for compounds active against tuberculosis. Screens of compounds showing growth inhibition of *Mycobacterium smegmatis*, an organism that is related to the more pathogenic and clinically important *Mycobacterium tuberculosis*, have led to the discovery of several compounds that have good potency against the latter [26]. These initial screens were performed without any notion of a predetermined cellular target in mycobacteria. However, once active compounds were found that showed visible growth inhibition, genome sequencing [3] and biochemical [14] studies revealed that some of them targeted the mycobacterial ATP synthase. This multienzyme complex is essential for metabolic energy production in the organism. Nevertheless, prior to these reports, this enzyme complex had not even been considered as a molecular target for drug development. Currently, the multienzyme complex is being utilized as a molecular screening target in order to obtain more knowledge concerning the detailed structure–activity relationships of these inhibitors.

15.5 COMPOUND ACQUISITION

Once an appropriate target has been identified and a suitable corresponding assay has been devised, there is, of course, a need for compounds to screen. Some drug development projects have screened several hundred thousand compounds against single targets. For example, a recent high-throughput screening campaign to find $K_V1.3$ channel blockers was reported, which screened 650,000 compounds at an initial concentration of 10 μM. The screen required over two thousand 384-well plates to run and was completed in a period of 17 days, obtaining 332 high-quality hits to be further pursued [15]. Obviously, making or acquiring such a large number of compounds for screening is a major undertaking.

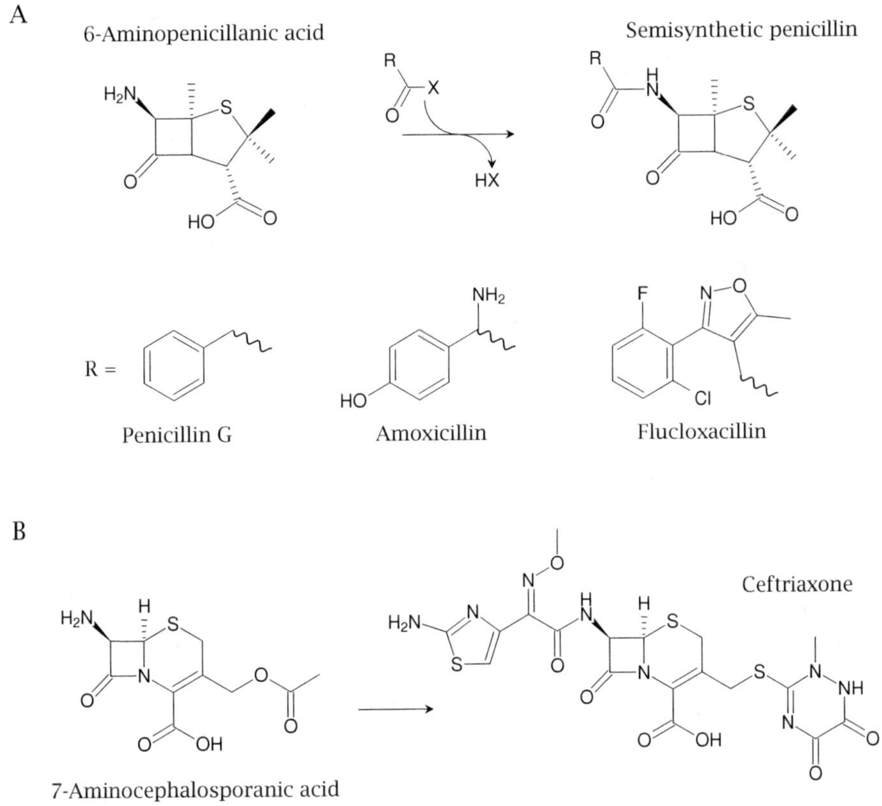

Figure 15.10 Production of β-lactam antibiotics as an example of semisynthetic preparation of novel drugs. **A:** Various penicillins are obtained through chemical acylation of 6-aminopenicillanic acid, which is obtained by fermentation. **B:** Cephalosporins are derived from 7-aminocephalosporanic acid, which is also obtained through fermentation. Semisynthetic strategies have led to compounds with superior antibacterial properties such as ceftriaxone.

Where do researchers obtain these compounds? There are various sources: natural products, new synthetic compounds available from various academic and drug company research projects, chemical libraries sold by pharmaceutical companies to other companies for testing against their own targets, or in-house pharmaceutical libraries. The latter typically consist of collections of compounds that have been developed over the years for several specific projects in a particular pharmaceutical company but are now stored away to become that company's private compound library useful for other in-house projects.

15.5.1 Natural compounds

The large number of existing compounds notwithstanding, there remains a need for new ones. One source of new compounds is the natural product approach. Microorganisms, plants, and animals produce compounds that can

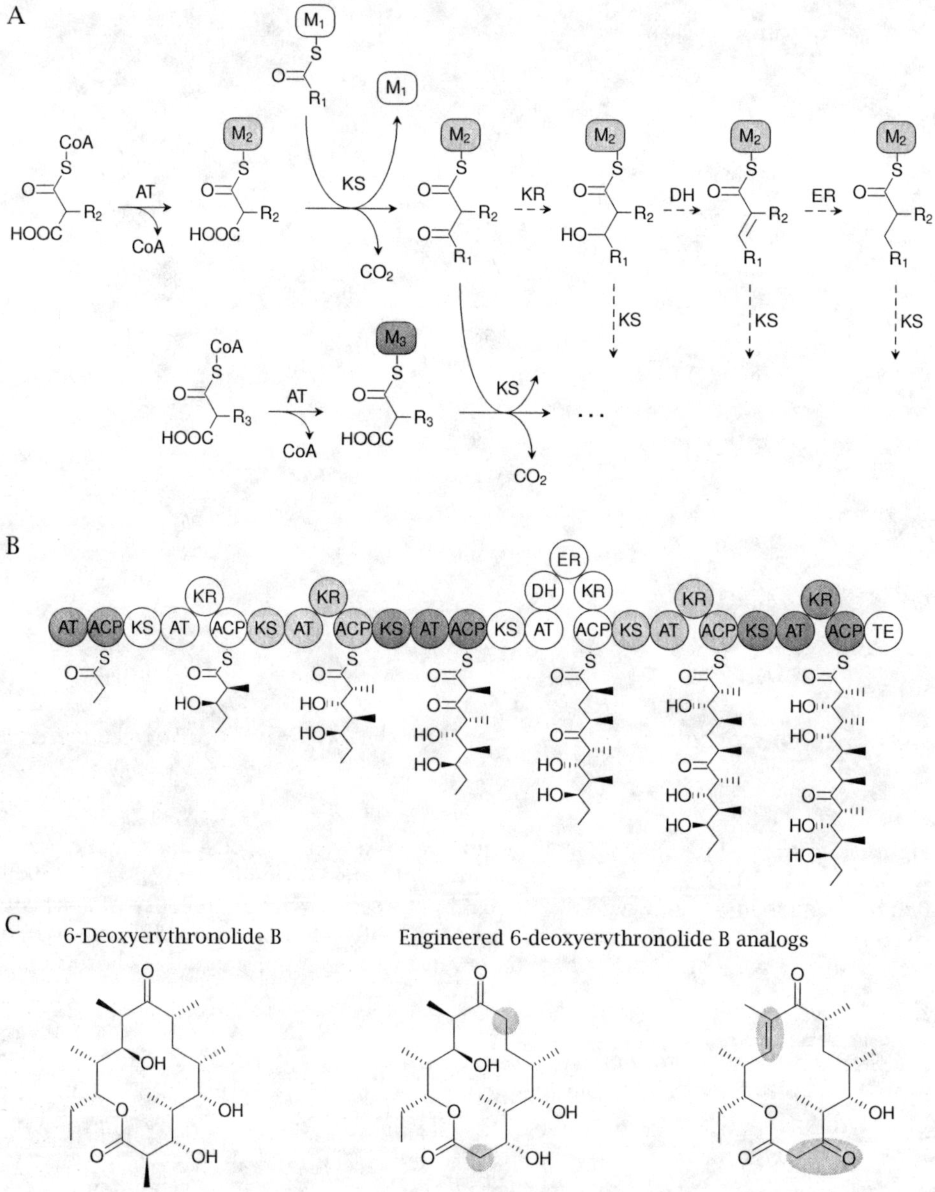

Figure 15.11 Genetic engineering of polyketide synthesis for the biosynthesis of novel compounds. **A:** The biosynthesis of polyketides is similar to that of fatty acids, with the following differences: Each C_2 subunit is incorporated by a separate enzyme module (indicated here as M_1–M_3), and the polymerization can occur at any stage of reduction of the β carbon. Enzyme activities: AT, acyltransferase; KS, ketosynthase; KR, ketoreductase; DH, dehydratase; ER, enoylreductase. **B:** Structure of 6-deoxyerythronolide B synthase. There are six separate modules, five of which contain one or more reducing enzyme activities. The final thioesterase (TE) releases and cyclizes the product. (ACP is acyl carrier protein.) **C:** 6-Deoxyerythronolide B is a precursor of erythromycin. The two derivatives shown, as well as many others, were obtained by replacing several modules of the synthase with ones from rapamycin polyketide synthase [16].

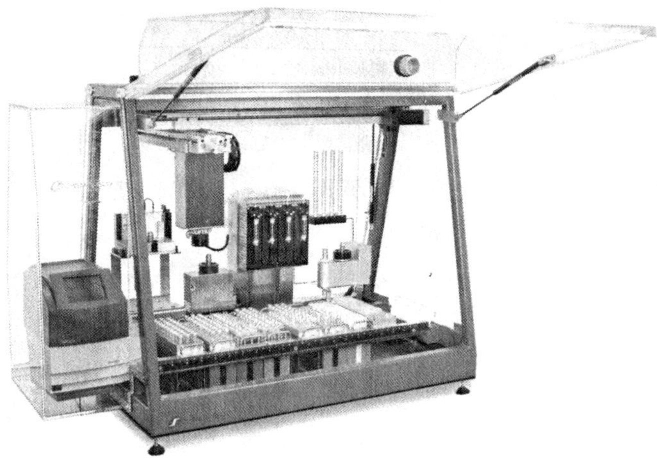

Figure 15.12 An example of an automated setup to robotically handle the microwave-catalyzed synthesis of numerous reactions and their subsequent automated purification. Shown here is a commercially available system, ChemSpeed SWave (ChemSpeed, Inc., Switzerland).

be isolated and screened against various biochemical targets. For example, a *Streptomyces* bacterial strain can be isolated from soil and grown in fermenters, and its secondary metabolites can then be fractionated, isolated, identified and screened for a particular bioactivity, such as activity against various cancer cell lines or infectious agents. Compounds that show promising activity are then studied as to their chemical structure, allowing scientists to undertake focused structure–activity investigations to optimize their activities.

To expand the molecular diversity of natural products, two powerful approaches are used. The first is that of *semisynthesis*, that is, the use of chemical modification of the natural product to form new analogs [11]. A key example is the chemical derivatization of the core β-lactam nucleus, 6-aminopenicillanic acid, which has yielded numerous derivatives with improved medicinal properties (Figure 15.10).

The second strategy to expand the repertoire of natural compound is based on molecular genetics. Once the genes responsible for the biosynthesis of a natural product have been isolated and characterized, they can be manipulated with recombinant methods, yielding redesigned biosynthetic pathways that produce modified versions of the original natural product [31,22]. An example is the production of new analogs of 6-deoxyerythronolide B [16,4], the biosynthetic precursor of the antibiotic erythromycin (see Figure 4.5). Polyketides such as 6-deoxyerythronolide B are produced by multienzyme complexes with highly modular structures, which can be rearranged and recombined to yield synthetic pathways for novel compounds in an almost algorithmic fashion (Figure 15.11).

15.5.2 Synthetic chemistry

The second major source of new compounds is chemical synthesis. Where possible, compounds are synthesized in high-throughput mode such as by robotic

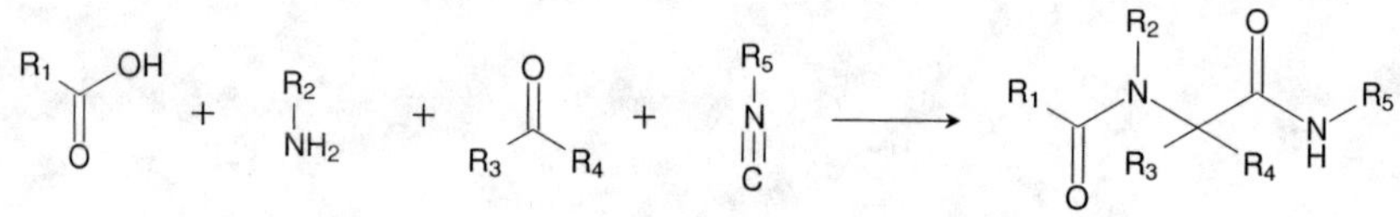

Figure 15.13 The Ugi reaction [37]. A carboxylic acid, an amine, a ketone or aldehyde and an isonitrile condense to form a bis-amide. Through combinatorial variation of the functional groups R_1–R_5, many different compounds can be prepared in parallel.

systems that allow for many different chemical reactions to be undertaken simultaneously in a parallel fashion [6]. Frequently the chemical reactions are enhanced by application of microwave heating that can also be done robotically (Figure 15.12).

Great care and planning is required in designing parallel synthetic reaction schemes that produce many new compounds with minimal technical complexity [42]. For example, a library of compounds can be generated very rapidly by using a series of reactions that have multiple components. An example of this is the Ugi chemical reaction that combines four components: an aldehyde, an amine, a carboxylic acid, and an isocyanide [10]. The aldehyde and amine first condense to form an imine, which then can react with the other two components (Figure 15.13). For each of these four components, several variants with different side chains can be used, resulting in a potentially large number of combinations. Some of the side chains introduced in the first reaction may contain modifiable groups that can then be used for further diversification. Synthetic strategies like this one are subsumed under the term *combinatorial chemistry* [17].

With both natural compound and synthetic libraries, compounds with reasonable bioactivity, or *hits*, are identified in preliminary screens. These compounds are then modified to try to develop a series of *lead compounds*, which improve on the initial hits with respect to synthetic accessibility, potential toxicity, and the physicochemical features that determine the so-called *ADME properties*, that is, absorption, distribution, metabolism and excretion. Many compounds are too hydrophobic to use directly, and these new agents will then need to be formulated for better pharmacokinetic properties. Chapter 14 elaborates on several of these topics.

In summary, drug discovery is an exciting field of research. It is an area requiring various kinds of expertise from computational and synthetic chemistry, to enzymology, cell biology, physiology, and clinical medicine. Perhaps some of you will find your future in this critically important career!

STUDY QUESTIONS

15.1 The compound shown below has been determined to have biological activity as a P-glycoprotein blocker, which may be useful in cancer chemotherapy (see Section 3.2.3). However, to become clinically viable, the structure needs to be modified so as to increase the affinity to the target.

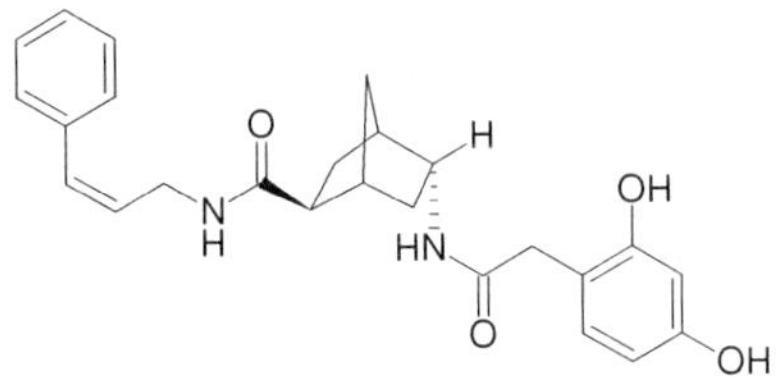

Based on the structure of the starting compound, suggest a strategy to generate a compound library that can be screened for derivatives with higher affinity.

15.2 Your pharmaceutical company is investigating inhibitors of a protease that accomplishes the following reaction:

Ala–Leu–Phe–Gly–Ile → Ala–Leu–Phe + Gly–Ile

Before synthesizing a complete compound library, your team wants to obtain a crystal structure of the enzyme with an inhibitor bound in the active site.

Propose two compound structures, likely to be reasonable inhibitors, that the chemists in your company should synthesize for use in the crystallography experiments.

15.3 Although computer modeling can be very effective in predicting the activity of a compound against a defined drug target, there are several factors that may detract from its accuracy and prove detrimental to an exclusively in silico approach to drug design. What might these be exactly?

REFERENCES

[1] E. J. Adie et al., "CypHer 5: a generic approach for measuring the activation and trafficking of G protein-coupled receptors in live cells," *Assay Drug Dev. Technol.* 1 (2003), 251–259.

[2] F. Allen, "The Cambridge Structural Database: a quarter of a million crystal structures and rising," *Acta Crystallogr. B* 58 (2002), 380–388.

[3] K. Andries et al., "A diarylquinoline drug active on the ATP synthase of Mycobacterium tuberculosis," *Science* 307 (2005), 223–227.

[4] G. W. Ashley et al., "Preparation of erythromycin analogs having functional groups at C-15," *J. Antibiot. (Tokyo)* 59 (2006), 392–401.

[5] S. Brinster et al., "Type II fatty acid synthesis is not a suitable antibiotic target for Gram-positive pathogens," *Nature* 458 (2009), 83–86.

[6] J. F. Cargill and M. Lebl, "New methods in combinatorial chemistry-robotics and parallel synthesis," *Curr. Opin. Chem. Biol.* 1 (1997), 67–71.

[7] T. Chen, ed., *A Practical Guide to Assay Development and High-Throughput Screening in Drug Discovery,* CRC Press, 2009.

[8] M. S. Chorghade, ed., *Drug Discovery and Development,* Wiley-Interscience, 2007.

[9] R. A. Copeland, *Evaluation of Enzyme Inhibitors in Drug Discovery,* Methods of biochemical analysis series, vol. 46, John Wiley & Sons, 2005.

[10] A Dömling and I Ugi, "Multicomponent reactions with isocyanides," *Angew. Chem. Int. Ed.* 39 (2000), p. 18.

[11] M. T. Hamann, "Enhancing marine natural product structural diversity and bioactivity through semisynthesis and biocatalysis," *Curr. Pharm. Des.* 9 (2003), 879–889.

[12] A. P. Kaplan and P. A. Bartlett, "Synthesis and evaluation of an inhibitor of carboxypeptidase A with a Ki value in the femtomolar range," *Biochemistry* 30 (1991), 8165–8170.

[13] F. D. King, *Medicinal chemistry: principles and practice,* 2nd ed., Royal Society of Chemistry, 2002.

[14] A. Koul et al., "Diarylquinolines target subunit c of mycobacterial ATP synthase," *Nat. Chem. Biol.* 3 (2007), 323–324.

[15] K. Liu et al., "High-throughput screening for Kv1.3 channel blockers using an improved FLIPR-based membrane-potential assay," *J. Biomol. Screen.* 15 (2010), 185–195.

[16] R. McDaniel et al., "Multiple genetic modifications of the erythromycin polyketide synthase to produce a library of novel 'unnatural' natural products," *Proc. Natl. Acad. Sci. U. S. A.* 96 (1999), 1846–1851.

[17] B. L. Miller, ed., *Dynamic Combinatorial Chemistry in Drug Discovery, Bioorganic Chemistry, and Materials Science,* John Wiley & Sons, 2009.

[18] B. Nagar et al., "Crystal structures of the kinase domain of c-Abl in complex with the small molecule inhibitors PD173955 and imatinib (STI-571)," *Cancer Res.* 62 (2002), 4236–4243.

[19] D. T. Newton, C. Creuzenet, and D. Mangroo, "Formylation is not essential for initiation of protein synthesis in all eubacteria," *J. Biol. Chem.* 274 (1999), 22143–22146.

[20] R. Numann and P. A. Negulescu, "High-throughput screening strategies for cardiac ion channels," *Trends Cardiovasc. Med.* 11 (2001), 54–59.

[21] R. H. Oakley et al., "The cellular distribution of fluorescently labeled arrestins provides a robust, sensitive, and universal assay for screening G protein-coupled receptors," *Assay Drug Dev. Technol.* 1 (2002), 21–30.

[22] I. Ojima, "Modern natural products chemistry and drug discovery," *J. Med. Chem.* 51 (2008), 2587–2588.

[23] K. Onodera, K. Satou, and H. Hirota, "Evaluations of molecular docking programs for virtual screening," *J. Chem. Inf. Model.* 47 (2007), 1609–1618.

[24] G. L. Patrick, *An introduction to medicinal chemistry,* 3rd ed., Oxford University Press, 2005.

[25] J. G. Robertson, "Mechanistic basis of enzyme-targeted drugs," *Biochemistry* 44 (2005), 5561–5571.

[26] J. C. Sacchettini, E. J. Rubin, and J. S. Freundlich, "Drugs versus bugs: in pursuit of the persistent predator Mycobacterium tuberculosis," *Nat. Rev. Microbiol.* 6 (2008), 41–52.

[27] R. H. Shoemaker, "The NCI60 human tumour cell line anticancer drug screen," *Nat. Rev. Cancer* 6 (2006), 813–823.

[28] R. B. Silverman, *The organic chemistry of drug design and drug action,* 2nd ed., Academic Press, 2004.

[29] L. K. Smith et al., "Monitoring of gene knockouts: genome-wide profiling of conditionally essential genes," *Genome Biol.* 8 (2007), R87.

[30] M. Steiner-Mosonyi et al., "The Pseudomonas aeruginosa initiation factor IF-2 is responsible for formylation-independent protein initiation in P. aeruginosa," *J. Biol. Chem.* 279 (2004), 52262–52269.

[31] W. Strohl, "The role of natural products in a modern drug discovery program," *Drug Discov. Today* 5 (2000), 39–41.

[32] Z. Su and J. F. Honek, "Emerging bacterial enzyme targets," *Curr. Opin. Investig. Drugs* 8 (2007), 140–149.

[33] J. R. Sufrin, D. A. Dunn, and G. R. Marshall, "Steric mapping of the L-methionine binding site of ATP:L-methionine S-adenosyltransferase," *Mol. Pharmacol.* 19 (1981), 307-313.

[34] C. J. Swanson et al., "Metabotropic glutamate receptors as novel targets for anxiety and stress disorders," *Nat. Rev. Drug. Discov.* 4 (2005), 131–144.

[35] G. Thomas, *Medicinal Chemistry: an Introduction,* 2nd ed., John Wiley & Sons, 2007.

[36] O. Trott and A. J. Olson, "AutoDock Vina: improving the speed and accuracy of docking with a new scoring function, efficient optimization, and multithreading," *J. Comput. Chem.* 31 (2010), 455–461.

[37] I. Ugi, "The α-addition of immonium ions and anions to isonitriles accompanied by secondary reactions," *Angew. Chem. Int. Ed.* 1 (1962), 8–21.

[38] M. D. Vaughan, P. B. Sampson, and J. F. Honek, "Methionine in and out of proteins: targets for drug design," *Curr. Med. Chem.* 9 (2002), 385–409.

[39] W. Wang et al., "Discovery of inhibitors of Escherichia coli methionine aminopeptidase with the Fe(II)-form selectivity and antibacterial activity," *J. Med. Chem.* 51 (2008), 6110–6120.

[40] C. Wermuth et al., "Glossary of terms used in medicinal chemistry (IUPAC recommendations 1998)," *Pure Appl. Chem.* 70 (1998), 1129-1143.

[41] C Wolff, B Fuks, and P Chatelain, "Comparative study of membrane potential-sensitive fluorescent probes and their use in ion channel screening assays," *J. Biomol. Screen.* 8 (2003), p. 533.

[42] J. Xiang et al., "Identifying tumor cell growth inhibitors by combinatorial chemistry and zebrafish assays," *PLoS One* 4 (2009), e4361.

ANSWERS TO STUDY QUESTIONS

2.1 The question states explicitly that further increases of A or B should not increase their effects any further. Such behavior can never be observed with reversible inhibitors. So, both A and B are irreversible.

Now, it is clear that, even at saturating concentration, drug A does not manage to knock out all receptors. This means there must be two different receptors that participate in the response, only one of which reacts with A. The second one responds to B, so that after application of both drugs, almost all receptors are gone. (Data from Ref. [20].)

2.2 For the inhibition by variable concentrations of two inhibitors A and B, we can derive (not shown, but analogous to Section 2.6.2)

$$Y = \frac{[\mathrm{L}]}{[\mathrm{L}] + K_\mathrm{L}\left(1 + \frac{[\mathrm{A}]}{K_\mathrm{A}} + \frac{[\mathrm{B}]}{K_\mathrm{B}}\right)}$$

Let's say that $[\mathrm{A}]_\mathrm{T}$ and $[\mathrm{B}]_\mathrm{T}$ are the concentrations of A alone and of B alone, respectively, necessary to reduce the receptor saturation Y to a certain threshold, Y_T. Then, in order for variable combinations of A and B to achieve the same Y_T, we must stipulate

$$\frac{[\mathrm{A}]}{K_\mathrm{A}} + \frac{[\mathrm{B}]}{K_\mathrm{B}} = \frac{[\mathrm{A}]_\mathrm{T}}{K_\mathrm{A}} \left(= \frac{[\mathrm{B}]_\mathrm{T}}{K_\mathrm{B}}\right)$$

from which we obtain

$$[\mathrm{B}] = \frac{K_\mathrm{B}}{K_\mathrm{A}}\left([\mathrm{A}]_\mathrm{T} - [\mathrm{A}]\right)$$

which describes the linear relationship between [B] and [A] depicted in the plot.

Deviations from linearity imply that two drugs do not act in the same manner. If the observed graph resembles the dotted line, then higher than expected dosages of B will be required to complement A; the drugs have a less than additive effect. If the graph resembles the dashed line, the drugs act *synergistically*—the combination is stronger than the sum of its parts.

2.3 A search on PubMed (`http://www.ncbi.nlm.nih.gov/pubmed`) for the phrase "quinine[ti] quinidine[ti] pharmacodynamics" turns up many interesting references.

3.1 An oligonucleotide would be degraded in the intestine and be unable to cross epithelial barriers, so only parenteral application would work. Assuming a molecular weight of 400 Da for each base, the total molecular weight would be 6000 Da, so that it should be able to cross capillary walls to reach the interstitial volume and to undergo glomerular filtration. It would be excluded by the blood-brain barrier and would not effectively cross cell membranes.

3.2 Ascorbic acid in large doses should acidify the urine. LSD contains two amines which, when protonated, should reduce reuptake of filtrated LSD by nonionic diffusion and so accelerate elimination.

3.3 **A:** According to Equation 3.12, the two parameters plotted here should be strictly proportional to each other. The observed scatter indicates that the tabulated data do not perfectly obey that equation. This might be due to either measurement errors, or to elimination kinetics of other than first order (remember that Equation 3.12 is based on first order kinetics). **B:** Protein binding should inhibit renal elimination. Indeed, most drugs with a high degree of protein binding show a low percentage of elimination in the urine. However, drugs that have a low degree of protein binding show considerable scatter, which suggests that their rates of metabolic inactivation vary greatly as well. **C:** Common textbook wisdom—faithfully conveyed in these pages also—is that polarity is a key determinant of oral availability. I was surprised to see that this relationship does not show up in the data. We have to consider though that these are calculated logP values, not experimental ones.

4.1 Drug L can be applied at concentrations substantially greater than necessary for the therapeutic effect, and so can remain effective even if a significant fraction is consumed because of enzyme induction. In contrast, drug S will not have that kind of reserve and will be more likely to dive below the therapeutic concentration.

4.2 Desaturation is an oxidative reaction, and cytochrome P450 enzymes are at least partly responsible [18]. Vinyl compounds react with nucleophiles, and toxicity due to glutathione depletion has been documented [11].

4.3 No. Unlike, for example, in phenylketonuria, the toxicity in glucose-6-phosphate dehydrogenase deficiency is *not* caused by accumulation of toxic metabolites to unusually high levels, which would be prevented by maternal-fetal equilibration, but instead arises from increased sensitivity to normal levels. It is possible for the same metabolite level to be harmless to the mother but toxic to the fetus [5].

5.1 (1) Thioperadine shows equal potency on all receptors, whereas clozapine shows significant differences. Possibly, thioperamide interacts only with conserved residues within the binding site, while clozapine interacts with a different set of residues that also includes some that are not conserved between the species in question. (2) The surprisingly large differences in affinity observed with clozapine suggest that its physiological effects should also differ significantly between different animals. Such differences might affect not only the receptor that is being targeted but also others that might then elicit side effects. Therefore, animals may lie.

5.2 The key finding is that the two compounds together achieve a higher degree of receptor activation than either alone. This is only possible if they bind to different sites, which has been substantiated by molecular experiments; GABA binds to the extracellular N-terminal domain, whereas CGP7930 binds to the heptahelical domain [3]. It also shows that GABA itself is only a partial agonist at this receptor.

5.3 Recall that G proteins interact with multiple receptors. The relatively small fluorescence change may reflect the fraction of labeled G protein that interacts with this particular receptor.

6.1 See for example Ref. [1].

6.2 After the cycle has been started by the release of transmitter, it is driven forward by transmitter removal. In the case of acetylcholine, the energy for the removal is supplied by enzymatic hydrolysis of the transmitter, which is exergonic (see Section 6.8.3). With other transmitters, it is supplied by the active transporters that scavenge them from the synaptic cleft.

6.3 While tetanus toxin affects neurons that inhibit motoneurons, botulinum toxin disrupts transmitter exocytosis in the motoneurons themselves. Tetanus and botulinum toxin each consist of two polypeptide chains. The heavy chain mediates uptake into motoneurons, and in the case of tetanus toxin, its retrograde axonal transport, whereas the light chain is the protease that cleaves synaptobrevin [4].

6.4 A dehydrated ion must be very tightly coordinated by the channel protein, or else dehydration will simply not occur, because the associated energy barrier is too large. This requires the diameter of the ion and that of the ion selectivity filter to be very closely matched, and therefore enables a high degree of selectivity. In contrast, the hydration shells of hydrated ions, while differing in size, are not clearly delineated, and therefore cannot be as accurately discriminated by the corresponding channels.

7.1 This finding suggests that trans-repression and translational regulation are activated by distinct conformations of the receptor. While the term doesn't seem to have been used in this context, this could be considered another example of agonist-specific coupling (see Section 2.2.6)

7.2 (1) The formation and dissolution of minerals is governed by the solubility product, that is, the product of the concentrations of their constituent ions in solution. In the case of hydroxyapatite, OH^- is one of the ions. When OH^- is consumed by acid, this lowers the solubility product and so promotes the dissolution of the mineral. In contrast, the concentration of fluoride is not affected by the pH, which keeps the solubility more stable. Note, however, that the protonation of phosphate (PO_4^{3-}) at low pH will lower the solubility product in both cases.

(2) Fluoride that is applied to the surface of the teeth cannot penetrate them effectively and therefore will affect only the surface itself. In contrast, fluoride that is ingested (with the tap water or otherwise) will participate in the deposition of the mineral itself—but only during formation of the teeth, that is, during early childhood. Once the permanent teeth are out, superficial application will be more effective.

For political background on and military implications of tap water fluoridation, see Ref. [8].

8.1 While NO can cross cell membranes easily, compounds such as *S*-nitrosocysteine or *S*-nitrosoglutathione would require some sort of facilitated transport. Thus, their distribution would likely be slower and less effective.

8.2 See for example References [24], [2] and [9].

9.1 Inhibition of FAAH will increase the levels of endocannabinoids, and inhibition of cyclooxygenase may divert arachidonic acid to lipoxygenase; therefore, cyclooxygenase inhibitors might indirectly raise the levels of endocannabinoids.

An interesting experimental study that deals with the effect of cyclooxygenase inhibition on endocannabinoid signaling is Ref. [19].

9.2 NO degradation occurs in the peroxidase site of NOS. It seems unlikely that NO consumption by cyclooxygenase would be limited by inhibition of the cyclooxygenase site, which is the target of most Cox inhibitors. If it were so limited, however, the reduced consumption of NO should detract from the inhibition of inflammation. See, however, Ref. [15] for an intriguing twist on this connection.

10.1 (1) Metformin is supposed to inhibit complex I of the respiratory chain, which should increase the level of NADH. If NADH were toxic, metformin should compound the pathogenic effect of the sorbitol pathway. Note, however, that the explanation of the metformin effect given here is not universally accepted, either. (2) The combined action of malate dehydrogenase (a), malic enzyme (b) and pyruvate carboxylase (c) would do the trick, at the expense of ATP:

$$\text{Oxaloacetate} + \text{NADH} \longrightarrow \text{malate} + \text{NAD}^+ \qquad \text{(a)}$$

$$\text{Malate} + \text{NADP}^+ \longrightarrow \text{pyruvate} + \text{CO}_2 + \text{NADPH} \qquad \text{(b)}$$

$$\text{Pyruvate} + \text{CO}_2 + \text{ATP} \longrightarrow \text{oxaloacetate} + \text{ADP} + \text{P}_i \qquad \text{(c)}$$

$$\text{NADH} + \text{NADP}^+ + \text{ATP} \longrightarrow \text{NAD}^+ + \text{NADPH} + \text{ADP} + \text{P}_i$$

10.2 Apart from the obvious answer—that is, to study the effect of both in vitro with purified enzymes—we could also learn something from clinical outcomes: If arginine itself exercises this inhibition, then the effect should be observable in patients with arginase defects. On the other hand, application of ornithine should have no beneficial effect. The opposite should be true if ornithine were the inhibitor.

10.3 We could try to increase the production of uric acid. The most direct way would be to inhibit purine salvage, for example, by inhibition of hypoxanthine-guanine phosphoribosyltransferase (HGPRT). However, enzyme defects for HGPRT cause pathology, which suggests that this therapy might be problematic. Also, the solubility of uric acid is limited, and increased excretion will likely lead to formation of urate kidney stones.

11.1 The interaction of carbenicillin with EDTA is synergistic, whereas that with polymyxin B is slightly antagonistic. Both EDTA and polymyxin disrupt the outer membrane of Gram-negative bacteria; in the case of EDTA, this occurs through the withdrawal of calcium ions that connect adjacent lipopolysaccharide molecules. This should facilitate the entry of carbenicillin and make it more effective. Therefore, synergism would be expected in both cases. It is not clear why the interaction with polymyxin B is (slightly) antagonistic rather than synergistic. However, mutual binding and neutralization have been observed with anionic β-lactams and cationic aminoglycosides in vitro; this seems possible with carbenicillin and the cationic polymyxin B, too.

11.2 Fosfomycin requires uptake by active transport, and flucytosine requires metabolic activation by the cell. Neither the glycerophosphate transporter (with fosfomycin) nor the deaminase (with flucytosine) is essential for the cell. Therefore, any mutation that disrupts these proteins will render the drugs ineffective, without jeopardizing cell viability. With any gene, the number of possible disruptive mutations is very large, and therefore the development of resistance is fast.

In contrast, penicillin or ketoconazole do not depend on active transport or metabolic activation and only interact with enzymes that are essential to the cell. Mutations that render these targets resistant to the drugs must therefore spare their interactions with the physiological substrates. The number of possible mutations with such discriminatory effects is far smaller, and therefore they are much less likely to occur within a limited period of time.

11.3 Amantadine and rimantadine reach the intracellular space and the endosome by nonionic diffusion. Protonation of the amino groups in the acidic lumen of the endosome then causes the drugs to accumulate within. The situation mirrors the accumulation of acetylsalicylic acid in the gastric epithelial cells (see Figure 3.4).

12.1 (1) Pentostatin is a transition state analog. (2) Inhibition of ADA will affect lymphatic cells and drive them into apoptosis, as is the case in the genetic enzyme defect (see Section 10.1.4). Susceptible tumors are lymphomas and lymphatic leukemias.

12.2 The different effects of tamoxifen on breast tissue versus bone tissue must be related to tissue-specific differences in the expression of transcriptional coregulators. It is likely that breast tumors, particularly when under selective pressure from tamoxifen therapy, accumulate mutations that shift the expression pattern from one that favors growth inhibition to one that favors cell growth. Fulvestrant would likely be less affected by such a shift.

12.3 If we assume that simultaneous targeting of different targets helps to avoid tumor resistance, we may expect the empirical process of selecting clinically useful antitumor agents to favor "dirty" drugs, that is, ones with multiple targets and modes of action.

13.1 See Ref. [17].

13.2 See Ref. [22].

13.3 See Ref. [7].

13.4 See References [10] and [23].

14.1 A carrier that remains buoyant evades the normal peristaltic gastric emptying process and stays within the stomach until it dissolves. Delivery from such a carrier can be protracted and sustained, which may provide an advantage for drugs that are rapidly eliminated. Gas-containing and gas-generating polymer carriers have been developed [16].

14.2 Both DNA and the cell surface are negatively charged. Cationic polymers bind and condense the DNA, and also mediate adsorption to the cell surface through electrostatic interaction. The adsorbed complex can then be taken up by endocytosis [6].

14.3 We can (1) replace L-amino acids by D-amino acids, (2) use cyclized peptides (these will be resistant to exopeptidases), (3) block the C- and N-termini by amidation or methylation, (4) insert unnatural side chains that sterically interfere with proteolysis, or (5) replace peptide bonds by other, isosteric groups. For an in-depth discussion, see for example Ref. [21].

While chemical modification is a viable and valuable strategy, some recent progress with the oral delivery of unmodified peptides is noteworthy as well [12].

14.4 It takes time for an antibody to reach the tumor cells after intravenous application. The shorter the half-life, the greater the proportion of the radionuclide that will decay during the time interval required for transport. This considerably increases the nonspecific toxicity [13].

14.5 They all are hydrophobic, hence they all can be applied through skin patches [14]. (Drugs: 1, estradiol; 2, scopolamine; 3, nicotine; 4, clonidine.)

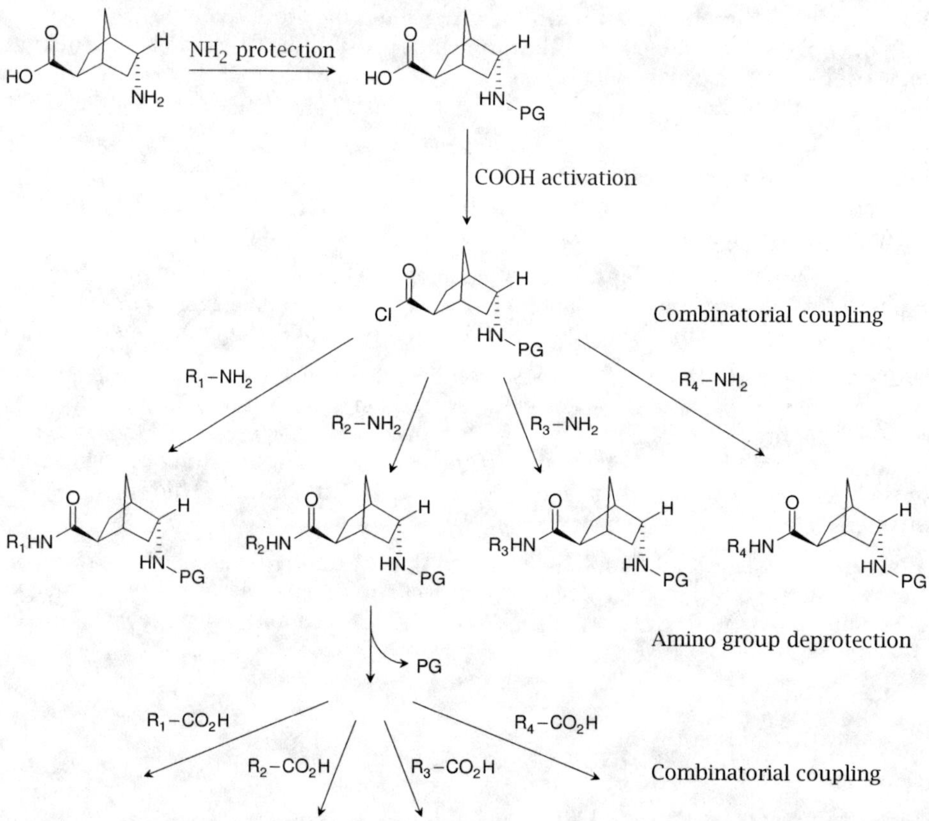

Figure A.15.1 Combinatorial synthetic strategy for creation of a compound library (Question 15.1). After attaching a protecting group (PG) to the amine, the free carboxylic acid is chemically activated as an acyl chloride. Various amine-containing molecules are then added in separate reaction vessels. Once these intermediates are synthesized, each is deprotected and then portioned into many other reaction vessels. In each of these, a different activated carboxyl acid-containing compound is attached.

15.1 The initial structure contains two amide bonds, which can be formed synthetically in a straightforward manner. We can break up the initial structure across these amide bonds and use the central piece with the bicyclic ring system as the core scaffold, and create the compound library using the synthetic strategy outlined in Figure A.15.1.

The number of different compounds in the library will be the product of the numbers of substituents used in each of the two synthetic steps. Additional variation can be achieved by replacing the starting scaffold with other, similar bifunctional molecules.

Note that amide (or peptide) bonds may be unstable during intestinal passage. Therefore, after isolating derivatives with sufficiently high affinity for the target, it may be necessary to replace the amide bonds, or sterically protect them in order to make the molecule suitable for oral application.

15.2 The catalytic mechanism of proteases resembles that of cholinesterase (see Figure 6.26). However, to make this problem a little easier, let us assume that the protease simply activates a water molecule to attack the specific peptide bond. The structure of an intermediate that would form in the active site is shown in Figure A.15.2.

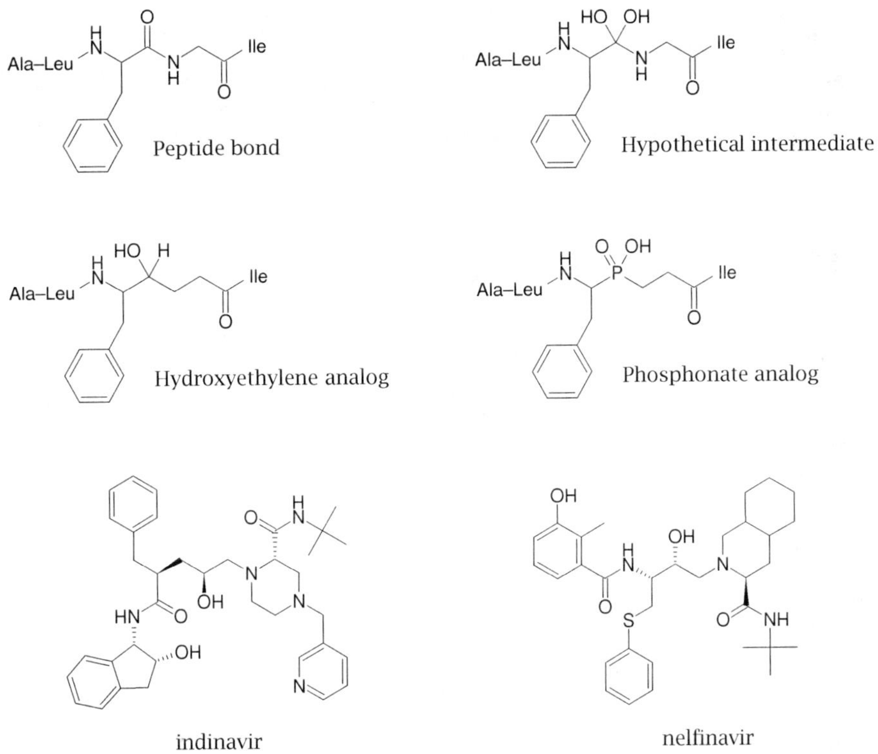

Figure A.15.2 Illustration for answer to question 15.2. Hydrolysis of the peptide bond will go through an unstable intermediate stage that contains two hydroxyls. Of the two transition state analogs shown, the hydroxyethylene compound has the correct positioning of the hydroxyl group as well as an sp^3-hybridized carbon, whereas the phosphonate has the correct number of non-hydrogen substituents. The compounds indinavir (Crixivan™) and nelfinavir (Viracept™) are potent HIV-1 protease inhibitors that employ the hydroxyethylene functionality.

As a rule, enzymes bind transition states very avidly. While the transition state as such is not stable and cannot be synthesized, it is possible to create analogs that are stable and resemble the transition state with respect to geometry and polarity, and as such are tightly bound by the enzyme as well. Such transition state analogs tend to be good inhibitors. Two transition state analogs of the protease are also shown in Figure A.15.2. Compare also Figure 15.2 in the chapter text.

15.3 The following list provides some plausible answers.

1. The protein structure used for molecular docking experiments may be inaccurate or have limited resolution. For example, a flexible loop or a series of water molecules may be missing from the crystal structure, but they may nevertheless be important determinants of ligand binding. This would likely lead to failure in obtaining accurate ranking orders for a series of docked compounds.

2. The calculated intermolecular potential between the compound and the receptor may not capture the true impact of various binding forces on the compound's affinity for the receptor. For example, how does one properly set up the contributions by electrostatic interactions, when the dielectric constant of the ligand binding site, and the ionization status of amino side chains, are not accurately known? Another potential

source of error are the entropic forces associated with changes in the solvation of both the ligand and the amino acid residues in the binding site.

3. If the ligand has a number of rotatable bonds, does the modeling experiment sample all possible conformations of the ligand? Conformational flexibility of the target protein tends to be even more difficult to treat appropriately.

REFERENCES

[1] J. S. Albert and W. G. R. Crampton, "The Physiology of Fishes," D. H. Evans and J. B. Claiborne, ed., (2005), Taylor & Francis, 429–470.

[2] M. Y. Ali et al., "Regulation of vascular nitric oxide in vitro and in vivo; a new role for endogenous hydrogen sulphide?," *Br. J. Pharmacol.* 149 (2006), 625–634.

[3] V. Binet et al., "The heptahelical domain of GABA(B2) is activated directly by CGP7930, a positive allosteric modulator of the GABA(B) receptor," *J. Biol. Chem.* 279 (2004), 29085–29091.

[4] A. T. Brunger and A. Rummel, "Receptor and substrate interactions of clostridial neurotoxins," *Toxicon* 54 (2009), 550–560.

[5] C. Corchia et al., "Favism in a female newborn infant whose mother ingested fava beans before delivery," *J. Pediatr.* 127 (1995), 807–808.

[6] X. Gao, K. Kim, and D. Liu, "Nonviral gene delivery: what we know and what is next," *AAPS J.* 9 (2007), E92–104.

[7] R. Garzon, G. Marcucci, and C. M. Croce, "Targeting microRNAs in cancer: rationale, strategies and challenges," *Nat. Rev. Drug. Discov.* 9 (2010), 775–789.

[8] P. George and S. Kubrick, *Dr. Strangelove: or, How I learned to stop worrying and love the bomb; a novel,* Corgi book, Bantam Books, 1964.

[9] L. Grossi, "Hydrogen sulfide induces nitric oxide release from nitrite," *Bioorg. Med. Chem. Lett.* 19 (2009), 6092–6094.

[10] A. L. Jackson and P. S. Linsley, "Recognizing and avoiding siRNA off-target effects for target identification and therapeutic application," *Nat. Rev. Drug. Discov.* 9 (2010), 57–67.

[11] M. Jurima-Romet et al., "Cytotoxicity of unsaturated metabolites of valproic acid and protection by vitamins C and E in glutathione-depleted rat hepatocytes," *Toxicology* 112 (1996), 69–85.

[12] M. A. Karsdal et al., "Lessons learned from the development of oral calcitonin: the first tablet formulation of a protein in phase III clinical trials," *J. Clin. Pharmacol.* 51 (2011), 460–471.

[13] M. J. Mattes, "Radionuclide-antibody conjugates for single-cell cytotoxicity," *Cancer* 94 (2002), 1215–1223.

[14] A. Naik, Y. N. Kalia, and R. H. Guy, "Transdermal drug delivery: overcoming the skin's barrier function," *Pharm Sci Technolo Today* 3 (2000), 318–326.

[15] M. J. Paul-Clark et al., "15-epi-lipoxin A4-mediated induction of nitric oxide explains how aspirin inhibits acute inflammation," *J. Exp. Med.* 200 (2004), 69–78.

[16] V. K. Pawar et al., "Gastroretentive dosage forms: a review with special emphasis on floating drug delivery systems," *Drug Deliv.* 18 (2011), 97–110.

[17] M. I. Recht, S. Douthwaite, and J. D. Puglisi, "Basis for prokaryotic specificity of action of aminoglycoside antibiotics," *EMBO J.* 18 (1999), 3133–3138.

[18] A. J. Sadeque et al., "Human CYP2C9 and CYP2A6 mediate formation of the hepatotoxin 4-ene-valproic acid," *J. Pharmacol. Exp. Ther.* 283 (1997), 698–703.

[19] L. E. Staniaszek et al., "Effects of COX-2 inhibition on spinal nociception: the role of endocannabinoids," *Br. J. Pharmacol.* 160 (2010), 669–676.

[20] I. Takayanagi et al., "Propylbenzilycholine mustard (PrBCM)-sensitive cholinoceptors and contractile response to partial agonist in guinea pig ileal muscle," *Jpn. J. Pharmacol.* 56 (1991), 151–158.

[21] J. Vagner, H. Qu, and V. J. Hruby, "Peptidomimetics, a synthetic tool of drug discovery," *Curr. Opin. Chem. Biol.* 12 (2008), 292–296.

[22] E. Wang et al., "Posttranscriptional regulation of protein expression in human epithelial carcinoma cells by adenine-uridine-rich elements in the 3'-untranslated region of tumor necrosis factor-alpha messenger RNA," *Cancer Res.* 57 (1997), 5426–5433.

[23] K. A. Whitehead, R. Langer, and D. G. Anderson, "Knocking down barriers: advances in siRNA delivery," *Nat. Rev. Drug. Discov.* 8 (2009), 129–138.

[24] W. Zhao et al., "The vasorelaxant effect of H_2S as a novel endogenous gaseous K_{ATP} channel opener," *EMBO J.* 20 (2001), 6008–6016.

INDEX

Page numbers in italics refer to figures or figure legends.

ABC transporters, *see also* blood-brain barrier, *46*, 45–47
  relationship to drug metabolism, 83
absorption, *see* drug absorption
acarbose, *241*, 241
acetaminophen
  inhibition of cyclooxygenase, 215–217
  metabolism, 78, *79*
N-acetyl-p-benzoquinone imine (NAPQI), 78, *79*, 217
acetyl-CoA, *see* drug acetylation
acetylation, *see* drug acetylation
acetylcholine, 148, *149*
  receptor agonists and antagonists, *149*
  role in NO release, 192
acetylsalicylic acid, *44*, 214–215
acetyltransferases, *see* drug acetylation
acid maltase deficiency, 233
acrolein, 317
acromegaly, 170
actinomycin D, 14
action potential
  definition, 112
  diffusion potential, 154–155
  voltage-gated channels, 116–119
active metabolites, 68
active solute transport, 44
acute urate nephropathy, 237–238
acyclovir, 288, *289*
adenine dinucleotide, 152
adenosine deaminase deficiency, *232*, 231-233
adenosine receptors, CNS, 152
adenoviruses, in gene therapy, 367
α-adrenergic receptors, 143
  autonomic nervous system, 160
β adrenergic receptors, 143–144
  antagonists, effect on HCN channels, 127
  signal amplification in cascades, 27
adrenocorticotropic hormone (ACTH), 176
aerosols, application of drugs as, 52
AG 957, *299*
agonist-specific coupling, *see also* receptor activation, 94
agonists, *see* receptor activation
agranulocytosis, 80
albendazole, 283
albumin, 2
  binding of drugs to, 15, 55
alcohol, in gout, 236
aldose reductase inhibitors, 239
aldosterone, *see also* mineralocorticoids
  biosynthesis, *177*
allantoin, 238
allopurinol, 236, *237*

allosteric ligands, *see also* receptor activation, 24
  G protein-coupled receptors, 91
amantadine, *287*, 287
amikacin, 272, 273
amino acid conjugation
  drugs, 82
  in urea cycle enzyme defects, 230
γ-aminobutyric acid, *see* GABA receptors
aminoglycoside antibiotics
  cellular uptake, 329
  enzymatic modification, *330*, 330
  structure, 324
p-aminohippurate, *48*, 49
  clearance, 60
5-aminosalicylic acid, 348
amiodarone, inhibition of thyroxine deiodination, 175
AMPA receptors, 135
amphetamine, 145, *146*
amphomycin, 265
amphotericin B, 277, *278*, 278
  liposomal delivery, 364–366
ampicillin, 268
anandamide, *see* endocannabinoids
androgens, 180–182
  antagonists, 181, 304
  use as drugs, side effects, 181
aneuploidy, in malignant tumors, *302*, 302
angina pectoris, 201
angiogenesis, in malignant tumors, 300
angiotensin, *9*
  activation of mineralocorticoid synthesis, 179
  converting enzyme, 8
annexins, 217
anorexia nervosa, as a cause of gout, 236
anovulation, 183
antagonists, *see* receptor activation
anthracycline antibiotics, *319*, 319
antibiotics
  bacteriocidal and bacteriostatic action, 265
  origin, 259–260
  selective toxicity and mode of action, 256–257
antibodies, humanized, 358, *359*
antibody conjugates, for drug delivery, 355–358
antimicrobial resistance
  and bacterial genetics, 263–264
  countermeasures, 259
  mechanisms, 259
  origin, 260
  prevention by combination therapy, 264–265
antiport, *see* active solute transport
antithrombin III, 251
apoptosis, 297, *298*, *299*, 299–301
  extrinsic pathway, 300
  induced by antitumor drugs, 303
aptamers, *see* RNA aptamers
aquaporins, 41, 168
arachidonic acid, 207, 208
N-arachidonyldopamine, *see* endocannabinoids
2-arachidonylethanolamine, *see* endocannabinoids
2-arachidonylglycerol, *see* endocannabinoids
arecoline, 150
aripiprazole, 23, *143*
aromatase, *306*
  inhibitors, in tumor therapy, 304
aromatic hydrocarbon receptor, role in enzyme induction, 83
β-arrestins, 102
arsphenamine, *5*, 5
artemisinine, 281
Ascaris lumbricoides, 284
asparaginase, in tumor therapy, 312–313
Aspergillus fumigatus infections, 258
asthma
  aggravation by cyclooxygenase inhibitors, 217
  glucocorticoid nebulizers, 52
  use of leukotriene receptor antagonists, 219
atenolol, *144*
atherosclerosis, 241–251
  clinical manifestations, 243–244
  pathogenesis, *243*, 242–244
atovaquone, 282
ATP:L-methionine *S*-adenosyltransferase, pharmacophore for inhibition, *383*
atropine, 3, *149*
  use in parathion poisoning, 19
autoimmune diseases
  Graves' disease, 173
  Hashimoto's disease, 173
  myasthenia gravis, 150
  treatment with glucocorticoids, 178
  treatment with methotrexate, 310
  type 1 diabetes, 238
autonomic ganglia, 160
autonomic nervous system, 160
  ganglion blockers, 150
aztreonam, *269*

bacampicillin, 42, *43*
bacitracin, 265
baclofen, 138, *139*
bacterial cell wall, 260
  effect on antibiotic susceptibility, 260–261
  kinetics of drug transport, 261–262
  structure, *261*
balsalazide, 348
barbiturates, 133
basal membranes, 39
Bay 57-1293, *288*, 289
benserazide, 350
benzbromarone, 235, *237*

benzodiazepines, 133
benztropine, *149*, 150
bicuculline, 133
bile acids
  amino acid conjugation, 82
  depletion with polymeric absorbers, 247
bilirubin, 228
bioavailability, 50
biotransformation, *see* drug metabolism
bisphosphonates
  in antiprotozoal therapy, 283
  in bone mineralization, 188
blood coagulation, *249*, 248–251
blood pressure, pharmacological control, 247
blood-brain barrier, 40, 348, *350*
  and microbial infections, 258
body fat, in drug distribution, 53
bone marrow transplantation
  adenosine deaminase deficiency, 232
  Gaucher disease, 234
bone metabolism, *184*, 184–185
bone mineralization, 183, *184*
bortezomib, 314, 316
breast cancer resistance protein (BCRP), 45, *47*
5-bromo-deoxyuridine, 310
bromo-chloro-nitrosurea (BCNU), 317, *318*
  targeted delivery with Gliadel®, *361*, 360–361
bromocriptine, 142, *143*

C-peptide, 241
caffeine, 152
calcitonin, 186
calcitriol, 186, *187*
calcium
  intra- and extracellular concentrations, 113
  regulation by hormones and vitamin D, 185–188
calcium channels
  biochemical signaling, 119
  L-type, 158–159
    antagonists, 125
    heart rhythm, 157
  T-type
    antagonists, 125
    epilepsy, 124
    heart rhythm, 157
calcium-sensing receptor, 185
calicheamicins, antibody conjugates for cancer therapy, 355, *357*
calmodulin, 119
  role in regulation of nitric oxide synthase, 195
cancer, *see* malignant tumors
canfosfamide, 79, *80*
cannabis, reward circuit, 141
capsaicin, 121
carbamazepine, *122*, 123
carbamoylcholine, *149*
carbapenems, 269
carbenicillin, conversion to prodrug, 345
carbidopa, *350*, 350
carboxypeptidase A, femtomolar inhibitor, *376*
cardiac insufficiency, 124
cardiovascular disease, *see* heart disease, atherosclerosis
carindacillin, *344*, 345
caspases, *299*
caspofungin, 277
catechol-*O*-methyltransferase, *147*
  inhibitors, 145–146
catecholamines
  biosynthesis, *141*
  degradation, *147*
CB 1954, 81, *82*
CD20, conjugated antibodies against, 358
cefotaxim, *269*
cefoxitin, *269*
ceftobiprole, *269*, 270
cell cultures, and pharmacokinetics, 38
cell cycle, *301*, 301
  check points, 302
β cells (pancreas)
  autoimmune destruction, 238
  destruction by streptozotocin, 305
cellular senescence, 300
cephalosporins, 267, 268
Chain, Ernst, 14
cheese reaction, 145
chirality
  and drug solubility, 347
  and pharmacological activity, 16
chloramphenicol, 14, 272, 324
  interaction with ribosomal RNA, *328*, 328
chloride channels, 128, *see also* $GABA_A$ receptor, glycine receptor
chloride, intra- and extracellular concentrations, 113
chloroquine, 280–281
cholecalciferol, *see* vitamin D
cholesterol
  intestinal uptake, 244–246
  pathogenesis of atherosclerosis, 242
  statins, 246
  transport and metabolism, *245*
  transport by lipoproteins, 244
cholesterol ester transport protein (CETP), *245*
choline acetyltransferase, 148
cholinesterase, 148
  catalytic mechanism, *151*
  inhibitors, *151*, 150–152
chromosomal translocation, *308*
chronic myeloid leukemia, 307
cidofovir, *288*, 288
cimetidine, 7, *8*
cinacalcet, 185
ciprofloxacin, *274*, 274
cisplatin, 317, *318*
clavulanic acid, 270, *271*

clearance, 59
creatinine, 60
inulin, 59–60
clomiphene, *183*, 183
clonidine, *144*
clopidogrel, 249
Clostridium difficile, 267
clozapine, 142, *143*
coagulation factor VIII, 2
coagulation factors, 248
γ-glutamyl-carboxylase, *250*, 250
cocaine, 145, *146*
local anesthesia, 121
colchicine, *237*
combination therapy
in malignant tumors, 296
in microbial infections, 264
competitive inhibition, 18, 30
conjugation reactions, 75–83
effect on drug activity, 75
cooperative receptor activation, 32–33
cortisol, *see also* glucocorticoids
biosynthesis, *177*
covalent inhibition, 31
creatinine clearance, 60
curcumin, 219
cyclic guanosine monophosphate (cGMP), signaling mechanisms, 195–197
cyclic nucleotide-gated channels, 120
cyclins, 301
cyclizine, *8*
cyclooxygenase, 210
inhibitors, 211–215
inhibitors, type 2-selective, 214
isoforms, 213–214
reaction mechanism, 211
cyclophosphamide, 317, *318*
cycloserine, 265, *268*
NMDA receptor, 138
CYP3A4, 84
CypHer 5 assay for GPCR activation, *378*, 379
cys-loop receptor family, 129
cystic fibrosis, treatment with PTC124, 336
cytochrome P450, 71–73
eicosanoid biosynthesis, 219
inhibition by drugs, 73
metabolism of NO-releasing drugs, 202
mode of action, *71*, 71
reactions, 72, 73
substrate binding, *72*
transcriptional induction, 84
cytochrome P450 reductase, 71, 74
cytosine arabinoside, 313, *314*
cytotoxic cancer drugs, types, 309

dasatinib, 307
daunorubicin, *319*
7-dehydrocholesterol, *187*, 188
dehydroepiandrosterone, 180
biosynthesis, *177*
deoxycytidine kinase, role in adenosine deaminase deficiency, 233
dermal drug application, 52
desmopressin, *169*
dexamethasone, 178, *179*
DHPR (dihydropyridine receptor), *see* calcium channels, L-type
diabetes insipidus, 168, 238
diabetes mellitus, 238–241
effects on metabolism, 239
long-term complications, 239
symptomatic, 241
type 1 and type 2, 238
diabetic coma, 239
diaphorase, 74, *82*
diazoxide, 125
diclofenac, 213
didanosine, *288*
diethylamine, fast block of Na_V channels, 123
diethylcarbamazine, 283
diethylstilbestrol, *183*
diffusion potential
ion concentration gradients, 113
physical basis, 154–155
Digitalis glycosides
Na^+/K^+-ATPase inhibition, 116
use in cardiac insufficiency, 124
digitoxin, 116
enterohepatic cycling, *77*
digoxin, 116
1,4-dihydro-*N*-methylnicotinic acid moiety, in prodrugs, 351, *352*
dihydrofolate reductase, 310, *312*
species-specificity of inhibitors, 256
dihydropteroate synthase, 13, 263
dihydropyridine receptor, *see* calcium channels, L-type
dihydrotestosterone, 180
dimethylphenylpiperazinium, *149*
diphtheria toxin, antibody conjugates for cancer therapy, 356
dissociative anesthesia, 138
distal tubule, 57
distribution, *see* drug distribution
DNA repair, 299
DNA replication
and antimicrobial therapy, 273–274, 287–289
relationship to cell cycle, 301
tumor therapy, 313, 320
DNA topoisomerase
and cytosine arabinoside, *314*
bacterial, 274
inhibitors, tumor therapy, 313–314, *315*, 320
DNA-alkylating drugs, *318*, 316–320
docetaxel, 315
docosahexaenoic acid, 223
docosanol, 286
L-DOPA (levodopa), 348

DOPA decarboxylase, 348
inhibitors, *350*, 350
dopamine, *350*
prodrugs, 348
dopamine receptors, 141–142
agonists and antagonists, *143*
dose–effect relation
and signaling cascades, 25–28, 33
receptor overexpression, 28
doxorubicin, *319*, 320
liposomal delivery, 354
doxycycline, 283
drug absorption, 38
drug acetylation, 79–81
aminoglycoside antibiotics, *330*
drug antitarget, 126
drug application, routes, 49–53
drug design, 12
drug discovery, 373
parallel synthesis, 387–388
sage advice on, 11
strategies, 11–14
drug distribution, 38, 53–56
compartments, 53, *54*
kinetics, 55
drug efflux pumps, *see also* ABC transporters
bacterial, 262
drug elimination, 38
biliary secretion, 77
kidneys, 56–63
kidneys, variability, 62–63
drug metabolism, *see also* cytochrome P450, conjugation reactions
overview, *68*, 67–69
drug screening
animal experiments, 380
cell culture, 384
combinatorial chemistry, 388
compound acquisition, 384–388
computational, *381*, 381–383
displacement assays, 375
enzyme activity assays, 375–376
fluorescence assays, 379–380
natural compounds, 385–387
phenotypic, 384
drug side effects, and beneficial effects, 28
drug targets, 4–7
selection, 373
validation, 373–374
drug toxicity, *see also* specific drugs
and beneficial drug effects, 28
and drug metabolism, 69
drug transport, *see also* transport proteins
anatomical barriers, 38, *39*
cell membranes, 41, *42*
role of molecular size, 43
drugs
from natural sources, examples, 3
molecular size, 1–3
origin and preparation, 3–4
similarity to physiological ligands, 7
Duchenne muscular dystrophy, treatment with PTC124, 336

EC_{50}, 28
Echinococcus granulosus, 284
ecstasy (drug), 146
EDRF, *see* nitric oxide
edrophonium, *151*
efficacy, 28
Ehrlich, Paul
arsphenamine (Salvarsan), 4
blood–brain barrier, 40
eicosanoids, 207
biosynthesis, 208
eicosapentaenoic acid, 223
elacridar, *47*
electron crystallography, 129
eliglustat, 234
elimination, *see* drug elimination
enalaprilate, *9*, 10
endocannabinoids, 219–223
biosynthesis, 220, *221*
inactivation, 220
inhibition of synaptic transmission, 220–221
receptors, 219–220
endorphins, 10, 152
endothelia, drug transport across, 39
endothelium-derived relaxing factor, *see* nitric oxide
enfuvirtide, *286*, 286
enhanced permeability and retention effect, 353, *354*
enkephalins, 152
entacapone, 146
entero hepatic cycling, 69
enzyme defects, 227–235
and fetoplacental circulation, 227
principles of therapy, 228
enzyme replacement therapy
acid maltase, 233
adenosine deaminase, 232
glucocerebrosidase, 234
epilepsy
ethosuximide, 124
$GABA_A$ receptors, 133
ion channels, 122–124
ketogenic diet, 239
retigabine, 124
epinephrine, 8, 142–144
epithelia, drug transport across, 39
eplerenone, 179, *180*
epoxide hydrolase, 72
14,15-epoxy-eicosatrienoic acid, 219
EPR effect, *see* enhanced permeability and retention effect
ergocalciferol, *see* vitamin D
ergosterol, *278*
binding of amphotericin B, 278

ergotamine, 4
erythromycin, *72*, 272
analogs, genetically engineered, *386*, 387
essential genes, and antimicrobial therapy, 257, 374
estradiol, *see also* estrogens, *305*
delivery across the blood–brain barrier, 351, *352*
estrogens, 182–183
ethanol
$GABA_A$ receptor, 134
reward circuit, 141
ethinylestradiol, *183*
ethosuximide, 124
etomidate, *139*
etoposide, *299*
excitable cells, types of, 112–113
excitatory postsynaptic potential (EPSP), 127
excitotoxicity, 138
exemestane, *306*
EXP7711, 346
extracellular signal-regulated kinase (ERK), *103*

factor VIII, 2
false transmitters, 144
farnesylpyrophosphate synthetase, inhibition by bisphosphonates, 188
fat tissue, in drug distribution, 53
fatty acid synthesis, as antimicrobial target, 374
ω-3 fatty acids, 223
fatty acyl amide hydrolase, 220
favism, 75, *76*
ferredoxin, 263
fibrin, *249*
fibrinogen, *249*
fibrinolysis, 248, 251
finasteride, *183*
first-pass effect, 49, 69
flavin-containing monooxygenases, 73
Fleming, Alexander, 14
Florey, Howard, 14
fluconazole, 277, 279
flucytosine, 277, 279
fludrocortisone, 179, *180*
fluoroquinolones, 274
5-fluorouracil, 309–310, *311*
fluoxetine, *146*
flutamide, *183*
foam cells, 242, 244, *245*
follicle-stimulating hormone (FSH), 180, 181
fomivirsen, 290, 332
formylmethionine deformylase, *see* peptide deformylase
foscarnet (phosphonoformic acid), *288*, 289
fosfomycin, 265, *268*
fructose, in gout, 236
full agonists, 22
Fura 2, 379
fusidic acid, 272, *274*

G protein-coupled receptors
agonist-specific coupling, 94
antibodies as agonists and antagonists, 91–92
assay for activation
$\gamma^{35}S$, 100
based on endocytosis, *378*, 379
constitutive activity, 95
drugs acting on, 88
in synaptic transmission, 128
roles of conserved sequence motifs, 89–90, 94
structure, *90*
types of ligands, 88–89
G proteins, 98–101
activation, *99*
effector mechanisms, *101*
functions of α subunits, 99–100
functions of $\beta\gamma$ subunits, 100
G_0 phase, variability of duration, 301
GABA receptors, 138
agonists and antagonists, *139*
$GABA_B$ receptors, 138
$GABA_C$ receptors, 138
$GABA_A$ receptor, 133–134
ethanol, 134
ganciclovir, 290
ganglion blockers, 150
Gaucher disease, *234*, 233–235
gemtuzumab ozogamicin, 355, *357*
gigantism, 169
glaucoma, 150
glioblastoma (brain cancer), 360
glomerular filtration, *57*, 57–58
protein binding and, 55
rate, 59–60
glomerulus, *see* glomerular filtration
glucocerebrosidase deficiency, *see* Gaucher disease
glucocorticoid response elements, 176
glucocorticoids, 176–179
analogs and antagonists, *179*
anti-inflammatory effects, *178*, 178
induction of gluconeogenesis, 176
inhibition of $cPLA_2$, 217
non-transcriptional signaling, 176–178
gluconeogenesis, transcriptional induction by cortisol, 176
glucose, transport, 239
glucuronidation, 76–77
gludopa, *350*, 350
glutamate receptors, 374–375
agonists, *375*
ionotropic, 135, 138
metabotropic, 138
γ-glutamyl-carboxylase, *250*, 250
glutathione
conjugation, 78–79
glucose-6-dehydrogenase deficiency, 75, *76*
protein-*S*-nitrosylation, 198

glycine receptor, 134
  activation by endocannabinoids, 220
glycine receptors, 138–139
  agonists and antagonists, *140*
glycine, synaptic reuptake, inhibitors, 139
glycopeptide antibiotics, 266, *267*
goiter, 174
Goldman equation, 155
gonadal steroids, 179–183
gonadotropin releasing hormone (GnRH), 181
  analogs, in prostate cancer, 359
gonococci (*Neisseria gonorrhoeae*), 268
gout
  drugs, *237*
  pathogenesis, 235–236
  pharmacotherapy, 236–237
GPIIA/IIIB receptor, *249*, 250
Graves' disease, 173
griseofulvin, 277
growth hormone, 169–170
  illicit use, 170
  regulation by glucose, 170
guanethidine, *144*
6-guanidinonaltrindole (6-GNTI), *97*, 97
guanine, preferential alkylation at N7, 319
guanylate cyclase (sGC), 195

Haemophilus influenzae, 268
hairy cell leukemia, 305
half-life, drug elimination, 61
haloperidol, 141, *143*
Hashimoto's disease, 173
Hata, Sahachiro, 4
hay fever, 148
heart
  disease, 124–127
  excitation-conduction system, 157
  rhythm, generation, 156–157
hemophilia, 2
heparin, 251
hepatitis C virus, protease inhibitors, 290
HER2/neu
  inhibition with antibodies, 308
  inhibition with ribozymes, 338
heroin, 42, *43*
herpes simplex virus, 288, 289
hexamethonium, 43, *44*
high density lipoprotein (HDL), *245*
histamine receptors, 7, 146–148
HIV
  fusion with host cells, *286*, 286
  inhibitors of TAR RNA, *334*, 333–335
  protease inhibitors, *12*
    development, 12
    drug delivery, 73, 347
hormonal contraception, 182
5-HT_3 receptor, *see* serotonin receptors
HU-308, *223*
human chorionic gonadotropin (hCG), 180, 182
human immunodeficiency virus, *see* HIV
human ether-à-go-go-related gene (hERG) channels, 126
6-Hydroxy-L-DOPA, *143*
hydroxyapatite, 183, 185
hydroxylamines, metabolism, 79
11β-hydroxysteroid dehydrogenase, 178, *180*
hygiene, 255
hyperparathyroidism, 185
hyperpolarization-activated, cyclic nucleotide-gated (HCN) channels, 120, 126
hypersensitivity type I, 148
hypertension, L-type calcium channel blockers, 125
hyperthyroidism, 173–174
hypothalamus, 167
hypothyroidism, 173

^{131}I
  antibody conjugates, cancer therapy, 356–358
  in hyperthyroidism, 174
  in thyroid cancer, 305
idoxuridine, 310
imatinib, 307
  molecular docking, *381*
  penetration of blood–brain barrier, *47*
imipenem, *269*, 269
imipramine, *146*
immunoliposomes, 358
immunosuppression, *see also* autoimmune diseases
  and antimicrobial therapy, 265
indomethacin, 213
infectious diseases, socio-economic context, 255
influenzavirus
  fusion with host cells, 287
  life cycle, *285*
  M_2 protein, blockade by amantadine, *287*
  neuraminidase, 290
infusion, *see* intravenous drug application
inhalational anesthetics, *see* narcotic gases
inhibition
  competitive, 30
  covalent, 31
Inhibitory postsynaptic potential (IPSP), 128
insulin
  aggregation, *363*
  application methods, 362
  hexamer structure, *364*
  inhalable, 363
  lispro and aspart, 363, *364*
  pumps, 363
  secretion and turnover, 361
  slow- and fast-acting preparations, 362
insulin-like growth factor 1 (IGF-1), 169
interferon-α, use in hairy cell leukemia, 305
interferons, antiviral therapy, 290–291
interstitial space, 53

intracellular volume, 53
intramuscular drug application, 51
intravascular volume, 53
intravenous drug application, 50–51
inulin clearance, 59–60
inverse agonists, *see* receptor activation, 23
ion channels, 5, 114–116, *see also* calcium channels, chloride channels etc.
  block, fast, slow and use-dependent, 122
  electrophysiology, 155–156, 380
  fast block and slow block, *123*
  fluorescence assays, *379*, 379–380
ionotropic receptors, structural families, 129
ipratropium bromide, *149*, 150
iptakalim, 125
irreversible inhibition, 19
ischemic heart disease, 124
isoflurane, *139*
isoniazid (INH)
  acetylation, fast and slow, 80
  mode of action, 263, *264*
isoprenaline, *see* isoproterenol
isoproterenol, 8, *144*
isouramil, *76*
ivabradine, 126
ivermectin, *140*, 283

kainate receptor, 135
kanamycin, enzymatic modification, *330*
kernicterus, 228
ketamine, 138
ketoconazole, *72*, 279
  inhibition of human cytochrome P450, 73
ketogenic diet, 239
α-ketoglutarate, organic anion transporters, 48
kidney function
  overview, 57–59
  perfusion, 57
  quantitative aspects, 59–63
  tubular secretion, *48*

L-DOPA (levodopa), *350*
labetalol, stereoselective receptor binding, 16
β-lactam antibiotics, 267–271
  mode of action, 267–268
  semisynthesis, 267, *385*, 387
  side reactions, 268
β-lactamases, 270–271
  inhibitors, *271*, 270–271
Lambert-Eaton syndrome, 152
lapatinib, 308
law of mass action, 17, 30
LDL receptor, *245*
letrozole, *306*
leukotrienes, 207, 218–219
leuprolide acetate, 359
levodopa, 348, *see* L-DOPA
lidocaine, 123, 126
Ligand-gated channels, 6

linezolid, 272
liposomes, for drug delivery, 353–354, *355*
  targeting with antibodies, 358
lipoxin A_4, 218
lipoxins, 207, *see also* eicosanoids
5-lipoxygenase, 218–219
lipoxygenases, 210
lithium, 1, *2*
  inhibition of iodide transport, 175
LNB-7TM receptors, 95
long-term potentiation and -depression, 135
loop of Henle, 57, 58
losartan, *10*, 10, *346*
low density lipoprotein (LDL), 244
  modification of and atherosclerosis, 243
luteinizing hormone (LH), 180, 181
Lyell's syndrome (toxic epidermal necrolysis), 80

macugen, 338
magnetic resonance imaging (MRI), 64
malaria, 279
  chemotherapy, 280–283
  parasite species, 280
  tetracyclines, 279
malathion, *151*
malignant tumors
  angiogenesis, 300
  causes of diversity, 296
  cell aneuploidy, *302*, 302
  chemotherapy, 304–320
  disinhibition of cell growth, 297–298
  genetic instability, 299–301
  increased mutation rate, 300
  induction vs. progression, 300
  metastases, 295
maribavir, 290
mass action law, 17
MDR-1, 45, *46, 47*
mecamylamine, 43, *44*
mechanosensitive channels (TRPV4), 121
mechlorethamine, 316, *318*
melarsene oxide, *280*
melphalan, 316, *318*
membrane potential, *see also* action potential
  diffusion potential, 154–155
menstrual cycle, 182
mesalazine, 348
metazoans, 279
metformin, 240
  mode of action, *242*
  structure, *241*
methacrylic acid copolymers (Eudragit®), *344*, 344
methicillin, *269*, 270
methimazole, *174*
methionine aminopeptidase, *377*, 376–378
methotrexate, 310, *312*
α-methyl-DOPA, *144*
N-methyl-4-phenyl-tetrahydropyridine, *147*

methylenedioxyamphetamine (ecstasy), *146*
N,N'-methylenetetrahydrofolate, 310
methyltestosterone, *183*
metrifonate, 283
metronidazole, 263, 281
mexiletine, 126
mibefradil, 125
miconazole, *306*
microbes, classification, 256, *257*
microtubules, 315
mifepristone
  abortion, 182
  tumor therapy, 304, *305*
miglustat, 234
miltefosine, 281
mineralocorticoid receptor, 179
  activation by cortisol, 178
mineralocorticoids, 179
minoxidil, 125
Mississippi mud, 267
mitotane, 304, *307*
mitotic spindle, *317*
moenomycin, 266
molecular docking, *381*, 381-382
molsidomine, 203
monday headache, 201
monoamine oxidase (MAO), 140
  drug metabolism, 73
  inhibitors, 145
monoamine transmitters, 140-148
  biosynthesis, *141*
  degradation, *147*
  inhibitors of transport, 144-145
  modulatory function, 140
montelukast, 219
morphine, *10*, 10, 152
MRI (magnetic resonance imaging), 64
MRSA, *see Staphylococcus aureus*
mucosal drug application, 53
mucous membranes, drug transport across, 39
muramyl-transpeptidase, 267
murein biosynthesis, *266*
muscarine, *149*
muscarinic acetylcholine receptors
  agonists and antagonists, 150
  autonomic nervous system, 160
muscimol, *139*
myasthenia gravis, 150
Mycobacterium leprae, 274
Mycobacterium smegmatis, 384
Mycobacterium tuberculosis, 255, 261, 263, 264, 274

NADH:quinone oxidoreductase, 74, *82*
Na^+/K^+-ATPase, 45, 116
NAPQI, *see N*-acetyl-p-benzoquinone imine
NAR, *see* nicotinic acetylcholine receptor
narcotic gases, 134
  combination with muscle relaxants, 149

neonatal jaundice, 228
nephron, *57*, 57
Nernst equation, 154
neuromuscular blockade
  muscle relaxants, 148-149
  succinyl-bis-choline, 149
Neurotransmitter receptors, classification, 128-129
neutral antagonists, *see* receptor activation, 23
nevirapine, *288*
N-hydroxyarginine (NOHA), 192
nicotine, 148, *149*
  delivery with cyclodextrin, 366
  reward circuit, 141
nicotinic acetylcholine receptor, 129-133, 148
  acetylcholine binding site, 132
    cation-π interaction, 133
  antagonists (muscle relaxants), 148-149
  autonomic nervous system, 160
  desensitization, 133
  structure, 129
  subunit composition, 131
  trapping in open state, *131*
nifedipine, 125
nisin, 276
nitazoxanide, 281
nitrate tolerance, 202
nitric oxide
  bactericidal activity, 200
  binding to hemoglobin, 197
  enzymatic release from prodrugs, 202
  molecular properties, 191
  protein-*S*-nitrosylation by, 197-199
nitric oxide synthase, *194*
  inhibitors, 204
  isoforms, 193
  reaction mechanism, 192
nitroglycerine, 201
nitroimidazole drugs, mode of action, 263
nitroprusside, 203
NMDA receptor, 135
  cycloserine, 138
NO, *see* nitric oxide
nonionic diffusion, 42, *44*
nonsense mutations
  in inherited diseases, 335
  suppression with PTC124, *335*
norepinephrine, 8, *20*, 142-144
norgestrel, *183*
NOS, *see* nitric oxide synthase
NTBC (*p*-phenylpyruvate dioxygenase inhibitor), *229*, 229
nuclear hormone receptors, 5
  roles in enzyme induction, 84
  strucure and function, 164, *166*
  types, 165
5-nucleotidase, 313
5'-nucleotidase, *314*

oblimsen, 332
octanol-water partition coefficient, 41
odansetron, 134, 142
oligonucleotide drugs
  antisense, 3
  aptamers, 338
  aptamers, *in vitro* selection, *337*
  fomivirsen, 332
  with modified backbones, *333*
oltipraz, 85
oncogenes, 306
oncoproteins, 306
opioid receptors, 153
  heteromer-selective agonist, 97
opoids, reward circuit, 141
oral drug application, 49–50
organic anion and cation transporters, 47–49
  substrate specificity, 48
organic anion transporters (OAT), 83
Org 29535, *140*
orphan receptors, 87
orthosteric ligands, *see* receptor activation
oseltamivir, 290
osmotic engine, for drug delivery, 359, *360*
osteoblasts, *184*
osteoclasts, *184*
osteoporosis, 184
  estrogen therapy, 182
ouabain, 116
22-oxacalcitriol, 188
oxacillin, 270
oxytocin, *169*, 167–169

P-glycoprotein, 45, *46*
P2X receptors, 135
p53, 300
  gene therapy of tumors, 367
paclitaxel, 315
pancreatitis, 217
pancuronium, *149*
Paracelsus, 1
parasympathetic nervous system, 160
parathion, 19
parathyroid hormone, 185–186
parenteral drug application, 51
Parkinson's disease, 141, 348
  benztropine, 150
  MAO inhibitors, 145
paromomycin, 324
  interaction with ribosomal RNA, 325–326, *327*
partial agonists, 23
passive transport, 45
patch clamp method, 155
pegvisomant, 170
penicillin
  allergy, 148
  discovery, 14
  mode of action, 267
  semisynthesis, 385
penicillin G, 268, *269*
  antimicrobial spectrum, 268
  elimination, 62
penicillin-binding proteins, 266, 267
Penicillium notatum, 14
pentylenetetrazole, 133, *139*
peptide deformylase, 277, *377*, 377
peptide nucleic acids, *333*
peptide transmitters, 152
peroxynitrite
  and protein-*S*-nitrosylation, 198
  bactericidal activity, 200
pertussis toxin, inactivation of G$\alpha_{i/o}$ proteins, 101
PET (positron emission tomography), 64
pharmacogenetics, 68
pharmacokinetics, 37
  in antimicrobial therapy, 257–258
pharmacophore, 16, 382, *383*
phase I reactions, 70
phase II reactions, 70
phencyclidine, 138
phenobarbital, metabolism, *70*, 70
phenol, slow block of Na$_V$ channels, 123
phenoxybenzamine, *20*, 20
phenylacetate, conjugation with glutamine, *83*
phenylalanine
  degradation pathway, *229*
  hydroxylase, 228
phenylbutyrate, 230
phenylethylamine, 145
phenylketonuria, 228–229
  prevalence, 227
  treatment in pregnancy, 228
phenytoin, *122*, 123
pheochromocytoma, 21
phosphate, regulation by hormones and vitamin D, 185–188
phosphatidylserine, 248
3′-phosphoadenosine-5′-phosphosulfate (PAPS), 77
phospholipase A$_2$, 208
  inhibitors, 217
physostigmine, *151*
pilocarpine, *149*, 150
piperacillin, 268
piroxicam, 213
  delivery with β-cyclodextrin, 366
pituitary gland, 164–166
  hormones, 167
plasmids, in antimicrobial resistance, 264
plasmin, 248, 251
plasminogen, 248, 251
plasmocytoma, 314
Plasmodium falciparum, 279
platelets, *see* thrombocytes
PML-RARA oncoprotein, 308
polarity, of molecules, role in membrane transport, 41